ENVIRONMENTAL SCIENCE, ENGINEERING AND TECHNOLOGY

ESTUARIES

CLASSIFICATION, ECOLOGY AND HUMAN IMPACTS

ENVIRONMENTAL SCIENCE, ENGINEERING AND TECHNOLOGY

Additional books in this series can be found on Nova's website under the Series tab.

Additional E-books in this series can be found on Nova's website under the E-book tab.

MARINE BIOLOGY

Additional books in this series can be found on Nova's website under the Series tab.

Additional E-books in this series can be found on Nova's website under the E-book tab.

ENVIRONMENTAL SCIENCE, ENGINEERING AND TECHNOLOGY

ESTUARIES

CLASSIFICATION, ECOLOGY AND HUMAN IMPACTS

STEPHEN J. JORDAN

EDITOR

NOVA

Nova Science Publishers, Inc.

New York

QH
541.5
.E8
E845
2012

NOTICE TO THE READER

The Publisher has taken reasonable care in the preparation of this book, but makes no expressed or implied warranty of any kind and assumes no responsibility for any errors or omissions. No liability is assumed for incidental or consequential damages in connection with or arising out of information contained in this book. The Publisher shall not be liable for any special, consequential, or exemplary damages resulting, in whole or in part, from the readers' use of, or reliance upon, this material. Any parts of this book based on government reports are so indicated and copyright is claimed for those parts to the extent applicable to compilations of such works.

Independent verification should be sought for any data, advice or recommendations contained in this book. In addition, no responsibility is assumed by the publisher for any injury and/or damage to persons or property arising from any methods, products, instructions, ideas or otherwise contained in this publication.

This publication is designed to provide accurate and authoritative information with regard to the subject matter covered herein. It is sold with the clear understanding that the Publisher is not engaged in rendering legal or any other professional services. If legal or any other expert assistance is required, the services of a competent person should be sought. FROM A DECLARATION OF PARTICIPANTS JOINTLY ADOPTED BY A COMMITTEE OF THE AMERICAN BAR ASSOCIATION AND A COMMITTEE OF PUBLISHERS.

Additional color graphics may be available in the e-book version of this book.

LIBRARY OF CONGRESS CATALOGING-IN-PUBLICATION DATA

Estuaries : classification, ecology, and human impacts / editor, Steve Jordan.
    p. cm.
 Includes bibliographical references and index.
 ISBN 978-1-61942-083-0 (hardcover)
 1. Estuarine ecology. 2. Estuarine area conservation. 3. Estuaries. 4. Estuarine ecology--United States. 5. Estuarine area conservation--United States. 6. Estuaries--United States. I. Jordan, Stephen J.
 QH541.5.E8E845 2011
 577.7'86--dc23
                    2011043195

Published by Nova Science Publishers, Inc. + *New York*

B3243576

CONTENTS

FOREWORD

I was introduced to the science of estuaries as a graduate student in the early 1980s, studying the ecology of oyster populations in Chesapeake Bay. To undertake this research, I needed to learn not only about oyster biology, but also about the unique physical and chemical dynamics of estuaries. Further, to understand the relevance of my work, I needed to gain some knowledge of the oyster fishery, its management, and the underlying laws and policies. This experience, along with other opportunities for interdisciplinary study and research, has served me well in my career. It is hard to imagine ecosystems more conducive to such diverse learning and investigation than estuaries.

Working in the field of estuarine ecology taught me other lessons. One was that the traditional experimental model of scientific investigation, where one variable is isolated and all others controlled, is not a productive one for most ecological research. My oysters demonstrated interactive responses to changes in water temperature and concentrations of suspended solids (Jordan 1987), so that investigating either of these variables in isolation would have given at best an incomplete model – more likely an incorrect one. A second valuable lesson was that much of ecological theory, circa 1980, relied too much on cogitation and too little on data. From these and other experiences, I developed my personal style of research: observe, then analyze, then theorize; a model that is the reverse of many scientific investigations. In systems as complex as estuaries, often we don't know what model or hypothesis to test until we've had a thorough encounter with data. Or at least this was the case before the past decade or so. These days, so much data and scientific information have accumulated that I find myself asking younger scientists to define their model or hypothesis and then justify why they need to collect new data. It seems the era of exploration is giving way to a new era of data mining, modeling and synthesis.

All natural systems have infinite complexity for the committed reductionist, but estuaries manifest great complexity even at superficial levels of investigation. They are highly dynamic over brief spans of time and fine spatial scales. They are integrators of terrestrial, atmospheric, oceanic, and riverine influences, and subject to perturbations arising from any of these sources. Estuaries are centers of human populations, commerce, and biological productivity, contributing to and receiving from society far more than would be predicted by their modest geographic footprints. They are receptors, processors, and transporters of materials; producers of great quantities of seafood; and places of major concern for the ecological, sociological, and economic consequences of habitat loss, pollution, overexploitation of natural resources, and rising sea levels. Estuaries are especially sensitive to such influences, delivering feedback to society in the forms of diminished fisheries, impairment of recreation and tourism, and losses of other services and amenities. These losses are reflected in the many restoration and management programs that have been established for specific estuaries around the world, and the substantial body of research and modeling that they have stimulated.

This book reflects the interdisciplinary nature of estuarine research, ranging from physical science to socio-economics. Although wide-ranging, it is not fully comprehensive, nor is any other recent volume of which I am aware. Having taught estuarine ecology to graduate students for several years, I once had thoughts of writing or editing a textbook to replace the estimable, but dated, *Estuarine Ecology* (Day et al. 1989). This need remains unfulfilled, although I learned recently that a new edition of *Estuarine Ecology* is in preparation. Our volume is not intended to be a textbook; as directed by the publisher, its principal audience is the scientific community. Nevertheless, the authors and I have tried to make it more comprehensive and readable than a mere collection of scientific articles, so that it might serve as a text for graduate students until something better comes along.

I am writing this essay on the shores of Santa Rosa Sound, a ravishingly beautiful coastal lagoon on Florida's Gulf of Mexico coast, part of the Pensacola Bay estuarine system. Every time I step from my office to the outdoors, I'm reminded that estuaries are not just about the science, management, and policy we write about in this book. They are places where the beauty and joy of nature strike us, demand our attention, and refresh us.

REFERENCES

Day, J.W., Jr., Hall, C.A.S., Kemp, W.M. & Yañez-Arancíbia, A. (1989). *Estuarine Ecology*. Wiley & Sons, New York.

Jordan, S.J. (1987). *Sedimentation and Remineralization Associated with Biodeposition by the American Oyster* Crassostrea virginica *(Gmelin)*. Ph.D. Thesis, University of Maryland, College Park.

PREFACE

Estuaries are minor geographical features at the global scale, but have major importance for society and the world's economies. This book, with case studies from three continents, offers a perspective on estuaries as ecosystems that are intimately linked to society in both positive and negative ways. As focal points of trade, industry, energy production, fisheries, and tourism, estuaries provide a bounty of services to society; they are also centers of human population and vulnerable to the accompanying environmental stresses. This book covers a wide range of topics, from the definition and classification of estuaries, to chemical and physical properties, ecological processes, habitat values, fisheries, and tools for improving management and policy.

Chapter 1 - Estuaries, although minor geographical features at the global scale, have major importance for society and the world's economies. This chapter introduces estuaries by presenting an overview of definitions, origins, physical, chemical and ecological attributes, and the interactions of estuarine ecosystems with human activities, including attempts to restore degraded estuaries to a more desirable ecological condition.

Chapter 2 - Estuarine scientists have devoted considerable effort to classifying coastal, estuarine and marine environments and their watersheds, for a variety of purposes. These classifications group systems with similarities—most often in physical and hydrodynamic properties—in order to explain how estuaries respond to stressors. The authors evaluated four selected classification schemes for their effectiveness in (1) relating nitrate + nitrite concentrations to nitrogen load estimates, and (2) relating chlorophyll a concentrations to nitrogen loads and concentrations. Strong within-class similarities would be useful for developing numeric nutrient criteria and reference conditions in estuaries. Classifications were analyzed using a nationwide water quality data set. Of the classifications tested, the geographically focused Ecoregion classification seemed to provide the most robust grouping of estuaries on the basis of similarity of responses. Regionally-based classifications may account for factors such as land use, population density, and geographic attributes that within-estuary physical and hydrological variables do not represent. The authors offer recommendations for selecting and improving classification approaches, both for researchers developing classification tools, and for resource managers who use them. Our ability to manage estuarine environments will ultimately benefit from a classification scheme that provides standard terminology for various types of estuaries, guidelines for organizing data, facilitation of mapping and modeling efforts, and support for decisions.

Chapter 3 - Salt marshes are a very important part of the estuarine ecosystem, with an important role within the biogeochemical cycles, being areas of high primary production, which also contribute importantly as shoreline stabilizers. Besides, periodical tidal flooding of salt marshes also causes the transport of significant quantities of pollutants, which tend to accumulate in the marsh ecosystem. Therefore, salt marshes are considered to be important sinks namely of heavy metals. Their important role has been recently admitted by the inclusion of these ecosystems in the European Water Framework Directive (WFD). Multiple services of wetlands and its value are already well known. The Tagus estuary is the largest estuary in Western Europe, covering an area of approximately 320 km^2. The estuary has a vast inner bay with extensive intertidal areas from which about 15% of these areas are salt marshes. The Tagus estuary is a mesotidal system with a mean tidal range of 2.4 m, varying from about 1 m at neap tides to about 4 m at spring tides. This chapter compares the evolution of salt marsh plant structure over the past five decades in two Tagus salt marshes with different anthropogenic influences. The results show that in both salt marshes the differences in plant cover were accompanied by significant reductions in total plant biomass. Below/aboveground biomass ratios have been increasing, reflecting a higher investment in roots than in photosynthetic organs. There was a general decrease of total N content in all studied species, although shoot and root N concentrations were almost the same between years. These results suggest a lower N availability and/or a stress situation.

Chapter 4 - Successful marsh restoration requires recreating conditions to ensure proper ecosystem function. One approach to monitor restoration success is using a sentinel species as a proxy integrator of salt marsh function. The gulf killifish (*Fundulus grandis*, Baird and Girard) is a good candidate sentinel, as it provides an important trophic link between marsh surface and subtidal habitats. Accordingly, herein an *in-situ* approach is presented for assessing marsh function using reared and wild post-larval gulf killifish. Three field experiments conducted in the spring of 2002 and 2003 assessed post-larval growth and survival within cylindrical acrylic microcosms. Marginally significant differences in mean growth rate were apparent between marsh surface and subtidal habitats at Weeks Bayou, MS in 2002, with higher growth in subtidal habitats (6.2% d^{-1}) versus surface habitats (3.9% d^{-1}). Survival rates of 44.4% were equivalent within both habitats. Using the same approach, restored and reference sites at the Chevron-Pascagoula refinery, MS were compared in 2003. Survival rates were notably low for both restored (18.75%) and reference sites (0.0%) during the first experiment of 2003 (May). Reference mortality most likely reflected high sediment porosity. Survival was considerably higher at the restored site (76.2%) than at the relocated reference site (29.8%) in a second 2003 experiment (June); however, growth was poor at both sites (0.1 and –0.47% d^{-1}, respectively). Diet analysis showed that enclosed fish were feeding successfully. Estimates of infaunal abundances and daily growth of wild fish were comparable between sites. Growth rates of enclosed fish followed a trajectory more similar to laboratory fish than to wild fish, perhaps reflecting that reared fish within enclosures continued to grow slowly in the field. A number of valuable lessons from this study point to how an effective multipronged approach for using early stages of fundulids as sentinels of marsh restoration can be further developed.

Chapter 5 - Estuaries provide unique habitat conditions that are essential to the production of major fisheries throughout the world, but quantitatively demonstrating the value of these habitats to fisheries presents some difficult problems. The questions are important, because critical habitats in estuaries are threatened worldwide by development of the coastal zone, climate change, and pollution, among other stressors. This chapter begins by introducing estuarine fisheries, their importance, and the reasons for their prominence. Next the authors discuss the challenges of quantification, followed by three case studies of relationships between fisheries production and habitat structure and function. The authors propose that holistic, integrated approaches to research, modeling, and monitoring can lead to solutions beneficial to our understanding, to better management of coastal habitats, and to the sustainability of fisheries.

Chapter 6 - The accelerating increase of atmospheric carbon dioxide (CO_2) and its implications for climate change are of immense concern. The Hooghly estuary, along with the luxuriant mangroves of Sundarban is one of the world's unique ecosystems. Recent studies of this estuarine system reveal the potentially important role of the estuary in regulating atmospheric CO_2. Phytoplankton fix gaseous CO_2 from the atmosphere into organic carbon through photosynthesis; the carbon is then transformed through subsequent steps in the food chain. Mangrove litter, transported by tidal action, is the key contributor of soil organic carbon (SOC) to the estuary. Following standard techniques, the authors quantified several forms of carbon, including litter carbon, SOC, soil inorganic carbon (SIC), total aqueous organic carbon (TOC), dissolved aqueous organic carbon (DOC), aqueous particulate organic carbon (POC), aqueous dissolved inorganic carbon (DIC), and carbon content of phytoplankton and zooplankton. To study relationships of the various forms of carbon to physical and chemical factors, the authors measured litter biomass, soil salinity, soil pH, redox potential, soil temperature, water temperature, dissolved oxygen, salinity and pH. Statistical correlation analyses were then employed to assess how relationships between carbon forms and physical-chemical factors might contribute to regulating the organic carbon cycle. From this study, the authors have worked out a conceptual model of organic carbon cycling dynamics, and gained insight into mechanisms of transformation of different forms of carbon and the fundamental seasonal processes associated with carbon cycling in the estuarine system.

Chapter 7 - Nutrients in precipitation, in the groundwater of two subwatersheds and in two adjacent estuaries on Cape Cod, Massachusetts were measured regularly year-round from 1988 to the present. This 22-year data set captured low frequency (climate scale) equilibrium changes in nutrient levels, as well as transient changes brought about by an unexpected NO_3 increase to one of the subwatersheds-estuary systems. Both estuaries exhibit low background levels of NO_3 (< 0.2 mg/L or ~3 μM), which increase 6-fold over 3–5 year cycles. Dissolved oxygen concentration remains at 70–90% saturation, summer to winter. The annual average NO_3 concentration in one subwatershed-estuary system was correlated strongly with regional driving forces (low frequency precipitation and water table height), but was not correlated with anthropomorphic sources. Nitrate in the second subwatershed-estuary system exhibited the same correlation with regional driving forces until 2004, when a large, unexpected (20-fold) increase in groundwater NO_3 concentration occurred in this estuary, apparently introduced by an event in the up-gradient watershed. The correlation with regional driving forces disappeared and, after a year's delay, high frequency (3–4 month) NO_3 variability appeared during the winter collapse of the biological NO_3 demand; this cycle lasted for four years. The estuary then returned to its pre-2004 NO_3 level, even with the groundwater NO_3 input remaining high. The transient intervening high frequency behavior is consistent with "regime shift" concepts in their pre-shift stage; if so, the groundwater NO_3 input conditions for a full eutrophic regime shift remain to be found. The generalized case would be the following: in a temperate estuary seasonal NO_3 dynamics are statistically independent of long-term trends and cycles, much like weather is distinct from climate or turbulence from mean flow. The seasonal time scale then can be separated from long- term climate events. However, sudden events can destroy this scale separation and its simplification, with summer algae dynamics then leading to winter NO_3 variability after the collapse of the seasonal biological NO_3 demand. Estuary dynamics can bring about a return to pre-transient conditions, even with continued high NO_3 input, but this latter process is not yet well understood.

Chapter 8 - The European Water Framework Directive (WFD) considers water management from a wide perspective, looking for the prevention of any future deterioration of water bodies, as well as the protection and improvement of the state of marine ecosystems, in order to obtain "a good status" of water bodies. According to the WFD, good status of water bodies is obtained when concentrations of the priority substances in water, sediment and biota are below the established Environmental Quality Standards (EQSs). All the European Union member states must implement management plans in their river basins, including monitoring programs. In this chapter the authors compare two estuaries (Bilboa and Urdaibai in northern Spain) that have experienced very different human impacts. They are evaluated by reporting information on (1) concentrations and spatial distribution of trace metals and organic contaminants in sediments along both estuaries since 2002; (2) tools for finding potential sources of micro-contaminants (PCB, PAH) in some of the estuary compartments; (3) relationships between contaminant concentrations and particle size parameters or organic matter content; and, (4) the relevance of total organic carbon (TOC) distribution in the sediment as the principal pathway of transport or persistence of metals between inorganic compartments.

Chapter 9 - Estuaries are usually highly affected by pollutant discharges of both urban and industrial origin. Environmental risk assessment studies in estuaries most often include trace elements, bacause of their high toxicity and regular presence in these environments. The fate of trace elements in an estuary is clearly influenced by changes in physico-chemical properties of the estuary as mediated by tidal effects. Rigorous monitoring programs are necessary for the adequate study of the spatial distribution and evolution in time of trace elements in estuarine waters. These programs should include factors such as salinity, sampling depth, proximity to pollution inputs and sampling season. In this work the authrs describe a monitoring program carried out in the estuary of the Nerbioi-Ibaizabal River (metropolitan area of Bilbao, Basque Country) to investigate the fate of several trace elements in water. This estuary is located in one of the most populated areas (~1 million inhabitants) on the Bay of Biscay, with numerous industrial, urban and recreational activities. The exploitation of local iron at the end of the 19th century and ship-building in the bay also have had clear influence on the water quality of the estuary. Surface and deep (in contact with the sediment) water samples were collected approximately every three months from January 2005 to October 2010 (22 sampling campaigns) at eight different points of the estuary (four in tributary rivers, one in an enclosed dock area, two in the main channel, and one in the mouth of the estuary). The sites were sampled at low and high tides. Physico-chemical parameters such as temperature, redox potential, dissolved oxygen (DO) and conductivity were measured in situ by means of a multi-parameter probe. The values found were typical from a highly stratified estuary; the average dissolved oxygen concentration was ~65% of saturation. Concentrations of Al, As, Cr, Cu, Fe, Mn, Ni and Zn, measured simultaneously in the samples by inductively coupled plasma/mass spectrometry (ICP/MS), were comparable to those of similar estuaries in Europe, with the exception of As, Cu and Fe, which presented significantly higher values. Concentrations of most of the metals at most of the sampling sites significantly increased with time over the period investigated, a trend that was especially noticeable in samples collected at high tide at the bottom. Decreasing trends were found systematically in time series of pH, whatever the sampling site and the characteristics of the samples (deep, surface, high tide and low tide). Possible sources of each metal to the estuary were also investigated.

Chapter 10 - A newly developed ecosystem model—the first model describing ecological connectivity consisting of both benthic-pelagic and central bay-tidal flat ecosystem coupling while simultaneously describing the vertical micro-scale in the benthic ecosystem—was developed and applied to Tokyo Bay. The model permits prediction and evaluation of the effects of environmental measures, such as tidal flat creation or restoration, sand capping, dredging, and nutrient load reduction from rivers, on the hypoxic estuary from the perspectives of (1) the whole estuary composed of temporal-spatial mutual linkage of benthic-pelagic or central bay-tidal flat ecosystems (holistic approach), and (2) each biochemical and physical process contributing to oxygen production and consumption (elemental approach). The model outputs demonstrated significant ecosystem responses as follows. First, the oxygen consumption in the benthic system during summer was quite low due to low concentrations of dissolved oxygen (DO), i.e. hypoxia, although reduced substances, Mn^{2+}, Fe^{2+}, and S^{2-}, accumulated in the pore water. This result demonstrates the importance of using oxygen consumption rate at high DO concentrations as the index of hypoxia potential. Second, simulations of both tidal flat creation and nutrient load reduction decreased the anoxic water volume and mass of detritus in Tokyo Bay. Simulated creation of tidal flats, however, led to a greater biomass of benthic fauna, whereas nutrient load reduction led to lower biomass of benthic fauna compared to the existing situation. This result clarifies the differences between contrasting goals: (1) a bountiful ocean, non-hypoxic and with rich production of higher trophic level biology, vs. (2) a clear ocean, non-hypoxic and with low levels of particulate organic matter, and also the distinction between a bountiful ecosystem and higher water quality. Lastly, in the simulation, reproducing reclaimed tidal flats, as existed in the earlier Tokyo Bay system, prevented an increase of oxygen consumption potential (hypoxia potential) and a decline of higher trophic production leading to red tides, compared to the existing Tokyo Bay system with reclamation of tidal flats. This result demonstrates higher ecosystem tolerance of the earlier Tokyo Bay to red tide, and the tidal flat's function of keeping an optimum ecological balance resilient to environmental perturbation.

Chapter 11 - Coastal waters in the United States include estuaries, bays, sounds, coastal wetlands, coral reefs, intertidal zones, mangrove and kelp forests, seagrass meadows, and coastal ocean and upwelling areas. These coastal areas encompass a wide diversity of ecosystems that result from the tidal exchanges between freshwater rivers and saline ocean water that occur within coastal estuaries. Coastal habitats provide spawning grounds, nursery areas, shelter, and food sources critical for the survival of finfish, shellfish, birds, and other wildlife populations that contribute substantially to the economic health of the Nation. Section 305(b) of the Clean Water Act requires that the U.S. Environmental Protection Agency (EPA) report periodically on the condition of the nation's coastal waters. As part of this process, coastal states provide valuable information about the condition of their coastal resources to EPA; however, because the individual states use a variety of approaches for data collection and evaluation, it has been difficult to compare this information among states or on a national basis. In 1999—to better address questions about national coastal condition—EPA, the National Oceanic and Atmospheric Administration (NOAA), and the U.S. Fish and Wildlife Service (FWS), agreed to participate in a multiagency effort to assess the condition of the nation's coastal resources. The agencies chose to assess condition using nationally consistent monitoring surveys to minimize the problems created by compiling data collected using multiple approaches. The results of these assessments were compiled periodically into a series of *National Coastal Condition Reports* (NCCR). This series of reports contains the most comprehensive ecological assessment of the condition of U.S. coastal bays and estuaries. This chapter describes the planning, execution and communication of the National Coastal Assessments.

Chapter 12 - The ability of coasts to withstand floods and sea-level rise is investigated using a probabilistic Bayesian method. Fundamental principles of recently developed probabilistic computational methods for monotonic models are presented. Computational efficiency of these models and their coupling with a family of Monte Carlo (MC) methods are investigated for coastal estuaries and inter-connecting coasts. These integrated state-of-the-art computational modeling techniques are used in strategic planning and engineering management decision tools to evaluate impacts of storm hazards on estuaries and coasts in development of flood protection plans and assessment of damage to infrastructures.

Chapter 13 - This chapter introduces decision analysis concepts with examples for managing fisheries. Decision analytic methods provide useful tools for structuring environmental management problems and separating technical judgments from preference judgments to better weigh the prospects from decisions. First, an introduction to decision analysis methods will be given. To illustrate the concepts, a decision context of restoring and sustaining fisheries for the Natural Resource Damage Assessment process after the 2010 Gulf of Mexico oil spill is investigated and objectives derived. The fundamental objectives, measures for the achievement of fundamental objectives, and the means to achieve the fundamental objectives are selected. Additional topics important to decisions such as weighing trade-offs with decision-makers or stakeholders, considering uncertainty, and the value of information for monitoring data and other information that can assist in making a decision also are introduced and discussed. The decision analysis field offers tools that can integrate ecological and socioeconomic research to better understand problems and create improved opportunities. The processes described in this chapter might be useful ones for assessing restoration and recovery tasks as well as for providing more rigorous understanding of the opportunities available to better manage fisheries in the Gulf of Mexico and beyond.

Chapter 14 - Estuaries are subject to stresses that originate from intensive human activities in the proximal estuarine environment and in watersheds that drain to the coastal zone. These stresses are summarized here, with examples from this book and the scientific literature. To sustain the many benefits that estuaries provide to society, and to restore those that have been lost, will require better decisions than have been made in the past. Management and policy will need to be more proactive and less reactive, deal with upstream causes to relieve downstream effects, and account for a full range of ecosystem services in the economic balance of costs and benefits.

GLOSSARY OF SCIENTIFIC NAMES
AND TECHNICAL TERMS

Acanthus ilicifolius	Sea holly, or holly mangrove; a small coastal shrub native to India and Sri Lanka
Acipenseridae	Sturgeons; family name
Acipenser oxyrhincus desotoi	Gulf sturgeon; a subspecies of sturgeon; tributary rivers, estuaries and marine waters of the northeastern Gulf of Mexico
Alosinae	Shad and river herrings; sub-family name, family Clupeidae
ANCOVA	Analysis of covariance; a statistical technique that tests for differences in a dependent variable across two or more classes, while controlling for the the effect of a continuous covariate; a combination of analysis of variance with regression analysis
Angiosperms	Flowering plants
Anguillidae	Eels; family name
Anoxia	Absence of oxygen; in water, a non-detectable concentration of dissolved oxygen
Anthropogenic	Caused by humans
Atlantic bluefin tuna	See *Thunnus thynus*
Autotroph(ic)	"Self-feeding", i.e. capable of producing organic matter from CO_2 and light or chemical energy; applies to plants, algae, and some microorganisms
Avicennia alba	Mangrove native to India, southeast Asia and the western Pacific, known as *api-api putih* in Singapore
Avicennia marina	White or grey mangrove, widespread from Africa to New Zealand
Bayesian analysis	A method for determining the probabilities of outcomes given prior knowledge or beliefs and newly observed data; based on Bayes' theorem
Benthic organisms	Animals that live in or on the sea bottom
Beta distribution	The beta distribution is a family of continuous probability distributions defined on the interval (0, 1) parameterized by two positive shape parameters, typically denoted by α and β
Biodiversity	The variety of life on earth, including animals, plants,

	microorganisms, their habitats and genes; may refer to the varieties of life in a specified area
Biological criteria	Expected values of biological indicators (such as fish and invertebrate communities) in the absence of ecological impairment
Biogeochemical, biogeochemistry	Biogeochemistry is the study of chemical cycles, e.g., carbon, nitrogen, sulfur, that are either driven by or have an impact on biological activity
Biological productivity	The rate of production of biological material by living entities
Biomass	Mass of living biological matter; applied to species, communities of organisms, and specific areas or habitats
Biotic community	A group of interacting organisms within a habitat
Carbon cycle	The transformation of inorganic carbon (CO_2, CH_4) to organic carbon and back to inorganic carbon through biogeochemical processes, e.g., photosynthesis, respiration, decomposition and remineralization
Cephalorynchus hectori	Hector's dolphin; a small dolphin that inhabits coastal waters of New Zealand
Ceriops decandra	A mangrove native to southeast Asia
Chemosynthesis	Production of organic matter from inorganic carbon (CO_2 or CH_4) and water using chemical energy (rather than light); energy-yielding substrates for chemosynthesis include hydrogen sulfide (H_2S), ammonium (NH_4^+), methane (CH_4) and reduced iron (FeII)
Clupeidae	Herrings; family name
Cohorts	A group of organisms of the same age; the term is often used in fishery science
Coriolis force	An apparent force resulting from the Earth's rotation that deflects moving objects (such as water currents) eastward in the northern hemisphere and westward in the southern hemisphere
Crassostrea virginica	The eastern oyster, native to western Atlantic and Gulf of Mexico estuaries
Cynoscion nebulosus	Spotted seatrout; also speckled trout, spotted weakfish; estuaries of the eastern and southeastern U.S, also Mexico; an important recreational species
Derris trifoliata	A climbing, leguminous plant of southeast Asia and southwest Pacific islands; source of the pesticide rotenone
Detritus	Non-living particulate organic matter; base of important food chains in estuaries
Eastern oyster	See *Crassostrea virginica*
Ecosystem	The complex of a community of living organisms with its environment, functioning as an ecological unit
Ecosystem services	The goods and services supplied to humans as a result of ecosystems and ecological processes, e.g., food, timber, clean water
Effluent	Something that flows out, as water from waste streams and waste

	treatment systems; often associated with water pollution
Elicitation	The process or act of drawing out information or behavior (see Chapter 13)
Epifauna	Animals living on surfaces; in estuaries, includes oysters, mussels, barnacles, and encrusting bryozoans, among many others
Euryhaline	Capable of physiological tolerance for a range of salinity; typical of many estuarine animals
Eutrophication	An increase in the production of organic matter within an ecosystem; often associated with nutrient pollution in estuaries
Excoecaria agallocha	Milky mangrove; a toxic plant of coastal southern Asia and Australia
Farfantepenaeus aztecus	Brown shrimp; inhabit temperate and tropical shallow marine and estuarine waters worldwide; an important fishery species in the Gulf of Mexico and elsewhere
Fjord	A coastal water body formed in a glacial valley (see Chapter 1)
Frustule	The two-valved siliceous (glassy) shell of a diatom
Fundulids	Fishes of the family Fundulidae; mostly small, estuarine fishes, known generally as killifishes or topminnows
Fundulus grandis	Gulf killifish; a tropical killifish (see Fundulids) of eastern Florida and the Gulf of Mexico; brackish to fresh water, preferring marshes, seagrass beds and canals
Fundulus heteroclitus	Mummichog (see Fundulids); a fish of marine to brackish water in the western Atlantic, Canada to northeastern Florida; mainly in salt marshes and tidal creeks
Fundulus parvipinnis	California killifish (see Fundulids); southern California and Baja California (Mexico); shallow bays, estuaries and marshes
Gag grouper	See *Mycteroperca microlepis*
Gaussian error function	The normal (Gaussian or bell-shaped) frequency distribution of deviations from predicted values of a model or function
Grey mangrove	See *Avicennia marina*
Gulf killifish	See *Fundulus grandis*
Gulf sturgeon	See *Acipenser oxyrhincus desotoi*
Halimione portulacoides	Sea purslane; a salt-tolerant edible shrub of salt marshes and coastal dunes; Eurasia and parts of Africa
Halophyte	A plant tolerant of saline conditions, e.g., mangroves, salt marsh grasses, seagrasses; typically with special physiological adaptations for excreting salt
Harpacticoid copepods	Small (near-microscopic), mostly benthic crustaceans, inhabiting sediments; important food source for estuarine fishes and invertebrates
Hector's dolphin	See *Cephalorynchus hectori*
Hypoxia	Deficient in oxygen; hypoxic waters are defined generally as having dissolved oxygen <2 mg L^{-1} (125 μM); sometimes <3 mg L^{-1} (188 μM)

Infauna	Animals that live within sediments; burrowing animals such as worms and clams
Juncus gerardii	Saltmeadow rush, also blackgrass, black needle rush, saltmarsh rush; native to North America, also occurs in western Europe and central Asia
Kemp's ridley sea turtle	See *Lepidochelys kempii*
Lepidochelys kempii	Kemp's ridley sea turtle; Atlantic Ocean and Gulf of Mexico; critically endangered
Litopenaeus setiferus	White shrimp; U.S. Atlantic coast through Gulf of Mexico; important fishery species
Macroalgae	Colonial algae that form macroscopic aggregations or plant-like structures, e.g., sea lettuce (*Ulva* spp.), kelp (Laminariales)
Macrobenthic	Belonging to the larger species of benthic infauna (see Benthic organisms and Infauna); animals retained on a 0.5 or 1 mm screen
Macrotidal	Having a large tidal range, >4m
Mangrove	Any of several genera and species of shrubs and small trees that live in semi-aquatic, coastal habitats; tropical environments worldwide
Meiofauna	Animals small enough to pass through a 0.5 mm seive; usually refers to benthic fauna
Mesocosm	Partially controlled experimental systems intermediate in scale between fully controlled laboratory systems and uncontrolled field sampling; usually designed to measure interactive effects of organisms and their environment
Mesotidal	Having a medium tidal range, 2–4 m
Meroplankton	Planktonic early life stages of animals, e.g., larvae of fishes and benthic invertebrates
Metapopulation	A group of spatially separated sub-populations of a particular species that interact in some way (usually reproductively)
Microcosm	A relatively small, laboratory scale, experimental system (see Mesocosm); also a micro-scale representation of a macroscopic system
Microhabitat	The smallest scale of a habitat (≤ 1 m^2); may include environmental factors such as sediment grain size, plant species and density, and immediate predator-prey interactions
Microorganisms	Organisms of microscopic size including bacteria, fungi, viruses, etc.
Microtidal	Having a small tidal range, <2m
Mineralization	Conversion of organic matter to inorganic compounds, e.g., organic carbon to CO_2, organic nitrogen to NH_4^+
Mixohaline	Of water with salinity greater than fresh water and less than marine water; salinity = 0.5–30
Monotonic	A term in logic; a monotonic model assumes that a hypothesis holds over all possible observations of a parameter; e.g., most

	birds fly, so if it is a bird, it flies; a non-monotonic model would relax this obviously false premise
Monte Carlo methods	Methods for estimating uncertainty in mathematical models by repeated random draws from frequency distributions of data or model parameters
Morone americana	White perch; a semi-anadromous fish of western Atlantic estuaries; also introduced into some freshwater bodies, including the North American Great Lakes; a popular recreational and commercial fish in U.S. mid-Atlantic states; considered invasive in the Great Lakes
Morone saxatilis	Striped bass; a sport and commercial fish of the western Atlantic and Gulf of Mexico coasts; spawns in freshwater reaches of estuaries; also introduced to lakes
Moronidae	Family of temperate basses (see *Morone americana* and *Morone saxatilis*)
Mummichog	See *Fundulus heteroclitus*
Mycteroperca microlepis	A reef fish ranging from North Carolina through the Gulf of Mexico; commercially and recreationally important; juveniles inhabit seagrass beds in estuaries
Nekton	Free-swimming aquatic animals; contrast with Plankton
Nematodes	Roundworms; small, mostly near-microscopic inhabitants of most of earth's environments; extremely abundant and diverse in estuarine sediments; a prominent component of the Meiofauna (see above)
Ontogenetic movement	Movement or migration of animals as a feature of their development and life history, e.g., juvenile salmon migrate from their natal streams through estuaries into the ocean, later returning as mature adults to spawn
Otolith	A bony structure within the ears of fish; otoliths have concentric growth rings that can be used for determining the age of the fish
Oyster reefs	Aggregations of shells from living and dead oysters; the reefs include sediments and other epifauna; somewhat analogous to coral reefs
Penaeid shrimp	Penaeidae is a diverse family of prawns ("shrimp") that includes many species of commercial value
pH	The negative logarithm of hydrogen ion concentration in water; pH 7 means $H^+ = 10^{-7}$ M; a pH of 7 is neutral ($H^+ = OH^-$), <7 is acidic ($H^+ > OH^-$), and >7 is basic ($H^+ < OH^-$)
Photosynthesis	Conversion of CO_2 and water to organic matter using energy from sunlight
Phytoplankton	Single-celled, photosynthetic algae that float freely in water
Plankton(ic)	Passively floating or weakly swimming (usually minute) animal and plant life of a body of water
Porteresia coarctata	Tropical, salt tolerant wild rice; native to salt marshes in southeast Asia
Primary production	Production of organic matter from inorganic matter by

	photosynthetic and chemosynthetic organisms (see Photosynthesis and Chemosynthesis)
Recruitment	The process of becoming a member of a functional population; used in fisheries for the process of younger fishes growing to harvestable size (recruitment to the fishery) or reproductive competence (recruitment to the spawning stock); may mean growth to any life stage or functional state from an earlier stage
Red drum	See *Sciaenops ocellatus*
Redox potential	Reduction potential or oxidation-reduction potential; chemical potential to contribute electrons (oxidation) or to accept electrons (reduction) expressed in millivolts; in estuaries, usually refers to a bulk property of sediment or water rather than to a chemical species as in chemistry; reducing environments (negative redox potential) are anoxic, produce sulfides from sulfate, etc.
Residence time	The average time for a substance to pass through a compartment, e.g., the residence time of freshwater in an estuary; reciprocal of turnover time
Respiration	Production of CO_2 from organic matter that is used as an energy source in metabolism
Rhizo-sediment	A shallow sediment zone that contains roots and rhizomes of aquatic plants
Ria	An estuary formed from a former river mouth that has been submerged by rising sea level (see Chapter 1)
Salt wedge estuaries	Estuaries dominated by river flow, with relatively stable stratification (see Chapter 1)
Sarcocornia fruticosa	A succulent, salt tolerant coastal plant (glasswort); North America, Eurasia, Australia, Chile
Sarcocornia perennis	Perennial glasswort, a succulent salt tolerant coastal plant; widely distributed in coastal habitats
Sciaenidae	A family of fishes that includes drums, croakers, and weakfish; worldwide in warm temperate and tropical waters; many species inhabit estuaries
Sciaenops ocellatus	Red drum; also redfish; Coastal waters of the western Atlantic Ocean and Gulf of Mexico; important recreational and commercial species
Secondary production	Production of biomass by organisms that feed on primary producers or detritus; e.g., copepods, benthic infauna
Sentinel species	Species that are sensitive or prominent indicators of environmental effects, especially pollution
Silt	Granular mineral particles of intermediate size (3.9–62.5 μm in effective diameter), larger than clay and smaller than sand
Spartina maritima	Small cordgrass; a salt marsh plant native to western Europe and parts of Africa
Spotted seatrout	See *Cynoscion nebulosus*
Thunnus thynnus	Northern (or Atlantic) bluefin tuna; Atlantic Ocean and

	Mediterrean Sea; valuable commercial and sport fishery species; possibly endangered from overfishing
Tidal asymmetry	Deviations from even distributions (in space or time) of tidal flow that result from the Coriolis force, variations in topography and other physical irregularities; often refers to lateral (cross-estuary) gradients in tidal currents
Upwelling	Movement of a water mass upward from depth toward the surface; upwelling waters tend to be cold, high in nutrients, and low in dissolved oxygen
White shrimp	See *Litopanaeus setiferus*
Zooplankton	Animals that spend their entire lives as plankton, e.g., copepods; contrast with Meroplankton

In: Estuaries: Classification, Ecology and Human Impacts
Editor: Steve Jordan

ISBN: 978-1-61942-083-0
© 2012 Nova Science Publishers, Inc.

Chapter 1

INTRODUCTION TO ESTUARIES

Stephen J. Jordan
U. S. Environmental Protection Agency, Gulf Breeze, FL, US

ABSTRACT

Estuaries, although minor geographical features at the global scale, have major importance for society and the world's economies. This chapter introduces estuaries by presenting an overview of definitions, origins, physical, chemical and ecological attributes, and the interactions of estuarine ecosystems with human activities, including attempts to restore degraded estuaries to a more desirable ecological condition.

Keywords: Estuary classification, hydrology, fisheries, water quality, nutrients, pollution, restoration

INTRODUCTION

Estuaries are unique environments, sharing characteristics with oceans, lakes, and rivers, yet very different from any of them. Thus, estuaries collectively occupy a special niche in the geographical, hydrological, and ecological sciences, even though there is great diversity among individual estuaries. As we learn from Kurtz & Hagy (Chapter 2), much theoretical and empirical effort has been applied to classifying estuaries for various purposes, but a universally satisfactory classification has been elusive.

Beyond conveying a particular fascination for scientific inquiry, estuaries have distinctive relationships to society and economic development. Most of the world's seaports are in estuaries, where protected harbors connect with agriculture, industry, and transportation. Long before steam power, ships used the energy of incoming and outgoing tides to convey them from the ocean to estuarine ports and back. The estuaries of the Thames River (London), Hudson River (New York), Sacramento River (San Francisco Bay ports), Susquehanna River (Baltimore and Norfolk ports in Chesapeake Bay), and Haihe River (port of Tianjin on Bohai Bay, China) are home to notable examples of the world's great ports.

Estuaries support a large proportion of the world's fisheries because of their high biological productivity, critical habitat functions, and accessibility, with enormous economic and social consequences (Jordan & Peterson, Chapter 5). They also contribute to the quality of life for coastal residents and visitors in a variety of other ways, among them as centers of trade and employment, places for recreation and aesthetic enjoyment, and sites of distinctive cultures and communities.

Given all of these benefits and more, it is not surprising that estuaries, their watersheds, and adjacent coastal areas are centers of human population, and that coastal populations are growing at a faster rate than the world's population. There is little need to elaborate here that human occupation of the coastal zone, coupled with intensive uses of coastal and estuarine resources, has created problems of pollution (Bartolomé, Chapter 8; Gredilla et al., Chapter 9; Sohma, Chapter 10), overuse, habitat destruction (Jordan & Peterson, Chapter 5; Sohma, Chapter 10) and general environmental and ecological degradation (Martínez et al. 2007).

This chapter is intended to be a broad overview of estuaries. For more complete information, the interested reader should turn to the other chapters in this book, to the references therein, to the primary literature available through libraries and on-line searches, and to a smattering of books. In the last category, Day et al. (1989) and Hobbie (2000) are recommended as starting points.

DEFINITIONS AND PHYSICAL CHARACTERISTICS OF ESTUARIES

What is an estuary? The word estuary comes from the Latin *aestuarium*, literally "place of the tide." Webster (1975) defines estuary as "a water passage where the tide meets a river current; *esp*: an arm of the sea at the lower end of a river." Probably the most widely recognized definition is from Pritchard (1967): "*a semi-enclosed coastal body of water, which has a free connection with the open sea, and within which sea water is measurably diluted with freshwater derived from land drainage.*" Although we can accept the last of these as the formal definition of an estuary, in practice, there are exceptions. The key elements of Pritchard's (1967) definition are "semi-enclosed" and "measurably diluted with freshwater." The influence of some large estuaries, however, expressed as significant dilution of ocean water, along with high nutrient, chlorophyll, and suspended solids concentrations, extends far beyond their geomorphic boundaries into the open ocean. The Amazon estuary (Figure 1) and Chesapeake Bay are notable examples, where the oceanic zones of estuarine influence are known as estuarine plumes (Bezzera 2009, Guo & Valle-Levinson 2007). As for measurable dilution with freshwater, some coastal lagoons, e.g., Laguna Madre on the western coast of the Gulf of Mexico (Quammen & Onuf 1993), Lagoa de Arauama, Brazil (Kjerfve et al. 1996) and several others, have little freshwater inflow; evaporation generally exceeds freshwater inflows from rivers and precipitation, so that these bodies of water are often hypersaline, i.e. saltier than the ocean's salinity of 35, at times much saltier (salinity >50). Even water bodies that have no open connection with the sea and consist entirely of freshwater have been classified, studied, and incorporated into policy as "estuaries," as in the North American Great Lakes (Herdendorf 1990, http://lsnerr.uwex.edu/). For the remainder of this chapter, I adopt a convenient definition: *estuaries are semi-enclosed coastal water bodies that connect with the ocean and are influenced by ocean tides.* This definition includes hypersaline lagoons and tidal freshwater rivers, but excludes estuarine plumes and lakes.

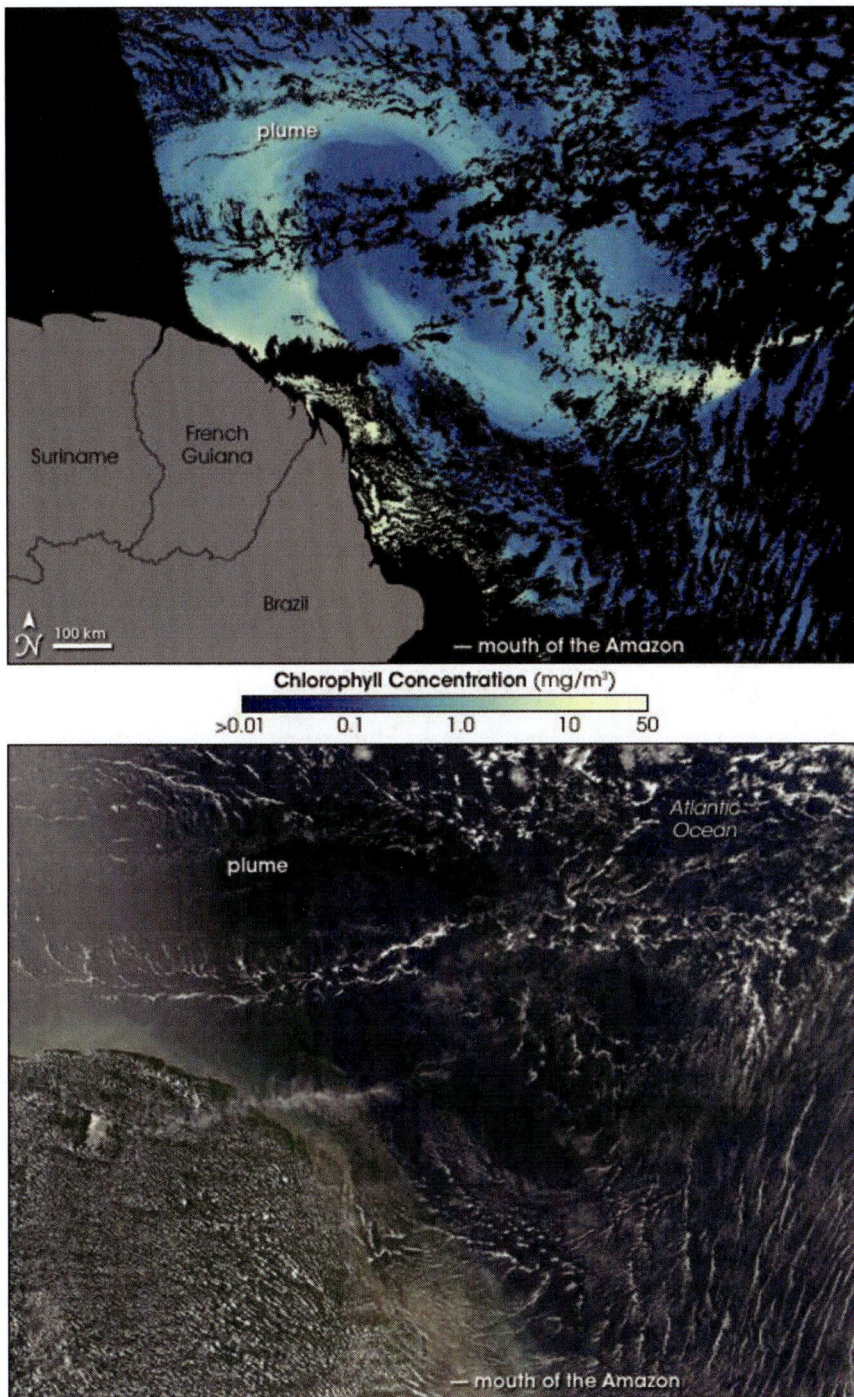

Figure 1. The Amazon estuarine plume extending into the Atlantic Ocean. Upper panel: chlorophyll concentrations; lower panel, natural color satellite image. Images from NASA, http://earthobservatory.nasa.gov/IOTD/view.php?id=7021 (April 21, 2011).

Estuaries originate from four distinct geologic processes: (1) drowning of river mouths by rising sea level and land subsidence (e.g., Chesapeake and Delaware Bays) – these estuaries are known internationally as *rias*; (2) formation of coastal declivities such as fault valleys by tectonic forces (e.g., San Francisco Bay); (3) formation of bars or barrier islands by sediment deposited along ocean shores, partially enclosing river mouths or other coastal waters (e.g., several small estuaries along the North American Pacific coast, many coastal lagoons); and (4) carving of coastal valleys by glaciers, forming *fjords* when the glaciers retreat (e.g., Puget Sound in the northwestern U.S., Sognefjord, Norway). These major classes of estuaries differ greatly in depth, tidal ranges and currents, stratification, biological attributes, and vulnerability to human influences.

Coastal lagoons tend to be the shallowest estuaries, with characteristic depths of 1 to a few m. Fjords are the deepest, up to ~1000 m; rias and tectonic estuaries have intermediate depth profiles (~5–100 m). Tidal ranges extend from microtidal (cm) to macrotidal (up to 10 m or more, with extremes >20 m). Tidal frequencies may be diurnal (one tidal oscillation, or one high and one low tide per day) or semi-diurnal (two tidal oscillations per day). In some estuaries, the magnitude of wind-driven changes in water levels can be larger than the astronomical tides.

Stratification occurs when water masses with different densities form more or less distinct layers. Density in estuaries and marine waters is determined by salinity and temperature: cold, saline water is denser than warmer, fresher water. Although salinity usually is the dominant component of the density function in estuaries, temperature gradients also can be important in the physical dynamics of stratification and mixing. Some estuaries are well-mixed (Figure 2A), as in coastal lagoons where shallow depths, winds, ocean currents, and lack of freshwater inflows tend to maintain a homogenous vertical profile of density. Partially mixed estuaries (Figure 2B) have zones of stratification and zones of mixing, where tidal energy, or sometimes the force of freshwater inflow from rivers, disrupts the density strata. Salt wedge estuaries (Figure 2C), where river flow dominates tidal forces, are characterized by a relatively stable lens of freshwater overlying a stable lens of ocean water. The freshwater layer thins seaward as the salt layer thickens. Episodic mixing occurs in salt wedge estuaries, driven by friction (shear) between the density layers (Geyer & Farmer 1989). Deep fjords (Figure 2D) tend to have very stable stratification, with a persistent layer of fresh or brackish water overlying very dense (cold and saline) deeper water (e.g., Gibbs et al. 2000). Exchange between a typical fjord and the ocean is impeded by a sill—the terminal moraine of the former glacier—at the mouth of the fjord.

The unique flow dynamics known as estuarine circulation arise from the combination of density gradients, tides, and river inflows, particularly in partially mixed estuaries. In two-layered estuarine circulation, fresher, less dense water flows seaward in the surface layer, while denser, saline water flows landward in the bottom layer (Figure 2B). These are net flows after accounting for oscillating tidal flows. Three-layered circulation can develop in lateral branches of estuaries (sub-estuaries) where both surface and bottom flows are landward, i.e. fresher water from the upper estuary and more saline water from the lower estuary both enter from the mouth of the sub-estuary flowing landward. In this case, a third layer develops at mid-depth flowing seaward, thereby conserving momentum (Chao et al. 1996). In estuaries where evaporation exceeds the rate of river inflow (e.g., Laguna Madre), salinity increases from seaward to landward. These systems are known as negative estuaries because of the reverse salinity gradients (Figure 2E).

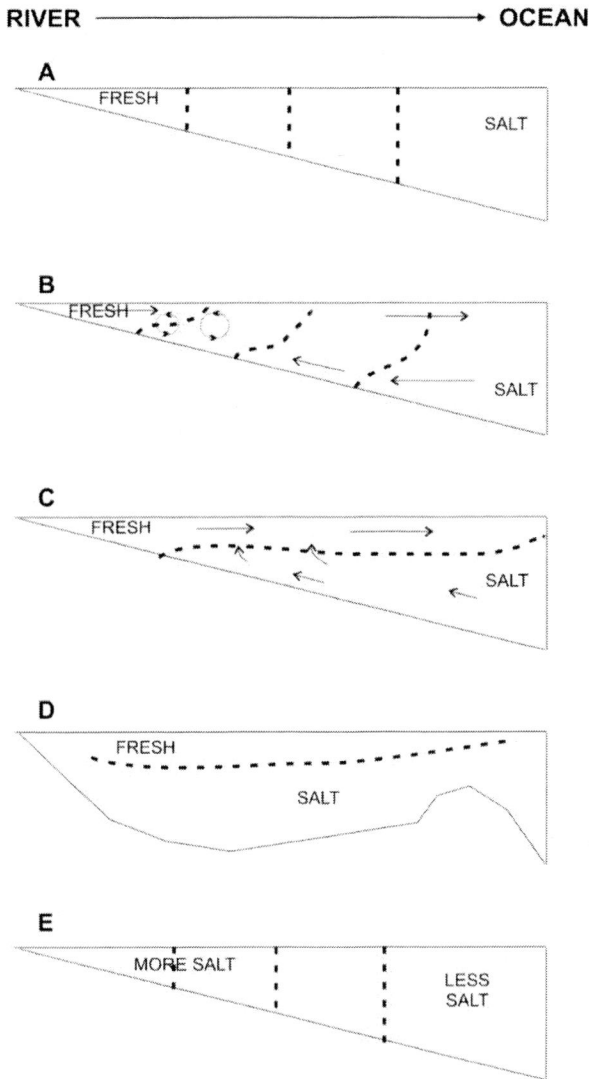

Figure 2. Five types of estuaries. A: well-mixed; B: partially mixed; C: salt wedge; D; fjord; E: negative estuary. Arrows indicate net flows of fresh and saline water and mixing. Vertical dashed lines indicate salinity gradients.

In some estuaries, wind stress, tidal asymmetry (spatial and temporal), the Coriolis force, and geomorphology contribute to significant density and flow variations in the lateral dimension (Dyer 1974). Upwelling of oxygen-depleted bottom water along one shore of the estuary, accompanied by downwelling along the opposite shore, is an example of lateral variation, where wind stress in the direction of the downwelling shore is the driving force. These dynamics can have negative consequences for aquatic life exposed to oxygen-depleted water (Breitburg et al. 1994, Reynolds-Fleming & Luettich 2004).

Estuaries are sedimentary environments. Large quantities of particulate matter, both inorganic and organic, enter estuaries from rivers, from the ocean, from land-based sources and shoreline erosion, and as the result of internal primary production. Much of this particulate matter sinks to the bottom, becoming bedded sediment, where it may be buried more or less permanently or resuspended, either to be re-deposited in the estuary or exported to the ocean. In general, estuaries accumulate sediment, with the consequence that they become shallower over time (Schubel & Kennedy 1984). Estuaries that are home to seaports and marinas typically need continual dredging to maintain navigational channels with sufficient depth for shipping and boating. The study of Chesapeake Bay sedimentation by Officer et al. (1984) is a good introduction to sedimentation processes in estuaries.

CHEMISTRY OF ESTUARIES

The highly dynamic chemistry of most estuaries is dominated by variations in salinity, ranging from near 0 to 35 or greater. Salinity gradients influence the concentrations and behavior of other chemical constituents, directly or indirectly: dissolved metals, organic contaminants, nutrients, and biochemicals interact with salinity through a variety of processes. The ionic strength of estuarine water, chiefly determined by salinity, affects the chemical activity, solubility, and bioavailability of metals such as Cd, Cr, Cu, Pb and Zn, although other factors that vary over estuarine gradients, such as pH, suspended particle concentrations, and organic matter complicate these relationships (Windom et al. 1983, Turner 1996, Turner et al. 2002). Complex bio-organic compounds—proteins, humic acids, etc.—are precipitated ("salted out") from solution at higher salt concentrations, contributing to suspended solids and sedimentary particulate fractions. Similar processes apply to organic pollutants (e.g., PAH, PCB) and some metals, one of the reasons why estuaries have been characterized as pollutant traps or filters (Schubel & Kennedy 1984). For more comprehensive accounts of the chemistry of dissolved and suspended matter in representative estuaries, see Gredilla et al. (Chapter 9) and Zwolsman & van Eck (1999).

Nitrogen and P, major nutrients for phytoplankton, play enormous roles in the chemistry and ecology of estuaries. In undisturbed aquatic systems, N tends to be the most limiting nutrient for phytoplankton production in marine waters, whereas this role is taken by P in freshwater systems (other nutrients, including Fe, Si, and dissolved inorganic C, along with light, can be limiting in some instances). The dynamic gradients of fresh to marine water in estuaries can result in primary production being limited by N, P or both in different reaches of the same estuary or in the same reaches at different times (Fisher et al. 1999). Most of the world's estuaries are enriched in N and P from anthropogenic sources (Seaver, Chapter 7); fertilizers and human and animal wastes contribute excess amounts of N and P to waterways that flow into estuaries. Atmospheric pollution from fuel combustion, industry, and agricultural sources is responsible for substantial loads of N to estuaries, by direct deposition to the estuary and by runoff of atmospherically deposited N from watersheds. By supporting excessive primary production, N and P can cause eutrophication, i.e. an increase in the rate of supply of organic matter (Nixon 1995), with the implication that organic matter accumulates at higher rates than it can be processed through the food chain (zooplankton, filter feeders, fishes, etc.) along trophic pathways leading to desirable outputs. The surplus is consumed by microbial communities, where respiration depletes dissolved oxygen (DO), leading to hypoxia and anoxia (Sohma, Chapter 10). Hypoxic and anoxic water are unsuitable as habitat for most life forms except some microbes. These conditions also drastically alter biogeochemical processes. In the absence of DO, bacterial metabolism converts oxidized chemical species such as NO_3, SO_4, and CO_2 to reduced species, including N_2, N_2O, H_2S, and CH_4. Both N_2O and CH_4 are greenhouse gases; H_2S is toxic to aerobic life. Decomposing organic matter is a source of NH_4, which is nitrified by bacteria to NO_3 in the presence of O_2; this is a major route for N recycling in estuaries.

In the presence of O_2, inorganic P in sediments is often bound in insoluble Fe-oxyhydroxide complexes, making P unavailable for primary production. In anoxic estuarine sediments, Fe is reduced and often bound to sulfides, preventing reoxidation. These processes can cause a significant release of PO_4, which then becomes available as a nutrient for phytoplankton (Jordan et al. 2008). These processes, combined with significant rates of denitrification in estuarine sediments, contribute to the prevalence of N limitation of primary production in estuaries.

Many other elements and compounds, some beneficial, some harmful, and some both, depending on concentration and availability, play important roles in estuarine ecosystems. Calcium, for example, is particularly important for shell formation in mollusks, many of which, including clams and oysters, are major contributors to fisheries in estuaries around the world. Supplied by rivers as a product of land erosion, Si as silicic acid ($Si(OH)_4$) is essential for construction of the external covering (frustule, or exoskeleton) of diatoms, phytoplankton that support a substantial portion of the food chain in many estuaries (Tréguer et al. 1995).

BIOLOGY AND ECOLOGY OF ESTUARIES

The great diversity of life in estuaries results from the coincidence of marine, freshwater, and estuarine species. Despite this diversity, fewer animal species are permanent residents in estuaries than in oceans or rivers, and those that are endemic to estuaries tend to be exceptionally hardy. Estuaries in general are harsh environments for aquatic life, where physiological systems must be adapted to rapidly varying salinity, temperature, light, turbidity, currents, and other physical and chemical properties of the environment.

Organic matter, the material of life, is produced in estuaries by phytoplankton, macroalgae, vascular aquatic plants, and autotrophic bacteria from inorganic carbon, using sunlight (or chemical energy) and nutrients. In addition, rivers, direct runoff, shoreline erosion, and point source effluents deliver organic matter from terrestrial and upland aquatic sources. Primary production and detritus from these sources are the base of food webs (Figure 3). Estuarine food chains may span up to eight trophic levels, from primary producers and detritus to zooplankton, meroplankton, benthic animals, forage fish, and top predators (Baird and Ulanowicz 1989). Some estuaries achieve extraordinarily high levels of organic productivity: Mencken (1940) famously referred to Chesapeake Bay as an "immense protein factory" because of the enormous quantities of crabs, oysters, fish, and waterfowl it produced for human consumption in the late 19th century. Much of the reason for the high productivity of Chesapeake Bay and other estuaries is related to their nutrient trapping and recycling properties. Shallow water, organic-rich sediments, dense beds of aquatic plants, and active microbial communities in surficial sediments (Ulanowicz and Baird 1999) supply key elements of structure and function for nutrient retention, recycling, and provision, that in turn support high levels of production at all trophic levels. By the late 20th century, Mencken's "protein factory" was producing less for human consumption and more (apparently) for microbial communities. Bay grasses, waterfowl, oysters, crabs, and edible fish populations declined, while large portions of the estuary became anoxic during summer. The reasons for these changes are too complex to elaborate upon in this overview, but chiefly involve interactions of excessive nutrient loading, intensive fishing pressure, habitat destruction, and other stresses, mostly related to human population growth and development. A key point is that none of these stresses alone was sufficient to cause the array of changes. For example, Breitburg et al. (2009) describe how interactions of fishing pressure and eutrophication negatively affect fish and shellfish production in estuaries.

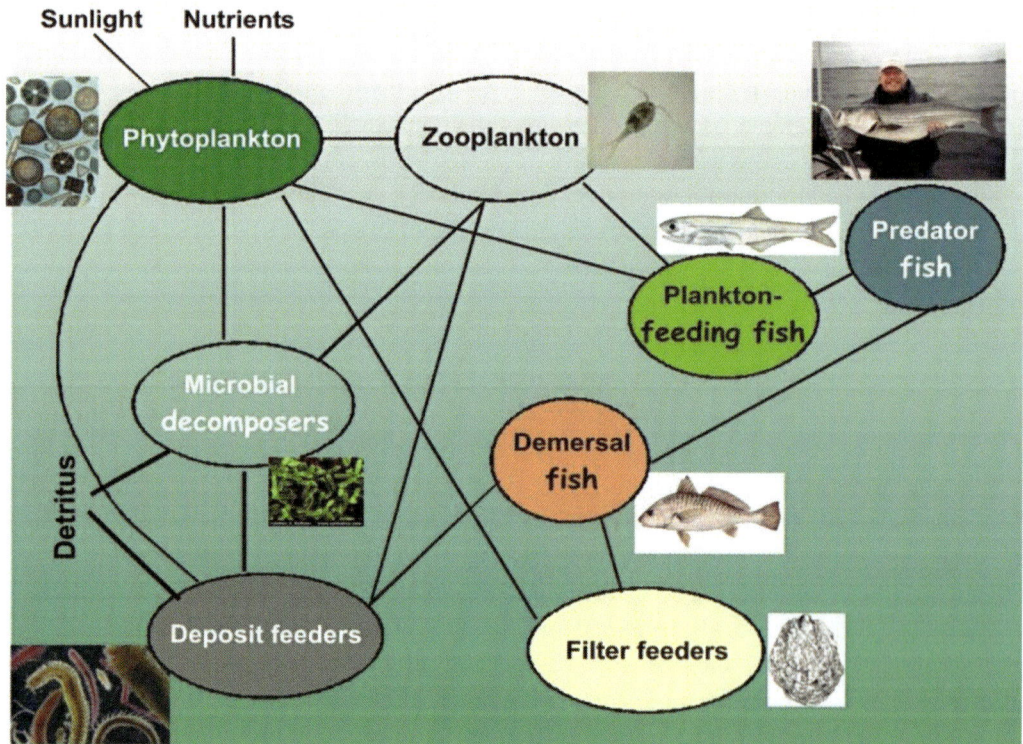

Figure 3. Idealized estuarine food web.

Estuarine ecology can be viewed from the perspective of communities of organisms, including aquatic plants, microbes, phytoplankton, zooplankton, benthic macrofauna, and fishes. A community perspective typically gives a broader, more balanced view of the ecosystem than a focus on individual species (Jordan & Smith 2005). Submersed aquatic vegetation (SAV) communities (seagrasses, bay grasses, etc.) have been studied and monitored extensively because of their habitat value for estuarine animals (Vivian et al., Chapter 4) and their sensitivity to eutrophication (e.g., Avery & Johansson 2008). Benthic macrofauna have received the most attention at the community level, with emphasis on the infauna, i.e. the worms, mollusks, crustaceans and other invertebrates that live within the sediments, as opposed to the epifauna that live on hard surfaces such as shell reefs and artificial structures. Soft-bottom benthic communities have been used extensively as indicators of pollution, hypoxia, and other stresses (e.g., Engle & Summers 1998). Communities of fishes have received less attention in estuaries than in lakes and streams, but nevertheless have value as indicators of ecosystem condition, especially at broad spatial scales (Lewis et al. 2007, Jordan et al. 2010a).

At a yet higher level of organization, estuaries can be understood as integral ecosystems. Jordan & Vaas (2000) combined indicators of phytoplankton, zooplankton, SAV, benthic, and fish communities with environmental variables (nutrients, DO, etc.) into a holistic index of ecosystem integrity for Chesapeake Bay. Whole-ecosystem simulation models have been employed to elucidate relationships that would not otherwise be recognized (Sohma, Chapter 10).

None of the ecological indicators, knowledge, and insight described so briefly in this section would be possible without reference to extensive bodies of data, spanning many estuaries over many years. In some cases, these data have been collected by one or a few individuals (Seaver, Chapter 7); some have been generated by large-scale systematic monitoring programs (Summers et al., Chapter 11). Systematic, ecosystem-based, sustained monitoring is critical to understanding, managing, and conserving ecosystems and the resources and services they supply to society (Jordan et al. 2010b). This ideal has been approached in some estuaries for limited periods of time, but never fully realized.

Many other aspects of estuarine ecology are covered in this book, at various levels of detail. Vivian et al. (Chapter 4) report on the value of salt marsh restoration for a sentinel fish species in estuaries of the Gulf of Mexico. Caçador and Duarte (Chapter 3) investigate interactions of N, metal pollution, and properties of salt marsh grasses. Mukherjee et al. (Chapter 6) develop an organic carbon budget for mangrove forests and associated biota in a South Asian estuary.

ESTUARIES AND SOCIETY

Finally, estuaries cannot be fully understood or managed successfully without attention to human factors. The tension between economic development and environmental sustainability is perhaps more evident in estuaries and their watersheds than in most other ecosystems. Historically, simultaneous uses of estuaries for food production, waste disposal, shipping, and recreation were sustainable when the uses did not exceed the capacity of the ecosystem to sustain them. Some uses, within limits, actually increase carrying capacity by increasing primary and secondary production. For example, increasing the supply of limiting nutrients (N and P) delivered to estuaries (via wastewater, contaminated runoff, and atmospheric deposition) can stimulate fishery production, up to a point (Jordan & Peterson, Chapter 5; Houde & Rutherford 1993). But as human populations grow and migrate to the coasts, making more and more demands—intentional and unintentional—on estuarine ecosystems throughout the world (U.S. EPA 2006), the demands exceed system capacity and become unsustainable. The results, illustrated in several chapters of this book, are widespread environmental degradation and losses of beneficial uses of estuaries.

Past abuses cannot be undone, but if society has the will, those that persist can be curtailed and future insults can be prevented. Restoration programs have been established for many estuaries. In the U.S., National Estuary Programs (NEPs) have supported comprehensive planning for management, restoration, and monitoring of 28 estuarine ecosystems spanning 19 coastal states and Puerto Rico.[1] Among the ten most frequent concerns expressed by the NEPs (Figure 4), habitat loss and alteration ranked highest.

The Chesapeake Bay Program (CBP), a precursor to the NEPs, has established an array of quantitative and qualitative goals, including nutrient load reductions, fishery and habitat restoration targets, and commitments to "develop, promote and achieve sound land use practices" (CBP 2000). This major program has achieved mixed results since its inception in 1983; in 2009, "bay health," based on indicators of water quality, habitat conditions, the food web, and fish and shellfish abundance, was only 45% of the desired value (CBP 2009). Much earlier in the history of estuary management (1960–1980), the estuary of the Thames River (UK) was restored from a thoroughly degraded condition to one of the world's least polluted estuaries, with dramatic increases in fish abundance and diversity (Andrews 1984). Most of the Thames restoration was accomplished by means of improved treatment and management of sewage and industrial wastes.

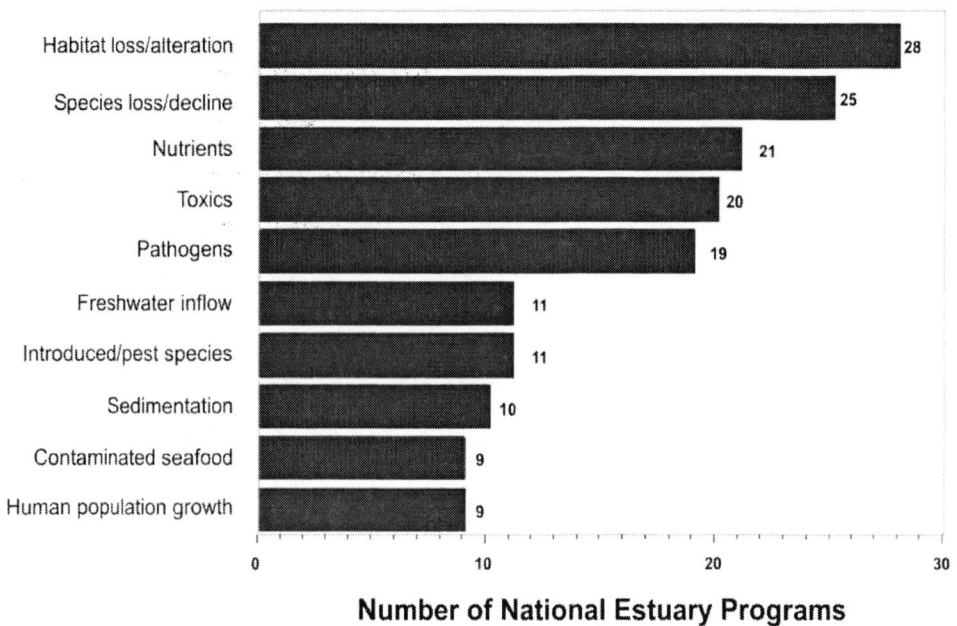

Number of National Estuary Programs

Figure 4. The ten most frequent environmental concerns for National Estuary Programs. Adapted from U.S. EPA (2006). Note: "Pathogens" refers to bacterial contamination inferred from indicator organisms such as fecal coliform and *Enterococcus* spp. bacteria.

The massive amounts of effort and money devoted to restoring estuaries are evidence that people greatly value them and the ecosystem services they provide. Nonetheless, it is problematic that the necessary levels of attention seldom are given to conserving and protecting coastal ecosystems that are not (yet) severely degraded. In recent years, ecologists, economists, and social scientists have begun to work together to understand the values of natural resources and ecosystems before they are lost to present and future generations. Estuaries, with their intricate ties to the land, the ocean, and human populations, should be

---

[1] http://water.epa.gov/type/oceb/nep/index.cfm.

excellent testing grounds for learning and for educating people about ecosystem services and the paths to a sustainable environment.

DISCLAIMER

The views expressed in this chapter are those of the author and do not necessarily reflect the views or policies of the U.S. Environmental Protection Agency. I appreciate thoughtful reviews by Jim Hagy and Jan Kurtz. This is contribution 1432 from the Gulf Ecology Division, and a product of EPA's Ecosystem Services Research Program.

REFERENCES

Andrews, M.J. (1984). Thames Estuary: pollution and recovery. In: P.J. Sheehan, D.R. Miller, G.C. Butler & Ph. Boudreau (Eds.), *Effects of Pollutants at the Ecosystem Level*. John Wiley & Sons, London, pp. 195-228.

Avery, W. & Johansson, R. (2008). Tampa Bay interagency seagrass monitoring program: A synopsis of seagrass trends from 1997–2006. Tampa Bay Estuary Program, http://www.tampabay.wateratlas.usf.edu/upload/documents/TBEP_04_08_Seagrass_Tran sect_Synthesis.pdf. Accessed May 24, 2011.

Bezzera, M.O. (2009). Current characterization at the Amazon estuary. *Geophys. Res. Abs. 11*, EGU2009-13373.

Baird, D. & Ulanowicz, R.E. (1989). The seasonal dynamics of the Chesapeake Bay ecosystem. *Ecol. Monogr. 59*, 329–364.

Breitburg, D.L., Steinberg, N., DuBeau, S., Cooksey, C. & Houde, E.D. (1994). Effects of low dissolved oxygen on predation on estuarine fish larvae. *Mar. Ecol. Prog. Ser. 104*, 235–246.

Breitburg, D.L. Craig, J.K., Fulford, R.S., Rose, K.A., Boynton, W.R., Brady, D., Ciotti, B.J., Diaz, R.J., Friedland, K.D. Hagy J.D. III, Hart, D.R., Hines, A.H., Houde, E.D., Kolesar, S.E., Nixon, S.W., Rice, J.A., Secor, D.H. & Targett, T.E. (2009). Nutrient enrichment and fisheries exploitation: interactive effects on estuarine living resources and their management. *Hydrobiologia 629*, 31–47.

CBP (2000). *Chesapeake 2000*. Chesapeake Bay Program, http://www.chesapeakebay. net/content/publications/cbp_12081.PDF

CBP (2009). *Bay Barometer*, Chesapeake Bay Program, http://www.chesapeakebay. net/content/ publications/cbp_50513.pdf.

Chao, S-Y., Boicourt, W.C. & Wang, H.V.C. (1996). Three-layered circulation in reverse estuaries. *Cont. Shelf Res. 16*, 1379–1397.

Day, J.W., Jr., Hall, C.A.S., Kemp, W.M. & Yañez-Arancíbia, A. (1989). *Estuarine Ecology*. Wiley & Sons, New York.

Dyer, K.R. (1974). The salt balance in stratified estuaries. *Estuar. Coast, Mar. S. 2*, 273–281.

Engle, V.D. & Summers, K. (1999). Determining the causes of benthic condition. E*nviron. Monit. Assess. 51*, 381–387.

Fisher, T.R., Gustafson, A.B., Sellner, K., Lacouture, R., Haas, L.W., Wetzel, R.L., Magnien, R., Everitt, D., Michaels, B & Karrh, R. (1999). Spatial and temporal variation of resource limitation in Chesapeake Bay. *Mar. Bio. 133*, 763–778.

Geyer, W.R. & Farmer, D.M. (1989). Tide-induced variation of the dynamics of a salt wedge estuary. *J. Phys. Oceanog.* 19, 1060–1072.

Gibbs, M.T., Bowman, M.J. & D.E. Dietrich. (2000). Maintenance of near-surface stratification in Doubtful Sound, a New Zealand fjord. *Estuar. Coast. Shelf S. 51*, 683–704.

Guo, X. & Valle-Levinson, A. (2007). Tidal effects on estuarine circulation and outflow plume in the Chesapeake Bay. *Cont. Shelf Res. 27*, 20-42.

Herdendorf, C.E. (1990). Great Lakes estuaries. *Estuaries 13*, 493–503.

Hobbie, J.E. (2000). *Estuarine Science: A Synthetic Approach to Research and Practice.* Island Press, Washington, D.C.

Houde, E.D. & Rutherford E.S. (1993). Recent trends in estuarine fisheries: Predictions of fish production and yield. *Estuaries 16*, 161-176.

Jordan, S.J., Lewis, M.A., Harwell, L.M. & Goodman, L. R. (2010a). Summer fish communities in northern Gulf of Mexico estuaries: indices of ecological condition. *Ecol. Indicators 10*, 504–515.

Jordan, S. J., Hayes, S.E., Yoskowitz, D., Smith, L.M., Summers, J.K., Russell, M. & Benson, W.H. (2010). Accounting for natural resources and environmental sustainability: linking ecosystem services to human well-being. *Environ. Sci. Technol. 44*, 1530–1536.

Jordan, S.J. & Vaas, P.A. (2000). An index of ecosystem integrity for northern Chesapeake Bay. *Environ. Sci. Policy 3*, S59–S88.

Jordan, T.E., Cornwell, J.C., Boynton, W.R. & Anderson, J.T. (2008). Changes in phosphorus biogeochemistry along an estuarine salinity gradient: The iron conveyer belt. *Limnol. Oceanogr. 53*, 172–184.

Kjerfve, B., Schettini, C.A.F., Knoppers, B., Lessa, G. & Ferreira, H.O. (1996). Hydrology and salt balance in a large, hypersaline coastal lagoon: Lagoa de Araruama, Brazil. *Estuar. Coast Shelf S. 42*, 701-725.

Lewis, M., Jordan, S., Chancy, C., Harwell, L., Goodman, L. & Quarles. R. (2007). Summer fish community of the coastal northern Gulf of Mexico: characterization of a large-scale trawl survey. *Trans. Am. Fish. Soc. 136*, 829–845.

Martínez, M.L., Intralawan, A., Vázquez, G., Pérez-Maqueo, O., Sutton, P. & Landgrave, R. (2007). The coasts of our world: Ecological, economic and social importance. *Ecol. Econ. 63*, 254–272.

Mencken, H.L. (1940). *Happy Days,* Alfred A. Knopf, New York, p. 55.

Nixon, S.W. (1995). Coastal marine eutrophication: A definition, social causes, and future concerns. *Ophelia 41*, 199–219.

Officer, C.B., Lynch, D.R., Setlock, G.H., & Helz, G.R. Recent sedimentation rates in Chesapeake Bay. In: V.S. Kennedy (Ed.), *The Estuary as a Filter*, pp. 1–11. Academic Press, Inc., Orlando, FL.

Pritchard, D.W. (1967). What is an estuary: physical viewpoint. In G.H. Lauff (Ed.), *Estuaries*, AAAS Publication no. 83, Washington, DC, pp. 3–5.

Quammen, M.L. & Onuf, C.P. (1996). Laguna Madre: seagrass changes continue decades after salinity reduction. *Estuaries 16*, 302–310.

Reynolds-Fleming, J.V. & Luettich, Jr., R.A. (2004). Wind-driven lateral variability in a partially mixed estuary. *Estuar. Coast. Shelf S. 60*, 395–407.

Schubel, J.R. & Kennedy, V.S. (1984). The estuary as a filter: introduction. In: V.S. Kennedy (Ed.), *The Estuary as a Filter*, pp. 1–11. Academic Press, Inc., Orlando, FL.

Tréguer, P., Nelson, D.M., Van Beenekom, A.J., DeMaster, D.J., Leynaery, A. & Quégioner, B. (1995). The silica balance in the world ocean: a reestimate. *Science 268*, 375–379.

Turner, A. (1996). Trace-metal partitioning in estuaries: importance of salinity and particle concentration. *Mar. Chem. 53*, 27–39.

Turner, A., Martino, M. & Le Roux, S.M. (2002). Trace metal distribution coefficients in the Mersey estuary, UK: Evidence for salting out of metal complexes. *Environ. Sci. Technol. 36*, 4578–4584.

Ulanowicz, R.E. & Baird, D. (1999). Nutrient controls on ecosystem dynamics: the Chesapeake mesohaline community. *J. Mar. Sys. 19*, 159–172.

U.S. EPA (2006). *National Estuary Program Coastal Condition Report*. U.S. Environmental Protection Agency, EPA-842/B-06/001 Washington, D.C.

Webster (1975). *Webster's New Collegiate Dictionary*. G. &. C. Merriam Co., Springfield, MA, USA.

Windom, H., Wallace, G., Smith, R., Dudek, N., Maeda, M., Dulmage, R. & Storti, F. (1983). Behaviour of copper in southeastern United States estuaries. *Mar. Chem. 12*, 183–193.

Zwolsman, J.J.G. & van Eck, G.T.M. (1999). Geochemistry of major elements and trace metals in suspended matter of the Scheldt estuary, southwest Netherlands. *Mar. Chem. 66*, 91–111.

In: Estuaries: Classification, Ecology and Human Impacts   ISBN: 978-1-61942-083-0
Editor: Steve Jordan   © 2012 Nova Science Publishers, Inc.

Chapter 2

CLASSIFICATION FOR ESTUARINE ECOSYSTEMS: A REVIEW AND COMPARISON OF SELECTED CLASSIFICATION SCHEMES

Janis C. Kurtz and James D. Hagy III
U. S. Environmental Protection Agency, Gulf Ecology Division, 1 Sabine Island Dr.,
Gulf Breeze, FL, US

ABSTRACT

Estuarine scientists have devoted considerable effort to classifying coastal, estuarine and marine environments and their watersheds, for a variety of purposes. These classifications group systems with similarities—most often in physical and hydrodynamic properties—in order to explain how estuaries respond to stressors. We evaluated four selected classification schemes for their effectiveness in (1) relating nitrate + nitrite concentrations to nitrogen load estimates, and (2) relating chlorophyll a concentrations to nitrogen loads and concentrations. Strong within-class similarities would be useful for developing numeric nutrient criteria and reference conditions in estuaries. Classifications were analyzed using a nationwide water quality data set. Of the classifications tested, the geographically focused Ecoregion classification seemed to provide the most robust grouping of estuaries on the basis of similarity of responses. Regionally-based classifications may account for factors such as land use, population density, and geographic attributes that within-estuary physical and hydrological variables do not represent. We offer recommendations for selecting and improving classification approaches, both for researchers developing classification tools, and for resource managers who use them. Our ability to manage estuarine environments will ultimately benefit from a classification scheme that provides standard terminology for various types of estuaries, guidelines for organizing data, facilitation of mapping and modeling efforts, and support for decisions.

INTRODUCTION

Estuaries are highly valued ecosystems providing a variety of services, including high biological productivity and important coastal habitats. These ecosystems are subject to escalating environmental pressures related to local population growth, global-scale sea level rise, and storm risk (Weiss et al. 2011). Protecting estuaries from these pressures is challenging in part because of the physical, chemical and biological diversity of estuaries. To protect and understand our coastal resources, scientists and environmental managers use classification to assist in a variety of ways. A classification system organizes data about ecological systems into a logical scientific framework, characterizing them based on their particular properties (Jay et al. 2000). Classification provides a lexicon for describing habitats, communities and ecosystems that can be used to inventory and map their extent and distribution. In order to help address environmental management objectives, a classification may also be predictive. For example, class membership may predict an ecological attribute or function, such as the nature of an expected ecological response to an environmental stressor. Management decisions required for conservation, protection and restoration are informed and supported by classification.

Early classification methods generally were based on physical themes: geomorphology and physiography, hydrography (water circulation, mixing, and stratification), salinity and tidal characteristics, sedimentation, and ecosystem energetics (reviewed by Kennish 1986). Pritchard (1967) placed estuaries into the four geomorphic classes discussed in Chapter 1: drowned river valleys, fjord-type, lagoon-type or bar built, and those produced by tectonic processes. Stommel & Farmer (1952) divided estuaries into four types based on stratification: well mixed, partially mixed, fjord-like, and salt wedge (Jordan, Chapter 1). This simple typology was the basis for numerous modifications (Ippen & Harlemann 1961, Simpson & Hunter 1974, Fischer 1976, Prandle 1986; Nunes Vaz & Lennon 1991). In addition to geomorphic classes, Pritchard (1967) placed estuaries into four classes based on circulation regime: type A or salt wedge, type B or partially mixed (moderately stratified), type C or vertically homogenous with a lateral salinity gradient, and type D or sectionally homogenous with a longitudinal salinity gradient. Hansen & Rattray (1966) devised a classification based on two dimensionless parameters expressing stratification (the relative salinity difference between surface and bottom water), and circulation, as the ratio of net longitudinal surface flow velocity to the mean cross-sectional velocity of freshwater discharge. In the Hansen & Rattray (1966) scheme, Type 1–3 estuaries differ in increasing magnitude of the circulation parameter, with subtypes a (well mixed) and b (partially stratified), e.g., a Type 1a estuary would be well-mixed with high relative river flow. Highly stratified estuaries with little circulation (fjord-like) were designated Type 4. These classification systems lacked consideration of some important forcing functions (wind, multiple freshwater discharges) that influence circulation and stratification, and do not express complex dynamic patterns, such as three-layered or reverse estuarine circulation, that occur episodically in some estuaries.

Salinity also has been used to divide estuaries into sections based on the average salinity in a section. In the Venice system (Venice System 1959), six distinct zones were recognized: limnetic or freshwater (salinity <0.5), oligohaline (0.5–5), mesohaline (5–18), polyhaline (18–30), euhaline (30–40), and hyperhaline (> 40).

Odum & Copeland (1974) developed a hierarchical, functional description of coastal ecosystems based on dominant forcing functions. This classification used characteristic energy sources, storage, and flows of estuaries and coastal ecosystems, including wetlands, to define six major classes: natural stressed systems, natural tropical systems with high diversity, natural temperate ecosystems with seasonality, natural Arctic systems with sea ice stress, emerging systems associated with man, and subsystems that organize extensive areas as a result of migrating organisms (as in the huge seasonal influxes of juvenile fish and invertebrates from coastal oceanic spawning areas to estuaries). Most of the classes were divided into more specific subclasses. The stated purpose of Odum & Copeland's (1974) work was to develop a classification system. ''... that groups together estuaries with similar responses to disturbance, planning, or management.'' Their work presaged present needs to develop stressor-based (disturbance) classifications for managing coastal ecosystems.

Although a variety of classification schemes has been developed for coastal habitats, no system has become universally accepted, perhaps because no classification system serves all purposes. One source of differences is scale, encompassing both extent and resolution. A classification scheme that is national in extent may utilize lower spatial resolution and more broadly defined categories than a scheme that is applied locally. Regional, more localized classifications may highlight more refined differences among water bodies or habitats that are not important at the national scale. Habitat classifications may employ a particularly small scale (several meters), delineating features such as seagrass beds or reefs (Madden et al. 2005). The most widely used habitat classification system was developed for wetlands by Cowardin et al. (1979), but the wetlands focus of that system does not adequately address estuaries.

The scale of a classification scheme may be determined in part by the scale of available data (Valentine et al. 2005). For example, information such as sediment grain size and composition, biological community composition, and water quality may be key descriptors applied to development of a classification (Madden et al. 2004b), but comparable datasets may not exist at a large scale (e.g., national). Conversely, data such as bathymetry, tidal amplitude, and land use are available at large scales, potentially driving uses of these measures in large scale classifications (NOAA 2003). To address problems associated with scale, classification frameworks may be organized in a nested hierarchy, initially dividing major groups or classes at a broad scale followed by division into subordinate groups or classes at progressively finer scales. Omernik's (1987) Ecoregion classification is an example of a hierarchical or nested classification with multiple Levels (I-IV) that vary in their scale of application.

As coastal and marine resource management issues expand to include sea level rise, freshwater inflow, wetland and barrier island protection and restoration, sediment distribution, water quality, and wildlife habitat the role of classification is also expanding. Habitat management is emerging as an important consideration for regulating coastal and fisheries resources. For example, a classification scheme created in the contiguous western U.S. for NOAA's National Marine Fisheries Service was adopted and modified to produce maps of essential fish habitat (Greene et al. 2007). New approaches to managing and assigning values to ocean resources (ecosystem-based, or ecosystem services based management) require an expanded information base, including identification, delineation, and valuation of estuarine and marine habitats. Classification and refined habitat mapping should improve (1) how sites are chosen for conservation and restoration projects, (2) identification of appropriate stations for monitoring that encompass specific habitat features, and (3) extrapolating the results of modeling to sites with similar structure and function.

Our objectives in this chapter are to (1) identify and briefly describe classification schemes most pertinent for coastal, estuarine and marine environments and their watersheds, along with their applicability, benefits, and limitations; (2) perform a test of selected classification schemes for a specific case; and (3) provide recommendations for selecting and improving classification approaches, both for researchers developing classification tools and the resource managers who use them.

CLASSIFICATION SCHEMES FOR COASTAL ENVIRONMENTS

Beyond the early geophysical classifications discussed at the beginning of this chapter, many classification schemes have been developed for coastal, estuarine and marine environments and their watersheds. Key features, including objectives for which they were developed, geographic focus and scale, organization or hierarchy, and data sources of 16 classifications are described below. Many examples are based on the hierarchical wetlands classification developed by Cowardin et al. (1979). Many address coastal environments in North America, but several are for marine ecosystems in Europe. Additional classification frameworks are reviewed by Madden et al. (2004a, 2005), Kurtz et al. (2006), and Greene et al. (2007).

Cowardin et al. (1979)

Cowardin's national benchmark is a geographically comprehensive, nested-hierarchical classification scheme for wetland and deepwater habitats, based on ecological parameters. The U.S. Fish and Wildlife Service (USFWS) developed it to inventory wetland and deepwater habitats on a national scale. Subsequent modifications addressed upland, high-energy coastal, deepwater marine, and estuarine habitats, expanding the coverage of the original work (Dethier 1992, Greene et al. 1999, Allee et al. 2000, Kutcher et al. 2005, Greene et al. 2007).

Ecoregions

Omernik's (1987) Ecoregion mapping framework (Figure 1) uses landscape and climatic data layers, aggregating areas based on spatial covariance in vegetation type, climate, and geology. The framework is hierarchical; higher levels address smaller, more resolved spatial scale. Differences among regions are based on the physical and biological components that influence ecological relationships. The Ecoregion delineations have been modified by various groups, in particular for application to biological criteria and water quality standards development (EPA 2004a). Although this classification applies to land, rather than estuaries, it has relevance to estuaries via the classification of watersheds.

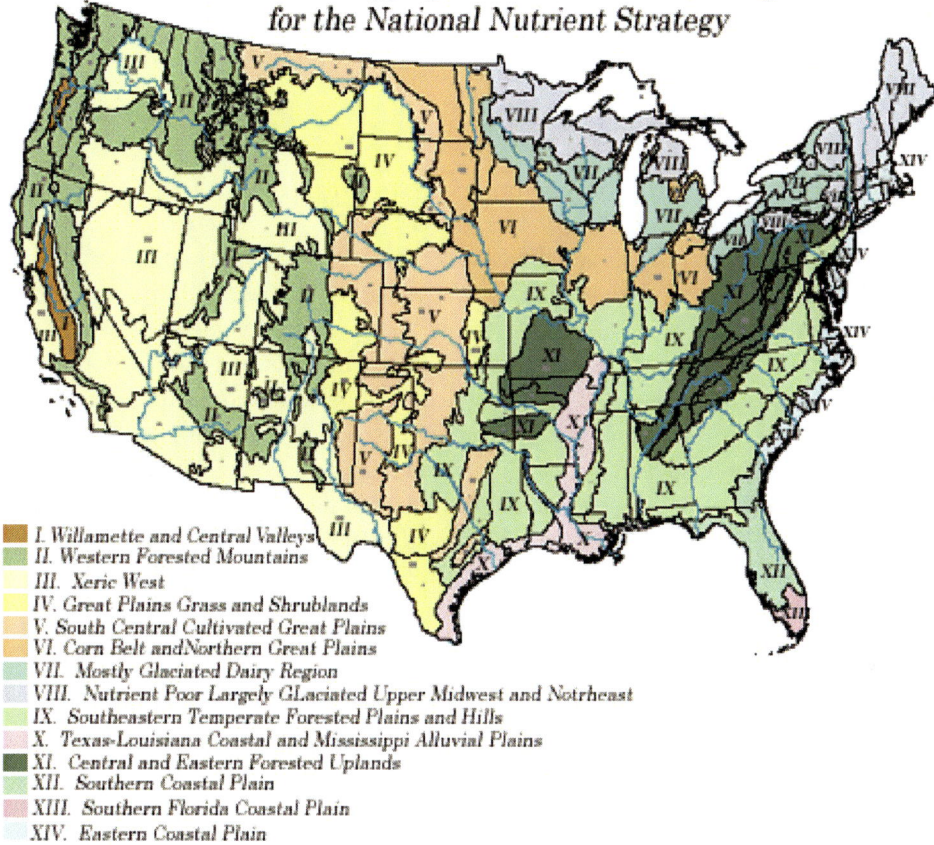

Figure 1. U.S. Ecoregions (Omernik 1987. http://www.epa.gov/wed/pages/ecoregions/na_eco.htm #Level%20III.).

Dethier (1992)

Dethier added "energy" as a classification variable, improving the applicability of the Cowardin et al. (1979) system to rocky substrates on the Pacific coasts of Washington. The energy descriptor includes designations such as "exposed," "moderately exposed," and "sheltered." This application demonstrates adaption of a regional approach to a smaller scale by considering locally important features. In this case, Dethier (1992) narrowed the original classification to focus on marine and estuarine systems, eliminating rivers and lakes from consideration.

Brown (1993)

Focusing on benthic habitats in Maine, Brown (1993) described a hierarchical classification organized by substratum, depth, energy level and salinity. Hierarchical levels of the classification include: system (biome types), subsystem (tidal regimes), class (substrata), subclass (energy levels), modifiers (as needed), diagnostic species, and common species. Descriptions of habitat types and associated species are provided as well as example applications.

National Estuarine Eutrophication Assessment (NEEA) Typology

Bricker et al. (1999) developed the NEEA typology, which was later modified (Smith et al. 2004; Figure 2) to classify the Nation's estuaries by susceptibility to nutrient over-enrichment. Addressing 138 U.S. estuaries, the classification employs measures of nutrient loading, dilution, flushing, and quantitative indices of estuarine susceptibility to nutrient pollution. Data sources and spatial units, defined by NOAA's Coastal Assessment Framework (NOAA 2003), are consistent and watershed-based, encompassing the estuary's entire watershed including land and water components (Engle et al. 2007). From an initial 90 parameters, NOAA selected five that explained most of the variability among classes, thereby grouping 138 estuaries into 10 classes using a statistical similarity index.

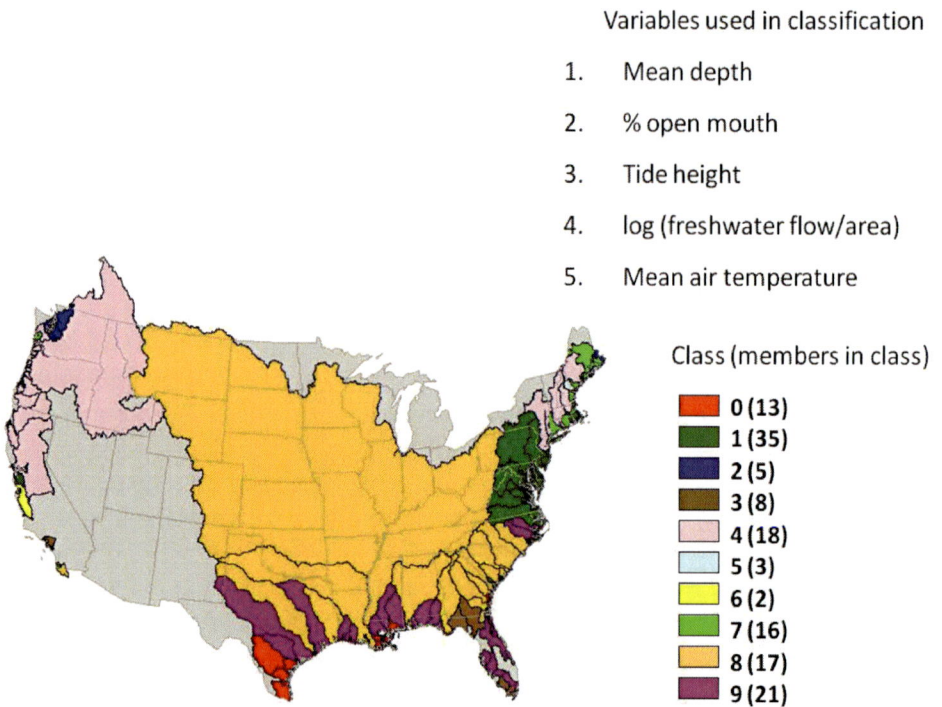

Variables used in classification

1.   Mean depth
2.   % open mouth
3.   Tide height
4.   log (freshwater flow/area)
5.   Mean air temperature

Class (members in class)

- 0 (13)
- 1 (35)
- 2 (5)
- 3 (8)
- 4 (18)
- 5 (3)
- 6 (2)
- 7 (16)
- 8 (17)
- 9 (21)

Figure 2. NOAA Typology (Bricker, et al. 2007).

Allee et al. (2000)

National in scope and comprehensive for coastal and marine systems, this classification encompasses the entire U.S. coastal margin from the landward extent of tidal influence to the outer edge of the continental shelf. Enhancements over Cowardin et al. (1979) include (1) wide applicability across a variety regions; (2) focus on marine and estuarine landscapes; and (3) linkages between geologic framework, energy and biology. This classification is not hierarchical; more refined habitats depend on subjective determinations rather than on standardized categories. This system does not distinguish between marine and estuarine categories, so that freshwater and brackish habitats of National Estuarine Research Reserve System (NERRS) sites are classified together, despite completely different species composition across salinity gradients (NERRS 2000).

Brown (2002)

Five major habitat types (Freshwater, Estuarine, Nearshore, Offshore and Oceanic Islands and Banks) are included in this design, developed to describe and define critical habitats for federally managed fishery species. The scheme does not provide a means for capturing habitat disturbances or gradients (NERRS 2000).

Madley (2002)

This classification, developed for coastal resource managers in Florida, focuses on estuarine and marine environments, including nearshore and neritic areas inhabited by corals, hard bottom, and seagrass communities. The boundaries of this hierarchical classification scheme are from the high tide line to the edge of the continental shelf. Data sources include Light Detection and Ranging (LIDAR), single beam sonar, and aerial photography. Although geologic structure, coastal complexity, and hydrodynamic features are not captured, the application of remote-sensing data and mapping may be relevant and useful.

BIOMAR

Developed for Britain and Ireland in the late 1990s and revised in 2004 (Connor et al. 2004) BIOMAR covers marine habitats from high tide (excluding salt marshes) seaward, excluding ocean habitats deeper than 80 m. Connor et al. (2004) focused on marine areas closely associated with the coast, but not on estuarine systems or upstream watersheds. The classification is readily accessible and web-based.[1] Five levels of classification are identified and a listing of habitat types within these levels is provided along with a detailed description of each type, distribution maps, color photographs and a glossary of terms.

The European Union Nature Information System (EUNIS) classification incorporates major themes of BIOMAR, expanded to cover aquatic and terrestrial habitats for most of Europe, including deeper water column habitats.[2] The full classification hierarchy is on-line with habitat identification keys, background, rationale, and a glossary of terms. The marine and coastal component is a small part of the overall classification scheme, which has eight

---

[1] http://www.jncc.gov.uk/MarineHabitatClassification.
[2] http://eunis.eea.europa.eu/index.jsp.

other main levels devoted to terrestrial habitats. In addition, the habitat types represented range from Northeast Atlantic (England and Ireland) to Mediterranean and Black Sea marine habitats (Davies et al. 2004).

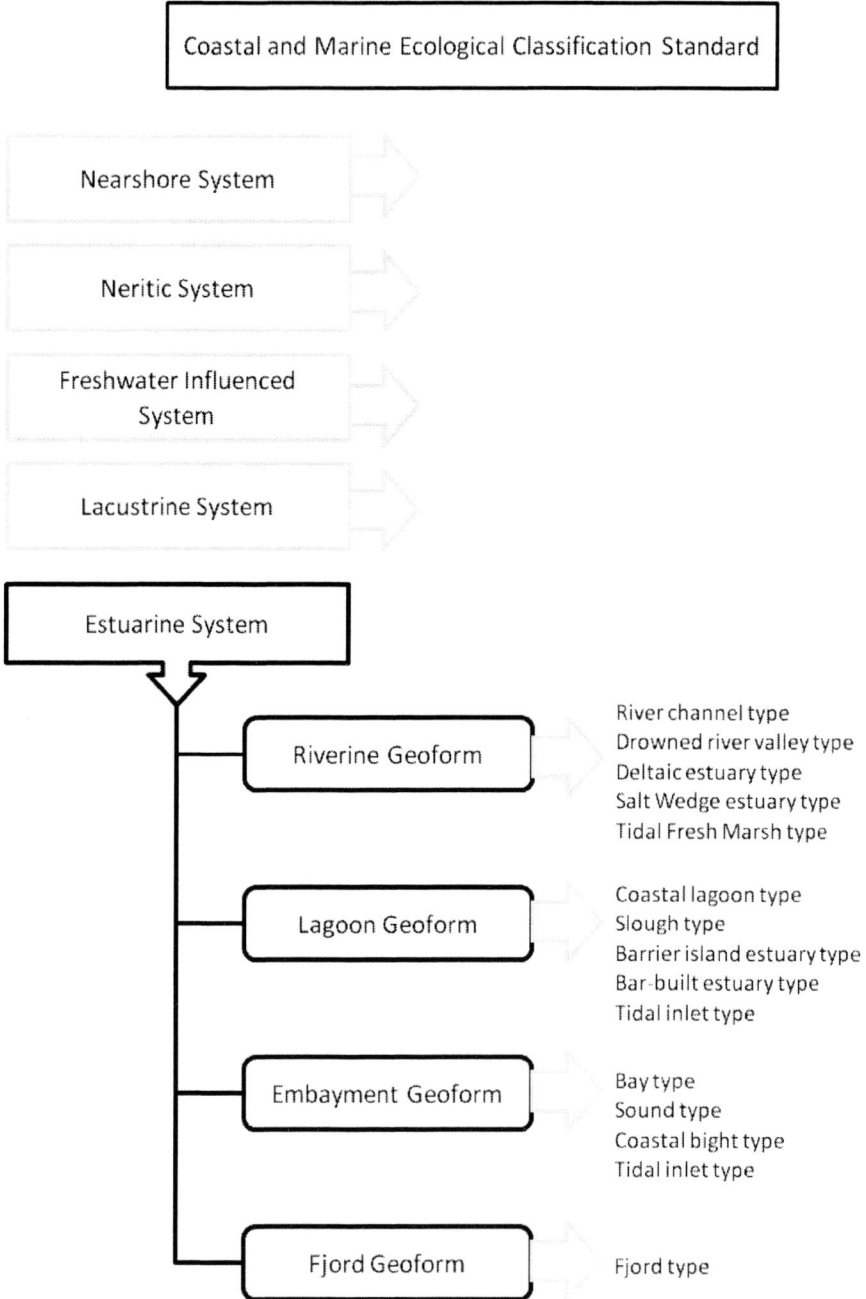

Coastal and Marine Ecological Classification Standard

Nearshore System

Neritic System

Freshwater Influenced System

Lacustrine System

Estuarine System

Riverine Geoform

River channel type
Drowned river valley type
Deltaic estuary type
Salt Wedge estuary type
Tidal Fresh Marsh type

Lagoon Geoform

Coastal lagoon type
Slough type
Barrier island estuary type
Bar-built estuary type
Tidal inlet type

Embayment Geoform

Bay type
Sound type
Coastal bight type
Tidal inlet type

Fjord Geoform

Fjord type

Figure 3. Hierarchical system, geoform and types of estuaries from the CMECS classification system. Each progressively lower level represents a finer habitat distinction (Madden et al. 2009).

EUNIS

Coastal and Marine Ecosystem Classification Standard (CMECS)

CMECS (Madden et al. 2005; Figure 3) was developed by NOAA, NatureServe, and other partners to provide a national framework for classifying the geological, physical, biological and chemical components of marine habitats including estuarine, coastal and ocean environments[3]. Version III (February 2009) of CMECS encompasses areas from the head-of-tides in the coastal zone to the deep ocean. Estuaries, wetlands, rivers, shorelines, intertidal and benthic zones and the entire water column from the shore to the deep ocean at scales of <1 m to >10 m are classified in a hierarchical organization. Five systems (nearshore, neritic, oceanic, estuarine, freshwater influenced and lacustrine) are further defined by five descriptors five descriptors (water column, geoform, surface geology, benthic biology and sub-benthos). CMECS is comprehensive for coastal and marine ecosystems, but does not yet address linkages to coastal uplands and watersheds.

National Estuarine Research Reserve System (NERRS)

The NEERS habitat classification scheme (Kutcher et al. 2008) facilitates measuring the magnitude and extent of habitat change in estuarine systems and linking observed changes to watershed land-use practices. The classification is ecologically and spatially explicit, with a nested hierarchy intended to allow effective use of existing data, and to inventory and classify all land-cover types comprehensively (Kutcher 2006). Largely based on Cowardin et al. (1979), the NERRS classification includes coastal uplands, wetlands, and nearshore estuarine habitats. It is organized in a four-level, nested, numerically coded hierarchy. Non-hierarchical categories for more detailed descriptions also are accommodated. The system requires data at two scales: 1) broad-scale, low resolution data for characterizing entire watersheds; and 2) detailed, fine-scale, high resolution data for characterizing specific properties of habitats (Walker et al. 2005). The NERRS scheme focuses on nearshore rather than deep water habitats.

Greene et al. (2007)

Greene's original classification was developed for deep seafloor habitats, with the goal of understanding and predicting spatial distribution of fish species. Revisions to date generally have been based on geomorphology, substrate type, textures produced by physical processes, and sessile biology primarily for deep water (>30 m) habitats. The classification scheme applies to seafloor habitat types throughout marine regions, from sub-arctic to tropical latitudes, and from shallow, intertidal regions and estuaries to abyssal plains. Seafloor mapping datasets (e.g., multi-beam bathymetry, backscatter intensity, and ground truth imagery) and analogous optical data, such as LIDAR, provide the foundation for applying this

---

[3] http://www.csc.noaa.gov/benthic/cmecs/index.html.

scheme. The hierarchy is largely based on spatial or mapping scale and emphasizes deepwater marine habitats as opposed to estuarine and coastal environments.

Valentine et al. (2005)

The goal of this scheme is to develop a practical method to classify marine sublittoral (subtidal continental-shelf and shelf-basin) habitats in northeastern North America. The classification approach is based on topographical, geological, biological and oceanographic attributes and natural and anthropogenic processes. Seabed substrate type, substrate dynamics and physical and biological features on the seabed are emphasized in this partially hierarchical classification of seafloor habitats. The classification was developed for the Gulf of Maine using multi-beam and sidescan sonar, video and photographic transects and sediment and biological sampling (i.e. benthic grabs). The highest level of classification is *themes*: topographical setting; seabed dynamics and currents; seabed texture, hardness and layering; grain size; seabed roughness; fauna and flora features; faunal association; human use and state of disturbance; and understanding of habitat recovery. Estuarine, water column and intertidal habitats are not considered.

Classification Framework for Coastal Systems

This non-hierarchical, multivariate classification (Figure 4; US EPA 2004a, Engle et al. 2007), groups coastal watersheds by similarities in physical and hydrologic characteristics that influence estuarine response to stressor loads. Similar to NOAA's Estuarine Classification (Bricker et al. 1999), and based on NOAA's Coastal Assessment Framework, this classification includes information on depth, salinity and water temperature, and discriminates on the basis of tidal prism volume and freshwater inflow, to integrate more direct measures of residence time and stratification into the classification framework.

Estuary Environment Classification

Hume et al. (2007) developed this framework for New Zealand's estuaries based on physical components and controlling factors such as climate and riverine conditions. Discriminations are made at several scales: the global scale, considering climatic and oceanic processes; a whole estuary scale based on hydrodynamic processes within the estuary; and finally a watershed or catchment scale taking into account land cover and geology. Data are primarily physical in nature and include depth, tidal currents, and wind-driven wave information among the factors. This multi-scale classification is designed principally for estuarine waters and defines eight categories of estuaries. Developed for New Zealand estuaries in collaboration with Australia, this classification is based on broad-scale hydrodynamic principles[4].

---

[4] http://www.naturalhazards.net.nz/tools/nzcoast/coastal/about/nz_estuarine_classification/estuary_types.

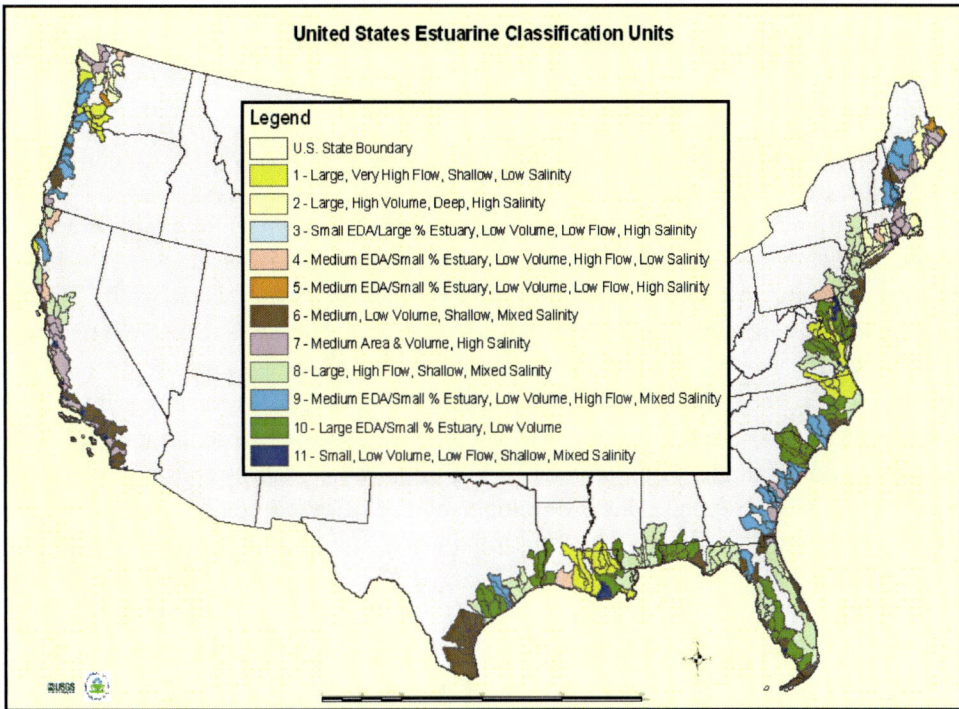

United States Estuarine Classification Units

Legend

- U.S. State Boundary
- 1 - Large, Very High Flow, Shallow, Low Salinity
- 2 - Large, High Volume, Deep, High Salinity
- 3 - Small EDA/Large % Estuary, Low Volume, Low Flow, High Salinity
- 4 - Medium EDA/Small % Estuary, Low Volume, High Flow, Low Salinity
- 5 - Medium EDA/Small % Estuary, Low Volume, Low Flow, High Salinity
- 6 - Medium, Low Volume, Shallow, Mixed Salinity
- 7 - Medium Area & Volume, High Salinity
- 8 - Large, High Flow, Shallow, Mixed Salinity
- 9 - Medium EDA/Small % Estuary, Low Volume, High Flow, Mixed Salinity
- 10 - Large EDA/Small % Estuary, Low Volume
- 11 - Small, Low Volume, Low Flow, Shallow, Mixed Salinity

Figure 4. Classification Framework for Coastal Systems (EPA, 2004a).

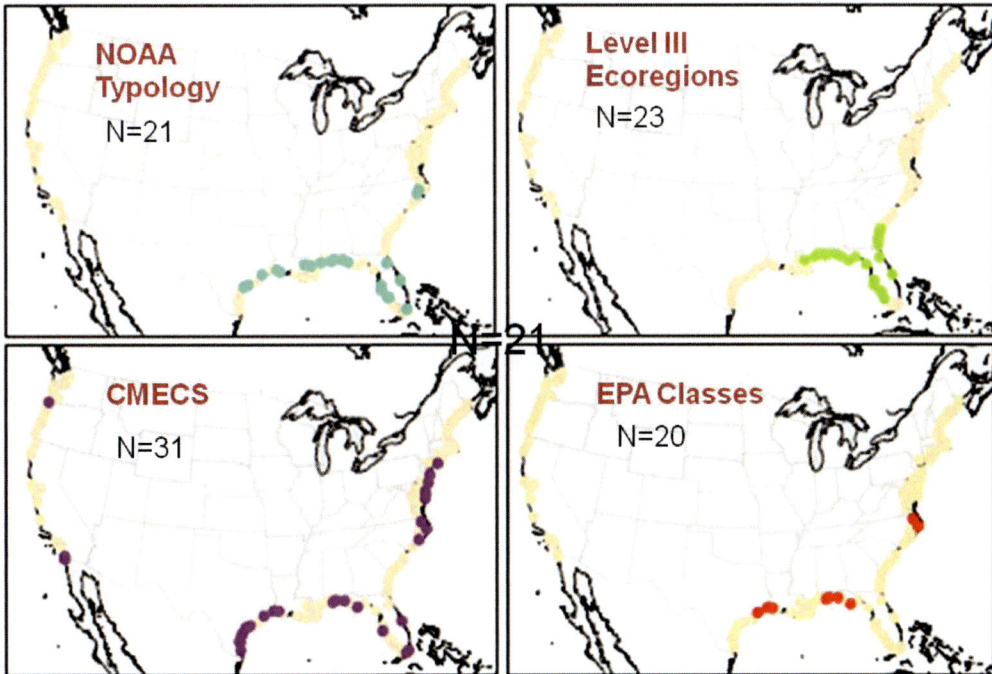

NOAA Typology
N=21

Level III Ecoregions
N=23

CMECS
N=31

EPA Classes
N=20

Figure 5. Classes that include Pensacola Bay and their members.

TESTING SELECTED HABITAT CLASSIFICATION SCHEMES: A CASE STUDY

Nutrient enrichment continues to be a major cause of impaired water quality in estuaries worldwide. In the U.S., the Clean Water Act (CWA; U.S. House of Representatives 2000) is a regulatory framework for water quality management that provides for establishment of water quality standards for pollutants. In the case of nutrients, effective management is hindered by the lack of numeric criteria. Criteria are the component of the water quality standard that expresses the concentration of a pollutant, that if achieved is expected to ensure that designated human and aquatic life uses of the water body will be supported. Numeric criteria development has been complicated by the reality that appropriate levels vary among estuaries. One approach to simplifying criteria development is to propose criteria for classes of estuaries, with a single value applied to the class rather than proposing criteria for each individual estuary. Beyond criteria development, classification could help establish expected conditions, optimal conditions, or ranges of water quality parameters that might be expected for a class of estuaries, so that condition can be assessed relative to a reference condition (EPA 2004b, 2008). Grouping similar estuaries can support more robust statistical inferences about reference water quality values, and thereby foster more effective management.

The objective of the analysis presented here was to evaluate the effectiveness of several classification schemes for the specific objective of maximizing similarity of water quality within classes, as would be useful for establishing nutrient criteria and reference conditions. Four classification schemes were analyzed on the basis of a nationwide water quality data set, focusing on the particular class that included Pensacola Bay, Florida, an estuary where approaches to numeric nutrient criteria have been considered (Hagy et al. 2008)

Methods

Four classification schemes, each based on physical and hydrological properties, were selected from those developed at the national scale, including Omernik's Ecoregions at hierarchical Level III (Figure 1), NOAA's Estuarine Typology (Figure 2), CMECS (Figure 3), and EPA's classification framework for coastal systems (Figure 4). Data to evaluate the schemes were compiled from the Environmental Monitoring and Assessment Program (Summers et al. 1995), the National Coastal Assessment (EPA 2008), the Florida Inshore Monitoring and Assessment Program (IMAP, Florida Department of Environmental Protection), the NOAA coastal assessment and decision-support (CA&DS) database, NEEA, and the NOAA Coastal Assessment Framework (CAF).

Linear regression analysis was employed to examine correlations between stressor (N load) and response (nitrate + nitrate and chl-a) variables within each classification. Each classification was tested by ANCOVA for its ability to explain the variance in annual mean nitrate + nitrate and chl-a concentrations after controlling for N loading as a covariate. All estuaries with complete data were tested in these analyses; load and concentration data were log-transformed to achieve approximately normal distributions of residual errors.

For a more focused test of the selected classifications, data were extracted for the class of estuaries in each classification system that included Pensacola Bay, Florida (Figure 5). At

Level III of the Ecoregions hierarchy, Pensacola Bay is in Ecoregion 12, which has 23 members. For the NOAA Estuarine Typology, Pensacola Bay is in Class 9 of 10, which includes 21 members nationwide. For CMECS (Madden et al. 2009) the system-geoform class of the hierarchy results in 4 classes of estuaries: Riverine, Lagoon, Embayment, and Fjords. Pensacola Bay falls within the Lagoon class, along with 31 other estuaries. By EPA's classification framework (Engle et al. 2007), Pensacola Bay is in a class described as having a large watershed area with a small percentage of the area as water, and with a relatively low volume of estuarine water. Twenty other estuaries fell into this class (Figure 5).

Across all 4 schemes, 56 estuaries were included in a class with Pensacola Bay; average class membership was 23. Only 3 estuaries were included with Pensacola Bay by all four schemes, illustrating that these classification approaches are far from interchangeable. The Ecoregion classification was mostly defined by geographical location (i.e., class members were likely to be adjacent), whereas CMECS was least defined by geography. The NOAA and EPA classifications were intermediate in this regard and similar in their grouping of estuaries.

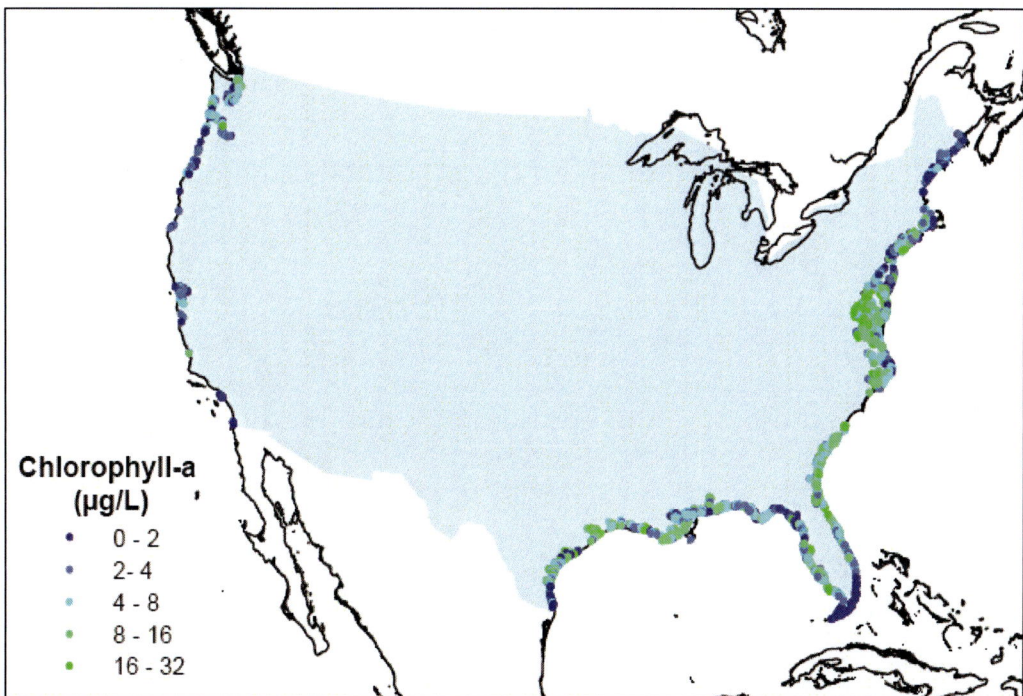

Figure 6. Chlorophyll a observations available for an estuary classification case study, including 3338 observations.

To evaluate which classification grouped estuaries so that classes had similar values for stressor and response parameters, nitrogen loading and chlorophyll-a (chl-a) concentration data were analyzed. From the datasets compiled, 3338 values were available for chl-a; nitrogen loading was available for 127 estuaries (Figure 6). Mean chl-a was determined for each class containing Pensacola Bay, and for all chl-a measurements (Figure 7).

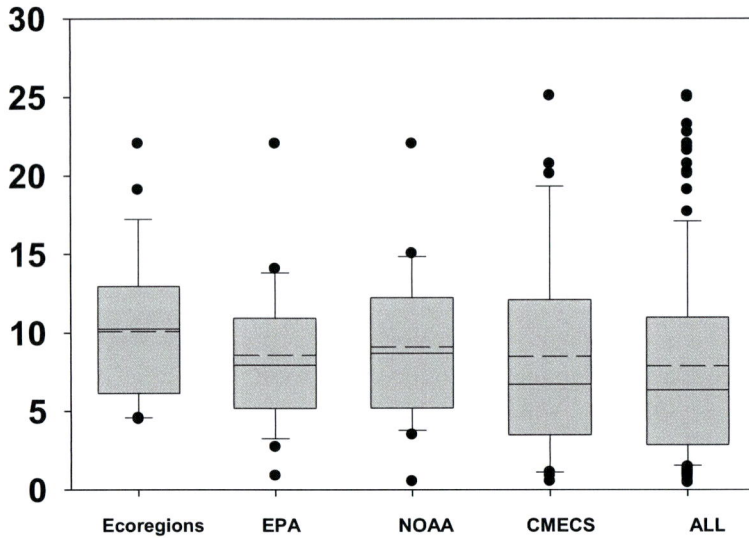

Figure 7. Box plots of chlorophyll-a (mg L^{-1}) by class and for all estuaries. Means are dotted lines, medians are solid lines, solid boxes are bounded by the 75th and 25th quartiles, respectively, dots are outliers, and brackets represent the 5th and 95th percentiles.

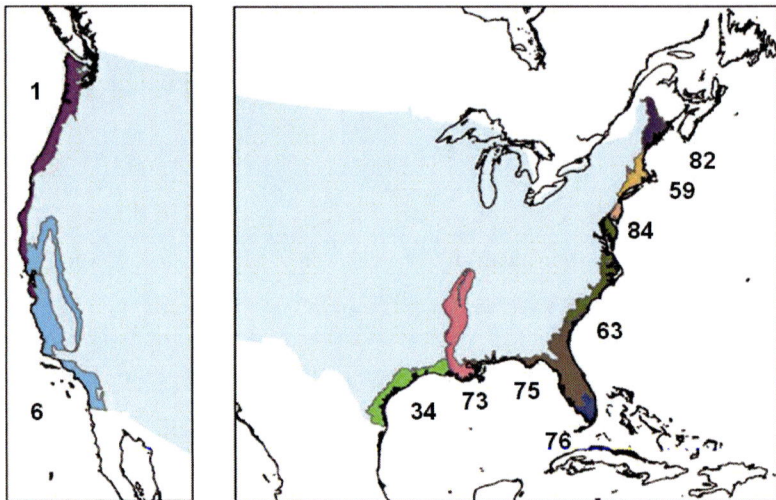

1 Coastal Range Pacific Northwest
6 South and Central California
34 Western Gulf of Mexico Coastal Plain
59 NE Coastal Zone
63 Mid-Atlantic Coastal Plain
73 Mississippi River Alluvial Plain
75 Southern Coastal Plain
76 South Florida Coastal Plain
82 Laurential Plains
84 Atlantic Coastal Pine Barrens

Figure 8. Ecoregion Level III, refinement of coastal classification into sub-regions.

Results

All of the classes containing Pensacola Bay had chl-a values slightly higher than those for all estuaries, but the classes were not very different from each other. Another comparison was made for chl-a measurements for all classes in a single classification. For example, Ecoregion Level III classes further refined into classes for coastal Ecoregions, i.e. Class 73 (Mississippi Alluvial Plain; Figure 8) showed greater differences, and the classification separated these differences relatively well (Figure 9).

Regression analysis of nitrate + nitrate and chl-a concentrations versus nitrogen loads by class for each classification system resulted in only one significant ($p < 0.05$) relationship: for the CMECS classification, there was a positive relationship between nitrate + nitrate and nitrogen load for river-dominated estuaries (Figure 10). Analysis of covariance generally indicated that either nitrate + nitrate, chl-a, or nitrogen loads differed between classes, but that classifications were at best weak predictors of load-response relationships (Tables 1 and 2). None of the classifications discriminated well on the basis of nitrate + nitrate concentrations; chl-a concentrations differed significantly among classes in the NOAA and Ecoregion classifications.

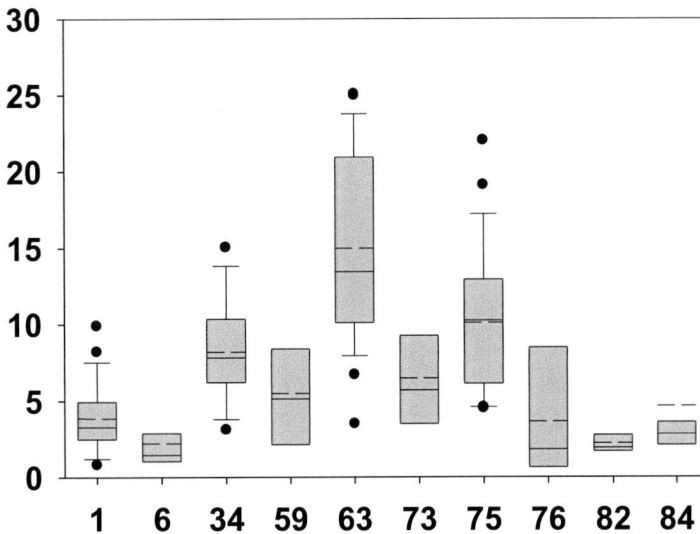

1 Coastal Range Pacific Northwest
6 South and Central CA
34 Western Gulf of Mexico Coastal Plain
59 NE Coastal Zone
63 Mid-Atlantic Coastal Plain
73 Mississippi River Alluvial Plain
75 Southern Coastal Plain
76 South Florida Coastal Plain
82 Laurential Plains
84 Atlantic Coastal Pine Barrens

Figure 9. Mean chlorophyll-a (mg L^{-1}) by estuary for Ecoregion Level III coastal classification.

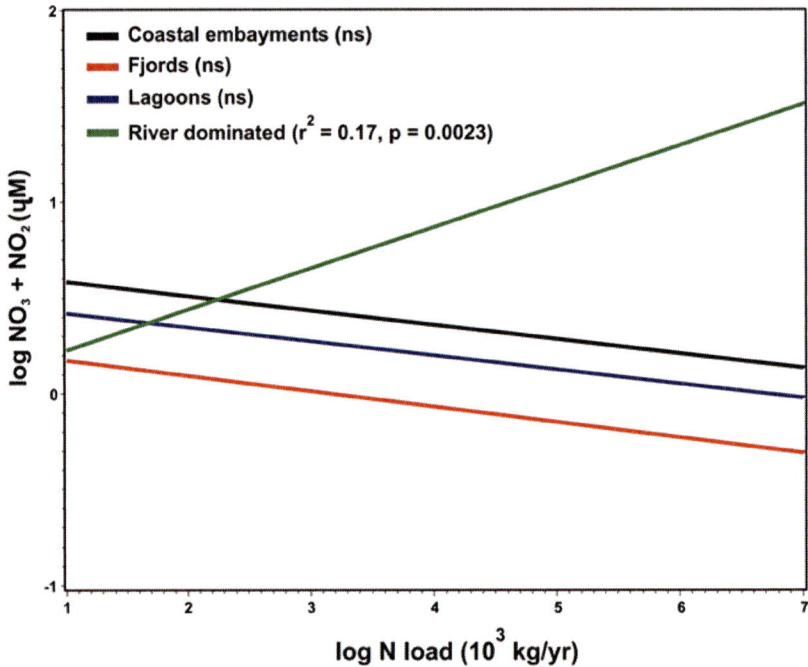

Figure 10. Relationship between N loading and Nitrate + Nitrite Concentrations for 4 types of estuaries from the CMECS classification.

Case Study Conclusions

None of the classifications evaluated was fully satisfactory for the purpose of discriminating between classes of estuaries on the basis of nitrogen load-response relationships. The CMECS is a descriptive classification system developed primarily to standardize nomenclature and facilitate mapping for coastal habitats, so it should not be expected to be predictive or diagnostic of stressor-response functions. Nevertheless, CMECS did demonstrate that among four classes of estuaries only river-dominated systems show a significant, although still quite variable, association between nitrogen loads and concentration responses. Regional approaches, specifically the Ecoregional classification, seemed to provide the most robust grouping of estuaries on the basis of similarity of responses. Regionally-based classifications may account (implicitly if not explicitly) for factors such as land use and population density that physical and hydrological variables do not represent. The EPA multivariate classification, developed specifically to facilitate prediction and diagnosis of stressor-response relationships in estuaries, appears to be less informative in this respect than the other classifications.

Although chl-a data were widely available on a national basis, there is a need for up-to-date, standardized nutrient loading data at all scales. In addition, data for other stressors and response parameters from national monitoring programs are important for understanding the condition of our waters, for determining trends, and for appropriate management.

Table 1. Analysis of covariance for effects of nitrogen loading and estuary classification system on $NO_3 + NO_2$ concentrations; N = 94; Nitrogen load and $NO_3 + NO_2$ concentration were log-transformed; F and p statistics are based on Type III sums of squares

Classification	Variable	F	p	r^2
NOAA		4.04	<0.0001	0.46
	N load	7.16	0.0091	
	Class	1.75	0.1105	
	N load x class	1.86	0.0875	
Ecoregion		2.34	0.0064	0.34
	N load	3.74	0.0569	
	Class	1.15	0.3421	
	N load x class	0.87	0.5489	
CMECS		6.53	<0.0001	0.35
	N load	0.00	0.9715	
	Class	0.58	0.6311	
	N load x class	2.26	0.0869	
EPA		1.89	0.0322	0.30
	N load	2.68	0.1056	
	Class	1.90	0.0724	
	N load x class	1.85	0.0806	

Table 2. Analysis of covariance for effects of nitrogen loading and estuary classification system on chl-a concentrations; N = 94; Nitrogen load and chl-a concentration were log-transformed; F and p statistics are based on Type III sums of squares

Classification	Variable	F	p	r^2
NOAA		5.39	<0.0001	0.53
	N load	2.71	0.1036	
	Class	8.16	<0.0001	
	N load x class	2.60	0.0183	
Ecoregion		11.07	<0.0001	0.71
	N load	0.00	0.9677	
	Class	4.70	0.0001	
	N load x class	2.85	0.0080	
CMECS		3.16	0.0051	0.20
	N load	0.00	0.9975	
	Class	0.89	0.4508	
	N load x class	0.49	0.6876	
EPA		5.83	<0.0001	0.57
	N load	0.09	0.7714	
	Class	1.82	0.0863	
	N load x class	0.79	0.6148	

Classification serves more than one purpose and can be used as a management tool, but there is room for further development of classifications for specific purposes, and a need for data to support their development and use. For the specific purpose of separating estuaries into classes with similar load-response relationships, or more broadly, similar relationships between watershed attributes and the condition of estuarine ecosystems, explicit analysis of data for individual estuaries may turn out to be more effective than classification. Computational, analytical, and modeling capabilities have grown enormously, along with the availability of data, since the early days of classification; for some applications these advances may have overtaken the need for classification.

RECOMMENDATIONS FOR SELECTING AND DEVELOPING CLASSIFICATION TOOLS FOR RESOURCE MANAGERS

Reviewing existing classification schemes is a valuable first step in determining which approaches or aspects of approaches are appropriate for classifying estuaries with specific objectives in mind. A brief, practical set of guidelines for successful selection, adoption and use of classification was developed for NERRs (Walker et al. 2005). Below, we discuss several of the issues to be considered before developing or adopting a classification scheme for a particular use.

Identify Objectives

Every classification tool is applied to accomplish a specific set of objectives (Kutcher 2006). To identify objectives, it is necessary to define the specific scientific or management needs or questions to be addressed by classification tools. Management needs should be clearly articulated to help define the type and level of classification required to achieve stated objectives. For example, classification may be used to facilitate project review, organize map products, assess change through time, design monitoring programs, plan for restoration, or develop conservation strategies. Identifying management objectives will influence the degree of resolution and detail needed in the classification and in the supporting data set.

Consider the Scale

In planning for classification, identifying the scale that is most important to classify, and the variety of elements or units that are encompassed by the selected scale to a level of detail that is practical and supported by data should be considered. Existing mapping resources and compilations of descriptive information can assist with these determinations. For example, preliminary planning efforts to site a submerged pipeline through an estuarine region may use coarse resolution survey data (e.g., remote sensing) to understand general sedimentary characteristics and distribution of seagrass, while fine-scale resolution of biological and physical characteristics is needed to assess and minimize potential environmental impacts to particular seagrass beds or oyster reefs. In many cases, both scales may be needed to develop a final plan.

Classification systems have been developed at the national scale (Figure 1), for coastal regions (Figures 2 and 4), for specific types of ecosystems (e.g., wetlands; Cowardin et al. 1979), deep-sea habitats (Greene et al. 1999), for specific regions such as northeastern North America (Valentine et al. 2005), or for discrete areas at finer scales (Walker, 2005). Nested hierarchal structures, beginning at the broad landscape level and transitioning to a finer habitat scale, lend utility and flexibility for various objectives, data sources and management applications.

Additionally, consideration may need to be given to political boundaries, as many estuaries and coastal waters share multiple state or national boundaries. These types of decisions cannot be made until the spatial area that will be used to develop the classification scheme is identified and all jurisdictions have had input into the extent to be included. Once the classification is developed it may be useful for wider application, but some constraints in the development stage will greatly facilitate progress. The value of a classification system may be enhanced if the scheme is widely accepted so that eventually a regional synthesis can be made, including synthesis between states. For example, the Gulf of Mexico Alliance has identified a common need among the Gulf States for development of a regional estuarine classification for use in nutrient management (GOMA 2008). Linking classification efforts in one state to other states in the regions may allow solutions to environmental problems in one state to be applied across the region.

Consider the Extent

The habitat coverage or range of the classification should also be considered. A framework focused on estuaries may encompass water bodies only, or may include coastal upland watersheds and ocean environments. Several existing schemes focus solely on the offshore, ocean environment, such as the Greene et al. (1999, 2007), Brown (2002) and Valentine et al. (2005). For example, Brown (2002) includes a supratidal zone, but falls short of capturing the upland habitats or even habitats on the coastal plain influenced by the spray zone. Both Valentine et al. (2005) and Greene et al. (1999, 2007) focus on seafloor habitats and do not detail the nearshore component.

The NERRS (Kutcher et al. 2005) and CMECS (Madden et al. 2005) classifications incorporate nearshore and upland habitat components. However, the NERRS approach may not include the level of detail that Coastal Zone Management programs need for offshore areas, in contrast to the Valentine et al. (2005) and Greene et al. (2007) classifications. If classification of wetland, estuarine and intertidal resources is desired, then a combination of the deepwater approaches with nearshore and upland approaches may be appropriate.

Classifying benthic habitat is of particular interest in Massachusetts (Lund & Wilbur 2007). A standard approach to classifying marine benthic habitats has not been established and endorsed by the scientific and management communities. The lack of a generally accepted classification scheme, encompassing benthic, water column and coastal environments, requires that resource managers either modify existing approaches for their study sites or develop new schemes.

Consider the Data

The classification scheme chosen should accommodate a variety of data sources, given that data describing estuaries and ecosystems take many forms, ranging from grab samples to sonar and satellite images. Locating data sources, and determining the density and spatial coverage of the combined data set is important to the development effort. Because classification is inherently a geospatial exercise, it is necessary to design and document an appropriate database structure. A review of existing data and a data gap analysis can assist with determining the specific scale or hierarchical levels of classification that can be incorporated into classification frameworks.

Know the Environments, Stressors, and Possible Impacts that May Be Observed

Estuaries and coastal environments are defined by a number of factors, including basin morphology, water quality, river inflow, tidal range, residence time, biological productivity, habitat differences, fisheries, circulation, etc. Coastal and fishery resource managers are frequently tasked with making decisions about uses of the estuarine and marine environment without sufficient knowledge of the impacts that may occur at various scales by proposed projects. For example, Massachusetts faces strong development pressures in coastal and estuarine areas from existing and proposed projects such as aquaculture, wind farms, pipeline and cable installations, deepwater ports, construction of docks and piers and sewage outfalls (Lund & Wilbur 2007). Development projects may change the transport of water, sediments and pollutants, such as nutrients and heavy metals impacting water and habitat quality. A classification system may assist with decisions about which estuaries or which sections of estuaries could be targeted for such projects so that risk can be minimized.

Stressors from a variety of sources (e.g., atmospheric deposition, groundwater, stormwater, etc.) will have effects on the estuaries and ecosystems of interest. The combination of anthropogenic influences and natural disturbances potentially disrupts and degrades ecosystem functions and values (Wilbur & Pentony 1999). Knowing which ecosystems are most resistant or susceptible to various stressors, which are more resilient to variations, and the factors that contribute to these traits may be important classification parameters to consider. Comprehensive maps of the estuarine and marine environment, and standardized classification of the habitats within these environments, provide resource managers with tools to assess the relative abundance of various habitat types.

Consider Biology

Incorporating biological community structure or ecological guild characteristics is not often undertaken in classification schemes. Biotic characteristics such as the presence and relative abundance of epifaunal communities and macroalgae beds are temporally variable and distribution may be patchy or ephemeral. Distributions of biological communities also are greatly altered by human impacts, are a substantial part of the societal and resource value of ecosystems, and frequently are the basis for regulatory strategies. CMECS uses biological information to modify or subdivide classes. The EUNIS approach to classifying biotopes (biological habitat class) may provide further guidance on incorporating biological resources into a habitat classification framework and could be further developed. Considerable data resources are required to integrate biological data into a classification framework, and objectives calling for finer scales of resolution may be the most amenable applications.

Consistent Nomenclature and Coding

Classified units require consistent names and coding systems to facilitate communication among the management, science and stakeholder communities. The nomenclature must clearly constrain the meaning of terms. An official glossary of terms should be agreed upon and implemented by all users to maintain consistency. Coding or naming conventions when used, may include numeric or letter sequences (or both). Although codes can be cumbersome and require effort to decipher, they are necessary shorthand for communication in situations where long text descriptions are inappropriate and where the information needs to be incorporated into a searchable database (Kutcher et al. 2005; Valentine et al. 2005). Supporting data should be organized in a manner that is consistent with the classification framework. In geographic information system (GIS) and spreadsheet formats, a coding system facilitates data organization, metadata development, and database queries (Kutcher et al. 2005).

Technical Advisory Group

A technical advisory group to provide input on the choice and implementation of a classification scheme may facilitate development and application of a classification intended for wide use. The advisory group should be composed of scientists and managers to ensure that the approach is valid and useful in a management context. Membership in this group should include a balance of expertise with knowledge of estuary processes, ecology, fishery resources, natural resource management, and mapping technology. Experts involved in the creation or implementation of relevant classifications should be engaged to ask and answer questions, discuss applicability, and to advise on the selection of an appropriate classification scheme and possible modifications.

Implementation

Technical staff who will implement classification tools should familiarize themselves with classification schemes, estuary function and structure, and mapping techniques, as well as data management. For our case study it was important to understand the ways that classes in each framework were defined, and how Pensacola Bay fit into these definitions based on its particular properties in order to place the Bay into the correct class for each framework. New users should identify areas where making class determinations or discrimination is uncertain, and where terminology or coding is unclear. Consistency in classifying elements and in use of coding and terminology is important for describing class membership.

The database that allowed us to perform the test was constructed from a number of sources and required careful design and management. Once data are identified as pertinent to the study objectives, those data should be incorporated into a searchable data management system. The data system will allow for subsequent dissemination of data, classification products and maps.

Comparative studies, such as the case study included in this chapter, are few. More work along these lines could foster wider acceptance and use of classification frameworks, as well as the development of better ones. A comparison of two large systems, the Delaware and Chesapeake Bays has shown that, although they are both drowned river mouth estuaries receiving waters from coastal plains, and geographically adjacent,[5] they respond to nutrient loads differently. The Delaware Bay has a greater nutrient load, but demonstrates less severe effects of nutrient enrichment, largely because of greater flushing rates or lower retention time (EPA, 2001). Studies should be conducted to evaluate classification schemes based on primary attributes important for determining the variability within and between classes. To be useful, classification should reduce variability of ecosystem-related measurements within specific classes and maximize the variability among classes (EPA, 2001).

Workshops to evaluate and discuss the application of the recommended classification framework, and the comparative studies that have been conducted may be useful. Environmental managers, researchers, policy makers and other stakeholders involved with the development, testing and application of relevant classifications should participate to evaluate and validate the classification and to identify next steps for wider implementation and incorporation of classification into environmental management procedures.

CONCLUSIONS

There is no single, standard classification approach currently endorsed by scientific and management communities, yet many are available. In order for an approach to be successful and useful on local, regional or national scales, users need to agree on and employ standardized terminology to classify and describe estuarine environments. Often these efforts are driven by available data, and it is important to understand the sources of data and incorporate them, but the foundation of a classification must be well grounded in ecological theory or basic science principles.

There are considerable benefits to adopting a standardized classification scheme into management paradigms for estuaries and ecosystems. Understanding the similarities and differences in types of estuaries and coastal waters, as well as the habitats within these systems, and how they respond to development pressures including pollutant and nutrient inputs is important to planning adaptive management strategies.

Multiple scales and expanded extent may be desired by managers, but sometimes simple is best. There is a tendency to aggregate classifications or extrapolate to broader scales or wider habitat types. The opposite is also true. Some broad scale classifications may be subdivided into finer and finer scales in a hierarchical manner. These efforts become much more credible when efforts are tested with appropriate data before they are applied to management problems.

Our ability to manage estuarine environments will ultimately benefit from a classification scheme that provides standardized terminology for different types, guidelines for organizing data, and facilitation of mapping and modeling efforts. Incorporation of the results of applications and comparative testing should guide efforts to modify and expand use of classification tools, the use of data bases that support them and will allow for adaptive management strategies that improve the condition of our estuaries.

---

[5] Chesapeake and Delaware Bays are connected by the Chesapeake and Delaware Canal, through which there are significant exchanges of tidal flows.

ACKNOWLEDGMENTS

We appreciate the review by Lisa Smith, and contributions to the chapter by S. J. Jordan. The views expressed in this chapter are those of the author and do not necessarily reflect the views or policies of the U.S. Environmental Protection Agency. This is contribution X1133 from the Gulf Ecology Division, and a product of EPA's Water Quality Research Program.

REFERENCES

Allee, R.J., Detheir, M., Brown, D., Deegan, L., Ford, R.G., Hourigan, T.F., Maragos, J., Schoch, C., Sealey, K., Twilley, R., Weinstein, M.P. & Yoklavich, M. (2000). *Marine and Estuarine Ecosystem and Habitat Classification*. NOAA Technical Memorandum NMFS–F/SPO–43.

Bricker, S.G, Clement, C.G., Pirhalla, D.E., Orlando, S.P. & Farrow, D.R.G. (1999). *National Estuarine Eutrophication Assessment: Effects of Nutrient Enrichment in the Nation's Estuaries*. National Oceanic and Atmospheric Adiministration, National Ocean Service, Special Projects Office and the National Centers for Coastal Ocean Science, Silver Spring, MD.

Bricker, S., B. Longstaff, W. Dennison, A. Jones, K. Boicourt, C. Wicks and J. Wocrncr 2007. Effects of Nutrient Enrichment in the Nation's Estuaries: A Decade of Change, National Estuarine Eutrophication Assessment Update. NOAA Coastal Ocean Program Decision Analysis Series No. 26. National Centers for Coastal Ocean Science, Silver Spring, MD.322pp. http://ccma.nos.noaa.gov/news/feature/Eutroupdate.html.

Brown, B. (1993). *A Classification System of Marine and Estuarine Habitats in Maine: An Ecosystem Approach to Habitats*. Maine Natural Areas Program. Department of Economic and Community Development, Augusta, ME.

Brown, S.K. (2002). *Our Living Oceans Benthic Habitat Classification System*. National Oceanic and Atmospheric Administration, National Marine Fisheries Service, Office of Science and Technology, Silver Spring, MD.

Connor, D.W., Allen, J., Golding, N., Howell, K., Lieberknecht, L. Northen K., & Reker, J. (2004). The national marine habitat classification for Britain and Ireland, version 04.05. Joint Nature Conservation Committee. http://www.jncc.gov.uk/pdf/04_05 _introduction.pdf

Cowardin, L.M., Carter, V., Golet F.C., & LaRoe, E.T. (1979). Classification of Wetlands and Deepwater Habitats of the United States. U.S. Fish and Wildlife Service, FWS/OBS–79/31. http://www.fws.gov/nwi/Pubs_Reports/Class_Manual/class_titlepg.htm

Davies, C., Moss D. & Hill, M.O. (2004). EUNIS Habitat Classification. European Environment Agency.

Dethier, M.N. (1992). Classifying marine and estuarine natural communities: an alternative to the Cowardin system. *Natural Areas J. 12,* 90–100.

Engle, V.D., Kurtz, J.C., Smith, L.M., Chancy, C, and Bourgeois, P. (2007) A Classification of U.S. Estuaries Based on Physical and Hydrologic Attributes. *Environ. Monit Assess* 129:397-412.

Fischer, H. 1976. Mixing and dispersion of estuaries. *Ann. Rev. Fluid Mech. 8*, 107–133.

Greene, H.G., Yoklavich, M.M., Starr, R.M., O'Connell, V.M., Wakefield, W.W., Sullivan, D.E., McRea, J.E. Jr. & G.M. Caillet. (1999). A classification scheme for deep seafloor habitats. *Oceanologica Acta 22,* 663–678.

Greene, H.G., Bizzarro, J.J., O'Connell V., & Brylinsky, C.K. (2007). Construction of digital potential marine benthic habitat maps using a coded classification scheme and their application. In H.G. Greene and B.J. Todd (Eds.), *Mapping the Seafloor for Habitat Characterization,* pp. 141–155. Geological Association of Canada, Special Paper 47.

Gulf of Mexico Alliance (GOMA). (2008). *Governors' Action Plan II for healthy and resilient coasts* 2009-2014.

Hagy, J.D. III, Kurtz, J.C. and Greene, R.M. (2008). An Approach for Developing Numeric Nutrient Criteria for a Gulf Coast Estuary. *U.S. Environmental Protection Agency, Office of Research and Development, National Health and Environmental Effects Research laboratory*, Research Triangle park, NC. EPA 600R-08/004 44 pp.

Hansen, D.V. & Rattray, M.J. 1966. New dimensions in estuary classification. *Limnol. Oceanog. 11*, 319–326.

Hume T., Snelder T., Weatherhead M. & Liefting, R. (2007). A controlling factor approach to estuary classification. *J. Ocean Coast. Manage. 50*, 905–929.

Ippen, A. & Harlemann, D. 1961. *Committee for Tidal Hydraulics Technical Bulletin 51,* U.S. Army Corps of Engineers, Washington, D.C.

Jay, D.A., Geyer, W.R. & Montgomery, D.R. (2000). An ecological perspective on estuarine classification. In: Hobbie, J.R. (Ed.), *Estuarine Science: A Synthetic Approach to Research and Practice,* pp. 149–176, Island Press, Washington, D.C.

Kennish, M. 1986. *Ecology of Estuaries: Physical and Chemical Aspects.* CRC Press, Boca Raton, Florida.

Kurtz, J.C., Detenbeck, N.D., Engle, V.D., Ho, K., Smith, L.M., Jordan S.J. & Campbell, D. (2006). Classifying coastal waters: current necessity and historical perspective. *Estuar. Coasts 29,* 107–123.

Kutcher, T.E., N.H. Garfield and K.B. Raposa. 2005 (draft). A recommendation for a comprehensive habitat and land use classification system for the National Estuarine Research Reserve System. National Estuarine Research Reserve, Estuarine Reserves Division. Draft report to NOAA/NOS/OCRM. Silver Spring, MD. 26 pp.

Kutcher, T.E. (2006). A comparison of functionality between two coastal classification schemes developed within the National Oceanic and Atmospheric Administration. *NERRS/ERD Proposal* (report to NOAA/NOS/OCRM), Silver Spring, MD.

Kutcher, T.E. (2008). *Habitat and Land Cover Classification Scheme for the National Estuarine Research Reserve System.* http://nerrs.noaa.gov/Doc/PDF/Stewardship/ NERRClassificationSchemeDoc.pdf, accessed May 18, 2011.

Lund, Katherine and Anthony R. Wilbur. 2007. Habitat Classification Feasibility study for Coastal and marine Environments in Massachusetts. Massachusetts Office of coastal Zone Management: Boston, MA. pp 31.

Madden, C.J., & Grossman, D.H. (2004a). A framework for a coastal/marine ecological classification standard. NatureServe, Arlington, VA. *Prepared for the National Oceanic and Atmospheric Administration, under Contract* EA-133C-03-SE-0275.

Madden, C.J., & Grossman, D.H. (2004b). *National coastal/marine classification pilot projects.* Draft final report. NatureServe, Arlington, VA.

Madden, C.J., Grossman, D.H. & Goodin, K.L. (2005). Coastal and Marine Systems of North America: Framework for an Ecological Classification Standard: Version II. NatureServe, Arlington, VA. http://www.natureserve.org/getData/CMECS/cm_pub.pdf

Madden, C.J. et al. (February 2009). Coastal and marine Ecological Classification Standard. Version III.2009.0213. *NatureServe,* Arlington, VA.

Madley, K. (2002). Florida system for classification of habitats in estuarine and marine environments (SCHEME). Report to US Environmental Protection Agency. Florida Fish and Wildlife Conservation Commission, Florida Marine Research Institute, *FMRI File Code* 2277-00-02-F.

National Estuarine Research Reserve System (NERRS). 2000. Evaluations of Marine and Estuarine Ecosystem and Habitat Classification. *NOAA Technical Memorandum* NMFS–F/SPO–43.

National Oceanic and Atmospheric Administration (NOAA). 2003. Coastal Assessment Framework. http://coastalgeospatial.noaa.gov/back_gis.html#caf.

Nunes Vaz, R.A. & Lennon, C.W. 1991. Modulation of estuarine stratification and mass transport at tidal frequencies. In: B.B. Parker (ed.), *Progress in Tidal Hydrodynamics,* John Wiley & Sons, New York, pp. 5–84.

Odum, H.T. & Copeland, B.J. 1974. A functional classification of the coastal systems of the United States. In: H.T. Odum, B.J. Copeland & E.A. McMahan (eds.), *Coastal Ecological Systems of the United States, Vol. I.* The Conservation Foundation, Washington, D.C., pp. 5–84.

Omernik, J.M. (1987). Ecoregions of the conterminous United States. *Annals Assoc. Am. Geogr. 77,* 118–125. (http://water.usgs.gov/GIS/metadata/usgswrd/ecoregion.html)

Prandle, D. 1986. Generalized theory of estuarine dynamics. In: J. van deKreek (ed.), *Physics of Shallow Estuaries and Bays*, Springer-Verlag, Berlin, pp. 42–57.

Pritchard, D. 1967. What is an estuary: Physical viewpoint. In: G. Lauff (ed.) *Estuaries,* Publication 83, American Association for the Advancement of Science, Washington, D.C., pp. 3–5.

Simpson, J. & Hunter, J.R. 1974. Fronts in the Irish Sea. *Nature 250*, 404–406.

Smith, S.V., Buddemeier, R.W., Bricker, S.B., Maxwell, B.A.,Pacheco, P., Mason, A., 2004. Estuarine typology: perturbations and eutrophication responses. In: ASLO/TOS 2004 Ocean Research Conference, Honolulu, Hawaii, February15–20, http://www.aslo.org/meetings/honolulu2004/files/aslotos-2004-abstracts.pdf

Stommel, H. & Farmer, H. 1952. On the nature of estuarine circulation. Reference Notes 52–51, 52–63, 52–88, *Woods Hole Oceanographic Institute,* Woods Hole, Massachusetts.

Summers, J.K., Paul, J.F. & Robertson, A. (1995). Monitoring the ecological conditions of estuaries in the United States. *Tox. Environ. Chem. 49*, 93.

U.S. Environmental Protection Agency (EPA). (2001). Nutrient Criteria Technical Guidance Manual. *Estuarine and coastal Marine Waters. U.S. Environmental Protection Agency,* Washington, D.C. EPA 822-B-01-003.

U.S. Environmental Protection Agency (EPA). (2004a). *Classification Framework for Coastal Systems.* U.S. Environmental Protection Agency, EPA 600/R–04/061, Washington, D.C.

U.S. Environmental Protection Agency (EPA). (2004b). *National Coastal Condition Report II,* US EPA, Office of Resarch and Development, Office of Water, Washington, DC. EPA-620/R-03/002.

U.S. Environmental Protection Agency (EPA). (2008). *National Coastal Condition Report III*. U.S. EPA, Office of Resarch and Development, Office of Water, Washington, DC. EPA/842-R-08-002http://www.epa.gov/nccr.

Valentine, P.C., Todd, B.J. & Kostylev, V.E. (2005). Classification of marine sublittoral habitats, with application to the northeastern North America region. *Am. Fish. Soc. Symp. 41*, 183–200.

Venice System. 1959. Final resolution of the symposium on the classification of brackish waters. *Arch. Oceanog. Limnol. 11*, 243–248.

Walker, S.P., Garfield, N.H., Kutcher, T.E. & Reed, T. (2005). *Recommended guidelines for adoption and implementation of the NERRS comprehensive habitat and land use classification system.* Technical document produced for pilot National Estuarine Research Reserves.

Weiss, J. L., J. T. Overpeck and B. Strauss. 2011. Implication of recent sea level rise science for low-elevation areas in coastal cities of the conterminous U.S.A. *Climatic Change.* 105: 635-645.

Wilbur, A.R. & Pentony, M.W. (1999). Human-induced nonfishing threats to Essential Fish Habitat in the New England region. In L. Benaka (Ed.), *Fish Habitat: Essential Fish Habitat and Rehabilitation. Am. Fish. Soc. Symp. 22*, 299–321.

In: Estuaries: Classification, Ecology and Human Impacts   ISBN: 978-1-61942-083-0
Editor: Steve Jordan   © 2012 Nova Science Publishers, Inc.

Chapter 3

TAGUS ESTUARY SALT MARSH STRUCTURE AND DYNAMICS: A HISTORICAL PERSPECTIVE

Isabel Caçador and Bernardo Duarte

Center of Oceanography, Faculty of Sciences of the University of Lisbon, Campo
Grande 1749-016, Lisbon, Portugal

ABSTRACT

Salt marshes are a very important part of the estuarine ecosystem, with an important
role within the biogeochemical cycles, being areas of high primary production, which
also contribute importantly as shoreline stabilizers. Besides, periodical tidal flooding of
salt marshes also causes the transport of significant quantities of pollutants, which tend to
accumulate in the marsh ecosystem. Therefore, salt marshes are considered to be
important sinks namely of heavy metals. Their important role has been recently admitted
by the inclusion of these ecosystems in the European Water Framework Directive
(WFD). Multiple services of wetlands and its value are already well known. The Tagus
estuary is the largest estuary in Western Europe, covering an area of approximately 320
km^2. The estuary has a vast inner bay with extensive intertidal areas from which about
15% of these areas are salt marshes. The Tagus estuary is a mesotidal system with a mean
tidal range of 2.4 m, varying from about 1 m at neap tides to about 4 m at spring tides.
This chapter compares the evolution of salt marsh plant structure over the past five
decades in two Tagus salt marshes with different anthropogenic influences. The results
show that in both salt marshes the differences in plant cover were accompanied by
significant reductions in total plant biomass. Below/aboveground biomass ratios have
been increasing, reflecting a higher investment in roots than in photosynthetic organs.
There was a general decrease of total N content in all studied species, although shoot and
root N concentrations were almost the same between years. These results suggest a lower
N availability and/or a stress situation.

INTRODUCTION: SALT MARSHES

Salt marshes have great ecological value, in terms of nutrient regeneration, primary production, habitat for wildlife, and as shoreline stabilizers. Periodic tidal flooding of salt marshes provides large quantities of pollutants to the marsh ecosystem. Thus, salt marshes are considered to be important sinks for pollutants (Caçador et al. 1993, Caçador et al. 1996, Davy 2000, Prange and Dennison 2000, Caçador et al. 2007a). The Water Framework Directive (WFD) requires that the member states of the European Union (EU) monitor the biological quality of all water bodies and elaborate management plans for the river basins, in order to solve water quality problems. The European Union (2000) Guidance Document No. 5 (Working Group 2.4, COAST, 2003) advises that, in transitional waters (TW), the intertidal areas from the highest tide limit to the lowest tide limit should be included in the monitoring program, thereby including salt marshes. The WFD mandates a reference condition for angiosperms (aquatic vascular plants), corresponding totally or nearly totally to undisturbed conditions, where there are no detectable changes in the angiosperm abundance related to anthropogenic activities.

The striking zonation of marsh plant communities has been explained as the product of competitively superior plants dominating in physically less-challenging habitats and displacing competitively subordinate plants (Bertness and Pennings 2000, Caçador et al. 2007b). Recent studies however, have suggested that both nutrient supply and thermal stress can influence this simple scenario by increasing nutrient availability, which may alleviate below ground competition for nutrients and lead to above ground competition for light, thus dictating competitive dominance among marsh plants. The evident zonation in salt marsh vegetation is now accepted to be the result of competitive advantages of superior plants in colonizing particular habitats with more favorable physical-chemical characteristics, leading to the drawback of less competitive species (Berteness and Pennings 2000; Caçador et al. 2009). External stresses driven by warming, such as nutrient balance disturbances (similar to eutrophication), may lead to the successful rise of less competitive species by alleviating belowground competition. Plant distribution and seasonal biomass variations have been studied for several decades in Tagus salt marshes, providing a baseline for species dynamics studies (Esteves-de-Sousa 1953, Melo et al. 1976, Catarino and Caçador 1981, Caçador 1987, Caçador et al. 2007b).

Salt marshes are natural sites for deposition of heavy metals in estuarine systems (Caçador et al. 1993, Doyle and Otte 1997, Williams et al. 1994, Caçador et al. 2009, Duarte et al. 2010). When located near polluted areas, these ecosystems receive large amounts of pollutants from industrial and urban wastes that either drift downstream within the river flow or are the direct result of waste dumping from nearby industrial and urban areas (Reboreda and Caçador 2007). When metals enter salt marshes they spread along with the tides and periodic floods and interact with soil and the biotic community (Suntornvongsagul et al. 2007). Most salt marsh plants accumulate large amounts of metals in their aerial and belowground organs (Caçador et al. 2000). Among these are the most common plants of southern European salt marshes: *Sarcocornia fruticosa, S. perennis, Halimione portulacoides* and *Spartina maritima*. Their ability to phyto-stabilize those contaminants in the rhizo-sediment is an important aspect of self-remediative processes and biogeochemistry in this ecosystem (Caçador et al. 1996, Sundby et al. 1998, Weis and Weis 2004).

THE TAGUS ESTUARY: A CASE STUDY

The Tagus estuary is one of the largest estuaries on the Atlantic coast of Europe, covering an area of 300 km^2 at low tide and 340 km^2 at extreme high tide (Figure 1). Surrounded by industries and cities, the estuary receives effluents from about 2.5 million inhabitants living in the Greater Lisbon area, together with the discharges from industries. These areas are usually covered with dense vegetation, which may act as a trap for estuarine sediment. Previous studies showed that Tagus salt marshes receive large quantities of anthropogenic metals and incorporate them into the sediments (Vale 1990, Salgueiro and Caçador 2007, Duarte et al. 2008, Caçador et al. 2009, Duarte et al. 2010).

The study reported here was carried out in two marshes (Pancas and Corroios) of the Tagus estuary (Figure 1). Both sites experience two episodes of tidal flushing each day. The marshes were chosen because of their contrasting situations: Pancas is a young salt marsh with extensive mudflats (800 ha) included in the Tagus Nature Reserve, whereas Corroios is older, smaller (400 ha) and located in the proximity of urbanised and industrial areas. The more abundant species colonising both marshes are *S. maritima*, *H. portulacoides*, and *S. fruticosa*.

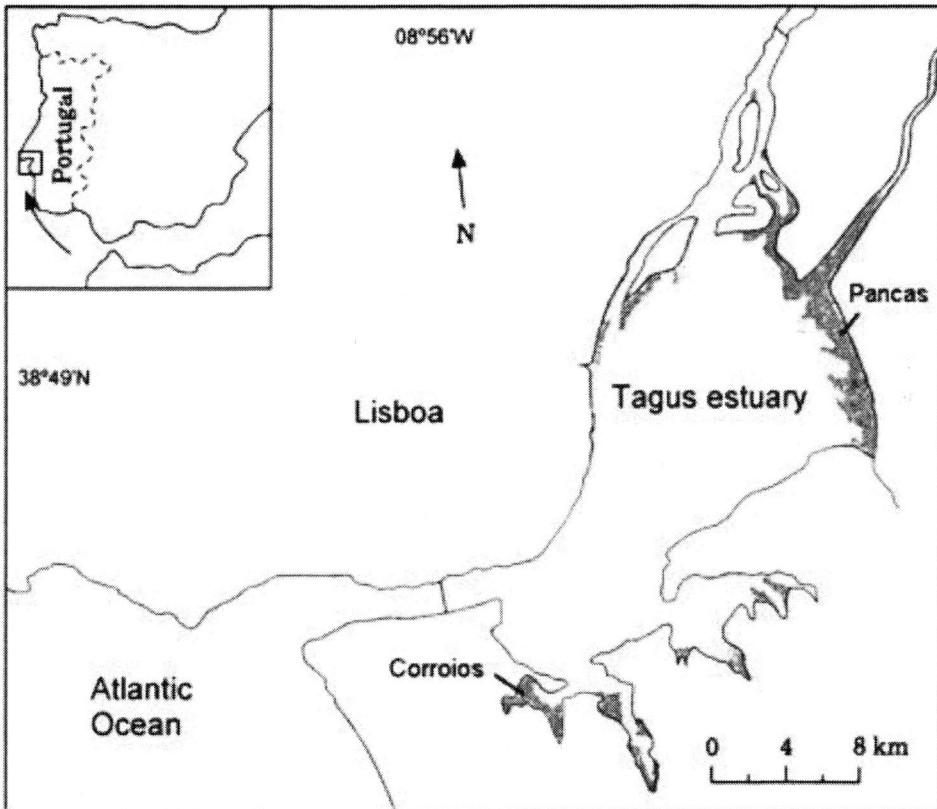

Figure 1. The Tagus Estuary sampling sites.

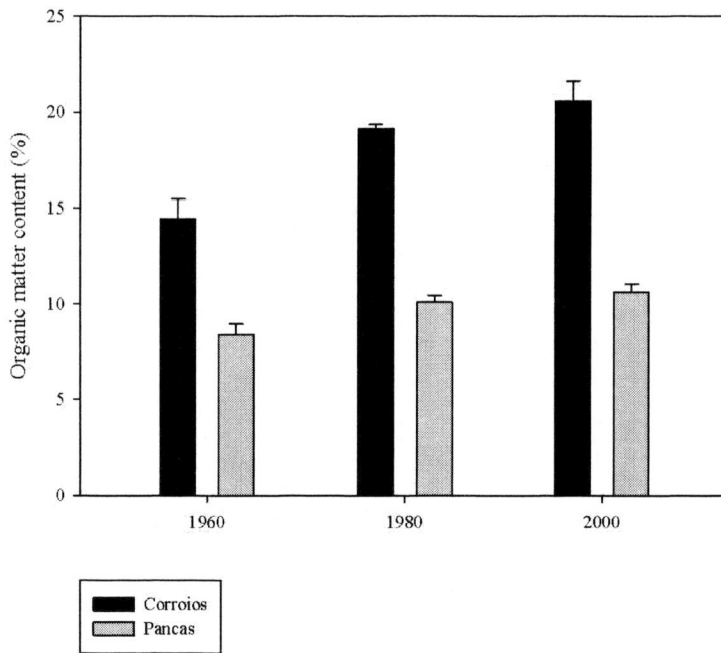

Figure 2. Organic matter content (%) in sediments of both salt marshes from 1960 to 2000.

PHYSICAL AND CHEMICAL CHARACTERISTICS OF THE SEDIMENTS

Silt and clay were the main constituents of the sediments of both salt marshes, with sand accounting for less than 5%. Previous works have shown that these salt marshes principally retain fine particles (Caçador et al. 2000). Sediment temperature was similar in the two marshes, varying seasonally between 15 and 26º C. Sediments were saltier in Corroios (23-47 psu) than in Pancas (13-19 psu) reflecting the salinity gradient along the estuary (Figure 2). In winter the salinity decreased pronouncedly in Corroios as a result of freshwater discharges reaching the lower estuary. Salinity in the *H. portulacoides* zone was higher than in *S. maritima*, most likely because of the longer periods that the *H. portulacoides* zone was exposed to the atmosphere during semi-diurnal tidal cycles; consequently evaporation rates were greater. The organic content of sediments has been increasing over the past four decades (Figure 2), although Corroios was much more enriched with organic matter than Pancas over the entire period.

EVOLUTION OF METAL CONTAMINATION IN TAGUS SALT MARSH SEDIMENTS

Salt marsh plants play an important role in the dynamics of the estuarine ecosystem. Plants act as sediment traps, facilitating the retention of suspended estuarine particulates with their associated metals, and influencing retention and accumulation processes of metals in salt marsh sediments (Salgueiro and Caçador 2007). The metal concentrations in salt marsh sediments are often related to contamination of the coastal environment. Through the analysis of sediment cores dated using ^{137}Cs isotopic signatures it is possible to determine the total concentrations of heavy metals at different periods of recent history (Salgueiro 2001).

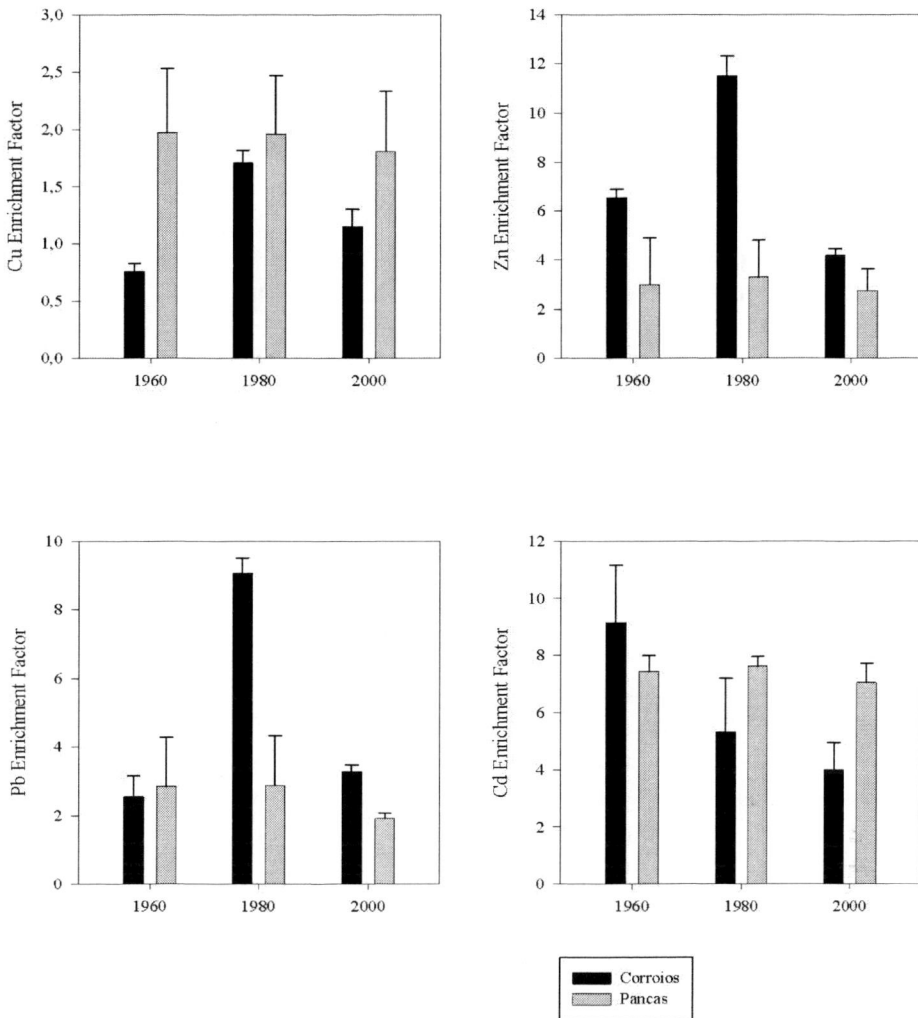

Figure 3. Heavy metal enrichment factors from 1960-2000 assessed from ^{137}Cs dated layers of salt marsh sediments.

Metal enrichment is more evident when observing ratios of metal concentrations in the sediments to their concentrations in the Earth's crust (Vale 1990; Figure 3). Tagus salt marshes were enriched substantially with the heavy metals Pb, Zn, Cd and Cu. Although concentrations of other heavy metals have been measured in these sediments (Duarte et al. 2008, 2009, 2010, Caçador et al. 2009) their enrichment factors were rather low compared to Pb, Zn, Cd, and Cu. As shown in Figure 3, Corroios salt marsh generally has higher enrichment factors, attributable to its proximity to large industrialized areas. Although total heavy metal concentrations or enrichment factors are important aspects to consider when evaluating sediment contamination, there are also some other factors that are often more important when we are referring to salt marsh plants. Halophytes take up heavy metals from sediments through their roots and alter the sediment biogeochemistry. This uptake is influenced by several factors such as: metal availability, root activity (Duarte et al. 2007), sediment biogeochemistry (Reboreda and Caçador 2007a), microorganism activity (Duarte et al. 2008, 2009) and plant species (Reboreda and Caçador 2007b, Caçador et al. 2009, Duarte et al. 2010).

Although Corroios showed the highest metal enrichment factors, it also had sediments with high organic matter content, thereby diminishing metal availability (Table 1). Organic compounds can establish strong chemical bonds with metals, reducing their availability (Caçador et al. 2000). Pancas salt marsh is a less impacted salt marsh, but it is also composed of sediments with lower organic matter content (Figure 2).

Considering that metal availability (as opposed to concentration) could influence plant diversity and coverage, it might be expected that the Pancas salt marsh plant community would have suffered more severe modifications than those in Corroios. This did not turn out to be the case, as illustrated and discussed later in this chapter.

NITROGEN CONCENTRATION IN SEDIMENTS

One of the greatest problems in coastal waters is eutrophication. Salt marshes import inorganic nutrients and export organic nutrients. As tidal water flows through salt marshes, plants, bacteria and algae produce or transform the organic matter of the food chain that supports fish and shellfish populations (Teal and Howes 2000). While salt marshes modify the forms of the principal plant nutrients N and P, some of the pathways result in removal of nutrients from biologically active systems. Flooding with seawater leads to an input of inorganic and organic substances into the marsh (Rozema et al. 2000). Nitrogen is removed primarily either by (1) being trapped in refractory organic matter that contributes to marsh maintenance through accretion, or (2) by loss to the atmosphere (as N_2) through denitrification. Coastal marshes tend to be nitrogen limited. With increasing nitrogen supply, marshes show greater primary productivity by both grasses and algae. Unlike some coastal systems, salt marshes can withstand very large additions of nitrogen without severe damage (Teal and Howes 2000). In addition, estimates of the size of the N-pool do not always indicate the availability of nitrogen to plants. The increasing soil nitrogen content with salt marsh age indicates that the salt marsh soil acts as a sink for N. Some other studies focused on the N-content of salt marsh compartments have been published (Caçador et al. 2007a, Sousa et al. 2008, Rozema et al. 2000).

Table 1. Available metal ($\mu g \cdot g^{-1}$) concentrations in sediments, determined by diethylene triamine pentaacetic acid (DTPA) extraction, according to Lindsay and Norvell (1978)

	Zn	Pb	Cu	Cd
Corroios	22.47 ± 8.31	15.60 ± 7.30	2.40 ± 0.36	0.20 ± 0.00
Pancas	43.83 ± 14.92	10.90 ± 3.49	2.03 ± 0.49	0.17 ± 0.06

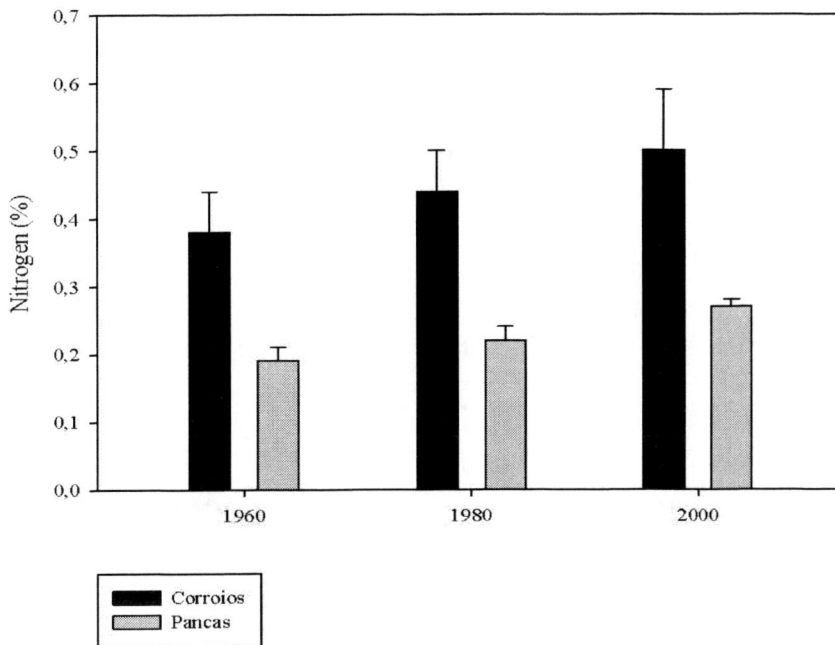

Figure 4. Nitrogen concentration (%) in the sediments of Corroios and Pancas salt marshes, from 1960 to 2000.

Table 2. Species composition in Corroios and Pancas salt marshes over the period of study

	Corroios				Pancas	
	1951	1977	1986	2003	1977	2003
S. fruticosa	X	X	X	X	X	X
S. perennis	X	X	X	X	X	X
S. marítima	X	X	X	X	X	X
H. portulacoides	X	X	X	X	X	X
Juncus maritimus	X	X			X	
Limonium vulgare	X				X	X
Spergularia spp	X					
Polygonum maritimum	X				X	X
Artemisia arborescens	X					
S. macrostachyum	X	X			X	X
Pucinelia maritima	X	X			X	X
Atriplex patula						X
Inula crithmoides		X	X	X	X	X
Salicornia nitens					X	X
Cyperus longus		X				

The amount of organic nitrogen transferred yearly to sediments in Pancas and Corroios is reflected in the vertical profiles of nitrogen concentrations in sediments. Depth variation of nitrogen differed considerably in both salt marsh sediments (Figure 4). Pancas sediments showed approximately 0.25% of nitrogen and decreased gradually with the depth/age. Corrios sediments exhibited a sub-surface enrichment in layers of higher root biomass that was 0.4%

N in Pancas and reached 0.5% in Corroios. This pattern was observed in all surveys and indicates retention of organic matter from sedimentation and the decay of root material. Where belowground biomass was higher and degradation rate slower (Corroios) the accumulation of nitrogen in sediments was more intense.

SALT MARSH PLANT STRUCTURE: DYNAMIC AND EVOLUTION

The most abundant plant species in both salt marshes are *S. maritima, H. portulacoides, S. fruticosa,* and *S. perennis* (Table 2).

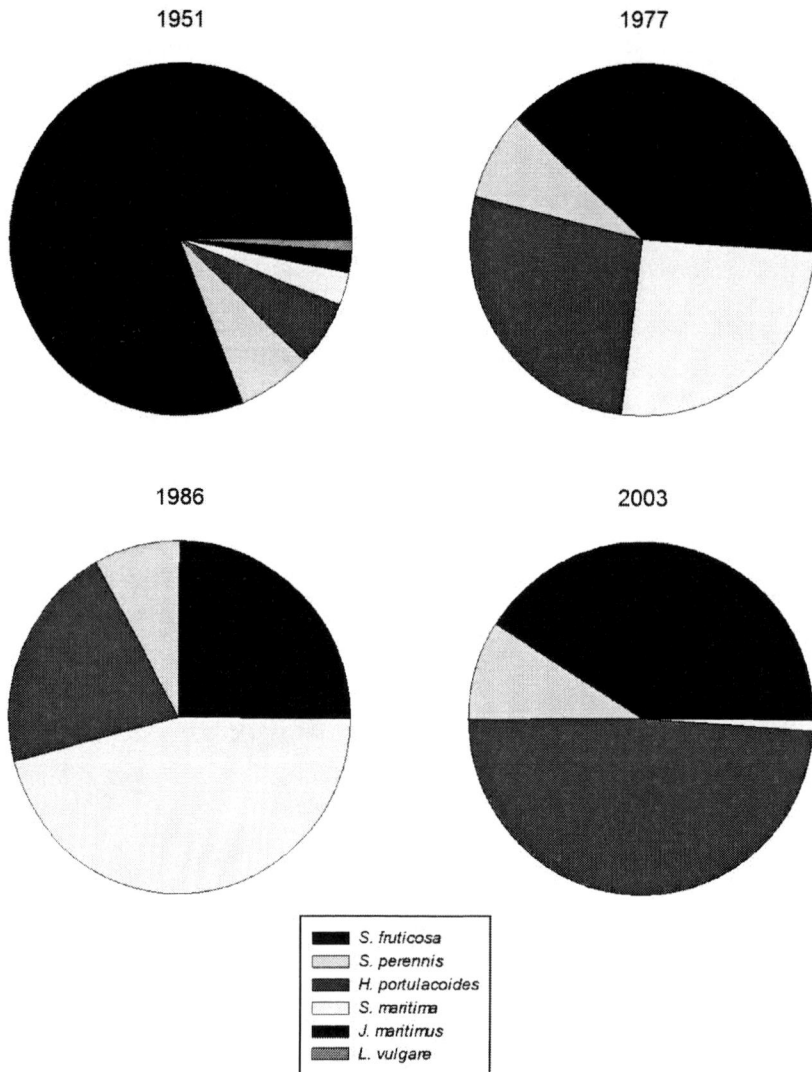

Figure 5. Species coverage (%) in Corroios salt marsh from 1951 to 2003.

1977

2003

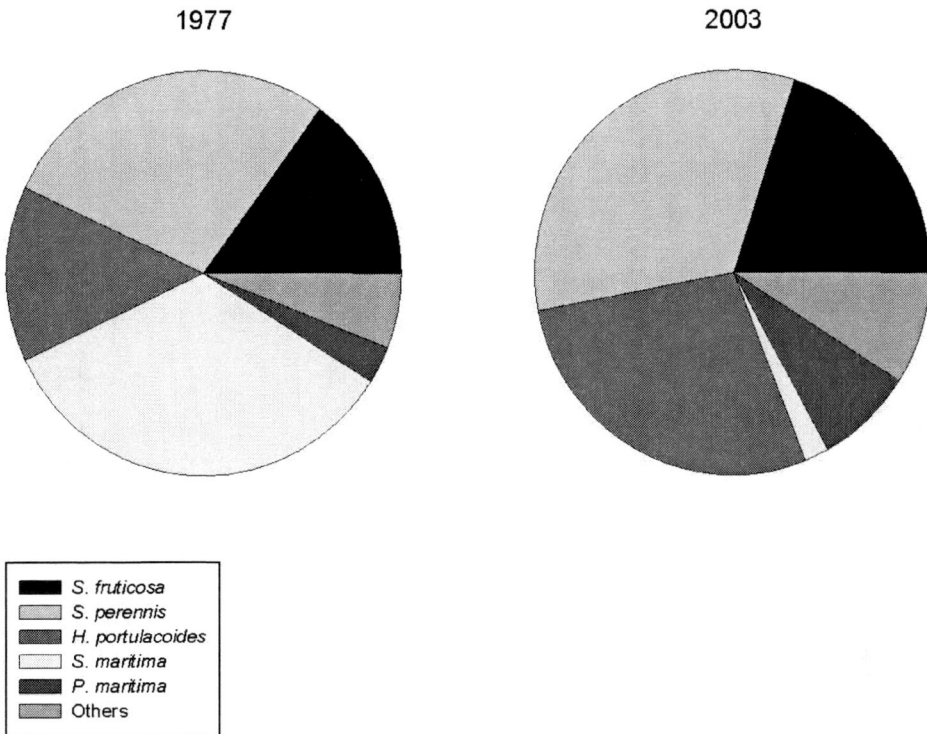

S. fruticosa
S. perennis
H. portulacoides
S. maritima
P. maritima
Others

Figure 6. Species coverage (%) in Pancas salt marsh in 1977 and 2003.

Comparing the results obtained in 1977 with recent data, the more dramatic change occurred in Corroios salt marsh, which lost about 50% of its plant species. Also, the coverages of the various species changed (Figure 5). Another difference occurred in the percentage of the area colonized by *S. fruticosa*, which changed from 13% in 1986 to 28% in 2003, and in the area colonized by *H. portulacoides*, changing from 24% in 1986 to 31% in 2003. Although Pancas salt marsh has conserved its plant diversity there occur also some modifications in plant cover. The more important modifications occur with *S. maritima*, which declined in area from 34% to 2%, and *H. portulacoides*, which increased in areal coverage from 14% to 28%. Another significant difference occurred in Pancas where *J. maritimus* colonized 18% of the total area (included in "Others" in 1977) and now has disappeared.

PLANT BIOMASS

Corroios salt marsh shows the highest values of plant biomass (Figure 7). For the three studied species, in both salt marshes, biomass decreased from 1980 to 2000. The evolution of biomass values in both salt marshes differed: In Pancas salt marsh, with the exception of *S. maritima*, there was a small decrease in biomass values, whereas Corroios salt marsh showed a higher reduction in biomass, mainly in the above ground parts.

Corroios

Pancas

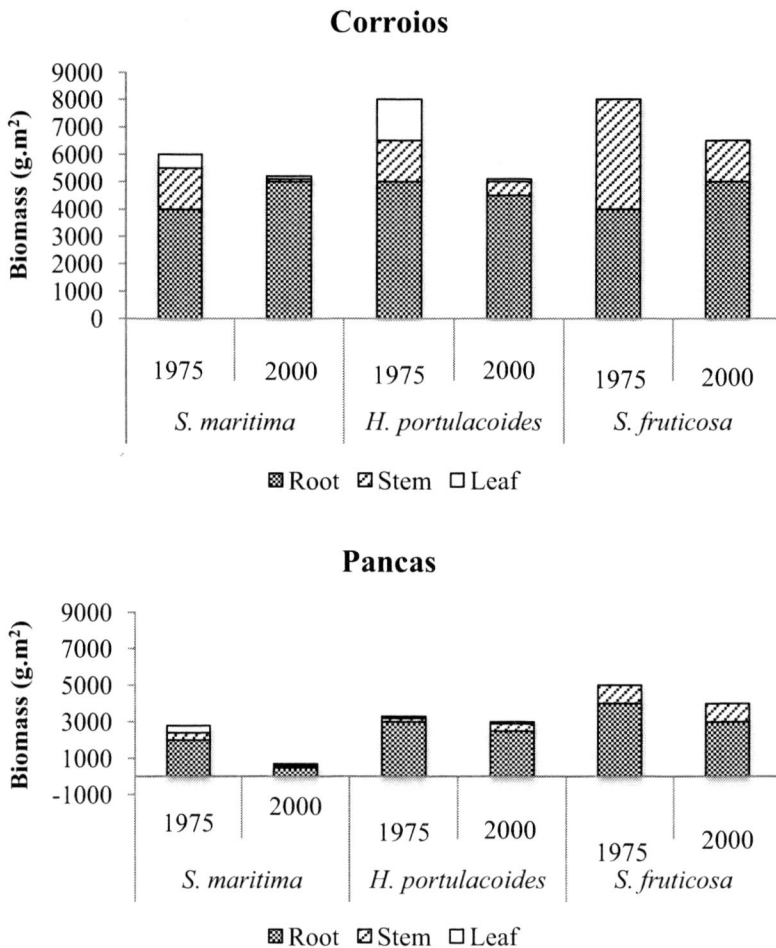

Figure 7. Biomass by species and organs in Pancas and Corroios salt marshes in 1975 and 2000.

In both salt marshes root biomass was higher than above ground biomass, ranging between 60 and 95 % of total biomass, and the differences were more pronounced in Corroios (Figure 8). When we compared the ratio of root biomass to total biomass, the three studied species had higher values in 2000 in Corroios salt marsh, but in Pancas salt marsh, with the exception of *S. maritima*, the values were higher in 1980.

NITROGEN CONCENTRATION IN PLANT BIOMASS

In both salt marshes N concentration was higher in leaves than in roots or stems of the marsh plants. The values were similar for the three studied species in Corroios salt marsh in 1975 and in 2000. In Pancas salt marsh an increase in N concentration in 2000 was evident (Figure 9).

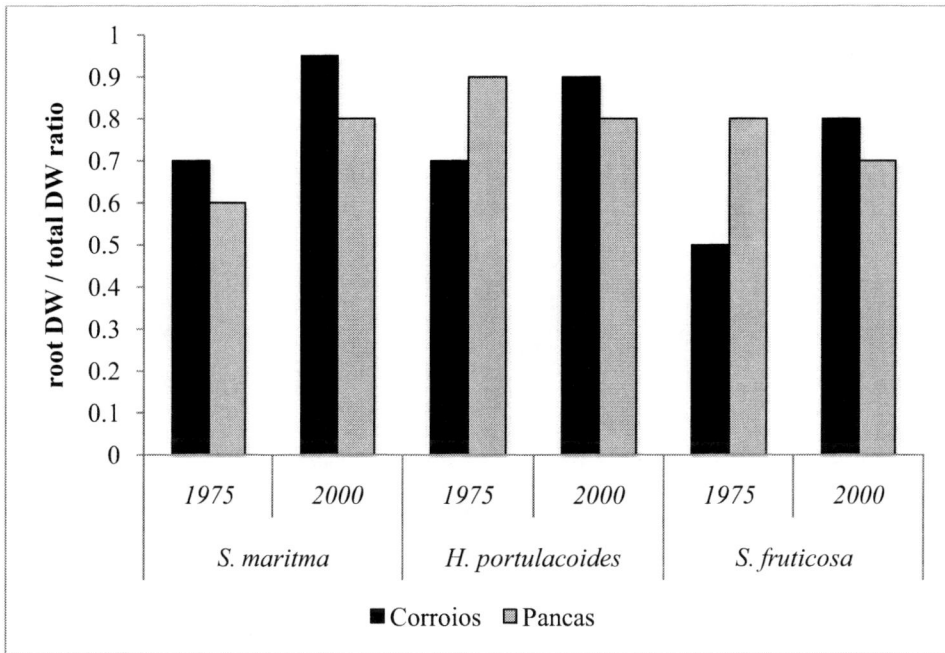

Figure 8. Root to total dry weight ratios in 1975 and 2000 in Corroios and Pancas for three species of marsh plants.

Spartina maritima in Corroios salt marsh and *H. portulacoides* in Pancas salt marsh increased the amount of biomass produced per unit of nitrogen absorbed (Berendse and Aerts, 1987), in other words, they increased their N efficiency from 1975 to 2000. The other species showed decreases in N efficiency over time in both salt marshes (Figure 10).

FINAL REMARKS

The results suggest contrasting trends in the dynamics of salt marsh vegetation in the two Tagus salt marshes. *Spartina maritima* helps to bind sediment, thus encouraging sedimentation, and has a large root system that helps to stabilize the sediment, thereby altering environmental conditions so that other species can invade the raised surface. The primary colonizers are lost, either as the result of their own effects on the physical environment, or through competitive interaction with other species able to invade the raised surface. (Long and Mason 1983). This was probably the reason for the more important changes that occurred in Pancas salt marsh. The pioneer species, *S. maritima*, was replaced by *H. portulacoides* and *S. fruticosa*.

Accretion rates in the two salt marshes are different (Caçador et al. 1996). In Pancas salt marsh vertical accretion rate has been approximately 2 cm yr^{-1} over 40 years; and about 1.3 cm yr^{-1} in Corroios salt marsh for the same period. Maximal accretion rates occur in younger marshes (Wijnen and Bakker 1999). An increase in marsh elevation also caused a decrease in the number of flooding events, which resulted in an increase in soil salinity.

An increase of N may increase the salt tolerance of some plant species (Rozema et al. 1985). Chenopods such as *H. portulacoides* will be able to tolerate more salt if more N is present in well-aerated sediment (Jensen 1985). This hypothesis may justify the loss of species diversity in Corroios and may explain the expanded coverage of *H. portulacoides* and *S. fruticosa*. Wijnen (1999), concluded that in older salt marsh sites with a larger soil-N content, more rapid succession will take place than at younger salt marsh sites. Species such as *H. portulacoides* may become the dominant species in this habitat as plant communities become less diverse. Plants that can produce more N-containing osmotic compounds such as proline and betaine when more N is present, like *H. portulacoides* (which accumulates methylated quaternary ammonium compounds for osmoregulation), were found to become established in marshes with high N levels (Wijnen 1999). Species such as *Juncus gerardii* do not accumulate N compounds for osmoregulation, and therefore they are not expected to compete successfully with other species when N levels increase. This is probably the reason for the disappearance of *J. maritimus* in Corroios salt marsh.

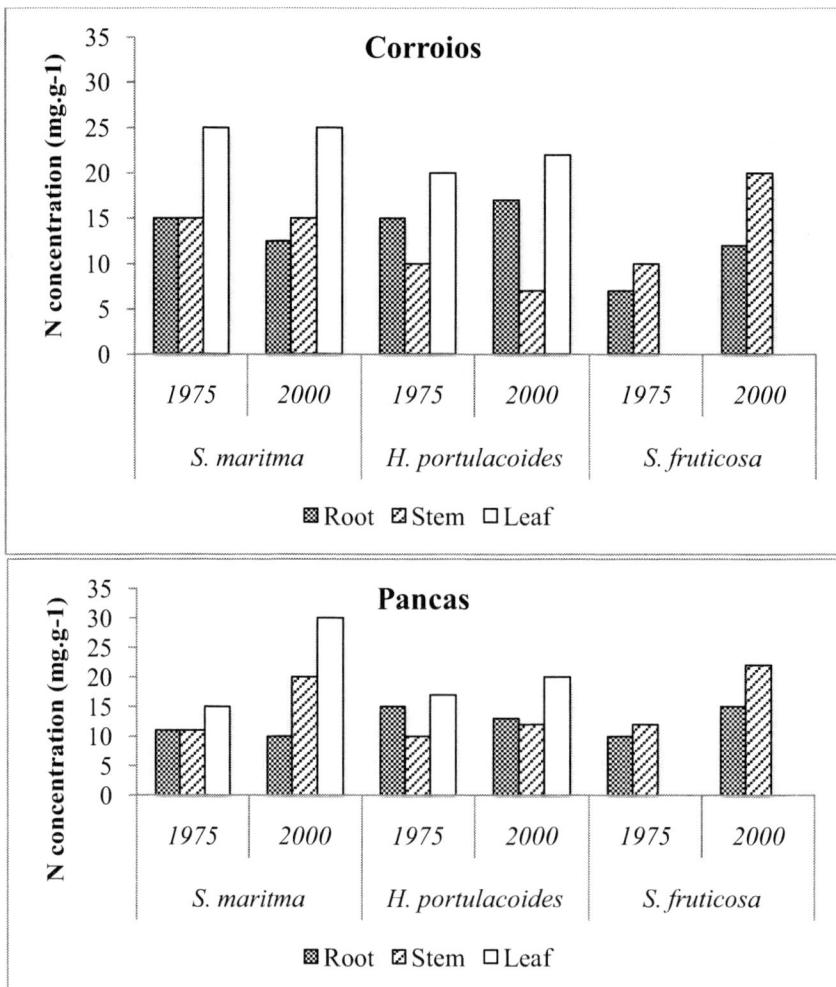

Figure 9. Nitrogen concentrations in roots, stems and leaves observed in the studied species, during the years of 1975 and 2000 in Corroios and Pancas salt marshes.

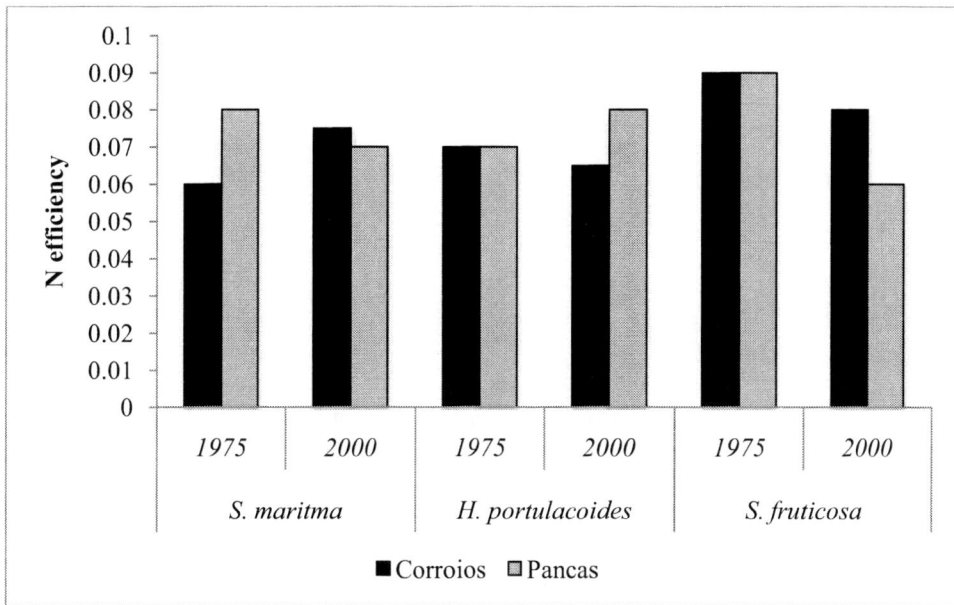

Figure 10. Nitrogen efficiency observed in three species of marsh plants, during the years of 1975 and 2000 in Corroios and Pancas salt marshes.

In salt marsh halophytes the quantity of below ground biomass is often much larger than the amount of above ground material (Gross et al. 1991, Caçador et al. 1999). Plants in physiologically stressed environments, such as salt marshes, have been assumed to have high below ground biomass (Waisel 1972, Groenendijk and Vink-Lieavaart 1987). In young salt marshes the physical and chemical conditions keep the plant populations young and the competition for nutrients to a minimum. This in turn minimizes the amount of root material required by the plant. In old salt marshes intense competition for nutrients results in increased allocation of biomass to root material (Gross et al. 1991). Using this hypothesis we are able to understand better the higher below-ground biomass in Corroios, an older marsh, when compared to Pancas, a much younger salt marsh. High root-shoot biomass common in halophyte species has been considered to be indicative of adaptive mechanisms, with a need for greater root surface under unfavorable soil conditions (Lana et al. 1991). Many factors, such as elevation, interstitial salinity, N availability, oxygenation of the root system, and anthropogenic factors, have been reported to have important roles in biomass production (Lana et al. 1991). In Corroios salt marsh a decrease in biomass and N use efficiency with an increase in the root biomass proportion, suggests an old and stressed ecosystem – N concentrations in Corroios are much the same over the period of study. Other kinds of stresses may be involved. Concentrations of Cu, Zn, Cd and Pb in the upper sediment are 3–12 times higher than pre-industrial metal levels, indicating that a substantial quantity of the anthropogenic metals is incorporated into the sediments in Corroios salt marsh, located near an industrial zone (Caçador et al. 1999).

REFERENCES

Berendse, F. and Aerts, R., 1987. Nitrogen-Use-Efficiency: A Biologically Meaningful Definition? *Functional Ecol. 1*, 293–296.

Bertness, M. & Pennings, S. (2000). Spatial Variation in Process and Pattern in Salt Marsh Plant Communities. In *Concepts and Controversies in Tidal Marsh Ecology,* Weinstein, M. & Kreeger, D.A. (eds.), Springer, Netherlands, pp. 39–57.

Caçador, I. (1994). *Acumulação e retenção de metais pesados nos sapais do Estuário do Tejo.* PhD Thesis, University of Lisbon, Portugal.

Caçador, I. (1987). Estrutura e Função das Manchas de Sapal no Estuário do Tejo. Tejo que futuro? Associação dos Amigos do Tejo (eds.), Lisboa, Portugal, pp: 131–138.

Caçador, I., Caetano, M., Duarte, B., & Vale, C. (2009). Stock and losses of trace metals from salt marsh plants. *Mar. Environ. Res.* 67, 75–82.

Caçador, I., Costa, A. & Vale, C. (2007a). Nitrogen sequestration capacity of two salt marshes from the Tagus estuary. *Hydrobiologia* 587, 137–145.

Caçador, I., Tibério, S. & Cabral, H. (2007b). Species zonation in Corroios salt marsh in the Tagus estuary (Portugal) and its dynamics in the past fifty years. *Hydrobiologia* 587, 205–211.

Caçador, I., Vale, C. & Catarino, F. (1996). Accumulation of Zn, Pb, Cu and Ni in sediments between roots of the Tagus estuary salt marshes, Portugal. *Est. Coast. Shelf Sci.* 42, 393–403.

Caçador, I., Vale, C. & Catarino, F. (1993). Effects of plants on the accumulation of Zn, Pb, Cu and Cd in sediments of the Tagus estuary salt marshes, Portugal. In: Vernet, J.-P. (ed). *Environment Contamination,* Elsevier Science Publisher B.V., Amsterdam, pp. 355–364.

Caçador, I., Vale, C. & Catarino, F. (2000). Seasonal variation of Zn, Pb, Cu and Cd concentrations in the root-sediment system of *Spartina maritima* and *Halimione portulacoides* from Tagus estuary salt marshes. *Mar. Environ. Res.* 49, 279–290.

Catarino, F. & Caçador, I. (1981). Produção de biomassa e estratégia do desenvolvimento em *Spartina maritima* e outros elementos da vegetação dos sapais do estuário do Tejo. *Boletim Sociedade Broteriana* 54, 384–403.

Davy, A. (2000). Development and structure of salt marshes: community patterns in time and space. In: Weinstein, M.P. & Kreeger, D.A. (eds.). *Concepts and Controversies in Tidal Marsh Ecology.* Springer, Netherlands, pp. 137–156.

Doyle, M. & Otte, M. (1997). Organism-induced accumulation of Fe, Zn and As in wetland soils. *Environ. Poll.* 96, 1–11.

Duarte, B., Caetano, M., Almeida, P., Vale, C.& Caçador, I. (2010). Accumulation and biological cycling of heavy metal in the root-sediment system of four salt marsh species, from Tagus estuary (Portugal). *Environ. Poll.* 158, 1661–1668.

Duarte, B., M. Delgado & I. Caçador. (2007). The role of citric acid in cadmium and nickel uptake and translocation, in *Halimione portulacoides. Chemosphere* 69, 836–840.

Duarte, B., Raposo, P., & Caçador, I. (2009). *Spartina maritima* (cordgrass) rhizosediment extracellular enzymatic activity and its role on organic matter decomposition and metal speciation processes. *Mar. Ecol.* 30, 65–73.

Duarte, B., Reboreda, R., & Caçador, I. (2008). Seasonal variation of Extracellular Enzymatic Activity (EEA) and its influence on metal speciation in a polluted salt marsh. *Chemosphere* 73, 1056–1063.

Esteves-de-Sousa, A. (1950). Acerca da sub-halosérie da região salgadiça litoral, entre Corroios e Talaminho. *Rev. Fac. Cienc* (Lisboa), Rev. Ciências Naturais.

European Union. (2000). Directive 2000/60/EC of the European Parliament and of the Council of 23 October 2000 establishing a framework for community action in the field of water policy. *Official J. European Communities* L327, 1–73.

Berendse, F. and Aerts, R., 1987. Nitrogen-Use-Efficiency: A Biologically Meaningful Definition? *Functional Ecology* 1, 293–296.

Groenendijk, A. & Vink-Lieavaart, M. (1987). Primary production and biomass on a Dutch salt marsh: emphasis on the belowground component. *Vegetatio 70*, 21–27.

Gross, M., Hardisky, M., Wolf, P., & Klemas, V. (1991). Relationship between aboveground and belowground biomass of *Spartina alterniflora* (smooth cordgrass). *Estuaries 14*, 80–191.

Lana, P., Guiss, C. & Disaró, S. (1991). Seasonal variation of biomass and production dynamics for above and belowground components of a *Spartina alterniflora* marsh in the euhaline sector of Paranaguá Bay (SE Brazil). *Est. Coast. Shelf Sci.* 32, 23–241.

Lindsay, W. and Norvell, W., 1978. Development of a DTPA soil test for zinc, iron, manganese and copper. *Soil. Sci. Soc. Am. J.* 42, 421–428.

Long, P. & Mason, C. (1983). *Saltmarsh Ecology.* Blackie, Glasgow.

Melo, A., Correia, A., Caçador, I., Ramos, M. & Catarino, F. (1976). Aspectos da estrutura e dinâmica das comunidades vegetais do Salgado de Corroios. Bases para a sua protecção. Liga Prot. Nat, Vol. XXII p. 41–49. Lisboa.

Prange, J. & Dennison, W. (2000). Physiological responses of five seagrass species to trace metals. *Mar. Poll. Bull.* 41, 327–336.

Reboreda, R. & I. Caçador. (2007a). Halophyte vegetation influences in salt marsh retention capacity for heavy metals. *Environ. Poll. 146*, 147–154.

Reboreda, R. & Caçador, I. (2007b). Copper, zinc and lead speciation in salt mar*sh sediments colonised by Halimione portulacoides and Spartina maritima. Chemosphere 69*, 1655–1661.

Rozema, J., Leedertse, P., Bakker, J. & Wijnen, H. (2002). Nitrogen and Vegetation Dynamics in European Salt Marshes. In: Weinstein, M.P. & Kreeger, D.A. (eds.). *Concepts and Controversies in Tidal Marsh Ecology,* Springer, Netherlands, pp. 469–491.

Rozema, J., Lupper, E. & Broekman, R. (1985). Differential response of salt marsh species to variation of iron and manganese. *Vegetatio 62*, 293–301.

Salgueiro, N. and Caçador, I. (2007). Short-term sedimentation in Tagus estuary, Portugal: the influence of salt marsh plants. *Hydrobiologia 587*, 185–193.

Salgueiro, N. (2001). *Sedimentação a curto e longo prazo nos sapais do estuário do Tejo.* Graduation Thesis, Faculty of Sciences of the University of Lisbon, Portugal.

Sousa, A., Lillebo, A., Caçador, I. & Pardal, M. (2008). Contribution of *Spartina maritima* to the reduction of eutrophication in estuarine systems. *Environ. Poll. 156*, 628–635.

Sundby, B., Vale, C., Caçador, I., Catarino, F., Madureira M.J., & Caetano, M. (1998). Metal-rich concretions on the roots of salt marsh plants: Mechanism and rate of formation. *Limnol. Oceanog. 43*, 245–252.

Suntornvongsagul, K., Burke, D., Hamerlynck, E., & Hahn, D. (2007). Fate and effects of heavy metals in salt marsh sediments. *Environ. Poll. 149*, 79–91.

Teal, J., & Howes, B. (2000). Salt marsh values: Restrospection from the end of the century. In: Weinstein, M.P. & Kreeger, D.A. (eds.). *Concepts and Controversies in Tidal Marsh Ecology*, Springer, Netherlands, pp. 9–19.

Vale, C. (1990). Temporal variations of particulate metals in the Tagus river estuary. *Sci. Total Environ. 97/98*, 137–154.

Waisel, Y. (1972). *Biology of Halophytes*. Academic Press, New York.

Weis, J. & Weis, P. (2004). Metal uptake, transport and release by wetland plants: implications for phytoremediation and restoration. *Environ. Internat. 30*, 685–700.

Wijnen, H. & Bakker, J. (1999). Nitrogen accumulation and plant species replacement in three salt marsh systems in the Wadden Sea. *J. Coast. Conserv. 3*, 19–26.

Wijnen, H. (1999). *Nitrogen dynamics and vegetation succession in salt marshes*. PhD Thesis. Groningen, Netherlands.

Williams, T., Bubb, J. & Lester, J. (1994). Metal accumulation within salt marsh environments: A review. *Mar. Poll. Bull. 28*, 277–290.

In: Estuaries: Classification, Ecology and Human Impacts   ISBN: 978-1-61942-083-0
Editor: Steve Jordan   © 2012 Nova Science Publishers, Inc.

Chapter 4

HABITAT FUNCTION OF A RESTORED SALT MARSH: POST-LARVAL GULF KILLIFISH AS A SENTINEL

Deborah N. Vivian

United States Environmental Protection Agency, Gulf Ecology Division, Gulf
Breeze, FL, US

Chet F. Rakocinski and Mark S. Peterson

Department of Coastal Sciences, The University of Southern Mississippi, Ocean
Springs, MS, US

ABSTRACT

Successful marsh restoration requires recreating conditions to ensure proper ecosystem function. One approach to monitor restoration success is using a sentinel species as a proxy integrator of salt marsh function. The gulf killifish (*Fundulus grandis*, Baird and Girard) is a good candidate sentinel, as it provides an important trophic link between marsh surface and subtidal habitats. Accordingly, herein an *in-situ* approach is presented for assessing marsh function using reared and wild post-larval gulf killifish. Three field experiments conducted in the spring of 2002 and 2003 assessed post-larval growth and survival within cylindrical acrylic microcosms. Marginally significant differences in mean growth rate were apparent between marsh surface and subtidal habitats at Weeks Bayou, MS in 2002, with higher growth in subtidal habitats (6.2% d^{-1}) versus surface habitats (3.9% d^{-1}). Survival rates of 44.4% were equivalent within both habitats. Using the same approach, restored and reference sites at the Chevron-Pascagoula refinery, MS were compared in 2003. Survival rates were notably low for both restored (18.75%) and reference sites (0.0%) during the first experiment of 2003 (May). Reference mortality most likely reflected high sediment porosity. Survival was considerably higher at the restored site (76.2%) than at the relocated reference site (29.8%) in a second 2003 experiment (June); however, growth was poor at both sites (0.1 and –0.47% d^{-1}, respectively). Diet analysis showed that enclosed fish were feeding successfully. Estimates of infaunal abundances and daily growth of wild fish were comparable between sites. Growth rates of enclosed fish followed a trajectory more similar to laboratory fish than to wild fish, perhaps reflecting that reared fish within enclosures continued to grow slowly in the field. A number of valuable lessons from this study point to how an effective multipronged approach for using early stages of fundulids as sentinels of marsh restoration can be further developed.

INTRODUCTION

Salt marshes are among the most productive ecosystems in the world (Kneib 2003), thus they are utilized by many aquatic and terrestrial species for breeding, feeding, and nursery functions. However, salt marsh ecosystems have been degraded by urbanization and industrialization in coastal regions of the USA (Partyka & Peterson 2008; Peterson & Lowe 2009), as evidenced by an almost 50% reduction in salt marsh habitat in recent history (Kennish 2001). This degree of habitat loss should have a substantial impact on estuarine function and the recruitment of estuarine-dependent fauna (Peterson 2003).

In view of an increasing focus on ecosystem services and efforts to restore them, it is essential to develop procedures to assess the success of restoration efforts. Success rests upon reproducing ecological conditions well enough that habitats function normally in the restored area. Assessments of multiple endpoints are also critical for a full evaluation of habitat function within a restored marsh (Rozas & Minello 2009). One endpoint might be to assess restoration success through the use of sentinel species as proxy integrators of various aspects of marsh function. Several studies have shown that resident marsh species show promise as sentinels of restored salt marshes (James-Pirri et al. 2001, Teo & Able 2003, Able et al. 2003; Rozas & Minello 2009).

The fundulids are a group of resident salt marsh fishes that primarily use the marsh surface as habitat for feeding, growth and survival (Weisberg & Lotrich 1982; Lopez et al. 2010). Fundulids also provide an important trophic link by relaying secondary production from the marsh surface to the estuary proper (Kneib 1997a). Although adults will move on and off the marsh surface, early stages of most fundulids reside in marsh surface microhabitats (Smith and Able 1994; Kneib 1997a). Because of the important trophic link that fundulids provide between the marsh surface and estuary proper, early stages of fundulids provide good candidate sentinels for monitoring marsh restoration success.

Two of the most abundant fundulids found on the Atlantic and Gulf of Mexico coasts of the United States respectively are the mummichog, *Fundulus heteroclitus,* and the gulf killifish, *F. grandis.* Although many studies have been done on the life history of *F. heteroclitus*, few studies have been conducted on the life history of *F. grandis.* Previous studies of the gulf killifish have focused mainly on its spatial distribution relative to the marsh surface (Peterson & Turner 1994), feeding (Rozas & LaSalle 1990), salinity tolerance (Crego & Peterson 1997), and reproduction (Hsiao & Meier 1989). As a cognate species of the mummichog along the Gulf of Mexico coast, the gulf killifish likely fulfills a similar trophic role in the salt marshes.

This chapter introduces an *in-situ* approach for assessing marsh function using reared, wild post-larval gulf killifish. We compared restored and reference salt marsh sites by (1) measuring growth and survival rates of reared post-larval gulf killifish within microcosms, and (2) quantifying growth rates of wild caught post-larval gulf killifish using otolith microstructure. Additionally, we examined the diets of wild and experimental fish along with infaunal abundances between sites to consider feeding success relative to site-specific growth and benthic function. Results provide a number of valuable lessons that point to how an effective approach for using resident fundulids as sentinels of restoration of marsh function can be further developed.

METHODS

Field

Assessments encompassed three separate field experiments. First, a pilot study was conducted in late spring 2002 within a 20 m section of Weeks Bayou, Ocean Springs, Mississippi (Figure 1A) to see whether laboratory reared post-larval fish could survive and grow within field microcosms, and to determine the appropriate size of fish for such studies. In late spring of 2003, two more enclosure experiments were conducted within a 10.1 ha restored salt marsh site located at the Chevron-Pascagoula refinery and at an adjacent reference marsh site located in the Bangs Lake portion of the Grand Bay National Estuarine Research Reserve (GB-NERR) (Figure 1B). The restored salt marsh was created in November 1985 by excavating a section of pine flatwood uplands and dredging a 1–2 m wide tidal creek to allow access to aquatic organisms and nutrient exchange (LaSalle 1996). The GB-NERR reference site represented comparable habitat at a site adjacent to the refinery. There was no site replication; the primary goal of these experiments was to evaluate the potential of using early stages of reared gulf killifish as sentinels. Reared post-larval gulf killifish used for the growth experiments were obtained by artificial fertilization. Parental gulf killifish were collected from Weeks Bayou with minnow traps during their peak spawning period and stripped of gametes. The resulting fertilized zygotes were hatched and reared at The University of Southern Mississippi, Gulf Coast Research Laboratory (Vivian 2005).

Cylindrical acrylic field microcosms (0.91 m high, 0.117 m diameter) were deployed to monitor *in-situ* field growth and survival of enclosed post-larval fish. Eight microcosm windows (0.762 cm diameter) were covered with mesh (1.0 mm) screens. Microcosms were inserted several centimeters into the sediment, leaving wells above the sediment to prevent enclosures from being completely drained at low tide (Figure 2) (Drury-McCall 2007). After fish were placed into microcosms, a 1.0 mm mesh screen was fixed over the tops to prevent escape during high tides (Figure 2). Microcosms were placed at stations located 5–12 m apart within marsh-edge (intertidal) and subtidal (permanently flooded) habitats (Vivian 2005).

Twenty-four hours prior to deployment, three randomly selected reared fish (12–15 mm TL) were weighed (g) using an OHAUS analytical balance and kept in one flow-through marked Ziploc® container for each enclosure within 182 L coolers for transporting to the field. The fish were weighed collectively for the first 2003 experiment. In the second experiment fish were individually marked with non-toxic acrylic paint (yellow, green, red) (Duffy-Anderson & Able 1999) so that they could be individually weighed (g) and monitored.

Experiments lasted from 16 to 23 days (Vivian 2005). At the completion of each experiment, field microcosms were carefully removed along with enclosed sediment and the contents sieved through a 1 mm mesh bag. The remaining contents were brought back to the laboratory where the fish were removed, weighed (g) and preserved in 95% ETOH. Specific growth rates (SGR) of the fish from enclosures were estimated by: SGR = ln (final weight (g)) – ln (initial weight (g)) / number of days (Wootton 1990).

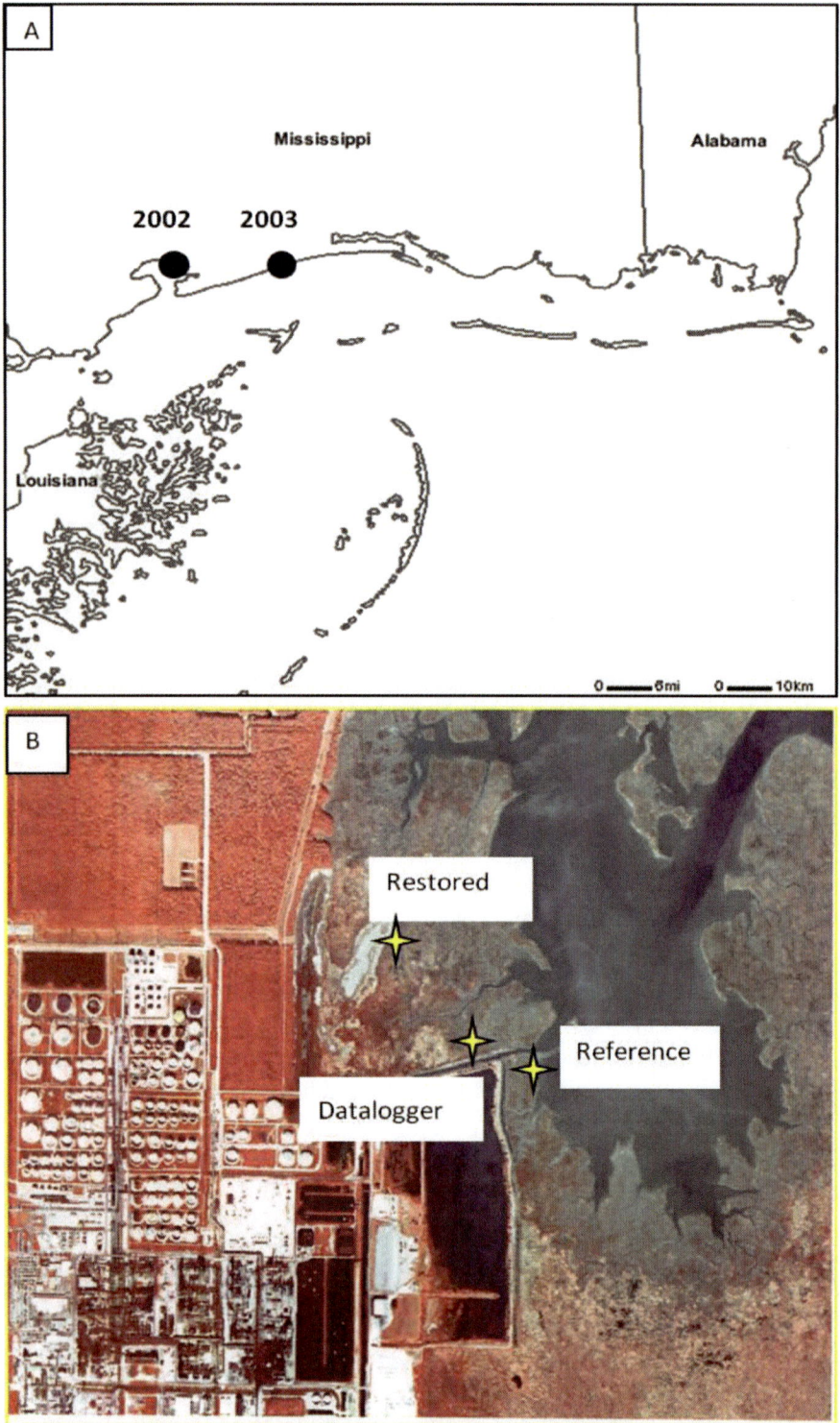

Figure 1. (A) Study site locations in 2002 and 2003; (B) the restored and reference site in the Chevron Pascagoula Refinery (Image obtained from Google Earth).

Figure 2. Left: a microcosm at low tide. Right: a microcosm at high tide.

Water depth (m) and dissolved oxygen (DO) (mg L^{-1}) were measured weekly both inside and outside the enclosures, and enclosures were inspected and freed of any fouling material on the mesh. Temperature (C), salinity (‰), and DO also were recorded with a YSI 85 meter. In 2003, a Datasonde IV datalogger was placed in the bayou that connects both the restored and reference sites (Figure 1B). Hourly measurements of T, S, DO, and depth (m) were recorded every 15 minutes for both experiments in 2003.

Infaunal abundance was sampled on three dates before and after the 2003 experiments (May 5, June 3, and June 19, 2003) to determine possible differences in food availability and benthic function between restored and reference sites. A 5.45 cm diameter PVC tube was used to extract 5 cm deep sediment cores (n = 93) near each enclosure. Cores were placed into Ziploc® bags and kept on ice.

Laboratory

Otolith microstructure was examined to compare wild post-larval gulf killifish to enclosed fish (by inferred growth rates) for all experiments. Several methods were used to collect wild fish during each experiment including pit traps, kick nets, and/or Breder traps (Vivian 2005). Wild post larval gulf killifish were placed on ice, brought back to the laboratory, and preserved in 95% ETOH. Later, fish were identified, enumerated, measured (mm, TL and SL), and placed in separate vials with 95% ETOH for later processing and examination of otolith microstructure.

Both left and right sagittal otoliths were removed from the wild-caught fish for ageing (Vivian 2005). The left sagittal otolith was used routinely to age fish; however, the right one was used when left ones were inadequate. Otoliths were removed from each fish, embedded in a resin block, and prepared following Siegfried & Weinstein (1989) and Secor et al (1992). Each otolith was transversely sectioned, sanded, and polished until the primordium and other daily growth increments became visible. The presence of daily otolith increments in *F. grandis* has not yet been validated, but has been documented for *F. heteroclitus* (Radtke & Dean 1982). Seventy percent of *F. heteroclitus* had two increments at hatching, which agrees with our observations of otoliths of reared *F. grandis*. Daily increments along the longest axis of the otolith from the edge to the primordium were counted using a compound 40x microscope. Age-length relationships based on the otolith microstructure data were determined and compared among sites.

In order to determine if fish within enclosures were feeding, stomach contents of enclosed and wild-caught fish were examined. Contents from the entire gut were removed using a dissecting microscope. Prey items were identified, counted, and lengths and widths measured (μm) using an ocular micrometer. Prey items were enumerated and grouped into nine categories. Harpacticoid copepods and unidentified larval chironomids were distinguished from the other major prey groups because of their high frequency in the diet. The miscellaneous category included flatworms and unidentified organisms.

In the laboratory, the benthic cores were placed into 10% buffered formalin with Rose Bengal. Benthic organisms retained by a 0.63 μm sieve were removed, identified, and counted. Ten percent of the core samples (n = 10) were resorted for quality assurance and control. Percentage contributions of taxa and their $\log_{10}$ abundances ($\log_{10} N + 1$) were determined for each site and collection date.

Data Analysis

Growth rates of fish from surface and subtidal habitats in 2002 and from restored and reference sites in 2003 were compared with a Student's t-test (SPSS, Inc., v11.0, Chicago, IL). Age-length (days mm^{-1} SL) relationships of wild caught gulf killifish were estimated with linear regression and their slopes compared using a t-test (Zar 1984).

Percent diet similarity between wild and experimental fish was determined using the Qbasic program, Quantan Pro. Any differences in $\log_{10}$ infaunal abundances between collections and sites were determined using a 2-way ANOVA with site and date as main effects ($\alpha = 0.05$). The Shannon diversity (H') index also was used to compare the infaunal samples: $H' = -\Sigma (p_i (\ln p_i))$ where p_i is the proportion of the total number of organisms belonging to the ith taxon.

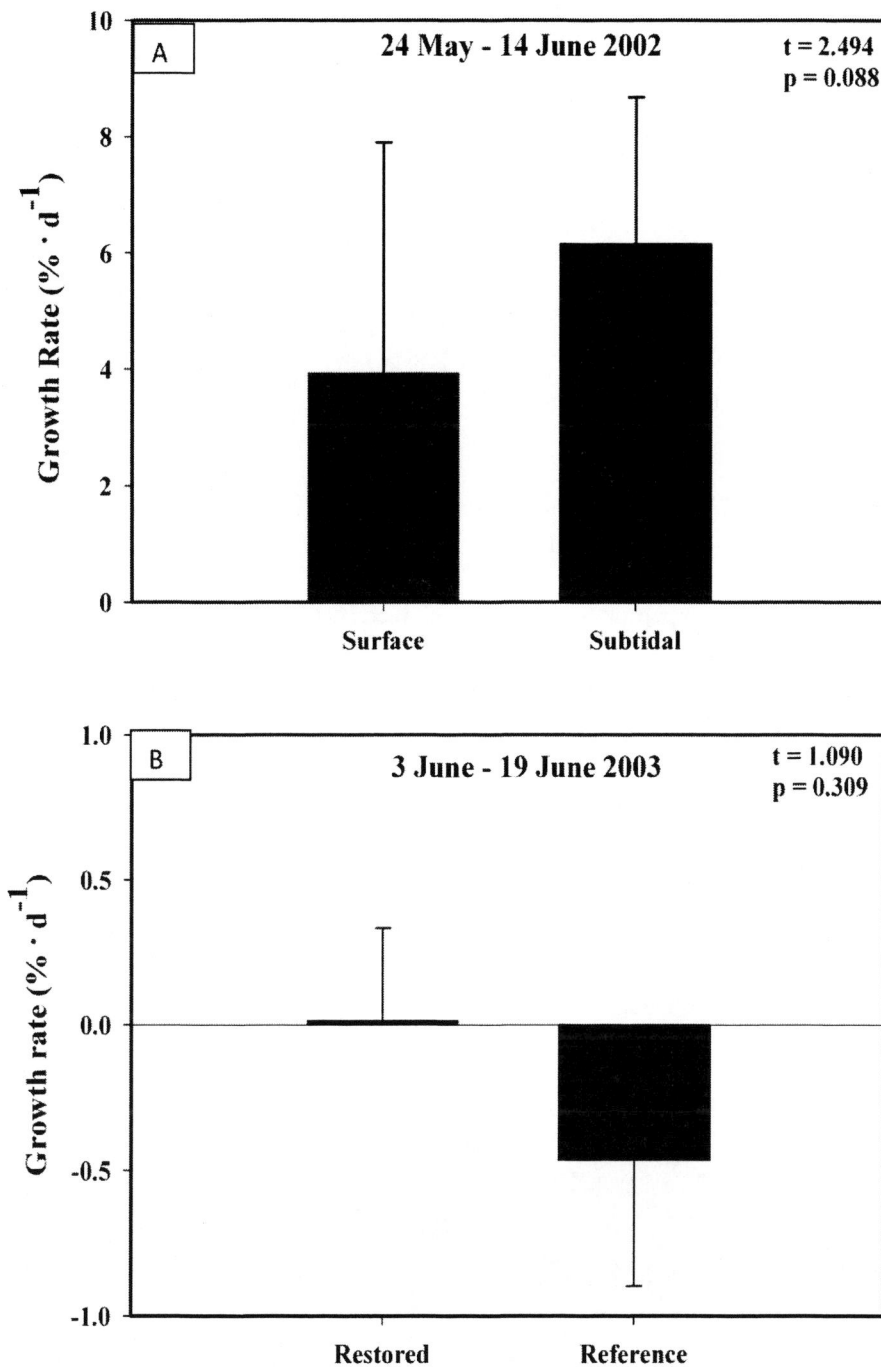

Figure 3. (A) Mean growth rates ($\pm$ SE) in 2002 in surface and subtidal habitats; (B) plot of the mean growth rate ($\pm$ SE) in the 2003 experiment at the restored and reference sites.

RESULTS

Survival and Growth

Although only 44% of fish deployed in 2002 were recovered, recovery rates were similar between marsh surface and subtidal habitats. Slightly larger fish (12–15 mm TL) were used for the 2003 experiments because of the 44% recovery rate in 2002. However, during the first 2003 experiment, enclosure wells did not hold water due to porous sediment, resulting in many enclosures draining at low tide. Thus, recovery rates were 0.0 % for the reference site and 19% for the restored site for the first experiment in 2003. Subsequently, reference enclosures were moved to a nearby intertidal creek for the second 2003 experiment. The recovery rate for the second experiment in 2003 was much higher, totaling 30 % at the reference site and 76 % at the restored site.

Mean growth rates in 2002 were 3.9% d^{-1} and 6.2% d^{-1} in marsh surface and subtidal habitats, respectively (Figure 3A). Growth rates were higher in the subtidal habitat (Student's t = 2.49, p = 0.09) Fish did not grow much in any of the 2003 experiments, and growth rates were not significantly different between the restoration and reference sites (Student's t = 0.87 p = 0.40) for the second experiment. Growth rates were somewhat positive at the restoration site (0.01% d^{-1}) and negative at the reference site (–0.47% d^{-1}; Figure 3B).

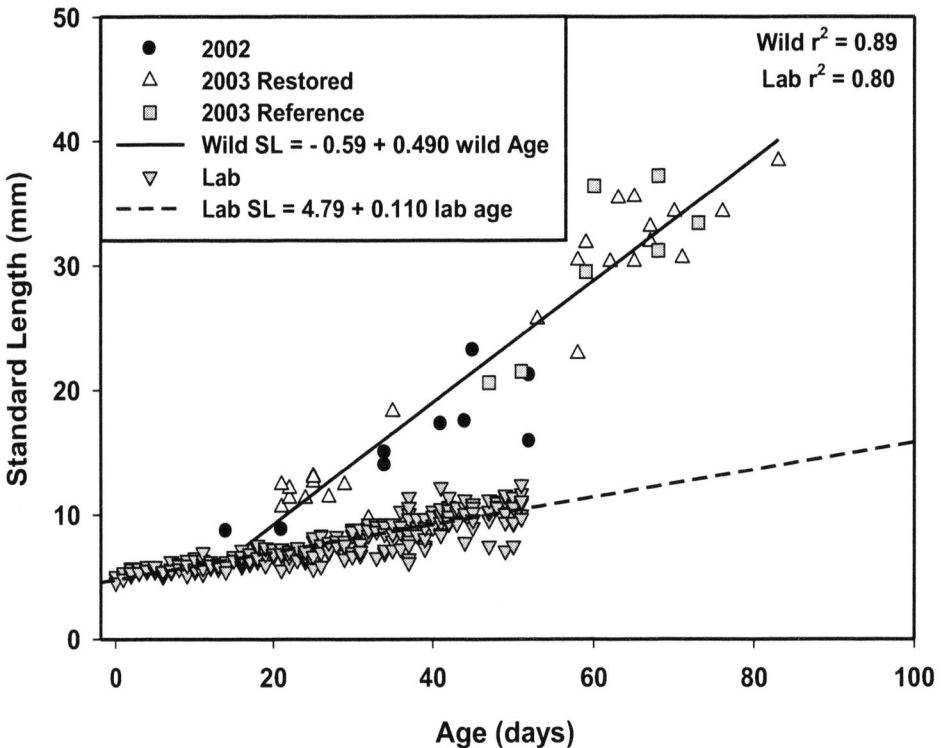

Figure 4. Linear regressions of age (d) vs. standard length (mm) of post-larval gulf killifish. Wild-caught fish and laboratory reared fish are represented by symbols. Dashed line depicts the growth trajectory for lab reared fish.

Environmental Measurements

Mean temperature and salinity ranged between 28–33 C and 13–19 ppt, respectively, during the study period (Table 1). Slightly lower DO levels were recorded at Weeks Bayou in 2002 than at the Chevron-Pascagoula refinery in 2003. Minimum DO values were sometimes low within the microcosms, but mean DO inside the microcosms was generally similar to ambient water (Table 1).

Growth of Wild-Caught Fish

A total of 56 wild gulf killifish were collected in 2002 and 2003 (Weeks Bayou n = 13, Restored n = 29, Reference n = 14). Mean SL (±SE) of wild fish was 14.8 (±1.3) mm for Weeks Bayou, 22.8 (±1.9) mm for the 2003 restoration site, and 29.9 (±2.2) mm for the 2003 reference site. There was a significant positive relationship between daily age and SL of wild caught fish (F = 397.23, p < 0.05, r^2 = 0.89) (Figure 4). The age-length relationship for reared fish at GCRL (1–51 d old; n = 266) was compared with that of the wild fish to determine whether lab-reared fish followed a different growth trajectory than wild-caught fish. Age-length relationships were significant for both reared and wild fish (Figure 4), but the slope for reared fish (0.11) was markedly and significantly shallower than for wild fish (0.49; t = 6.44; df = 309, two tailed p <0.001).

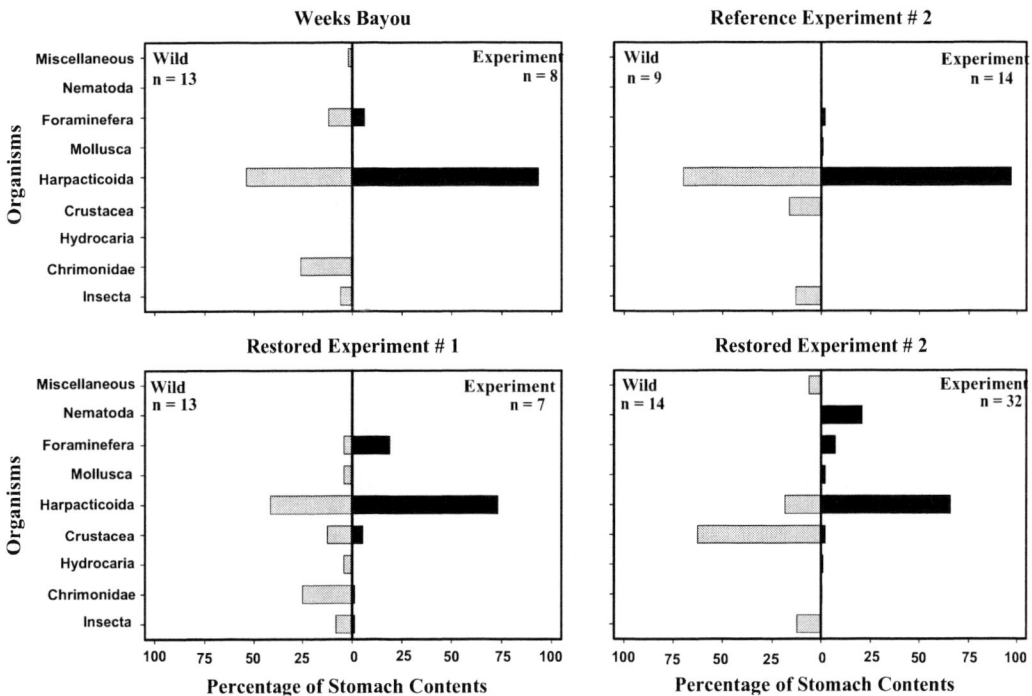

Figure 5. Percentage of each prey category in the diets of wild-caught (left side) and experimentally reared (right side) fish from 2002 and 2003 growth experiments.

Diet Analysis and Infaunal Abundance

The diets of enclosed and wild-caught fish comprised a total of 19 prey taxa, including mostly benthic organisms (Vivian 2005). A major diet component for both wild and enclosed fish was harpacticoid copepods (Figure 5). The diets of enclosed fish at the restored site (i.e. mollusks, nematodes, foraminifera, and crustaceans) were generally more diverse than those of enclosed fish at the reference site (Figure 5). Wild fish exhibited a somewhat more varied diet than enclosed fish, including more terrestrial insects and unidentified larval chironomids (Figure 5). Percent diet similarities between wild and enclosed fish were moderate in 2002 (0.50), high at the 2003 reference site (0.70), and low to moderate at the 2003 restored site (0.21–0.54). Both wild and enclosed fish mostly consumed food items < 1.0 mm in maximum width (Figure 6). Wild fish from the reference site consumed a wider range of prey widths; three of the largest fish caught had a mean prey width of > 1.0 mm due to the consumption of large prey such as *Palaemonetes* spp.

Table 1. Mean (± SE) water temperature, salinity and dissolved oxygen measured during the 2002 and 2003 growth studies via a handheld YSI and Datasonde. Minimum and maximum values are given in parenthesis. Dashed lines indicate that no readings were obtained

Date and Site	Temperature (°C)	Salinity (ppt)	DO (water) (mg/L)	DO (microcosm) (mg/L)
24 May – 14 June 2002 Weeks Bayou	27.7 ± 2.3 (22.6–35.7)	16.2 ± 1.1 (12.2–22.3)	4.22 ± 0.4 (3.2–5.4)	4.13 ± 0.5 (1.0–8.0)
5 May –28 May 2003 Reference Site	28.0 ± 0.8 (25.6–29.9)	16.4 ± 1.0 (13.3–19.1)	5.77 ± 0.5 (3.7–6.7)	---
5 May –28 May 2003 Restored Site	27.7 ± 1.2 (22.6–30.5)	18.8 ± 1.1 (15.3–22.3)	5.49 ± 0.5 (4.2–6.9)	5.44 ± 0.4 (0.3–9.8)
5 May – 28 May 2003 Datasonde	27.6 ± 0.1	17.0± 0.1	3.47 ± 0.03	---
2 June – 19 June 2003 Reference Site	31.3 ± 3.3 (28.0–34.6)	13.2 ± 1.0 (12.2–14.2)	5.72 ±1.1 (4.6–6.8)	5.49 ± 0.7 (0.7–13.2)
2 June – 19 June 2003 Restored Site	32.9 ± 2.8 (30–35.7)	13.4 ± 0.9 (12.5–14.3)	5.24 ± 0.7 (4.5–6.0)	6.53 ± 0.5 (2.2–12.5)
2 June – 19 June 2003 Datasonde	28.9 ± 0.9	13.6 ± 0.8	3.34 ±0.1	---

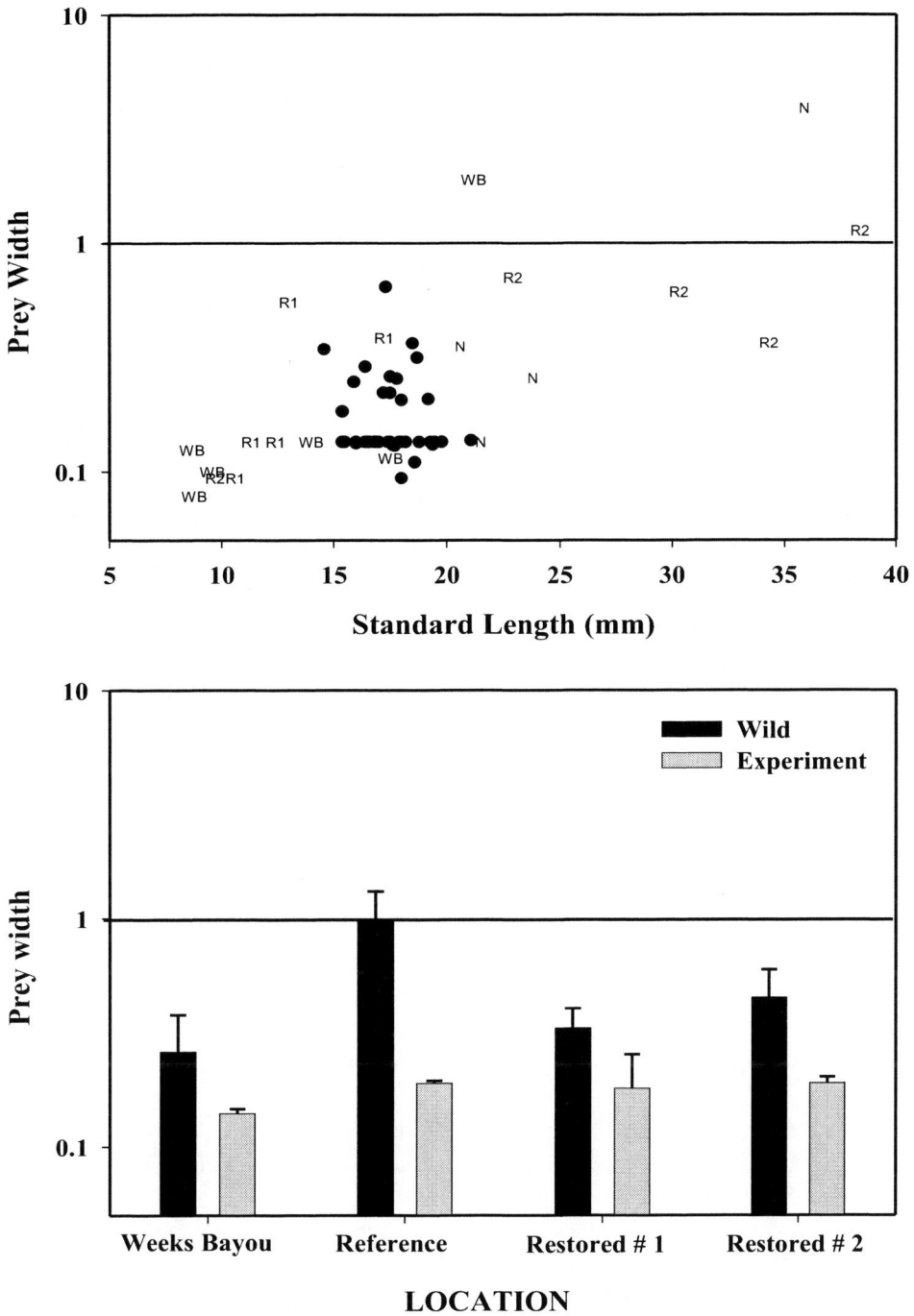

Figure 6. (Top panel) Mean prey width vs. standard length of gulf killifish. Data is plotted on common log coordinates for visualization. Circles represent experimental fish and letters represent wild-caught fish for respective site-events. WB = Weeks Bayou; N = reference site; R1 = restored 1; R2 = restored 2. (Bottom panel) Mean prey width (± 1 SE) by site-event for wild (black) and experimental (gray) fish. The horizontal line represents the mesh size of microcosm window screens.

A total of 26 nominal taxa occurred in the infaunal samples from both sites over all three events (Vivian 2005). A 2-way ANOVA failed to show any significant differences in mean total abundances of organisms between sites (reference vs. restored; F = 1.28, p = 0.26). There was no interaction effect between site or collection date (site x date; F = 0.53, p = 0.59), however, a significantly greater number of organisms was collected in May 2003 (F = 4.61, p = 0.01). The eleven most common groups of sediment-dwelling organisms were *Laeonereis culveri, Capitella capitata, Steblospio gynobranchiata,* Enchytraeidae, harpacticoid copepods, calanoid copepods, *Hargaria rapax, Grandierella bonnieroides,* Chironomidae, Nematoda, and Foraminifera. The diversity of infaunal organisms ranged higher at the restored site (1.3–1.8) than at the reference site (0.5–1.4). However, percent similarities of benthic samples were generally high between restored and reference sites (> 70%).

DISCUSSION

The utility of using early stages of fundulids as indicators of successful marsh restoration depends on recognizing how they perform as integrators of marsh function. Various trophic and biogeochemical processes might be expressed through the growth and survival of fundulids. Furthermore, early stages of fundulids should be tied more to marsh surface and benthic functions than adults. In the present study, differences in growth of enclosed fish were not detected between restored and reference sites. Enclosed fish did grow, albeit slowly. In this study, growth rates were 0–0.09 g Δw d^{-1} (2002) and –0.04–0.04 g Δw d^{-1} (2003). Reported growth rates of wild YOY *F. heteroclitus* in a restored marsh ranged from 0.04 to 0.08 g Δw d^{-1} (Teo & Able 2003) and in a natural marsh, YOY growth ranged from 0.03–0.14 g Δw d^{-1} (Kneib 1993). Although gulf killifish grew slower in the present study, the highest observed growth rates were within the range reported for *F. heteroclitus*. Importantly, similar growth rates of wild-caught fish in both 2002 and 2003 indicated that restored marsh function was sufficient for normal fundulid growth, despite the lower growth rates of reared fish within microcosms.

Survival rates of enclosed fish should provide another useful indicator of marsh restoration success, to the degree that they reflect trophic or biogeochemical processes. For example, differences in hydrology, sediment properties, or food availability could affect survival of enclosed fish. In the present study, survival rates were actually higher at the restored marsh than at the reference site in 2003. Even a complete lack of survival at the reference site in the first 2003 experiment likely reflected marsh function vis-a-vis poor availability of refuge habitat on the marsh surface. Due to the porous sediment, the enclosures were dewatered at low tide. Similar issues with using mesocosms to evaluate marsh restoration were reported by Rozas & Minello (2009). In contrast, the construction of the restored marsh and resulting elevation differences we observed (LaSalle 1996) probably explain the more frequent marsh inundation and higher survival within microcosms at the restored site. Kneib (1993) found that growth was positively associated with duration of tidal flooding. This factor may indicate that, whereas tidal inundation may enhance growth through increased food availability, early stages can still survive in vegetated marsh surface habitat that is less frequently inundated. Indeed, similar survival rates in surface and subtidal habitats and higher growth rates in subtidal habitat at Weeks Bayou in 2002 are consistent with this interpretation.

A multipronged approach should provide a robust way to monitor marsh restoration success. Using enclosures as one element of a multipronged approach to assessing restoration success carries both advantages and disadvantages. Enclosures isolate experimental subjects from density dependence and predation, thus precluding effects of such population level effects. On the other hand, other effects can be standardized through the use of a set number of individuals and the placement of enclosures. Using reared fish in enclosures also reduces extraneous variance in their responses because of their common early history and more similar genetic inheritance. For example, food quality and quantity as well as the density and parental sources of fish being reared can be controlled (Vivian 2005).

Effects of potential artifacts also must be considered when using microcosms to evaluate marsh restoration success. Enclosed fish are confined to specific areas, restricting their ability to move to other adjacent habitats to obtain food or refuge. Access to the marsh surface can be essential to the growth and survival of wild *F. heteroclitus* (Weisberg & Lotrich 1982). In California, adult *F. parvipinnis* can double food consumption while on the marsh surface and thus grow 20–44% faster when fish have access to both the marsh surface and to subtidal areas (Madon et al. 2001). However, Kneib (1993) concluded that *F. heteroclitus* larvae grow more rapidly at lower marsh elevations.

Wild fish, just as lab-reared fish, can provide useful diagnostic metrics. For example, in the present study it was unclear whether the growth response of enclosed reared fish was hampered by their prior growth history. They were only slightly larger than would be expected based on their former growth trajectory. Slow growth also has been observed for *F. heteroclitus* raised under laboratory conditions, where 60 d old fish only reached 12–15 mm TL (Sakowicz 2003). In the present study, fish were fed *ad libitum* to satiation once per day and food was continuously available between water changes. Thus, the quantity of food should have been sufficient for normal fish growth. However, the growth of *F. heteroclitus* is known to be rapid during the first year (Kneib & Stiven 1978). Relatively rapid growth of wild fish was also implied in the present study. Because wild fish would already be acclimated to prevailing conditions at study sites, their growth responses might be more sensitive. On the other hand, the use of reared fish obviates possible site-specific acclimation effects. Thus, a robust multi-pronged protocol might include the use of both wild and reared fish in enclosures. Metrics obtained from ambient wild fish can provide an additional element of a multipronged approach to assess marsh restoration success. In contrast to enclosure fish, metrics obtained from *in-situ* wild fish also reflect population-level differences associated with the density and predation effects referred to above.

Using other indicators of marsh function to relate with fish metrics can expand insights, extend functional assessments, reinforce interpretations, and inform how connections between function and metrics are expressed. Differences in diets and relative abundances of infaunal prey organisms were informative relative to possible trophic effects on fish growth and may also reflect biogeochemical differences in habitat function (Rakocinski & Zapfe 2005, Ferguson & Rakocinski 2008). James-Pirri et al. (2001) compared the diet of fish from restored and reference marshes in New England as an indicator of restoration success. They found that food availability in the restored marsh was sufficient within one year after restoration in some cases. In the present study, enclosed fish fed on similar types and sizes of prey as wild fish, and had a more diverse diet at the restored marsh. All fish consumed harpacticoid copepods as the major dietary component. Harpacticoid copepods have often been shown to be a common prey type for small estuarine fishes, especially for YOY fundulids (Kneib 1997b, Smith et al. 2000, James-Pirri et al. 2001). Harpactacoids are usually the second most abundant benthic organism in marshes (Hicks & Coull 1983). If the restored marsh is providing food resources similar to those of reference marshes for major resident and transient organisms, then similar benthic and planktonic communities probably occur at the sites.

Infaunal composition and abundance also proved to be useful indicators for evaluating marsh restoration success in the present study. Percent similarities of infaunal communities were high among all sites. Benthic taxa from the core samples in the restored marsh were similar to those from a restored marsh site located at Bayou Portage, Mississippi. McClelland et al. (1999) noted similar taxa, including *L. culveri, Streblospio* sp., *A. louisianum, G. bonnierodies*, and Hydrobiidae. Similarities in the benthic fauna at the Chevron restoration marsh were also observed by LaSalle (1996). Percent benthic similarities between restored and reference marshes in this earlier study were about 0.60, a value that is comparable to what we found in 2003. The most dominant organism reported by LaSalle (1996) was *Hargeria rapax*, as opposed to harpacticoid copepods. However, LaSalle used a 1.0 mm sieve which likely allowed harpacticoid copepods to pass through the mesh, whereas a 63 μm sieve was used in this study.

Better experimental designs incorporating early stages of fundulids to assess marsh restoration success should be executed on broader spatial and temporal scales. The spatial scale of microcosms used in the present study did not seem to limit the availability of food to post-larval fundulids. Relatively similar diets between experimental and wild-caught fish, especially at the reference site, indicated that the same types of food were available to enclosed fish. The mesh size of 1 mm in microcosm windows was large enough to allow passage of appropriate sizes of prey based on diet comparisons between enclosed and wild-caught fish. Microcosm height would not have restricted food availability during flooding tides either. At 15 mm TL, *F. heteroclitus* cease remaining motionless at the bottom, and fish larger than 60 mm TL utilize the entire water column (Sakowicz 2003). Thus, fish were likely able to utilize the entire microcosm height at high tide to obtain food. However possible differences in growth between intertidal and subtidal habitats observed in the 2002 experiment suggest that enclosures connecting marsh surface and subtidal habitats could be more effective.

A better experimental design should also include a range of restoration site ages to integrate across a gradient of restoration stages. The amount of time required for any marsh restoration site to reach functional equivalence can be variable. The average recovery time for restored marshes to approach reference conditions ranges from 5 to 20 years (Jones & Schmitz 2009; Borja et al. 2010). The present study was conducted about 18 years after completion of the Chevron-Pascagoula marsh restoration. Considerations of survival of reared post-larval fundulids, growth rates of enclosed reared fish, inferred growth of wild fish, diets of enclosed fish, benthic food availability, and benthic community composition in this study all indicated that the restored marsh at the Chevron-Pascagoula Refinery was functioning adequately after the 18-year period.

In conclusion, a number of valuable lessons from this study point to how an effective multipronged approach for using early stages of fundulids as sentinels of marsh restoration can be further developed. Comparisons of wild and reared fish within enclosures may help distinguish effects of prior growth history from other artifacts of using field enclosures. A combination of both enclosed and ambient wild fish could be used to distinguish population-level from other isolated effects on growth and survival. To further address such population-level differences, studies of wild fish could incorporate mark recapture methods to estimate densities and mortality rates of natural populations. Other proxies of marsh function also should be incorporated, such as represented by benthic infaunal sampling in this study. Differences in growth between intertidal and subtidal habitats in the first experiment suggest that enclosures might best be contiguous across these habitats. Finally, a better design should include a range of site ages to integrate across a gradient of restoration stages.

ACKNOWLEDGMENTS

We would like to thank the Chevron-Pascagoula refinery, particularly W. Calhoun, for allowing us to conduct our research at the refinery's restoration site, and the Mississippi-Alabama Sea Grant Consortium for partial funding. Special thanks to Dr. Thomas Grothues for his critical and helpful review; he contributed many insights. Thanks to B. Comyns for his lab facilities and training in otolith processing and C. Moncreiff for greenhouse space for rearing fish. We would also like to thank "Team Fundulus" for helping with fieldwork in sometimes trying conditions. This chapter is based on a thesis by Deborah N. Vivian presented to the Department of Coastal Sciences of the University of Southern Mississippi in fulfillment of her MS degree. The information in this document has been subjected to review by the National Health and Environmental Effects Research Laboratory and approved for publication. Approval does not signify that the contents reflect the views of the Agency, nor does mention of trade names or commercial products constitute endorsement or recommendation for use. This is contribution number 1422 from the Gulf Ecology Division.

REFERENCES

Able, K.W. (1999). Measures of juvenile fish habitat quality: examples from a National Estuarine Research Reserve. *Am. Fish. Soc. Symp. 22*, 143–147.

Able, K.W., Hagan, S.M. & Brown, S.A. (2003). Mechanisms of marsh habitat alteration due to *Phragmites*: response of young-of-the-year mummichog (*Fundulus heteroclitus*) to treatment for *Phragmites* removal. *Estuaries 26*, 484–494.

Borja, A., Dauer, D.M., Elliot, M. & Simenstad, C.A. (2010). Medium and long-term recovery of estuarine and coastal ecosystems: patterns, rates and restoration effectiveness. *Estuar. Coasts 33*, 1249–1260.

Breder, C.M. (1960). Design for a fry trap. *Zoologica 45*, 155–164.

Crego, G.J. & Peterson, M.S. (1997). Salinity tolerances of four ecologically distinct species of *Fundulus* (Pisces: Fundulidae) from the northern Gulf of Mexico. *Gulf Mexico Sci. 15*, 45–49.

Drury-McCall, D. (2004). Grass shrimp (*Palaemonetes* spp.) play a pivotal trophic role in enhancing *Ruppia maritima*. *Ecology 88*, 618–624.

Duffy-Anderson, J.T. & Able, K.W. (1999). Effects of municipal piers on the growth of juvenile fish in the Hudson River estuary: a study across the pier edge. *Mar. Biol. 133*, 409–418.

Ferguson, H.J. & Rakocinski, C.F. (2008). Macroinfaunal functional metrics for tracking marsh restoration: Implementing a practical approach. *Wetlands Ecol. Manage. 16*, 277–289.

Hicks, G.R.F. & Coull, B.C. (1983). The ecology of marine meiobenthic harpacticoid copepods. *Oceanogr. Mar. Biol Ann. Rev. 21*, 67–175.

Hsiao, S. & Meier, A. (1989). Comparison of semilunar cycles of spawning activity in *Fundulus grandis* and *F. heteroclitus* held under constant laboratory conditions. *J. Exp. Zool. 252*, 213–218.

James-Pirri, M.J., Raposa, K.B. & Catena, J.C. (2001). Diet composition of mummichogs, *Fundulus heteroclitus*, from restoring and unrestricted regions of a New England (USA) salt marsh. *Estuar. Coast. Shelf S. 53*, 205–213.

Jones, H.P. & Schmitz, O.J. (2009). Rapid recovery of damaged ecosystems. *PLoS One, 4*, e5653.

Kennish, M.J. (2001). Coastal salt marsh systems in the U.S.: a review of anthropogenic impacts. *J. Coast. Res. 17*, 731–748.

Kneib, R.T. (1993). Growth and mortality in successive cohorts of fish larvae within an estuarine nursery. *Mar. Ecol. Prog. Ser. 94*, 115–117.

Kneib, R.T. (1997a). The role of tidal marshes in the ecology of estuarine nekton. *Oceanogr. Mar. Biol. Ann. Rev. 35*, 163–220.

Kneib, R.T. (1997b). Early life stages of resident nekton in intertidal marshes. *Estuaries 20*, 214–230.

LaSalle, M.W. (1996). *Assessing the functional level of a constructed intertidal marsh in Mississippi.* Technical Report WRP-RE-15, US Army Engineer Waterways Experiment Station, Vicksburg, MS.

Lopez, J.D., Peterson, M.S. Lang, E.T. & Charbonnet, A.M. (2010). Linking habitat and life history for conservation of the rare saltmarsh topminnow (*Fundulus jenkinsi*): morphometrics, reproduction, and trophic ecology. *Endangered Species Res. 12*, 141–155.

Madon, S.P., Williams, G.D., West, J.M. & Zedler, J.B. (2001). The importance of marsh access to growth of the California killifish, *Fundulus parvipinnis*, evaluated through bioenergetics modeling. *Ecol. Model. 135*, 149–165.

Magnuson, J.J. (1991). Fish and fisheries ecology. *Ecol. Appl. 1*, 13–26.

Magnuson, J.J. (1962). An analysis of aggressive behavior, growth, and competition for food and space in medaka (*Oryzias latipes* (Pisces, Cyprinodontidae)). *Can. J. Zool. 40*, 313–363.

McClelland, J., LeCroy, S. & Heard, R. (1999). *An assessment of the macrobenthic community of a wetlands mitigation project near Bayou Portage, Hancock County, MS.* Report to Brown and Mitchell, Inc. Gulfport, MS.

Nemerson, D.M. & Able, K.W. (2004). Spatial patterns in diet and distribution of juveniles of four fish species in Delaware Bay marsh creeks: factors influencing fish abundance. *Mar. Ecol. Prog. Ser. 276*, 249–262.

Partyka, M.L. & Peterson, M.S. (2008). Habitat quality and salt marsh species assemblages along an anthropogenic estuarine landscape. *J. Coast. Res. 24*, 1570–1581.

Peterson, M.S. (2003). A conceptual view of environment-habitat-production linkages in tidal river estuaries. *Rev. Fish. Sci. 11*, 291–313.

Peterson, M.S. & Lowe, M.R. (2009). Implications of cumulative impacts to estuarine and marine habitat quality for fish and invertebrate resources. *Rev. Fish. Sci. 17*, 505–523.

Radtke, R.L. & Dean J.M. (1982). Increment formation in the otoliths of embryos, larvae, and juveniles of the mummichog, *Fundulus heteroclitus*. *Fish. Bull. 80*, 201–215.

Rakocinski, C.F. & Zapfe, G.A. (2005). Macrobenthic process indicators of estuarine condition. In S.A. Bortone [ed]. *Estuarine Indicators* (pp. 315–331), CRC Press, Boca Raton.

Rozas, L.P. & LaSalle, M.W (1990). A comparison of the diets of Gulf Killifish, *Fundulus grandis* Baird and Girard, entering and leaving a Mississippi brackish marsh. *Estuaries 13*, 332–336.

Rozas, L.P. & Minello, T.J. (2009). Using nekton growth as a metric for assessing habitat restoration by marsh terracing. *Mar. Ecol. Prog. Ser. 394*, 179–193.

Ryer, C.H. & Olla, B.L. (1996). Growth depensation and aggression in laboratory reared coho salmon: the effect of food distribution and ration size. *J. Fish Biol. 48*, 686–694.

Sakowicz, G.P. (2003). Comparative morphology and behavior of larval salt marsh fish *Fundulus heteroclitus* and *Cyprinodon variegatus*. M.S. Thesis, Rutgers University, New Brunswick, New Jersey.

Secor, D.H., Dean, J.M. & Laban, E.H.. 1992. Otolith removal and preparation for microstructural examination. *Canadian Journal Fishery and Aquatic Science, 117*, 17–59.

Siegfried, R.C. II & Weinstein, M.P. (1989). Validation of daily increment deposition in the otoliths of spot (*Leiostomus xanthurus*). *Estuaries 12*, 180–185.

Smith, K.J. & Able, K.W. (1994). Salt marsh tide pools as winter refuges for the mummichog, *Fundulus heteroclitus*, in New Jersey. *Estuaries 17*, 226–234.

Smith, K.J., Taghon, G.L. & Able, K.W. (2000). Trophic linkages in marshes: ontogenetic changes in diet for young-of-the-year mummichog, *Fundulus heteroclitus*, 221–237. In M.P. Weinstein & D.A. Kreeger (Eds.), *Concepts and Controversies in Tidal Marsh Ecology*. Kluwer Academic Publishers, the Netherlands.

Smith, M.E. & Fuiman, L.A. (2003). Causes of growth depensation in red drum, *Sciaenops ocellatus*, larvae. *Environ. Biol. Fish. 66*, 49–60.

Sogard, S.M. (1997). Size-selective mortality in the juvenile stage of teleost fishes: a review. *Bull. Mar. Sci. 60*, 1129–1157.

Teo, S.L.H. & Able, K.W. (2003). Growth and production of the common mummichog (*Fundulus heteroclitus*) in a restored salt marsh. *Estuaries 26*, 51–63.

Vivian, D.N. (2005). *Examining habitat function of a restored salt marsh using post-larval gulf killifish (*Fundulus grandis *Baird and Girard)*. M.S. Thesis, The University of Southern Mississippi, Hattiesburg, MS.

Weisberg, S.B. & Lotrich, V.A. (1982). The importance of an infrequently flooded intertidal marsh surface as an energy source for the mummichog *Fundulus heteroclitus*: an experimental approach. *Mar. Biol. 66*, 307–310.

Whitfield, A.K. & Elliott, M. (2003). Fishes as indicators of environmental and ecological changes within estuaries: a review of progress and some suggestions for the future. *J. Fish Biol. 64* (Supplement A), 229–250.

Wootton, R.J. (1990). *Ecology of Teleost Fishes*. Chapman and Hall, London.

Zar, J.H. (1984). *Biostatistical Analysis,* Second Edition, Prentice-Hall, Englewood Cliffs, N.J.

In: Estuaries: Classification, Ecology and Human Impacts   ISBN: 978-1-61942-083-0
Editor: Steve Jordan   © 2012 Nova Science Publishers, Inc.

Chapter 5

CONTRIBUTIONS OF ESTUARINE HABITATS TO MAJOR FISHERIES

Stephen J. Jordan

U.S. Environmental Protection Agency, Gulf Ecology Division, Gulf Breeze, FL, US

Mark S. Peterson

The University of Southern Mississippi, Department of Coastal Sciences, Ocean Springs, MS, US

ABSTRACT

Estuaries provide unique habitat conditions that are essential to the production of major fisheries throughout the world, but quantitatively demonstrating the value of these habitats to fisheries presents some difficult problems. The questions are important, because critical habitats in estuaries are threatened worldwide by development of the coastal zone, climate change, and pollution, among other stressors. This chapter begins by introducing estuarine fisheries, their importance, and the reasons for their prominence. Next we discuss the challenges of quantification, followed by three case studies of relationships between fisheries production and habitat structure and function. We propose that holistic, integrated approaches to research, modeling, and monitoring can lead to solutions beneficial to our understanding, to better management of coastal habitats, and to the sustainability of fisheries.

INTRODUCTION

The physical, chemical, and biological properties of the environment supply the structures and ecological functions that support life. Estuaries and near-coastal ecosystems feature a unique complex of attributes that are of critical importance to major fisheries, especially as nursery grounds for early life stages of fishes and invertebrates. Shallow water, aquatic vegetation, high primary productivity, and variable salinity are characteristic of most estuaries. In combination, these traits are favorable for fishery productivity, thus estuaries contribute far more to the world's fisheries than would be predicted simply based on their collective area (Houde & Rutherford 1993).

Many of the world's fisheries are heavily dependent on estuaries and their unique habitat features; for example, estuarine-dependent species comprise more than 50% of U.S. commercial fisheries landings (Houde & Rutherford 1993). Nearshore areas contributed over two million jobs and $117 billion to the U.S. economy in 2000; saltwater fisheries are worth more than $48 billion annually, mostly supported by estuarine-dependent fishes and decapods (U.S. Commission on Ocean Policy, 2004). Among fishes, the Sciaenidae (drums, croakers, seatrout, etc.) are major contributors to fisheries; they are ubiquitous inhabitants of estuaries as juveniles and adults. Moronids (striped bass, *Morone saxatilis* and white perch, *Morone americana*) are prominent fishery species in estuaries of the western Atlantic. Shads and river herrings (Alosinae), clupeids (e.g., anchovies and menhaden), eels (Anguillidae), some sturgeons (Acipenseridae), and salmonids make extensive use of coastal river and estuarine habitats. Most of the world's crab, shrimp, and oyster populations that support fisheries spend all or critical parts of their life cycles in estuaries.

Several properties of estuaries favor high productivity of fish and shellfish. Estuaries generally have high rates of primary production because (1) they are receptors for land-based nutrients, and (2) nutrient recycling is enhanced where shallow, organic-rich sediments support abundant communities of benthic fauna and microbes. Elevated primary production supports greater biomass production at higher trophic levels, at least up to the point where eutrophication causes hypoxia and degraded environmental conditions (Houde & Rutherford 1993). For diadromous species, estuaries are essential corridors between rivers and oceans, but many of these species also reside in estuaries for brief to extended periods of time (weeks to years) at some point in their life cycles. Seagrasses, salt marshes, oyster reefs, and soft sediments in shallow-water areas afford refuges or nurseries for juveniles of many fishery and non-fishery species (Ruiz et al. 1993).

Freshwater discharge into estuaries is a factor in the success of many estuary-dependent organisms. Freshwater inflows convey nutrients, trace elements, and detritus from the watershed, and in combination with tides, supply energy, mixing, and salinity gradients within the shallows of the estuary. Browder & Moore (1981), using a conceptual model, suggested that freshwater inflow influenced fishery production in five ways: (1) *transport of nutrients* needed to stimulate productivity of wetland vegetation, phytoplankton, and seagrasses, all providing food for juvenile fish and shellfish, either directly or through the food chain; (2) *transport of detritus*; the physical force of freshwater discharge flushes decaying wetland vegetation into tidal creeks and open waters, where it is processed by microorganisms into food for benthic organisms that are eventually consumed in the food chain; (3) *transport and deposition of sediments* needed to build, maintain and counteract erosion of tidal marshes; (4) *reduction of salinity* offers euryhaline larval stages protection from stenohaline marine predators; (5) *mixing and transport of water masses* provides oxygenation for decomposition and utilization of detritus, and transport of larval and postlarval stages throughout the estuary.

Despite recognition of the linkage between healthy coastal ecosystems and fisheries production (Heinz Center 2002, Crossett et al. 2004, U.S. Commission on Ocean Policy 2004), humans continue to degrade coastal habitats worldwide. Human activities, by altering landscapes and seascapes, alter coastal habitats in ways that can shrink their boundaries and weaken their supporting functions. It has been observed repeatedly that such alterations reduce the abundance and productivity of affected species or populations (Peterson & Lowe 2009). Nonetheless, it is difficult to determine how long term population success is related to habitat extent and condition. Key questions are whether particular habitats are truly essential, the extent to which alternative habitats can support populations, and how well organisms can adapt to major alterations in habitat extent and condition (Dantin et al. 2005).

There has been a promising shift from traditional definitions of nursery habitat based solely on structural features (e.g. marsh edge, open water, oysters, seagrass, etc.) toward recognition that habitats are spatially and temporally dynamic environments within the estuarine landscape (Peterson 2003, Stoner 2003, Browman & Stergiou 2004). Within this more comprehensive view, successful life histories of fish and invertebrates are linked directly to habitat diversity, quality and quantity, all nested within a framework of biotic and abiotic variability (Simenstad et al. 2000, Manderson et al. 2002, 2003, Peterson 2003, DeLong & Collie, 2004). Thus, one approach to assessing estuarine landscape mosaics is to treat the estuary holistically as an environment possessing both dynamic (short-term physical-chemical and biotic variability, salinity, prey fields) and stationary (long-term structural variability, sediment type, wetland context) components (Peterson 2003), each having its own influence on fisheries. The timing, positioning and the amount of overlap between the dynamic and the stationary components may determine the survival and growth rates of juvenile nekton as they pulse into the estuary. In this chapter, we examine some of the opportunities and difficulties associated with fishery production as a function of estuarine habitats. A vast array of factors influences the quantity and quality of the estuarine environment for fish and shellfish (Breitburg & Houde 2006); we discuss only a few. The focus is on the aspects that are most vulnerable to losses resulting from anthropogenic development of the coastal zone, effects of fishing, climate change, and rising sea level. In Figure 1, we present a simplified conceptual model, illustrating a holistic view of the fishery-habitat problem and a framework for integrated modeling.

Challenges for Quantifying Benefits of Estuarine Habitats

Although much is known about the value of estuarine habitats to fisheries, the substantial body of relevant research has been focused more on hypothesis testing and estimation at site-specific scales than on quantifying the composite values of habitats to the fisheries. For many estuary-dependent fishery species, the spatial scales of wide-ranging populations (or metapopulations) and fisheries are regional rather than local (Jordan et al. 2009), a fact that connotes a scale mismatch between site-specific, short term research and the central problem of predicting the cumulative effects of habitat loss or gain. Connecting these scales, while confronting the inherent complexities and uncertainties, requires innovative approaches to modeling. In the case studies described later in this chapter, we offer examples of how models have been used to predict the effects of habitat change on estuary-dependent fisheries.

Rose (2000) discussed several of the challenges in quantifying how fish populations are affected by environmental quality, with examples of successful modeling approaches. Above all, models require adequate data. Commercial fisheries for estuarine-dependent species are reasonably well-documented, and in most cases, total landings, fishing effort, and ex-vessel values are available at various spatial scales, including entire regions. For some species these records span many decades. These data are essential (although not always sufficient) for estimating such parameters as stock biomass, fishing mortality, and recruitment. Fishery-independent monitoring data (typically collected by state agencies in the U.S. for inshore waters) also can be very useful for estimating and verifying model parameters (Jordan & Coakley 2004, Jordan et al. 2002, 2009). Two other types of data are essential: (1) the spatial extent and distribution of specific habitat types (e.g., seagrass beds, oyster reefs); (2) habitat dependencies based on densities (number or biomass per unit area) and survival rates of critical life stages by structural habitat type and other factors such as salinity. These basic elements can be enhanced with other information, such as habitat condition (e.g., seagrass density and species) and growth rates of the species of interest by habitat type.

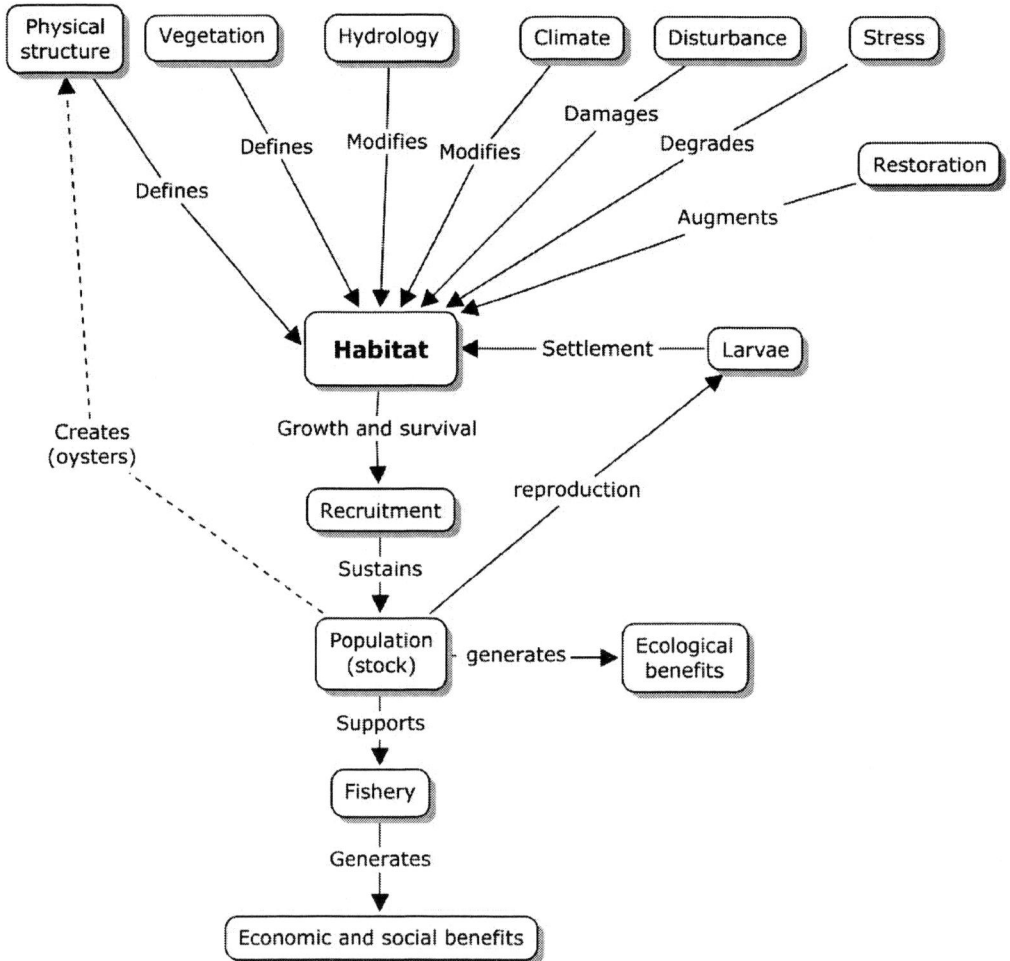

Figure 1. A conceptual overview of the relationships between estuarine-dependent fisheries and critical estuarine habitats. The dashed line indicates that among our examples, only oysters produce their own habitats. The diagram was produced with the use of Cmap tools, courtesy of the Institute for Human and Machine Cognition.

Given adequate data, the problem of quantifying the contributions of habitat to fishery stocks and harvest can be tractable. Although, as discussed by Rose (2000), there are uncertainties in every aspect of modeling fishery-habitat relationships, the largest source of uncertainty in our experience is, by far, temporal variability in recruitment, i.e. the numbers or biomass of animals added to the harvestable stock within a given time period (e.g., daily, monthly, yearly, depending on the species and the modeling questions). With such a large single source of uncertainty, the other sources tend toward insignificance, barring gross errors in model specification, assumptions, and parameterization. Individual-based models (Rose 2000), Monte Carlo methods (Jordan & Coakley 2004, Jordan et al. 2009), and Bayesian analysis (Kinas & Andrade 2007) can quantify major uncertainties within models, so that predicted outcomes can be expressed as probabilities, which may be more useful, and certainly less risky, than the (pseudo-) certainties generated by deterministic models.

Through the following case studies, from our research and the scientific literature, we explore the current state of the art in determining how losses, gains, and restoration of estuarine habitats affect the sustainability of estuary-dependent fisheries.

CASE STUDIES

Gulf of Mexico Crustacean Fisheries

There is a decades-long history of studies exploring the relationships between coastal habitats and major fisheries (reviewed by Peterson & Lowe 2009). In the Gulf of Mexico region, many of these studies have focused on the importance of salt marshes—which are extensive and widely threatened in the region—to penaeid shrimp, which support a dominant fishery in terms of biomass and economic value. In an early attempt to quantify the relationship, Turner (1977) conducted a statistical analysis of shrimp production as a function of latitude and area of salt marsh. Browder et al. (1985, 1989), employing remote sensing and geographical information systems, developed more sophisticated, spatially explicit models to examine the potential effects on the shrimp fishery of salt marsh loss along the Louisiana coast. Minello & Rozas (2002) estimated juvenile population sizes for brown shrimp (*Farfantepenaeus aztecus*), white shrimp (*Litopenaeus setiferus*) and blue crab (*Callinectes sapidus*) at the scale of a marsh complex (>400 ha). Haas et al. (2004) employed an individual-based model of brown shrimp biology in combination with habitat maps to investigate how habitat and behavioral factors affected growth and survival of juveniles; they found that salt marsh edge was a critical factor in determining differential survival rates. The scale of the Haas et al. (2004) study was 1-10,000 m^2; the authors also extrapolated their results to estimate effects of habitat change at the scale of the Louisiana coast (~5000 km^2 of saline marshes; Gosselink 1984).

Besides salt marshes, submersed aquatic vegetation (seagrasses) and unvegetated shallow (≤ 1 m) water have significant importance as habitats for juvenile shrimp and crabs (Minello et al. 2003, Dantin et al. 2005). With salt marshes, these habitats have been included, as mosaics of shallow water habitat, in studies of aggregate production of juvenile crustaceans within estuaries (e.g., Minello & Rozas 2002, Jordan et al 2009, O'Higgins et al. 2010).

In recent years, fishery-oriented population models have been used to explore the effects of habitat change for penaeid shrimp and blue crab at the scale of the U.S. Gulf of Mexico coast (Jordan et al. 2009; Jordan et al. in review).[1] The methods link dynamic fishery-based population models at the regional scale to habitat maps and research data at much finer scales. These models support quantitative predictions of the effects of habitat change at scales ranging from the Gulf of Mexico, to estuaries (Jordan et al. 2009), to sub-estuaries (O'Higgins et al. 2010).

This case study demonstrates a contrast between bottom-up models that simulate processes at the individual organism or local cohort scale, and top-down approaches. Although the former seem to have the potential for scaling-up ecosystem services and values over landscapes and seascapes to simulate cumulative effects for entire estuaries or regions (Brander et al. 2010), the challenge seldom has been met in practice. Top-down models can be applied readily to a range of spatial scales, but at the price of substantial uncertainty. In our crab and shrimp models, we have combined elements of bottom-up and top-down approaches (Jordan et al. 2009, Jordan et al. in review, O'Higgins et al. 2010). There is also a contrast between static models that make estimates either for a point in time, or an assumed equilibrium condition, and dynamic models that can be used to predict future outcomes and their probabilities for various scenarios of habitat change, climate change, etc., while accounting for stochastic variations in recruitment, mortality, fishing effort, and environmental factors. Among these, recruitment variability can be a major source of uncertainty, as illustrated in Figure 2 for Gulf of Mexico penaeid shrimp and Chesapeake Bay oyster populations. No model is perfect, but models help us to understand what is most important in complex systems. With this knowledge, and more holistic (i.e. less reductionist) approaches to quantification and prediction, we can prioritize the research needed to refine the models and achieve greater confidence in their results.

Oyster Reefs—Self-Sustaining Habitat

The Eastern oyster, *Crassostrea virginica*, supports major fisheries in estuaries from eastern Canada through the Gulf of Mexico. The shells created by oysters from calcium carbonate and organic matter not only shelter the oysters during their lives, but also supply habitats for settling oyster larvae. Although oyster larvae will settle and grow on almost any hard surface, they are most successful on clean oyster shells (Mackenzie 1996). Oysters growing on other oysters and the shells of dead oysters gradually build up reefs that not only are prime habitat for oysters, but also for many other invertebrates and some fishes (Coen et al. 2007, Hadley et al. 2010). Shells are added to the reef by recruitment and growth of oysters, and lost through disintegration, burial, and removal by fishing. These processes have been reported on extensively (e.g., Christmas et al. 1997, Powell et al. 2006) and modeled (Powell & Klinck 2007, Smith et al. 2005). Shell reefs are vital to sustaining oyster fisheries, and play important roles in estuarine ecosystems. Shell resources are actively managed and supplemented in some areas (MacKenzie 1996). In this case study, we focus on the virtually inseparable problems of quantifying oyster populations and their shell reef habitats, with examples from Chesapeake and Delaware Bays.

---

[1] It is difficult, if not impossible, to quantify the fished populations at finer scales, because both the fisheries and the invertebrate populations that support them overlap local and state boundaries.

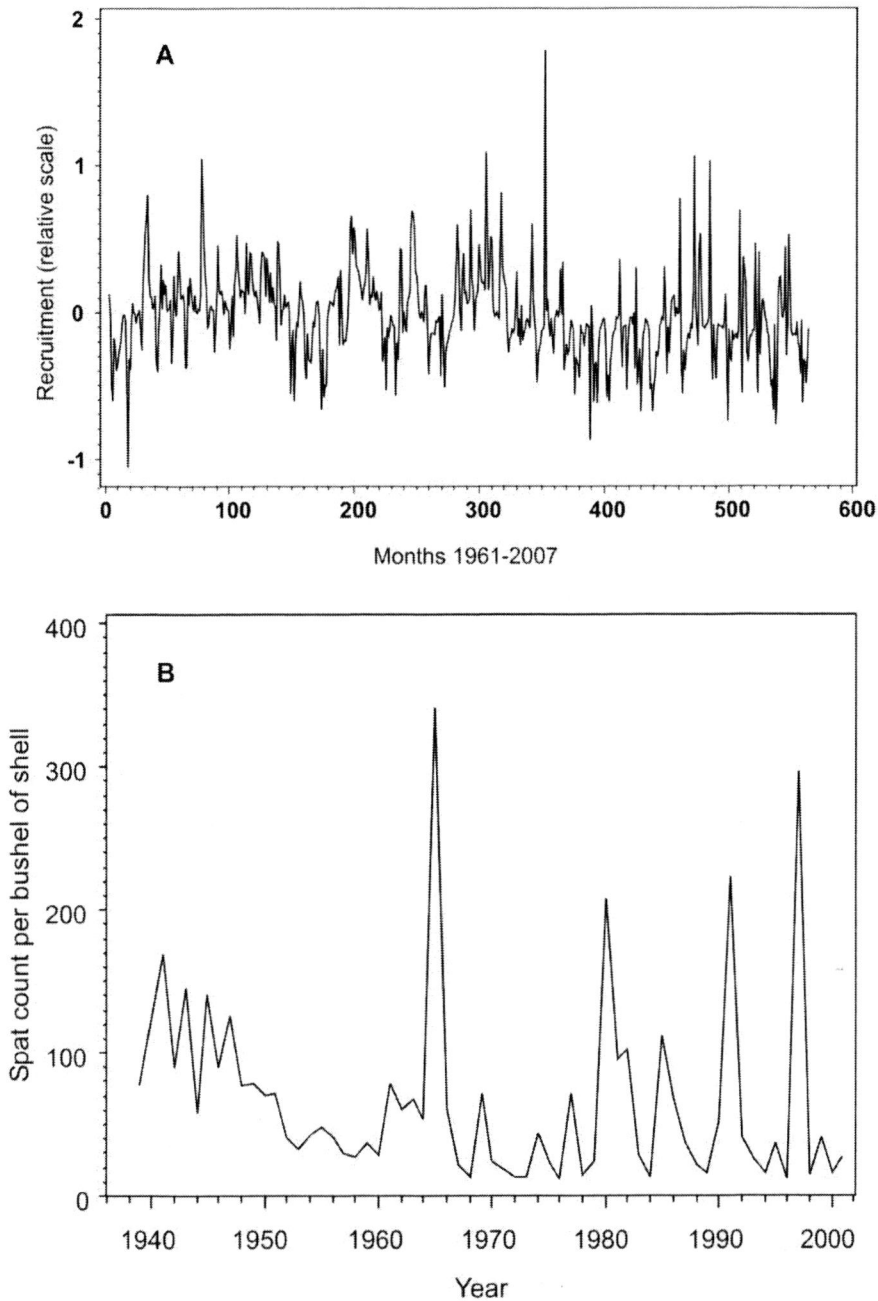

Figure 2. Examples of recruitment variability in fisheries. A: Modeled monthly recruitment of penaeid shrimp (3 species combined) in the U.S. Gulf of Mexico (S. J. Jordan, unpublished). B: Measured mean annual oyster settlement ("spat set") on reefs in the Maryland portion of Chesapeake Bay (data from the Maryland Department of Natural Resources). Spat set is an indicator of the general magnitude of future recruitment (3-4 years later).

Because oysters make their own habitat, there is strong feedback between the dynamics of oyster populations and the dynamics of oyster habitat (Mann & Powell 2007). When oyster populations decline, as in much of the range of *C. virginica* in recent decades (largely because of the combined pressures of parasitic diseases and fishing) the quantity and quality of shell habitat is gradually diminished (Powell et al. 2006). Reduced habitat reduces the probability of larval settlement and recruitment, which reduces the potential for shell deposition, which leaves the reef more vulnerable to sedimentation, in a vicious cycle. In northern Chesapeake Bay (Maryland), once a major oyster producing region, even massive plantings of dredged fossil shells could not curtail habitat loss in the face of high disease pressure, high sedimentation rates, and continued fishing (Smith et al. 2002b, 2005, Mann & Powell 2007). High sedimentation rates can be a major threat to the habitat, especially where oysters have been depleted by harvest and diseases (Smith et al. 2005). Urban and agricultural development of coastal watersheds is implicated in increased sedimentation, because of (1) runoff from impervious surfaces, farms, and disturbed land, and (2) *in situ* biological production that is enhanced by excess nutrient loading (Langland & Cronin 2003).

Surveys of oyster reefs in Chesapeake Bay have been conducted sporadically since the late 19th Century, with a variety of mechanical, acoustic, and visual observations (Smith et al. 2001, 2002a, 2002b, Smith & Jordan 1993, Mann et al. 2003). Some of these surveys were designed to delineate areas of oyster habitat by detecting shell bottoms and (more recently) their condition. Other surveys were designed primarily to obtain biological information about the oyster populations.[2]

Quantifying oyster habitat and oyster populations in large estuaries such as Chesapeake and Delaware Bays is a challenging problem. Small patches of habitat of variable quality, with variable densities of large, small, live, and dead oysters, are scattered over vast areas, often in water too turbid for visual assessment of bottom conditions. It was not until the early 2000s that serious attempts were made to quantify and model the oyster resource in Chesapeake Bay at a regional scale (Jordan et al. 2002, Mann et al. 2003, Jordan & Coakley 2004). At about the same time, advances were made in quantifying and classifying the shell reefs with acoustic techniques (Smith et al. 2002a), but to date there is not a model that fully integrates the dynamics of the population with those of the habitat.

More progress has been made in Delaware Bay, where New Jersey scientists have conducted rigorous assessments of the oyster resource and shell habitat. These scientific assessments have guided fishery management toward long-term sustainability of the oyster fishery (Powell et al. 2009), and they include shell budgets in an effort to ensure sustainable habitat. The quantification problem is somewhat easier in Delaware Bay than in Chesapeake Bay, where the area of oyster habitat, although much diminished from its historical extent, is at least 4-5 times greater than in Delaware Bay (estimated from unpublished reports).

There is much interest in restoring thriving oyster populations to estuaries for the ecological services they provide (Hadley et al. 2010). Besides food and income from harvesting, oyster populations contribute to water clarity and nutrient cycling by filtering particulate matter from the water; the structure and complexity of oyster reef habitats support biodiversity, fish production, and other valuable ecological functions (Coen et al. 2007). In order to know the extent of services provided, and the return on investments in restoration, it is essential to quantify the oyster populations, the habitat, and aspects of oyster reef biotic communities, over the long term. Therefore, long-term monitoring, assessment, and modeling should be integral to any restoration program. The importance (and costs) of these commitments will increase as the scale of restoration increases. Oyster restoration faces many

---

[2] Key biological surveys were standardized in the late 20th century (e.g., Smith & Jordan 1993).

obstacles, including harvesting (legal and illegal), diseases, predators, sedimentation, and availability and cost of shell or other durable materials as foundations for reefs. Not the least of these is the extreme variation in recruitment that can occur over time (Figure 2) and space. Some of these constraints can be managed more effectively than others, but the goals of restoration must be self-sustaining oyster populations, which will sustain their own habitats. The best hope for achieving this outcome is adaptive management, informed by sound research and by holistic, integrated, quantitative, population-habitat assessments.

Subtropical/tropical Mangrove Linked Reef Fisheries

Mangrove communities have long been studied but there is also a long history of these biomes being viewed as relatively unimportant habitats in subtropical and tropical environments worldwide (Farnsworth & Ellison 1997, Valiela et al. 2001, Alongi 2007). To the contrary, there is clear evidence of the importance of a spatially-explicit linkage of mangroves to adjacent habitat types, like seagrass beds, coral reefs and mud flats, in supporting nekton community structure and function, survivorship and prey abundance (Mumby et al. 2004, Skilleter et al. 2005, Jelbart et al. 2007). Figure 3 illustrates the connectivity between mangroves, coral reefs and other habitats for tropical reef fishes. The loss or degradation of this connectivity reduces habitat quality. For example, Skilleter & Warren (2000) quantified the decrease in habitat structure associated with grey mangrove, *Avicennia marina*, due to development of a boardwalk, which led to a decrease in numerous macrobenthic species and reduced habitat quality for nursery and feeding sites. Finally, much of the destruction of natural and human resources by the 2004 tsunami in Indonesia would have been reduced if mangroves had not been removed (Kathiresan & Rajendran 2005).

However, recent quantitative research has shown the value of mangroves in supporting important ecological services (Manson et al. 2005, Dorenbosch et al. 2006, Nagelkerken 2009) provided by many taxa, but questions remain about the role mangroves play as important nursery habitat (see Sheridan & Hays 2003, Chittaro et al. 2005). Faunce & Layman (2009) suggest that much of this is unclear because of definitions, and the difficulty in determining production metrics of species by habitat types. Their review indicates that much of the evidence for mangrove nursery use remains at the assemblage and species-level. In a review of the mangrove forest-fisheries connection, Fry & Ewel (2003) noted that mangrove carbon (as evaluated by stable isotope analyses) did not strongly support the perceived connection (i.e. there was not a strong trophic linkage), but that a broader approach of habitat use and residency and food webs was a more productive approach to answering whether mangroves serve as important nurseries. Recent work supported this dual approach (Cocheret de la Moriniére et al. 2003; Lin et al. 2007); however, Nyunja et al. (2009) and Mendoza-Carranza et al. (2010) provide isotopic evidence that mangroves do contribute to carbon pools and food webs in subtropical and tropical ecosystems in Africa and Mexico, respectively. They pointed out clear differences in isotopic ratios among species, size classes and sources (i.e., mangrove-lined creeks versus lagoons with seagrass beds and associated riparian vegetation).

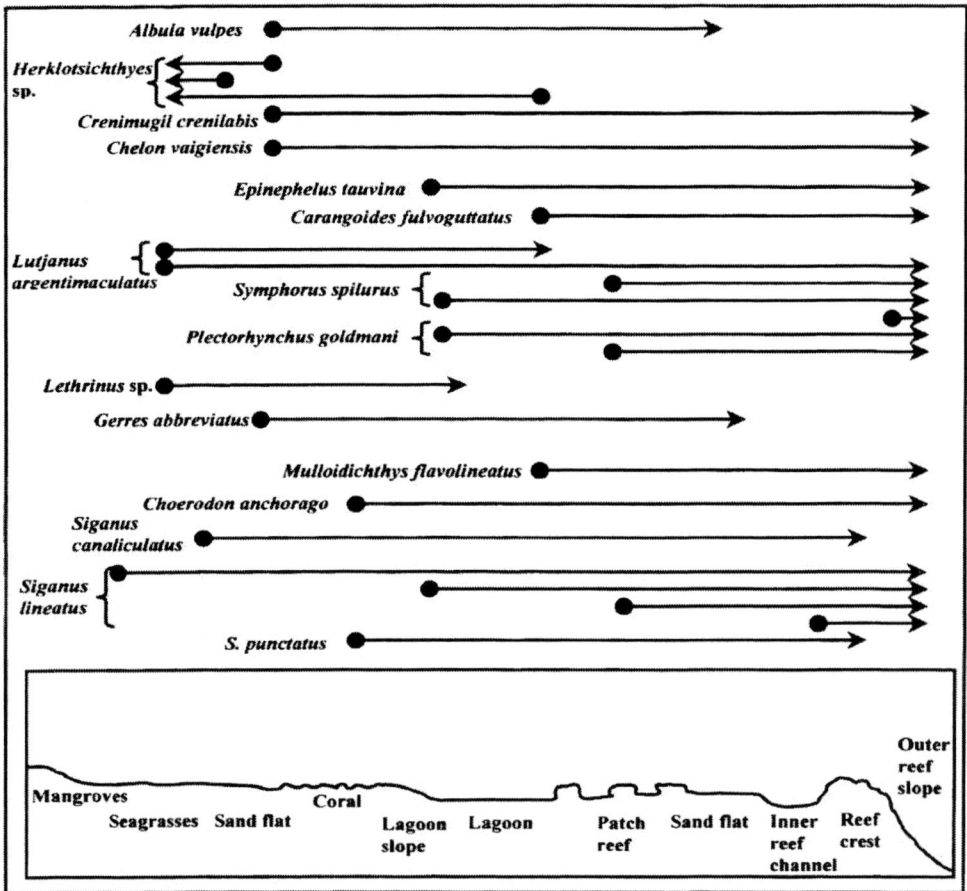

Figure 3. Spawning migrations of fish in Palau, Micronesia, from data collected through underwater observations by Johannes (1978). Arrows show the movement from usual habitat (•) to spawning sites (→). Actual distances travelled were not provided by Johannes (1978). Almost all species undertake a seaward spawning migration, with more than half of all species moving to the outer reef to spawn (the figure was reproduced from Pittman and McAlpine 2003 with permission from Elsevier). *Albula vulpes,* bonefish; *Herklotsichthyes* sp., herring (tropical); *Crenimugil crenilabis,* fringelip mullet; *Chelon (Liza) vaigiensis,* squaretail mullet; *Epinephelus tauvina,* greasy grouper; *Carangoides fulvoguttatus,* yellowspotted trevally; *Lutjanus argentimaculatus,* mangrove red snapper; *Symphorus (Symphorichthys) spilurus,* sailfin snapper; *Plectorhynchus goldmani (Plectorhinchus lineatus),* yellowbanded sweetlips; *Lethrinus* sp., emperor; *Gerres abbreviatus (erythrourus),* deep-bodied mojarra; *Mulloidichthys flavolineatus,* yellowstripe goatfish; *Choerodon anchorago,* orange-dotted tuskfish; *Siganus canaliculatus,* white-spotted spinefoot; *Siganus lineatus,* golden-lined spinefoot; *S. punctatus,* goldspotted spinefoot. Nomenclature from FishBase (http://www.fishbase.org).

It seems likely that tropical mangrove ecosystems act similarly to the salt marshes of temperate estuaries, performing much the same ecological functions, providing ecosystem services, and supporting fishery production. In a fisheries context, the large extents of shallow water in mangrove forests, with great complexity of habitat structure, are likely to be important refuges from large predators for juvenile fishes and invertebrates. With a much smaller body of research for mangroves than for salt marshes, however, these similarities remain mostly hypothetical, and the evidence is equivocal. As in our other case studies, holistic, integrated research and modeling are needed to quantify the value of mangroves as habitats supporting fishery production.

RESEARCH NEEDS

Our case studies have demonstrated the need for more integrated approaches to gain better understanding of fishery-habitat relationships. Plot-scale studies are needed to elicit the processes contributing to habitat dependence, long-term monitoring is required to document fisheries and fished populations, and models at appropriate scales are essential to quantifying and predicting the cumulative costs of habitat loss, or the values of habitat restoration. In general, although these three elements share common objectives, they are seldom coordinated in practice. Collaborations among estuarine ecologists, fisheries biologists, and modelers could be powerful means not only for better understanding, but also for more informed resource management.

We have made the case that more holistic views of habitats and their functions, bringing together structural, chemical, hydrologic, and dynamic attributes, are needed. This approach not only allows empirical quantification of habitat mosaics (Pittman et al. 2004; Peterson et al. 2007) but, coupled with movement patterns, choice experiments, and population dynamics of selected species, allows more comprehensive modeling of fishery-habitat systems. Modeling the influence of changes in habitat distribution and quality on population responses (e.g., Alderman et al. 2005) should foster better-informed and more holistic management actions.

Some specific research needs are suggested by the case studies. For modeling purposes, it is essential to know not only that particular habitats hold higher densities of animals (e.g., blue crabs in submersed aquatic vegetation beds), but also how these habitats contribute to the survival of critical life stages. The few applicable studies for blue crab and shrimp, for example, have used field tethering or laboratory mesocosms to investigate survival rates of early juveniles (Minello et al. 2003). Each of these methods involves somewhat unnatural conditions. An alternative approach avoids this problem: with sufficiently intensive field sampling, size frequency analysis of multiple cohorts can be used to estimate survival for early life stages. Where this technique is successful, much could be learned about the spatial and temporal dynamics of habitats and habitat dependence.

The oyster case study illustrates the need for integrated monitoring, modeling, research, and management that address oyster populations and their shell habitats simultaneously. Sustaining or restoring oyster populations as resources, both for fisheries and for their ecological benefits, depends on better quantification and understanding of the interactions between oysters, the habitats they create for themselves, and environmental factors. Recent developments in acoustic sea floor classification and imaging systems (e.g., Smith et al. 2002a) have the potential for habitat assessments that are more comprehensive, of higher resolution, non-destructive, and more informative than traditional techniques such as bottom trawls (using toothed dredges) and grab samples. Although much progress has been made by some programs, major uncertainties remain in modeling oyster populations. Among these, variability in recruitment success and disease-related mortality (Krantz & Jordan 1996) are topics for continued in-depth research. Integrating shell dynamics with oyster population dynamics presents another challenge for modelers and managers with interests in oyster conservation or restoration.

Although there have been many studies in recent years tracking and quantifying ontogenetic movements of fishes from mangroves through seagrass to coral reef systems and back (Nagelkerken 2009), it is clear that more research directed at eliciting the functions of mangrove habitats is needed. The data will be invaluable in protecting these coastal habitats from the severe impacts they have suffered resulting from coastal development (Alongi 2007). For example, detailed studies using stable isotope approaches are warranted given the variability in results (Fry & Ewel 2003; Nyunja et al. 2009; Mendoza-Carranza et al. 2010), about whether mangroves serve as important foraging grounds by contributing greater production to offshore stocks than other habitat types. Furthermore, approaches need to be developed to quantify mortality and growth of important commercial species that use mangroves early in life, and to track the newly-produced biomass across the landscape to coral reefs or other nearshore habitat types, for example, seagrass beds and mudflats.

CONCLUSIONS

Improved understanding of the relationships of estuarine habitats to major coastal fisheries requires integrated, holistic approaches to research, data collection, and modeling. Structural attributes of the habitat, environmental variability, biological interactions, dynamics of the fished populations, and measures of the fisheries (catch, effort, fishing mortality, dollar value, etc.) all should be considered if our goal is to sustain the fisheries and their contributions to society. In some cases, only a few of these aspects and associated parameters will turn out to be critical, thereby simplifying the problem. The importance of each parameter will depend on (1) the strength of its relationship to the population, and (2) its variability relative to the other parameters. Parameters with strong influence and large variability should be the most important targets for research.

Spatial and temporal scales also need careful consideration. We have shown in the shrimp and crab case study that the influence of changes in habitat requires attention at the regional scale of the fisheries and populations. Only then can habitat effects on the fisheries be quantified at local scales. Mangrove habitats in much of the world are associated with artisanal fisheries, which tend to operate over much smaller spatial areas than the extensive fisheries in our shrimp and crab example. In these cases, local-scale analyses may be appropriate. For example, if destruction or restoration of mangroves in a particular locality were correlated strongly with reduced or increased catches, respectively, it would be strong, though indirect, evidence for habitat dependence.

Habitat conditions, fished populations, and fisheries vary over a wide range of time scales. Sustainability requires that signals of long-term trends, past and future, need to be resolved from noisy shorter-term variability, especially unexplained (stochastic) variations in recruitment. Habitat loss can be an insidious threat to estuary-dependent fisheries if it occurs gradually and cumulatively, as when one after another bit of critical habitat is lost to shoreline armoring, dock construction, channelization, water pollution or other stresses and disturbances. Marginal effects on annual recruitment typically are not observed given the multiple climatic and anthropogenic signals affecting the baseline, yet they can accrue over years or decades until the fishery is in peril of irreversible loss. Models at the appropriate scales of space and time are essential, even though they may have large uncertainties.

ACKNOWLEDGMENTS

We thank Kevin Summers and William Fisher for preliminary reviews of the manuscript. The views expressed in this chapter are those of the authors and do not necessarily reflect the views or policies of the U.S. Environmental Protection Agency. This is contribution 1421 from the Gulf Ecology Division.

REFERENCES

Alderman, J., McCollin, D., Hinsley, S.A., Bellamy, P.E., Picton, P., & Crockett, R. (2005). Modelling the effects of dispersal and landscape configuration on population distribution and viability in fragmented habitat. *Landscape Ecol. 20*, 857-870.

Alongi, D.M. (2007). Mangrove forests: resilience, protection from tsunamis, and responses to global climate change. *Est. Coast. & Shelf Sci. 76*, 1-13.

Brander, L., Ghermandi, A., Kuik, O. Markandya, A., Nunes, P.A.L.D., Schaafsma, M. & Wagentendonk, A. (2010). Scaling up ecosystem services values: methodology, applicability and a case study. *Nota Di Lavoro 41.2010*, Fondazione Eni Enrico Mattei, Milan.

Breitburg, D. & Houde, E. (2006). Habitat, habitat requirements, and habitat management. Chapter 3, Element 3 In: (M. McBride & E. Houde (Eds.), *Fisheries Ecosystem Planning for Chesapeake Bay*. American Fisheries Society, Trends in Fisheries Science and Management 3, Bethesda, MD.

Browder, J.A., May L.N., Jr., Rosenthal, A., Gosselink, J.G. & Baumann, R.H. (1989). Modeling future trends in wetland loss and brown shrimp production in Louisiana using thematic mapper imagery. *Remote Sensing Environ. 28*, 45-59.

Browder, J. A. & Moore, D. (1981). A new approach to determining the quantitative relationship between fishery production and the flow of fresh water to estuaries. In R. D. Cross & D. L. Williams (Eds.), *Proceedings of the National Symposium on Freshwater Inflow to Estuaries Volume 1* (FWS/OBS-81/04, pp. 403-430). U. S. Fish and Wildl. Serv., U. S. Dept. of Interior, Washington, DC.

Browder, J.A., Bartley, H.A. & Davis, K.S. (1985). A probabilistic model of the relationship between marshland-water interface and marsh disintegration. *Ecol. Model. 29*, 245-260.

Browman, H..I, & Stergiou, K.I. (Eds.). (2004). Perspectives on ecosystem-based approaches to the management of marine resources. *Mar. Ecol. Prog. Ser. 274*, 269–303.

Chittaro, P.M., Usseglio, P., & Sale, P.F. (2005). Variation in fish density, assemblage composition and relative rates of predation among mangrove, seagrass and coral reef habitats. *Env. Biol. Fishes 72*, 175-187.

Christmas, J. F., McGinty, M. R., Randle, D. A., Smith, G. F. & Jordan, S.J. (1997). Oyster shell disarticulation in three Chesapeake Bay tributaries. *J. Shellfish Res. 16*, 115-123.

Cocheret de la Morinière, E., Pollux, B.J.A., Nagelkerken, I., Hemminga, M.A., Huiskes, A.H.L., & van der Velde, G. (2003). Ontogenetic dietary changes of coral reef fishes in the mangrove-seagrass-reef continuum: stable isotopes and gut-content analysis. *Mar. Ecol. Prog Ser. 246*, 279-289.

Coen, L.D., Brumbaugh, R.D., Bushek, D., Grizzle, R., Luckenbach, M.W., Posey, M.H., Powers, S.P. & Tolley, G.S. (2007). AS WE SEE IT: Ecosystem services related to oyster restoration. *Mar. Ecol. Prog. Ser. 341*, 303-307.

Cronin, T., Halka, J., Phillips, S, & Bricker, O. (2003). Estuarine sediment sources. In: Langland, M. & Cronin, T. (eds.). *A Summary Report of Sediment Processes in Chesapeake Bay and Watershed.* Water Resources Investigations Report 03-4123, U.S. Geological Survey, Denver CO, pp. 49-60.

Crossett, K. M., T. J. Culliton, P. C. Wiley, and T. R. Goodspeed. (2004). *Population trends along the Coastal United States: 1980–2008.* Washington, DC: NOAA.

Dantin, D.D., Fisher, W.S., Jordan, S.J. & Winstead, J.T. (2005). *Fishery Resources and Threatened Coastal Habitats in the Gulf of Mexico*, U.S. Environmental Protection Agency, EPA/600/R-05/051, Gulf Breeze, FL.

DeLong, A.K. & Collie, J.S. (2004). *Defining essential fish habitat: a model-based approach.* Rhode Island Sea Grant, Narragansett, RI.

Dorenbosch, M., Grol, M.G.G., Nagelkerken, I. & van der Velde, G. (2006). Seagrass beds and mangroves as potential nurseries for the threatened Indo-Pacific humphead wrasse, *Cheilinus undulates,* and Caribbean rainbow parrotfish, *Scarus guacamaia. Biol. Conserv. 129*, 277–282.

Farnsworth, E.J. & Ellison, A.M. (1997). The global conservation status of mangroves. *Ambio 26*, 328-333.

Faunce, C.H. & Layman, C.A. (2009). Sources of variation that affect perceived nursery function of mangroves, Chapter 11. In: I. Nagelkerken (Ed.), *Ecological Connectivity Among Tropical Coastal Ecosystems*, pp. 401-421, Springer Science + Business Media B.V.

Fry, B. & Ewel, K.C. (2003). Using stable isotopes in mangrove fisheries research – a review and outlook. *Isotopes Env. Health Stud. 39*, 181-196.

Haas, H.L., Rose, K.A., Fry, B., Minello, T.J. & Rozas, L.P. (2004). Brown shrimp on the edge: linking habitat to survival using an individual-based simulation model. *Ecol. Appl. 14*, 1232-1247.

Hadley, N.H., Hodges, M., Wilber, D.H. & Coen, L.D. (2010). Evaluating intertidal oyster reef development in South Carolina using associated faunal indicators. *Restor. Ecol. 18*, 691-701.

Heinz Center (2002). *The status of the nation's ecosystems.* The H. John Heinz III Center for Science, Economics, and the Environment, Cambridge University Press, Cambridge, U.K.

Houde, E.D. & Rutherford E.S. (1993). Recent trends in estuarine fisheries: Predictions of fish production and yield. *Estuaries 16*, 161-176.

Jelbart, J.E., Ross, P.M. & Connolly, R.M. (2007). Fish assemblages in seagrass beds are influenced by the proximity of mangrove forests. *Mar. Biol. 150*, 993–1002.

Johannes, R.E. (1978). Reproductive strategies of coastal marine fishes in the tropics. *Env. Biol. Fishes 3*:65–84.

Jordan, S.J. & Coakley, J.M. (2004). Long-term projections of eastern oyster populations under various management scenarios. *J. Shellfish Res. 23*, 63-72.

Jordan, S.J., Greenhawk, K.N., McCollough, C.B., Vanisko J., & Homer, M.L. (2002). Oyster biomass, abundance and harvest in northern Chesapeake Bay: trends and forecasts. *J. Shellfish Res. 21*, 733-741.

Jordan, S. J., O'Higgins, T. & Dittmar, J.A. (Submitted). Ecosystem services of coastal habitats and fisheries: multi-scale ecological and economic modeling. *Marine and Coastal Fisheries*.

Jordan, S. J., Smith, L. M. & Nestlerode J.A. (2009). Cumulative effects of coastal habitat alterations on fishery resources: toward prediction at regional scales. *Ecol. Soc.* [On-line serial] 14(1): 16. http://www.ecologyandsociety.org/vol14/iss1/art16/

Kathiresan, K. & Rajendran, N. (2005). Coastal mangrove forests mitigated tsunami. *Est., Coastal & Shelf Sci. 65*, 601–605.

Kinas, P.G. & Andrade, A. (2007). Bayesian statistics for fishery stock assessment and management: a synthesis. *Pan-Am. J. Aq. Sci. 2*:103-112.

Krantz, G. E. & Jordan, S.J. (1996). Management alternatives for protecting *Crassostrea virginica* fisheries in *Perkinsus marinus* enzootic and epizootic areas. *J. Shellfish Res. 15*:167-176.

Lin, H-J., Kao, W-Y, & Wang, Y-T. (2007). Analysis of stomach contents and stable isotopes reveal food sources of estuarine detritivorous fish in tropical/subtropical Taiwan. *Est., Coast. & Shelf Sci. 73*, 527-537.

Mackenzie, C.L, Jr. (1996). Management of natural populations. In: V.S. Kennedy, V. S., R.I.E. Newell & A. F. Eble (Eds.), *The Eastern Oyster*, Maryland Sea Grant College, College Park, pp. 707-721.

Manderson, J.P., Phelan, B.A., Meise, C., Stehlik, L.L., Bejda, A.J., Pessutti, J., Arlen, L., Deaxler, A., Stoner, & A.W. (2002). Spatial dynamics of habitat suitability for the growth of newly settled winter flounder *Pseudopleuronectes americanus* in an estuarine nursery. *Mar. Ecol. Prog. Ser. 228*, 227–239.

Manderson, J.P., Pessutti, J., Meise, C., Johnson, D., & Shaheen. P. (2003). Winter flounder settlement dynamics and the modification of settlement patterns by post-settlement processes in a NW Atlantic estuary. *Mar. Ecol. Prog. Ser. 253*, 253–267.

Mann, R. (2000). Restoring the oyster reef communities in the Chesapeake Bay: A commentary. *J. Shellfish Res.* 19, 335-339.

Mann, R., Jordan, S., Smith, G., Paynter, K., Wesson, J., Christman, M., Vanisko, J., Harding, J., Greenhawk, K. & Southworth, M. (2003). Oyster population estimation in support of the ten-year goal for oyster restoration in the Chesapeake Bay: Developing strategies for restoring and managing the Eastern oyster. In: *Chesapeake Bay Fisheries Research Program: Research Project Symposium*, NOAA Chesapeake Bay Office, Annapolis, MD, pp. 65-69.

Mann, R. & Powell, E.N. (2007). Why oyster restoration goals in the Chesapeake Bay are not and probably cannot be achieved. *J. Shellfish. Res. 26*, 905-917.

Manson, F.J., Longeragan, N.R., Skilleter, G.A. & Phinn, S.R. (2005). An evaluation of the evidence for linkages between mangroves and fisheries: A synthesis of the literature and identification of research directions. *Ocean. Mar. Biol. Ann. Rev. 43*, 485–515.

Mendoza-Carranza, M., Hoeinghaus, D.J., Garcia, A.M., & Romero-Rodriguez, A. (2010). Aquatic food webs in mangrove and seagrass habitats of Centla Wetland, a biosphere reserve in southwestern Mexico. *Neotrop. Ichthy. 8*, 171-178.

Minello, T.J. & Rozas, L.P. (2002). Nekton in Gulf Coast wetlands: fine-scale distributions, landscape patterns, and restoration implications. *Ecol. Appl. 12*, 441-455.

Minello. T.J., Able, K.W., Weinstein, M. P. & Hayes, C.G. (2003). Salt marshes as nurseries for nekton: testing hypotheses on density, growth and survival through meta-analysis. *Mar. Ecol. Prog. Ser. 246*, 39-59.

Mumby, P.J., Edwards, A.J., Arias-Gonzales, J.E., Lindeman, K.C., Blackwell, P.G., Gall, A., Gorczynska, M.I., Harborne, A.R., Pescod, C.L., Renken, H., Wabnitz, C.C.C. & Llewellyn, G. (2004). Mangroves enhance the biomass of coral reef fish communities in the Caribbean. *Nature 427*, 533-536.

Nagelkerken, I. (2009). Evaluation of nursery function of mangroves and seagrass beds for tropical decapods and reef fishes: patterns and underlying mechanisms. In: I. Nagelkerken (Ed), *Ecological Connectivity Among Tropical Coastal Ecosystems*, p. 357-399, Springer Science + Business Media B.V.

Nyunja, J., Ntiba, M., Mavuti, K., Soetaert, K. & Bouillon, S. (2009). Carbon sources supporting a diverse fish community in a tropical coastal ecosystem (Gazi Bay, Kenya). *Estuar. Coast. Shelf S. 83*, 333-341.

O'Higgins, T.G., Ferraro, S.P., Dantin, D.D., Jordan, S.J. & Chintala, M.M. (2010). Habitat scale mapping of fisheries ecosystem service values in estuaries. *Ecol. Soc.* [On-line serial] 15(4): 7. http://www.ecologyandsociety.org/vol15/iss4/art7/

Peterson, M.S. (2003). A conceptual view of environment-habitat-production linkages in tidal river estuaries. *Rev. Fish. Sci. 11*, 291-313.

Peterson, M.S. & Lowe, M.R. (2009). Implications of cumulative impacts to estuarine and marine habitat quality for fish and invertebrate resources. *Rev. Fish. Sci. 17*, 505-523.

Peterson, M.S., Weber, M.R., Partyka, M.L., & Ross, S.T. (2007). Integrating *in situ* quantitative geographic information tools and size-specific laboratory-based growth zones in a dynamic river-mouth estuary. *Aqua. Conserv.: Mar. Freshw. Ecol. 17*, 602–618.

Pittman, S.J. & C.A. McAlpine. (2003). Movements of Marine Fish and Decapod Crustaceans: Process, Theory and Application. *Adv. Mar. Biol. 44*:205-294.

Pittman, S. J., McAlpine, C.A., & Pittman, K.M. 2004. Linking fish and prawns to their environment: A hierarchical landscape approach. *Mar. Ecol. Prog. Ser. 283*, 233–254.

Powell, E.N., Ashton-Alcox, K.A. & Kraeuter, J.N. (2007). Reevaluation of eastern oyster dredge efficiency in survey mode: Application in stock assessment. *N. Am. J. Fish. Manage. 27*, 492-511.

Powell, E.N. & Klinck, J.M. 2007. Is oyster shell a sustainable estuarine resource? *J. Shellfish Res. 26*, 181-194.

Powell, E.N., Klinck, J.M., Ashton-Alcox, K.A. & Kraeuter, J.N. (2009). Multiple stable reference points in oyster populations: implications for reference point-based management. *Fish. Bull. 107*, 133-147.

Powell, E.N., Kraeuter, J.N. & Ashton-Alcox, K. (2006). How long does oyster shell last on an oyster reef? *Estuar. Coast. Shelf S. 69*, 531-542.

Rose, K.A. (2000). Why are quantitative relationships between environmental quality and fish populations so elusive? *Ecol. Appl. 10*, 367-385.

Ruiz, G.M., Hines, A.H. & Posey, M.H. (1993). Shallow water as a refuge habitat for fish and crustaceans in non-vegetated estuaries: an example from Chesapeake Bay. *Mar. Ecol. Prog. Ser. 99*, 1-16.

Sheridan, P.F. & Hays, C. (2003). Are mangroves nursery habitat for transient fishes and decapods? *Wetlands* 23(2), 449-458.

Simenstad, C.A., Brandt, S.B., Chambers, A., Dame, R., Deegan, L.A., Hodson, R., & Houde, E.D. (2000). Habitat–biotic interactions. In J.E. Hobbie (Ed.), *Estuarine Science: A Synthetic Approach to Research and Practice*, pp. 427–460, Island Press, Washington, DC.

Skilleter, G. A. & Warren, S. (2000). Effects of habitat modification in mangroves on the structure of mollusc and crab assemblages. *J. Exp. Mar. Biol. Ecol. 244*, 107–129.

Skilleter, G.A., Olds, A., Loneragan, N.R. & Zharikov, Y. (2005). The value of patches of intertidal seagrass to prawns depends on their proximity to mangroves. *Mar. Biol. 147*, 353–365.

Smith, G.F., Bruce, D.G. & Roach, E.B. (2002a). Remote acoustic habitat assessment techniques used to characterize the quality and extent of oyster bottom in the Chesapeake Bay. *Mar. Geodesy 24*, 1-19.

Smith, G.F., Bruce, D.G., Roach, E.B., Hansen, A., Newell, R.I.E. & A.M McManus. (2005). Assessment of recent habitat conditions of Eastern oyster *Crassostrea virginica* bars in mesohaline Chesapeake Bay. *N. Am. J. Fish. Mgmt. 25*, 1569-1590.

Smith, G., Greenhawk, K., Bruce, D., Roach, E. & Jordan, S. (2001). A digital presentation of the Maryland oyster habitat and associated bottom types in the Chesapeake Bay (1974-1983). *J. Shellfish Res. 20*, 197-206.

Smith, G. F. & Jordan, S.J. (1993). *Monitoring Maryland's Chesapeake Bay oysters*. Maryland Department of Natural Resources, Chesapeake Bay Research and Monitoring Division, CBRM-OX-93-3, Annapolis.

Smith, G.F., Roach, E.B. & Bruce, D.G. (2002b). The location, composition, and origin of oyster bars in mesohaline Chesapeake Bay. *Estuar. Coast. Shelf S. 56*, 391-409.

Stoner, A.W. (2003). What constitutes essential nursery habitat for a marine species? A case study of habitat form and function for queen conch. *Mar. Ecol. Prog. Ser. 257*, 275–289.

Turner, R.E. (1977). Intertidal vegetation and commercial yields of penaeid shrimp. *Trans. Am. Fish. Soc. 106*, 411-416.

U.S. Commission on Ocean Policy. (2004). An Ocean Blueprint for the 21st Century, Washington, DC.

Valiela, I., Bowen, J.L. & York, J.K. 2001. Mangrove forests: one of the world's threatened major tropical environments. *BioScience 51*, 807-815.

In: Estuaries: Classification, Ecology and Human Impacts   ISBN: 978-1-61942-083-0
Editor: Steve Jordan   © 2012 Nova Science Publishers, Inc.

Chapter 6

MECHANISMS OF TRANSFORMATION OF VARIOUS FORMS OF CARBON AND CYCLING PATHWAYS IN THE HOOGHLY ESTUARINE SYSTEM

Joyita Mukherjee, Madhumita Roy and Santanu Ray[*]

Ecological Modelling Laboratory, Department of Zoology, Visva-Bharati University,
Santiniketan, India

Phani Bhusan Ghosh

Institute of Engineering and Management, Y-12, Sector-V, Saltlake City,
Kolkata, India

Angshuman Sarkar

Department of Statistics, Visva-Bharati University, Santiniketan, India

ABSTRACT

The accelerating increase of atmospheric carbon dioxide (CO_2) and its implications for climate change are of immense concern. The Hooghly estuary, along with the luxuriant mangroves of Sundarban is one of the world's unique ecosystems. Recent studies of this estuarine system reveal the potentially important role of the estuary in regulating atmospheric CO_2. Phytoplankton fix gaseous CO_2 from the atmosphere into organic carbon through photosynthesis; the carbon is then transformed through subsequent steps in the food chain. Mangrove litter, transported by tidal action, is the key contributor of soil organic carbon (SOC) to the estuary. Following standard techniques, we quantified several forms of carbon, including litter carbon, SOC, soil inorganic carbon (SIC), total aqueous organic carbon (TOC), dissolved aqueous organic carbon (DOC), aqueous particulate organic carbon (POC), aqueous dissolved inorganic carbon (DIC), and carbon content of phytoplankton and zooplankton. To study relationships of the various forms of carbon to physical and chemical factors, we measured litter biomass, soil salinity, soil pH, redox potential, soil temperature, water temperature, dissolved oxygen, salinity and pH. Statistical correlation analyses were then employed to assess how relationships between carbon forms and physical-chemical factors might contribute

---

[*] Corresponding author.

to regulating the organic carbon cycle. From this study, we have worked out a conceptual model of organic carbon cycling dynamics, and gained insight into mechanisms of transformation of different forms of carbon and the fundamental seasonal processes associated with carbon cycling in the estuarine system.

Keywords: Estuary, conceptual model, carbon, mangrove litter

INTRODUCTION

The Hooghly-Matla estuarine complex, a positive mixohaline, net heterotrophic, tropical, river delta estuary (Ray 1987), occupies an important place among global ecosystems for its luxuriant mangrove vegetation (Bhunia 1979). Estuaries are semi-enclosed zones where river water mixes with sea water (Gattuso et al. 1998; Jordan, Chapter 1). They serve as reservoirs of carbon and play a critical role in the global carbon cycle. Climate change at global and regional scales is largely regulated by the carbon cycle (Kulinski & Pempkowiak 2008). Therefore, study of carbon cycles, as a whole or in part, has been an area of research emphasis for the past few decades.

The hydrosphere, as a large reservoir of carbon, plays a very important role in the global carbon cycle (Joos et al. 1996). Air-water CO_2 exchange, which is governed by a number of physical parameters and processes, has a vital part in this cycle. Estuaries are generally considered to be net heterotrophic systems and often act as sources of CO_2 to the Earth's atmosphere (Smith & Hollibaugh 1993, Frankignoulle et al. 1998, Biswas et al. 2004). In Europe, estimates of annual estuarine CO_2 emissions correspond to 5–10% of anthropogenic emissions (Ahad et al. 2008). Increasing CO_2 content of the atmosphere causes grave concern because of the climatic changes which may result from it (Bolin et al. 1979). The surface water of Sundarban (1,781 km^2) has a CO_2 flux of 314.6 mmol m^{-2} day^{-1} (Biswas et al. 2004). The magnitude of CO_2 flux is balanced mainly by two processes – photosynthetic uptake and respiratory release, which in turn are dependent upon various factors such as allochthonous input of labile organic matter, water residence time, sunlight availability, rates of community metabolism, temperature and nutrient load (Smith & Hollibaugh 1993, Howland et al. 2000, Wang & Veizer 2000, Abril et al. 2002, Ram et al. 2003, Ahad et al. 2008). Research has been done to show that seasonal variations can lead to estuaries shifting from net autotrophic to functioning as net heterotrophic (Kemp et al. 1992, Smith & Hollibaugh 1997, Howland et al. 2000, Ram et al. 2003).

Phytoplankton are the major producers of organic carbon in open waters and account for approximately 30% of the world's primary production (Falkowski 1980). These primary producers dominate the recycling of dissolved and potentially volatile carbon at the water-air interface (Adams & Caldeira 2008). In particular they act as sinks for CO_2. The uptake of CO_2 at the water surface has a positive correlation with estimates of primary production, surface irradiance, and phytoplankton biomass (Platt et al. 1990). Absorbed CO_2 in the form of organic carbon enters into the grazing food chain. From the food chain, the exudates and dead organisms contribute carbon to the sediment particulate organic carbon (POC) pool.

Dissolved organic carbon (DOC), as the potentially largest organic carbon pool, has drawn the attention of researchers. Previous work reveals the immense role of DOC as one of the major components of food webs (Packard et al. 2000), as a main source of energy for microbial metabolism (Tranvik 1992), and a major component of photosynthetic release (Ducklow & Carlson 1992). Changes in DOC concentrations along with the magnitude and proportion of autochthonous and allochthonous sources, temperature, depth, and season have been studied in various types of water bodies (Wetzel 2001, Sugiyama et al. 2004, Guo et al. 1995, Ribes et al. 1999, Dafner and Wangersky 2002, Zweifel et al. 1995, Mantoura & Woodward 1983). DOC forms complexes with metal ions, and can affect particle accumulation and sedimentation rates by controlling the surface charge of colloids and suspended matter in water through adsorption (Kretzschmar et al. 1997). Mineralization of DOC through microbial degradation (Jumars et al. 1989, Amon & Benner 1996) ultimately enriches the soil organic carbon pool. Export of particulate organic carbon (POC), another reservoir of the organic carbon pool, implies the net sequestration of CO_2 through sedimentation. Because any net increase in soil CO_2 emissions, perhaps in response to environmental changes, can influence global climate, identification of the factors that regulate soil respiration is critical in predicting ecosystem responses to global change (Ahn et al. 2009). A change of just 10% in soil organic carbon would be equivalent to all the anthropogenic CO_2 emitted over 30 years (Kirschbaum 2000).

Despite the important role of the carbon cycle in the Earth's biogeochemistry, studies from southern Asia have been few until recently. To understand the controlling factors and overall fate of organic carbon cycle in this system is challenging because estuaries are extremely dynamic both physically and biogeochemically. Thus, the objective of this study is to propose a holistic conceptual model explaining pathways of organic carbon cycling in the Hooghly estuarine system. For this purpose we studied various forms of carbon present in soil and estuarine water, and determined the processes of transformation from one form to another.

MATERIALS AND METHODS

Study Site

The Hooghly estuary, along with Sundarban, a vast expanse of forested coastal wetlands, is a globally unique ecosystem (Figure 1). The estuary is regarded as one of the best nursery grounds for a variety of shellfish and fin fishes, making the local economy and livelihood of residents very much dependent on this ecosystem. It extends over two countries, India (West Bengal state) and Bangladesh. The whole area is crisscrossed by dense network of rivers, canals and creeks (Ray & Straskraba 2001). Situated at the land-ocean boundary of the northeastern coast of the Bay of Bengal, it is a unique bioclimatic zone (Biswas et al. 2004). The meso-macrotidal Hooghly estuary (Biswas et al. 2010) has a wide mixing zone extending from Diamond Harbor to the mouth of the river. Sagar Island, located in the western sector of the estuary, is the largest delta island in this estuarine complex, lying between 21°56′–21°88′ N and 88°08′–88°16′ E (Figure 1). The island is about 144.9 km^2 in area, and is surrounded by the River Hooghly on the north and northwest and the River Mooriganga on the east (Mandal et al. 2009).

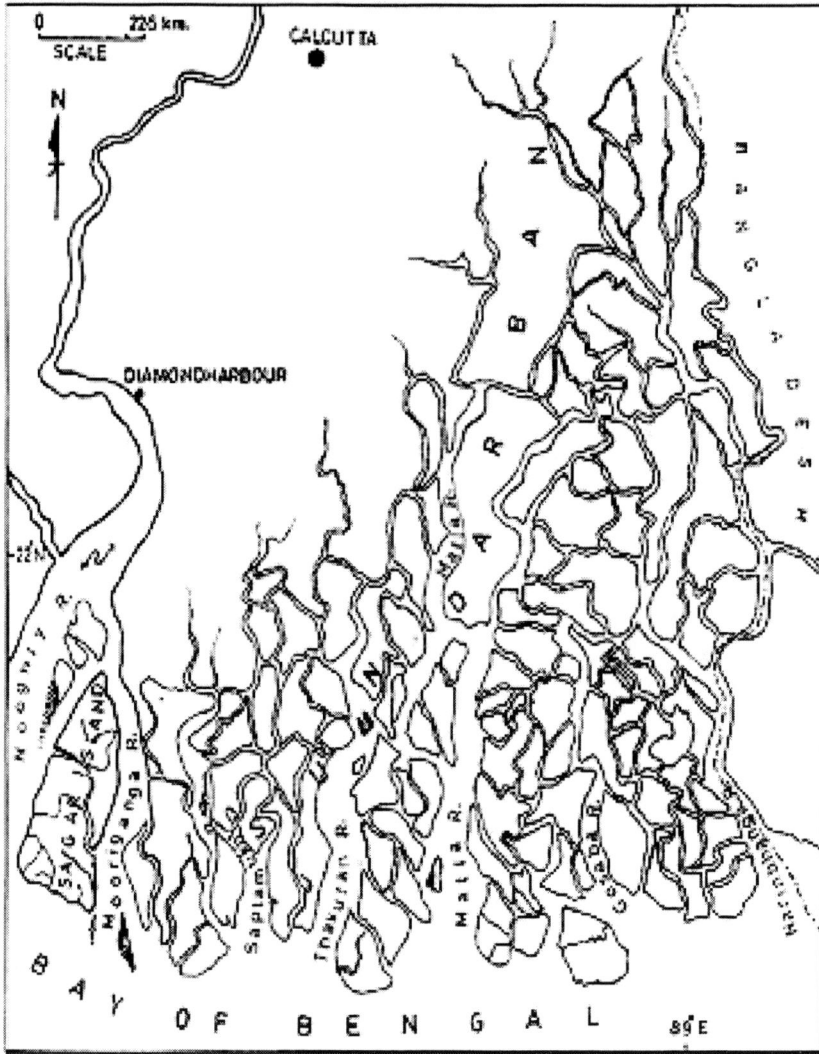

Figure 1. Hooghly estuary showing the Sagar Island study site, Sundarban, and portions of the Ganges delta.

This region is in the wet tropical climatic zone, with pronounced seasonal climatic changes. The seasons are characterized as pre-monsoon (March–June) with average high temperature ranging from 25–42 C and minimum precipitation; monsoon (July–October), when 70–80% of annual rainfall occurs, and post-monsoon (November–February), with cold weather (average 25^0 C) and negligible rainfall. The monsoon season is generally dominated by southwest winds. The average humidity is about 80% and more or less uniform throughout the year. *Avicennia marina* (grey mangrove) is the dominant species among the halophytes of Sagar island. *Avicenna alba*, *Porteresia coarctata*, *Excoecaria agallocha*, *Ceriops decandra*, *Acanthus ilicifolius* and *Derris trifoliate* are also present (Saha & Choudhury 1995).

Sampling and Experiment

Samples were collected from the mangrove forest floor of Sagar Island and the creeks that cross the island (Figure 1). Several experimental and survey efforts were conducted for one year (June 2009–May 2010) in the field to collect data for dissolved and particulate organic carbon (DOC, POC), water temperature, water pH, dissolved oxygen (DO), dissolved carbon dioxide ($CO_{2(aq)}$), soil organic carbon (SOC), soil temperature, salinity, etc.

Soil samples were collected from the field stations at a depth of ~8–10 inches (20–25 cm) in the mangrove forest floor, where tidal flow is encountered at monthly intervals. Soil temperature was measured in the field using a digital thermometer (EUROLAB-ST 9269). Soil samples were further analyzed by measuring organic and inorganic carbon content following standard methods (Greenberg et al. 1992). We measured pH monthly from a saturated soil–water paste (Gupta 2002). Soil salinity was derived from soil water extracts following the method of Gupta (2002), and calculated by the formula: salinity (ppt) = chlorinity (ppt) · 1.805 + 0.03 (Chattopadhyay 1998).

For collection of litter, several nets of known dimension (1m x 1m) were placed in the mangrove forest. Litter was collected monthly; litter biomass (m^{-2}) and mean litter biomass per kg of soil were calculated. Estuarine water samples were collected from the creeks at ~0.5 m depth for chemical analysis. Samples for DOC and POC were collected in glass bottles in the field and preserved on ice in the dark for transport to the laboratory. TOC in these samples was estimated with a TOC analyzer (Shimadzu). Water samples were passed through Millipore GF/F filters (0.45 μm pore size) and the filtrate used to determine POC. Total alkalinity (TA) of the samples was estimated by potentiometric titration (Greenberg et al. 1992) calibrated against standard sea water procured from the National Institute of Oceanography (NIO). Dissolved inorganic carbon (DIC) was estimated from TA. Dissolved free carbon dioxide ($CO_{2(aq)}$) was measured following a standard method (Greenberg et al. 1992). Bicarbonate alkalinity was calculated from pH and TA (Greenberg et al. 1992). Water temperature was measured digitally in the field (EUROLAB-ST 9269); DO was measured by Winkler's iodometric method. Water pH was measured with a pH meter (LUTRON-pH-206).

Plankton were collected from surface water using a plankton net during high tide, and preserved with Lugol's iodine solution (phytoplankton) or buffered formaldehyde (zooplankton). For quantitative analysis of phytoplankton, wet and dry weights were measured and phytoplankton carbon content was calculated following the literature (Jorgensen et al. 2000). For zooplankton carbon, the Sedgewick Rafter counting method was employed to obtain the number of organisms and the corresponding carbon content was estimated according to the literature (Friedler et al. 2003). Gross primary production and community respiration were measured in situ by the light and dark bottle method described by Strickland and Parsons (1972).

Statistical Analysis

Statistical analysis was done using S-PLUS 8.0 software. Multiple regression, correlation, and F statistics were used to assess relationships between environmental factors and the forms of carbon.

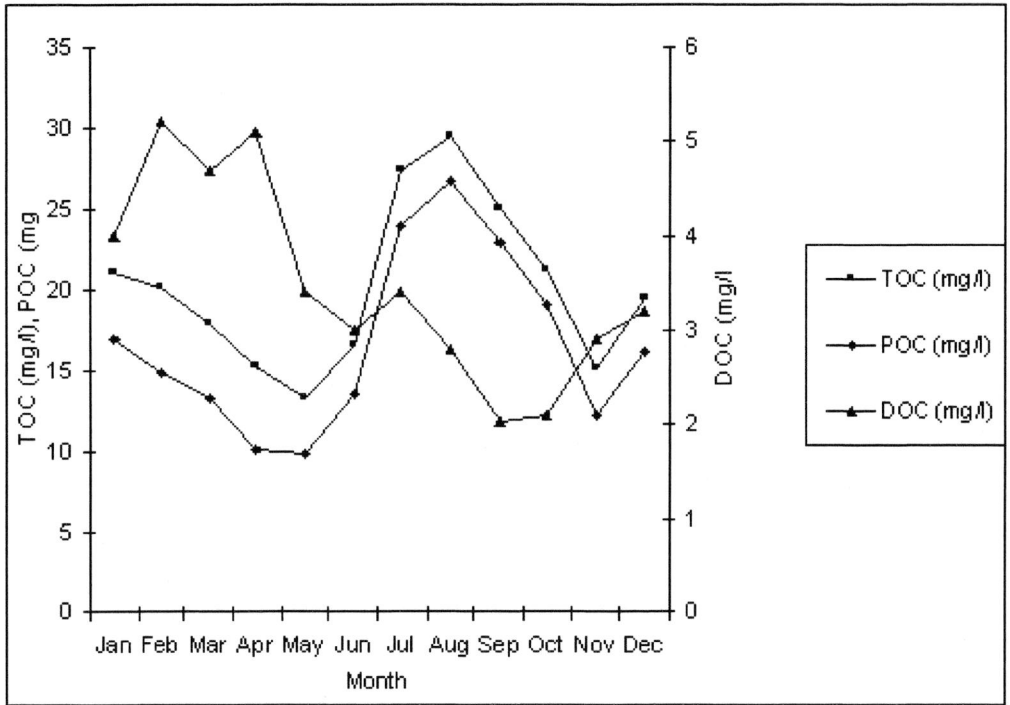

Figure 2. Fluctuation of TOC, POC and DOC during the study period.

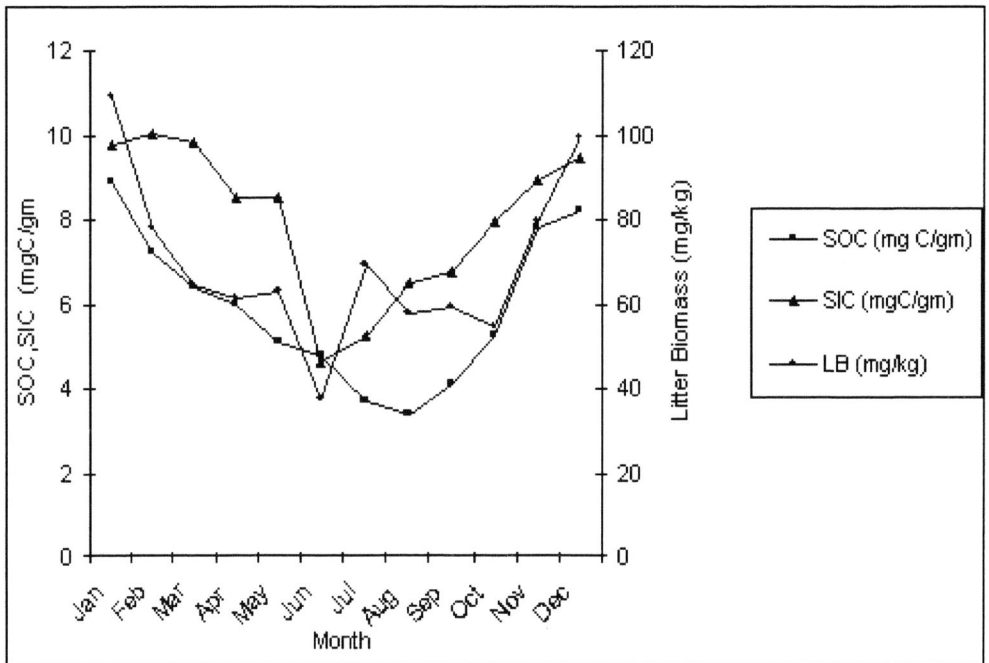

Figure 3. Fluctuation of SOC, SIC and LB during the study period.

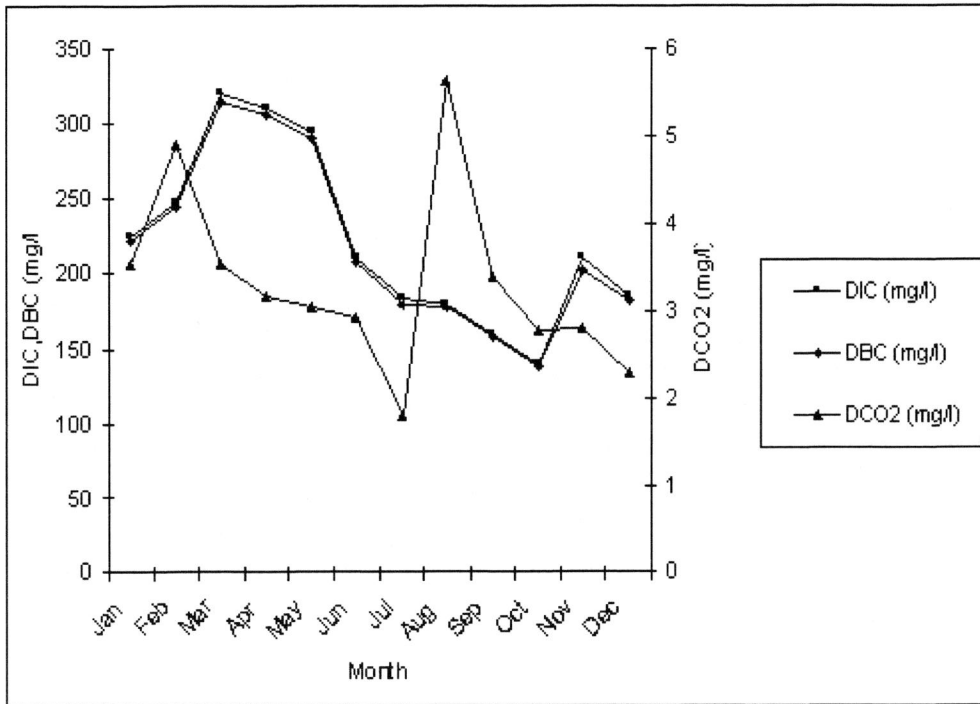

Figure 4. Fluctuation of DIC, DBC and $CO_{2(aq)}$ during the study period.

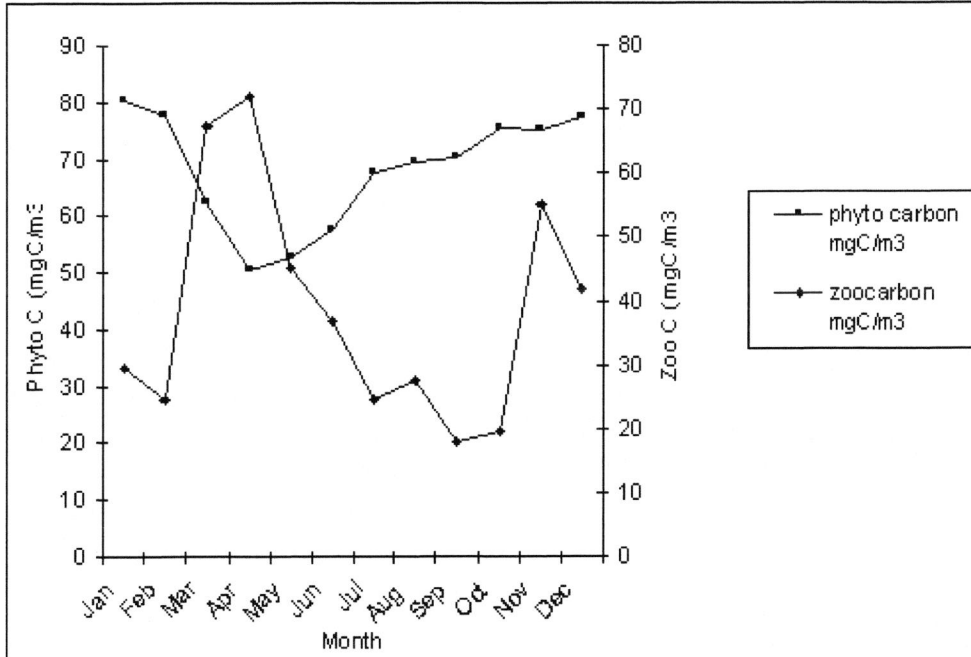

Figure 5. Fluctuation of phytoplankton carbon and zooplankton carbon during the study period.

RESULTS AND DISCUSSION

The Sundarban mangrove ecosystem provides detritus and nutrients to the adjacent Hooghly-Matla estuarine complex (Ray and Straskraba 2001). Highly productive mangrove ecosystems (2500 mg C m^{-2} day^{-1}) in the estuarine reaches of tropical rivers may be sources of nutrients to coastal waters, on one hand, and provide a sedimentary sink for nutrients on the other (Gonneea et al. 2004). The Ganges is one of the major rivers of India, where during the monsoon huge freshwater runoff carries large amounts of materials to the estuary. Mangrove wetlands along with their particulate and dissolved organic matter (POM and DOM) are potentially important sources of energy and nutrition to heterotrophic communities of surrounding estuarine and marine ecosystems (Odum & Heald 1975).

Deposition of organic matter in the form of litter (1603 g m^{-2} year^{-1}; Ghosh et al. 1990) from the Sundarban mangrove forest at the land–ocean boundary occurs throughout the year. Litter decomposition is a key process that facilitates the cycling of carbon and nutrients in forest ecosystems (Aerts 2006, Shiels 2006). Litter comprises new and old leaves, twigs, flowers and fruit. The breakdown of mangrove litter contributes a major role in maintaining the grazing food chain of the estuary through inorganic transformations, and leaf litter accounts for about 30–60% of primary production (Bunt et al. 1979). Carbon enters ecosystems through photosynthesis, of which a fraction is lost in autotrophic respiration. The net gain is the net primary production (NPP). In natural systems, this carbon is eventually transferred to the soil through litter fall, root turnover or death of individual plants, thus providing the substrate for the formation of soil organic carbon (Kirschbaum 2000).

Mangrove litter undergoes first degradation and then decomposition into dissolved inorganic nutrients, which are important for growth of phytoplankton and other higher plants, in turn stimulating the production of zooplankton and other aquatic fauna in the estuary. Higher values of litter biomass (LB) per kg soil are recorded during the post monsoon period and lower values are recorded during the monsoon (Figure 3).

Litterfall from mangroves varies seasonally and is one of the main sources of soil organic carbon (SOC). Death of flora and fauna (mainly macrobenthos, i.e. bivalves, gastropods, nemerteans, anemones, polychaetes and crabs) from adjacent mangrove forests also contribute to the SOC pool. SOC varies temporally and spatially as it is subjected to regulation by a number of factors like soil texture, tidal range, elevation, redox potential, bioturbation intensity, forest type, temperature and rainfall (Sebastian & Chacko 2006). Dynamics of SOC mineralization are strongly affected by temperature because soil microbial processes are a function of temperature (Insam 1990, Kirschbaum 1995, Winkler et al. 1996). So temperature is an important factor controlling both the amount and turnover time of SOC mineralization. Although a significant correlation between temperature and SOC mineralization is well established (Singh & Gupta 1977, Raich & Schlesinger 1992, Lloyd & Taylor 1994, Kirschbaum 1995, Kätterer et al. 1998), there is no uniform function to describe the relationship. Some authors qualitatively described the effect of temperature: at higher temperatures, the easily decomposable fraction was mineralized more quickly than at lower temperatures. Li-xia et al. (2006) showed that the dynamics of SOC mineralization followed a two-phase pattern: SOC was rapidly decomposed at the early incubation stages after which decomposition gradually slowed down to a comparatively steady rate. SOC produces CO_2 as the organic matter within soil undergoes microbial decomposition. The soil inorganic carbon (SIC) pool is enriched by SOC mineralization.

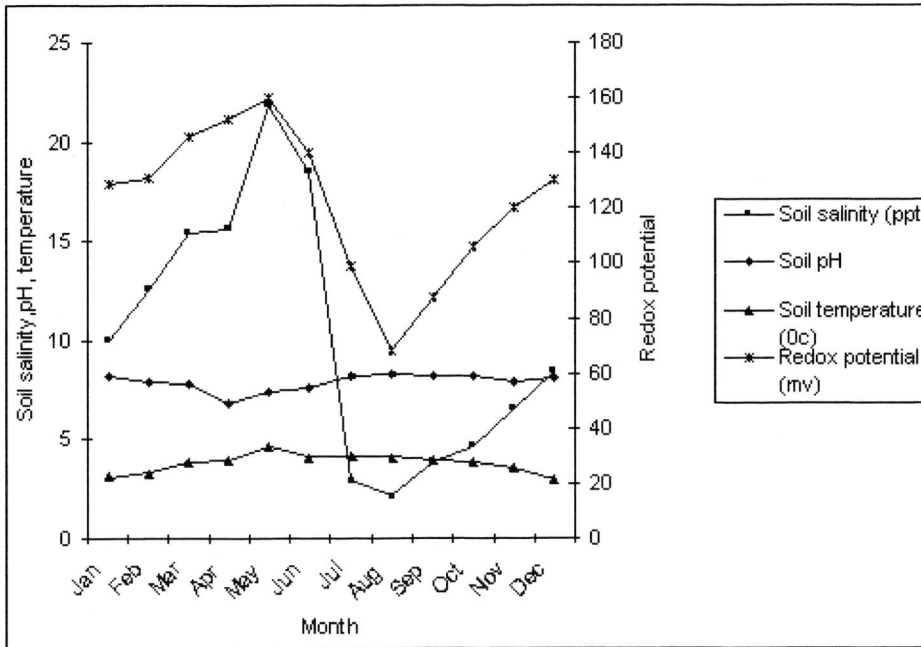

Figure 6. Fluctuation of soil salinity, pH, temperature and redox potential during the study period.

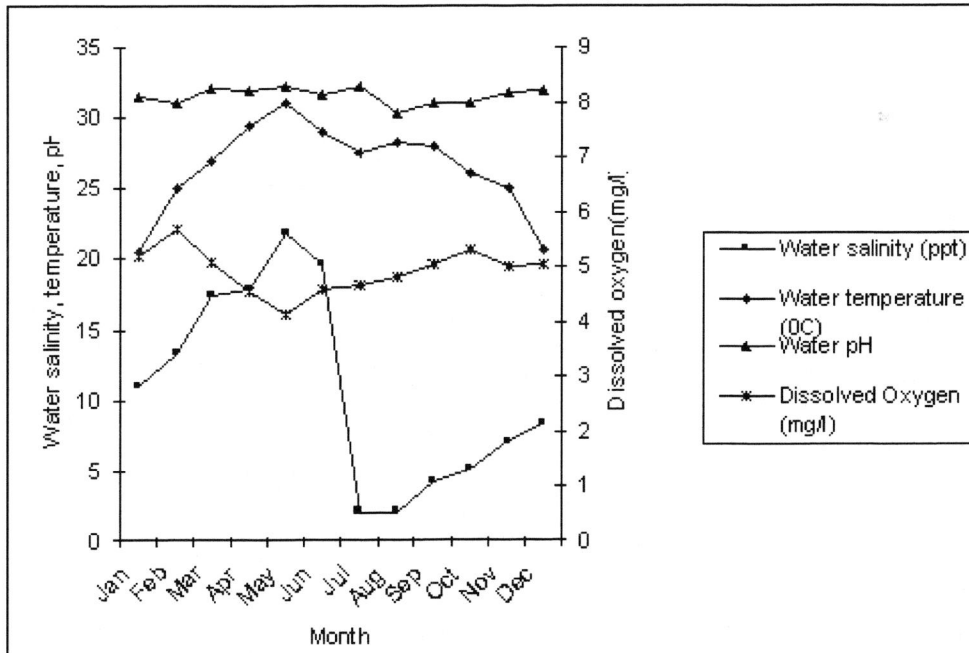

Figure 7. Fluctuation of water salinity, pH, temperature and dissolved oxygen during the study period.

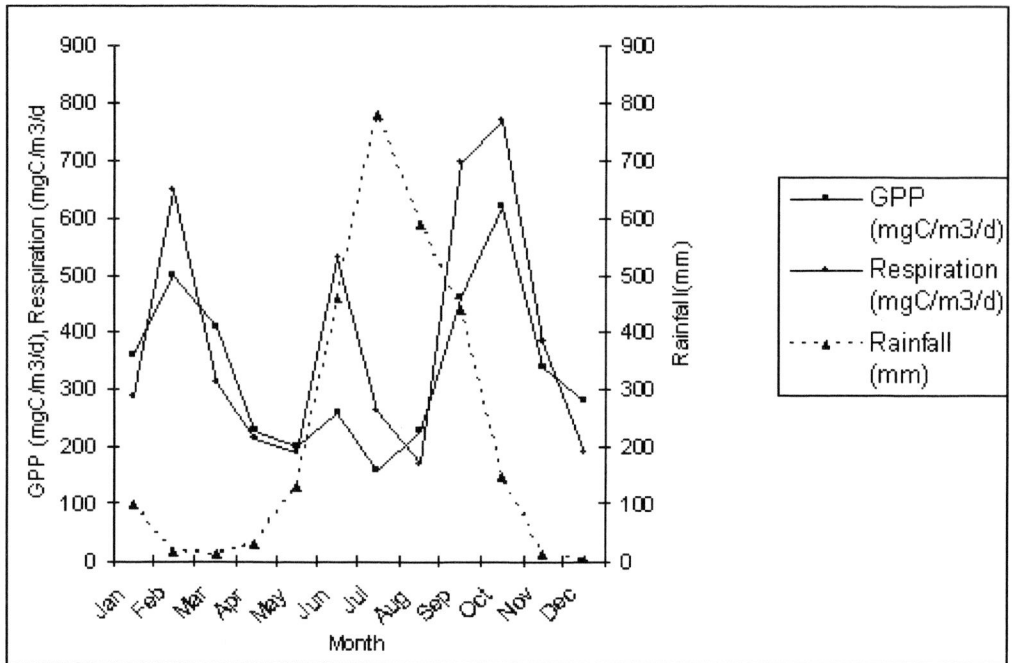

Figure 8. Fluctuation of GPP, respiration and rainfall during the study period.

Assessment of SOC is essential to understanding nutrient inputs into the surrounding mangrove water. In the present study SOC had higher values during the post-monsoon (7.2–8.9 mgC gm^{-1}) and lower values during the monsoon (3.4–5.25 mgC gm^{-1}; Figure 3). The reason for the seasonal differences may be that in the monsoon months following heavy rains nutrients are washed out from the soil. Mineralization of SOC to SIC is dependent on soil temperature and redox potential. Field observations show that soil temperature is lower during the post-monsoon (21.6–25.3) and higher during pre-monsoon (28.0–33.3; Figure 6). Higher values of soil temperature indicate higher microbial activity in pre-monsoon compared to post-monsoon (Gupta 2002). Oxygen diffuses through the soil, and the resulting aeration enhances mineralization. But this process is hindered when the pore spaces of soil are saturated with water and the aeration process is reduced. When soil is inundated with water, anaerobic conditions persist. In the Hooghly estuarine system, we observed that redox potential was lower during monsoon (+68 to +106mV), whereas higher values were observed during pre-monsoon (+140mV to +160mV), and post-monsoon values were intermediate (+120mV to +130mV; Figure 6). Lower values during monsoon coincide with heavy rainfall, which causes water logging. As a result oxygen penetration into the soil is dramatically reduced in this season. However, during pre-monsoon, higher temperature and lack of rainfall enhance the diffusion of oxygen in soil, which in turn increases redox potential. During the study period highest rainfall occurred in July and August and almost no rain fell in November and December (Figure 8). The pre-monsoon period generally remains dry.

Variations of SIC were studied throughout the year; the experimental results showed higher values in pre-monsoon (9.45–10.01 mgC gm^{-1}) and lower values in monsoon (4.61–6.73 mgC gm^{-1}; Figure 3). Soil salinity was higher during pre-monsoon (15.44–21.85 ppt) and much lower during monsoon (2.1–4.65 ppt; Figure 6), reflecting high rainfall. The pH is a very important property of soil as it determines the availability of nutrients, microbial activity and physical conditions of the soil. Soil pH is higher during monsoon (8.22–8.3) and lower during pre-monsoon (6.8–7.8; Figure 6).

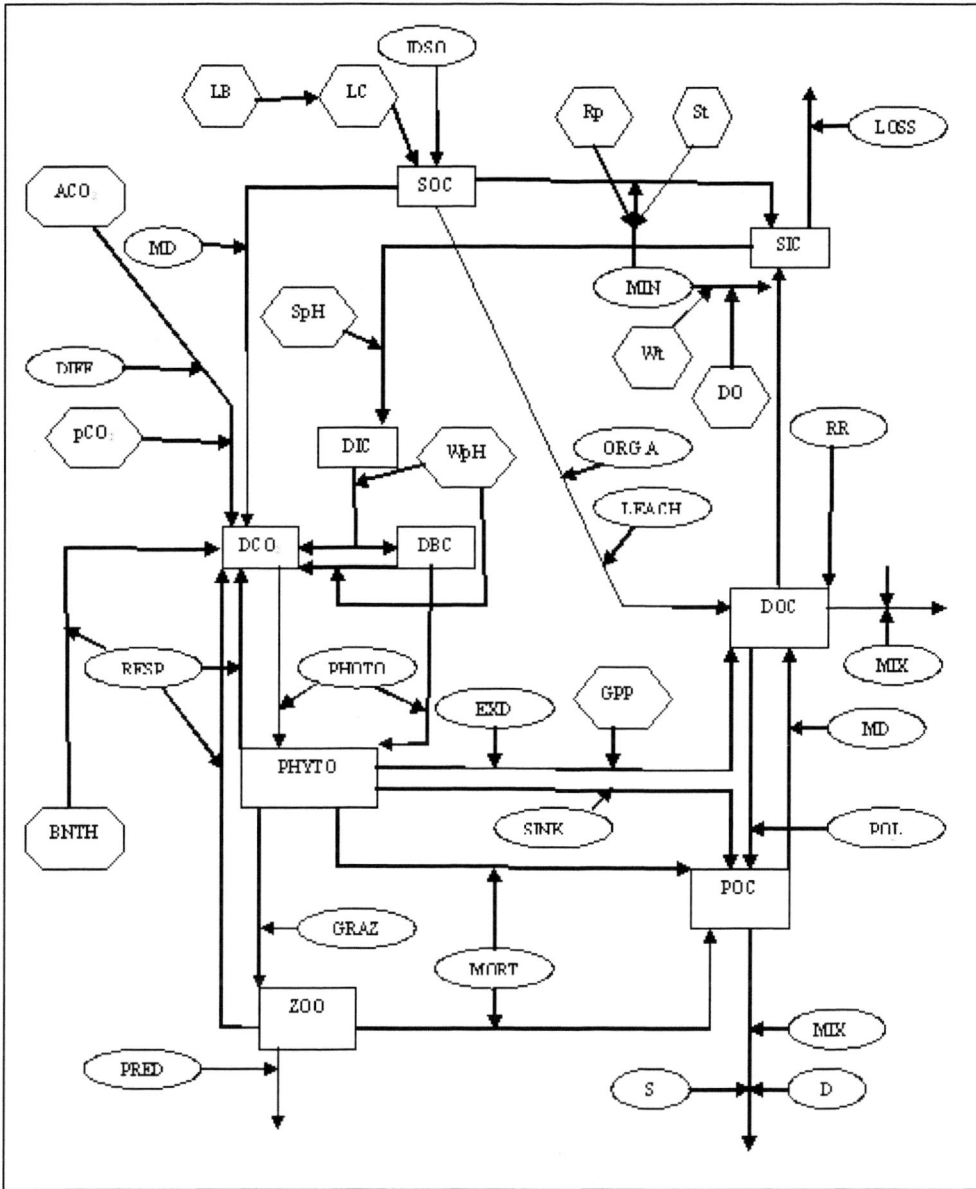

Figure 9. Conceptual diagram of pathways of organic carbon cycling in the Hooghly estuarine system.

The fractionation of DIC into $CO_2(aq)$, HCO_3^- and CO_3^{-2} in seawater is influenced by pH, alkalinity and temperature (Butler 1982; Takahashi et al. 1993). Field observations showed DIC concentrations were higher in pre-monsoon (210–320 mg L^{-1}) and lower in monsoon (140–183 mg L^{-1}; Figure 4). CO_2 (aq) and HCO_3^- are the portions of DIC available for phytoplankton uptake during photosynthesis. The $CO_2(aq)$ pool can be enriched by diffusion from the atmosphere, respiration, and as pH decreases, from the HCO_3^- pool. Field observations show that pH of water remains higher in post- and pre-monsoon than monsoon. Water pH varied from 7.80 to 8.30 throughout the year (Figure 7). As the pH of water and CO_2(aq) are inversely related, variations in CO_2(aq) were opposite those of pH, i.e. higher in monsoon (average 3.40 mg L^{-1}) and lower in pre-monsoon (average 3.17 mg L^{-1}) and post-monsoon (average 3.38 mg L^{-1}). Field observations showed that pH was lower in the month of August and September; accordingly CO_2(aq) values were higher in those months (Figure 4).

Some amount of CO_2 from air diffuses into water depending upon the partial pressure gradient of CO_2, and diffusion from water to air takes place if the partial pressure of CO_2 is greater in water than air. Because the Hooghly ecosystem is heterotrophic in nature, net diffusion from water to air takes place most of the year (Biswas et al. 2004, Mukhopadhyay et al. (2006). Only during post-monsoon when phytoplankton blooms peak is there net diffusion of CO_2 from air to water. Respiration by the biotic community of the ecosystem contributes to the pool of CO_2(aq). Degradation of SOC in the presence of oxygen releases free CO_2, which contributes to the CO_2(aq) pool. A part of SIC is converted to CO_2(aq). The conversion is dependent upon soil pH; under acidic conditions SIC contributes to DIC and some of the DIC is converted to CO_2(aq). When soil is inundated with water anoxic conditions prevail at the soil-water interface. Under such circumstances, SIC is converted to free CO_2. Bicarbonate is a reservoir for CO_2(aq); HCO_3^- values were higher in pre-monsoon (average 279.24 mg L^{-1}), lower in monsoon (average 163.42 mg L^{-1}) and intermediate in post-monsoon (average 212.92 mg L^{-1}) (Figure 4). Phytoplankton use both CO_2(aq) and HCO_3^- for photosynthesis (Clark and Flynn 2000; Tortell and Morel 2002; Cassar et al. 2004) and then by contributing carbon to the food chain take part in the organic carbon cycle. Phytoplankton blooms occur in post-monsoon. Accordingly field results showed that phytoplankton carbon reached its peak value during post-monsoon (average 77.56 mgC m^{-3}), particularly in January-February, with lower values during pre-monsoon (average 55.74 mgC m^{-3}). Zooplankton carbon varied from 18.04 mgC m^{-3} to 72.16 mgC m^{-3} showing higher values in pre-monsoon and lower in monsoon (Figure 5). These observations are supported by previous work (De et al. 1991, Biswas et al. 2004, 2010). Because the Hooghly estuary is heterotrophic in nature, the GPP is less than community respiration, as reflected in our field survey data (Figure 8).

The TOC pool in water consists of dissolved (DOC) and particulate (POC) fractions. The dynamics of the DOC pool follow inputs and outputs from various sources. Litterfall, leaching, decomposition and detrital storage in a mangrove system are the main sources of organic carbon in estuarine water adjacent to mangroves (Steinke et al. 1993a,b, Dittmar and Lara 2001, Machiwa and Hallberg 2002). A fraction of phytoplankton primary production is released as exudate, which contributes directly to the DOC pool. Mangrove leaves that fall into water initially undergo rapid leaching of DOM, which includes DOC (Fell et al. 1975, Newell et al. 1984, Robertson 1988). Easily leachable organic matter from dead organisms or shaded leaves directly contributes to the DOC pool during the tidal cycle. Microorganisms act upon the dead organisms (phyto- and zooplankton) and decaying matter, ultimately enriching the DOC pool. Microorganisms also consume DOC (Alongi et al. 1989, Boto et al. 1989, Cole 1999, Murrel & Hollibaugh 2000). Boto et al. (1989) found a weak significant correlation between bacterial production and DOC concentrations in the sediment of a tropical mangrove forest in Australia. Organic acids like humic and fulvic acids, constituents of humus, remain in the SOC pool, which under anaerobic conditions contributes to the DOC pool. The availability of DOC in water depends on the amount of organic carbon leached from SOC. DOC is formed from POC by two processes: (1) microbial degradation of POC releases DOC, and (2) solubilization, which is dependent on pH; as pH increases some of the POC pool is dissolved into DOC. The Hooghly River is loaded with terrestrial runoff which carries a significant amount of DOC of allochthonous origin (terrigenous). Ground water discharge also enriches the DOC pool. Some amounts of DOC are mineralized (to DIC) depending upon the physicochemical environment. Dissolved oxygen and temperature are the main factors regulating this process.

The POC pool is generated mainly by polymerization of DOC, sinking of phytoplankton and excretion, dead and decaying parts of zooplankton and their fecal matter. Rivers and ground water also transport some (allochthonous) POC. A portion of POC is taken up by bacteria, a portion is contributed to sediment, and another portion, along with DOC, is exported to the ocean by tidal action. Tidal amplitude, tidal regime, and frequency of inundation determine rates of export (Twilley 1985, Holmer & Olsen 2002, Machiwa & Hallberg 2002).

Estimated DOC from field observations showed higher values in post-monsoon (2.9–5.2 mg L^{-1}) and lower values in monsoon (2.02–3.4 mg L^{-1}). POC concentrations were higher in monsoon (19.09–26.74 mg L^{-1}) and lower in pre-monsoon (9.90–13.55 mg L^{-1}). The pattern of TOC concentrations was similar to that of POC (Figure 2). In the present study, we observed that DO varied annually between 4.13 and 5.67 mg L^{-1}. Higher values were observed during January–February and September–October, with the highest in February (5.67 mg l^{-1}), whereas lower values were observed during May–July, with the lowest in May (4.13 mg L^{-1}; Figure 7). In the tropics, temperatures are higher during pre-monsoon (March–June) and lower during post-monsoon (November to February).

Maximum average water temperature during the period of study occurred in May (31 C), with a minimum in January (20.5 C). Microbial activity increases with the increase of temperature in pre-monsoon and decreases with the decline of temperature in post-monsoon.

Table 1. Correlation coefficients of soil organic carbon with litter biomass (LB) and other forms of carbon in soil and estuarine water ($p< 0.01$, * $p< 0.05$)**

LB	TOC	POC	DOC	SIC	DIC
0.793**	−0.489	−0.529*	0.562*	−0.533*	0.255

Table 2. Correlation coefficients of governing factors of soil system with soil organic carbon (SOC) and soil inorganic carbon (SIC) ($p< 0.01$, * $p< 0.05$)**

	Redox potential	Soil salinity	Soil temperature	Soil pH	Rainfall
SOC	0.486	0.207	−0.809**	−0.036	−0.807**
SIC	0.299	0.549*	0.805**	−0.621*	0.077

Table 3. Correlation coefficients of soil inorganic carbon with litter biomass (LB) and other forms of carbon in soil and estuarine water ($p< 0.01$, * $p< 0.05$)**

LB	TOC	POC	DOC	SOC	DIC
−0.815**	−0.329	−0.309	−0.160	−0.533*	.443

Considering all the biological and abiological factors and physical processes, we developed a conceptual model of pathways of organic carbon cycling (Figure 9).

STATISTICAL ANALYSIS

Correlation results show that SOC is highly correlated with LB, DOC, POC, TOC, SIC, and weakly correlated with DIC (Table 1). Multiple regression of SOC against the aforesaid factors resulted in an overall R^2 of 0.97 (F = 11.97, p < 0.05). Redox potential (R_p), soil salinity (S_s), soil pH (S_{pH}), soil temperature (S_t) and rainfall (R_f) are the principal physical factors controlling the soil system, thereby controlling conversion of one form of soil carbon to another. The results indicate that these five environmental factors are prominent enough to explain the fluctuations of SOC (Table 2), with R_p and S_s positively correlated with SOC, and S_t, S_{pH} and R_f negatively correlated. Multiple regression of SOC against these factors gave an R^2 of 0.93 (F = 15.78, p < 0.01).

Correlation (Table 2) and multiple regression indicate control of mineralization of SOC to SIC primarily by S_t and R_p. Multiple regression for SIC as a function of the abiotic environmental factors gave overall $R^2 = 0.85$ (F = 6.95, p < 0.05). SIC is correlated with S_s, S_t and R_p positively, and negatively with S_{pH}. R_f is not significantly correlated with SIC (Table 2). These results indicate that variation of the SIC pool is a function of S_s, S_t, R_p and S_{pH}, with S_t and R_p dominant. SIC has significant negative correlations with SOC and LB (Table 3). Multiple regression with SIC as the response versus LB, DOC, POC, TOC, SOC and DIC as explanatory variables) gave $R^2 = 0.92$ (F = 9.16, p<0.05). Therefore it can be concluded that the significant abiotic factors, along with SOC and LB, can explain fluctuations of SIC throughout the year.

The significant negative correlation between pH and $CO_2(aq)$ (Table 5) indicates that pH, as expected, is a key factor controlling $CO_2(aq)$, with $CO_2(aq)$ concentrations decreasing as pH increases.

TOC is negatively correlated with salinity and pH and positively correlated with rainfall (Table 4). POC and DIC are significantly correlated with TOC (Table 6). The strong correlation of POC with TOC may merely reflect that POC is a major component of TOC. The multiple regression R^2 for TOC as a function of LB, POC, DOC, SOC, SIC, and DIC is >0.99 (F = 254.7, p < 0.01). DIC and SOC are negatively correlated with POC (Table 7). Water pH and rainfall are positively correlated with POC, whereas salinity is negatively correlated (Table 4). The multiple regression R^2 for POC as a function of TOC, DOC, SOC, SIC, DIC, and LB is >0.99 (F = 305.5, p < 0.01). The multiple regression R^2 for DOC with the same explanatory variables is 0.91 (F = 11.45, p < 0.01), indicating that LB, POC, TOC, SOC, SIC and DIC are sufficient to explain the variation of DOC. Among the forms of carbon, DOC was most strongly correlated with DIC and SOC (Table 8). Correlation of DOC with abiotic environmental factors shows that salinity and rainfall are positively and negatively correlated with DOC, respectively (Table 4).

There was a significant negative correlation (–0.557) between phytoplankton carbon and zooplankton carbon. Field results showed that in post-monsoon when phytoplankton blooms occur, the zooplankton population is moderately high. Then grazing pressure upon phytoplankton causes a decline in the number of phytoplankton. Zooplankton reach peak abundance in pre-monsoon. Subsequently, zooplankton abundance declines because high predation pressure reduces phytoplankton, and sometimes a phytoplankton bloom results at the end of monsoon.

Table 4. Correlation coefficients between governing factors of the estuarine system and total organic carbon (TOC), dissolved organic carbon (DOC), and particulate organic carbon (POC) in water ($p< 0.01$, * $p< 0.05$)**

	DO	Water salinity	Water temperature	Water pH	Rainfall
TOC	0.277	−0.813**	−0.100	−0.568*	0.721**
DOC	0.267	0.460	−0.312	0.273	−0.424
POC	0.231	−0.852**	−0.081	0.586*	0.748**

Table 5. Correlation coefficients of governing factors of the estuarine system with dissolved inorganic carbon (DIC), CO_2(aq), and HCO_3^{-1} ($p< 0.01$, * $p< 0.05$)**

	DO	Water salinity	Water temperature	Water pH	Rainfall
DIC	−0.314	0.813**	0.291	0.518*	−0.494
CO_2(aq)	0.313	−0.027	−0.096	−0.754**	−0.010
HCO_3^{-1}	−0.309	0.815**	0.291	0.506*	−0.490

Table 6. Correlation coefficients of total organic carbon with other forms of carbon in soil and estuarine water ($p< 0.01$, * $p< 0.05$)**

POC	DOC	DIC	SOC	SIC
0.983**	−0.234	−0.631*	−0.49	−0.328

Table 7. Correlation coefficients of particulate organic carbon with other forms of carbon in soil and estuarine water ($p< 0.01$, * $p< 0.05$)**

TOC	DOC	DIC	SOC	SIC
0.983**	−0.390	−0.734**	−0.529*	−0.309

Table 8. Correlation coefficients of dissolved organic carbon with other forms of carbon in soil and estuarine water ($p< 0.01$, * $p< 0.05$)**

TOC	POC	DIC	SOC	SIC
−0.234	−0.390	0.669*	0.562*	−0.160

Table 9. Key to abbreviations used in Figure 9

ACO_2	Aerial carbon dioxide
BNTH	Benthos
BU	Bacterial uptake
D	Detritivory
DBC	Dissolved bicarbonate
DO	Dissolved oxygen
DOC	Dissolved organic carbon
DCO_2	Dissolved free carbon dioxide
DIC	Dissolved inorganic carbon
DIFF	Diffusion of CO_2 at the air water interface
EXD	Exudation of phytoplankton
GPP	Gross primary productivity
GRAZ	Grazing of zooplankton upon phytoplankton
IDSO	Input from death of soil organisms
LB	Litter biomass
LC	Litter carbon
LEACH	Leaching
LOSS	Loss from the system
MD	Microbial degradation
MIN	Mineralization process
MIX	Mixing of sea water with estuarine water
MORT	Mortality
ORG A	Organic acid
pCO_2	Partial pressure of CO_2
PHOTO	Photosynthesis
PHYTO	Phytoplankton
POC	Particulate organic carbon
POL	Polymerization of DOC to POC
PRED	Predation of fish
RESP	Respiration
Rp	Redox potential
RR	River runoff
S	Sedimentation
SIC	Soil inorganic carbon
SINK	Sinking of phytoplankton
SOC	Soil organic carbon
SpH	pH of soil
St	Soil temperature
WpH	pH of water
Wt	Water temperature
ZOO	Zooplankton

CONCLUSIONS

Field data and analytical results suggest that the cycling of organic carbon in Hooghly estuary, along with the adjacent mangrove forest, is an integrated process of the soil and the waters of the estuarine system. Litter biomass is transformed to SOC; S_t and R_f are the major environmental factors controlling the fluctuations of SOC. Mineralization of SOC to SIC is governed by microbial activity, which is regulated by S_t. S_s, S_{pH}, R_p, R_f are controlling factors in the soils. SOC is transformed to DOC through leaching of organic acid. It is likely from our study that R_f and W_s play important roles in regulating the estuarine water chemistry of the Hooghly system. Death of organisms in soil and water enriches the SOC and POC pools, respectively.

With the aid of a thorough literature survey and field observations results a conceptual diagram has been worked out. This conceptual model will certainly aid in understanding the nature of the carbon cycle in this estuarine region and in assessing the mechanisms of transformation of different forms of carbon in the estuarine system. It will help to understand the fundamental seasonal processes associated with carbon cycling in mangroves and the adjacent estuarine ecosystem. Further study is needed to transform our conceptual model into a mathematical one. By running the mathematical model the sensitive parameters and key factors for controlling the organic carbon cycling will be determined, thereby helping to take the proper management strategy for restoring this luxuriant ecosystem.

ACKNOWLEDGMENTS

The authors (JM, SR and AC) are grateful to their respective departments (Department of Zoology & Department of Statistics) of Visva-Bahrati University for giving all sorts of facilities to conduct this research. One of the authors (JM) is thankful to University Grants Commission (UGC), Govt. of India, New Delhi for providing RFSMS fellowship as financial support to carry out this work and Institute of Environmental Studies and Wetland Management, Saltlake, Kolkata and West Bengal Pollution Control Board for giving the opportunity to carryout experiments on their laboratory. Authors are also thankful to Mr. Jalad Gayen for his assistance in the collection of samples during field surveys for this work.

REFERENCES

Abril, G., Noguueira, M., Etcheber, H., Cabeçadas, G., Lemaire, E. & Brogueira, M.J. (2002). Behaviour of organic carbon in nine contrasting European estuaries. *Estuar., Coast. Shelf S. 54*, 241–262.

Aerts, R. (2006). The freezer defrosting: global warming and litter decomposition rates in cold biomes. *J. Ecol. 94*, 713–724.

Adams, E.E., & Caldeira, K. (2008). Ocean storage CO_2. *Elements, 4*, 319–324.

Ahad, J.M.E., Barth, J.A.C., Ganeshram, R.S., Spencer, R.G.M. & Uher, G. (2008). Controls on carbon cycling in two contrasting temperate zone estuaries: The Tyne and Tweed, UK. *Estuar. Coast. Shelf S. 78*, 685–693.

Ahn, M.Y., Zimmerman, A.R., Comerford, N.B., Sickman, J.O. & Grunwald, S. (2009). Carbon Mineralization and Labile Organic Carbon Pools in the Sandy Soils of a North Florida Watershed. *Ecosystems 12*, 672–685.

Amon, R.M.W. & Benner, R. (1996). Bacterial Utilization of Different Size Classes of Dissolved Organic Matter. *Limnol. Oceanogr. 41*, 41–51.

Alongi, D.M., Boto, K.G. & Tirendi, F. (1989). Effect of exported mangrove litter on bacterial productivity and dissolved organic carbon fluxes in adjacent tropical nearshore sediments. *Mar. Ecol. Prog. Ser. 56*, 133–144.

Bhunia, A.B. (1979). *Ecology of tidal creeks and mudflats of Sagar Island (Sunderbans) West Bengal*. Ph.D Thesis, Calcutta University.

Biswas, H., Mukhopadhyay, S.K., De, T.K., Sen, S. & Jana T. K. (2004). Biogenic controls on the air-water carbon dioxide exchange in the Sundarban mangrove environment, northeast coast of Bay of Bengal, India. *Limnol. Oceanogr. 49*, 95–101.

Biswas, H., Dey, M., Ganguly D., De T.K., Ghosh S. & Jana T.K. (2010). Comparative analysis of phytoplankton composition and abundance over a two-decade period at the land–ocean boundary of a tropical mangrove ecosystem. *Estuar. Coasts 33*, 384–394.

Bolin B., Degens, E.T., Duvigneaud, P. & Kempe, S. (1979). The global biogeochemical carbon cycle. In B. Bolin, E.T. Degens, S. Kempe, & P. Kenter (Eds.), *The Global Carbon Cycle* (1st edition, pp. 1–56), Chichester,UK, John Wiley & Sons.

Boto, K.G., Alongi D.M. & Nott, A.L.J. (1989). Dissolved organic carbon-bacteria interactions at sediment-water interface in a tropical mangrove system. *Mar. Ecol. Prog. Ser. 51*, 243–251.

Butler, N.B. (1982). *Carbon dioxide equilibria and their applications*. NewYork: Addison-Wesley Publishing Co., Inc.

Bunt, J.S., Boto K.G., & Boto, G. (1979). A survey method for estimating potential levels of mangrove forest primary production. *Mar. Biol. 52*, 123–128.

Cassar, N., Laws, E.A. & Bidigare, R.R. (2003). Bicarbonate uptake by Southern Ocean phytoplankton. *Global Biogeochem. Cycles 18* Online serial: GB2003, doi:10.1029/2003GB002116.

Chattopadhyay, G.N. (1998). *Chemical analysis of fish pond soil and water*. Daya Publishing House, New Delhi, India.

Clark, D.R., & Flynn, K.J. (2000). The relationship between the dissolved inorganic carbon concentration and growth rate in marine phytoplankton. *Proc. Royal Soc. London Series B, 267*, 953–959.

Cole, J.J. (1999). Aquatic microbiology for ecosystem scientists: new and recycled paradigms in ecological microbiology. *Ecoystems 2*, 215–225.

Dafner, E.V. & Wangersky, P.J. (2002). A brief overview of modern directions in marine DOC studies Part II—Recent progress in marine DOC studies. *J. Environ. Monit. 4*, 55–69.

Ducklow, H.W. & Carlson, C.A. (1992). Oceanic bacterial production. *Advances in Microbial Ecol. 12*, 113–181.

Dittmar, T. & Lara, R.J. (2001b). Driving forces behind nutrient and organic matter dynamics in a mangrove tidal creek in North Brazil. Estuarine. *Coast. Shelf Sci. 52*, 249–259.

De, T.K., Ghosh, S.K., Jana, T.K. & Choudhury, A. (1991). Phytoplankton bloom in the Hooghly estuary. *Indian J. Mar. Sci. 20*, 134–137.

Falkowski, P.G. (1980). *Primary Productivity in the Sea*. Plenum Press, New York.

Frankignoulle, M., Abril, G., Borges, A., Bourge, I., Canon, C., Delille, B., Libert, E. & Theate, J. M. (1998). Carbon dioxide emissions from European estuaries. *Science 282,* 434–436.

Fell, J.W., Cefalu, R.C., Masters, I.M. & Tallman, A.S. (1975). Microbial activity in the mangrove (*Rhizophora mangle*) leaf detritus system. In: G.E. Walsh, S.C. Snedaker & H. J. Teas (Eds.), *Proceedings of International Symposium on Biology and Management of Mangroves, Honolulu, 1974,* vol. II (pp 23–42), University of Florida, Gainesville Florida, 23–42.

Friedler, E., Juanico, M. & Shelef, G. (2003). Simulation model of wastewater stabilization reservoirs. *Ecol. Eng. 20,* 121–145.

Gattuso, J.P., Frankignoulle, M., & Wollast, R. (1998). Carbon and carbonate metabolism in coastal aquatic ecosystems. *Ann. Rev. Ecol. Systematics 29,* 405–434.

Greenberg, A.E., Clesceri, L.S. & Eaton A.D. (1992). *StandardMethods for Examination of Water and Wastewater.* American Public Health Association, Washington, D. C., USA.

Ghosh, P.B., Singh, B.N., Chakraborty, C., Saha, A., Das, R.L. & Choudhury, A. (1990). Mangrove litter production in the tidal creek of Lothian island, Sunderbans, India. *Indian J. Mar. Sci. 19,* 292–293.

Gonneea, M.E., Paytana, A. & Herrera-Silveira, J. (2004). Tracing organic matter sources and carbon burial in mangrove sediments over the past 160 years. *Estuar. Coast. Shelf S. 61,* 211–227.

Guo, L., Santschi, P.H. & Warnken, K. (1995). Dynamics of dissolved organic carbon (DOC) in oceanic environments. *Limnol. Oceanogr. 40,* 1392–1403.

Gupta, P.K. (2002). *Methods in Environmental Analysis: Water, Soil and Air.* Agrobios (India), Jodhpur, India, 203–217.

Holmer, M. & Olsen, A. B. (2002). Role of decomposition of mangrove and seagrass detritus in sediment carbon and nitrogen cycling in a tropical mangrove forest. *Mar. Ecol. Prog. Ser. 230,* 87–101.

Howland, R.J.M., Tappin, A.D., Uncles, R.J., Plummer, D.H. & Bloomer, N. J. (2000). Distributions and seasonal variability of pH and alkalinity in the Tweed Estuary, UK. *Sci. Total Environ. 251–252,* 125–138.

Insam, H. (1990). Are the soil microbial biomass and basal respiration governed by the climatic regime? *Soil Biol. Biochem. 22,* 525–532.

Joos, F., Raynaud, D. & Wigley, T. (1996). Radiative forcing of climate change. In: *Climate Change 1995,* (pp 76–86), Cambridge University Press, Cambridge, UK.

Jorgensen, L.A., Jorgensen, S.E. & Nielsen, S.N. (2000). *ECOTOX, Ecological Modelling and Ecotoxicology.* Elsevier, Netherlands.

Jumars, P.A., Penry, D.L., Baross, J.A., Perry, M.J. & Frost, B. W. (1989) Closing the microbial loop : dissolved carbon pathway to heterotrophic bacteria from incomplete ingestion, digestion and absorption in animals. *Deep-Sea Res. 36,* 483–495.

Kätterer, T., Reichstein, M., Andrén, O. & Lomander, A. (1998). Temperature dependence of organic matter decomposition: a critical review using literature data analysed with different models. *Biol. Fert. Soils 27,* 258–262.

Kemp, W.M., Sampou, P.A., Garber, J., Tuttle, J., Boynton, W.R. (1992). Relative roles of benthic versus planktonic respiration in the seasonal depletion of oxygen from bottom waters of Chesapeake Bay. *Mar. Ecol. Prog. Ser. 85,* 137–152.

Kirschbaum, M.U.F. (2000). Will changes in soil organic carbon act as a positive or negative feedback on global warming? *Biogeochem. 48*, 21–51.

Kirschbaum, M.F. (1995). The temperature dependence of soil organic matter decomposition, and the effect of global warming on soil organic C storage. *Soil Biol. Biochem. 27*, 753–760.

Kretzschmar, R., Hesterberg, D., Sticher, H. (1997). Effects of adsorbed humic acid on surface charge and flocculation of kaolinite. *Soil Sci. Soc. Am. J. 61*, 101–108.

Kulinski, K. & J. Pempkowiak. (2008). Dissolved organic carbon in the southern Baltic Sea: Quantification of factors affecting its distribution. *Estuar. Coast. Shelf S. 78*, 38–44.

Li-xia, Y., Jian-jun, P. & Shao-feng, Y. (2006). Predicting dynamics of soil organic carbon mineralization with a double exponential model in different forest belts of China. *J. Forestry Res. 17*, 39–43.

Lloyd, J. & Taylor, J.A. (1994). On the temperature dependence of soil respiration. *Functional Ecol. 8*, 315–323.

Machiwa, J.F. & Hallberg R.O. (2002). An empirical model of the fate of organic carbon in a mangrove forest partly affected by anthropogenic activity. *Ecol. Model. 147*, 69–83.

Mandal, S., Ray, S., Sarkar, A. & Ghosh, P.B. (2009). Degradation of Mangrove Litter and Its Contribution As Dissolved Inorganic Nitrogen to the Adjacent Estuary of Sagar Island, Sunderban Mangrove Ecosystem, India. In: J. N. Metras (Ed.), *Mangroves Ecology, Biology and Taxonomy* (1st edition, Chapter 9, pp. 1–19) Nova Science Publishers.

Mantoura, R.E.C. & Woodward, E.M.S. (1983). Conservative behavior of riverine dissolved organic carbon in the Severn estuary: Chemical and geochemical implications. *Geochim. Cosmochim. Acta 47*, 1293–1309.

Mukhopadhyay, S.K., Biswas, H., De, T.K. & Jana, T.K. (2006). Fluxes of nutrients from the tropical river Hooghly at the land–ocean boundary of Sunderbans, NE Coast of Bay of Bengal. India. *J. Mar. Sys. 62*, 9–21.

Murrell, M.C. & Hollibaugh, J.T. (2000). Distribution and composition of dissolved and particulate organic carbon in northern San Francisco Bay during low flow conditions. *Estuar. Coast. Shelf Sci. 51*, 75–90.

Newell, S. Y. (1984). Carbon and nitrogen dynamics in decomposing leaves of three coastal marine vascular plants of the subtropics. *Aq. Botany 19*, 183–192.

Odum,W.E.; Heald, E.J. 1975. The detritus based foodweb of an estuarine mangrove community. In: L.E. Cronin (Ed.). *Estuarine Research,* (pp. 265–286), Academic Press, NewYork, USA.

Packard, T., Chen, W., Blasco, D., Savenkoff, C., Vezina, A.F., Tian, R., Amand, L. St., Roy, S.O., Lovejoy, C., Klein, B., Therriault, J.C., Legendre, L. & Ingram, R.G. (2000). Dissolved organic carbon in the Gulf of St. Lawrence. *Deep-Sea Res. 47*, 435–459.

Platt, T., Sathyendranath, S. & Ravindran, P. (1990). Primary production by phytoplankton: analytic solutions for daily rates per unit area of water surface. *Proc. Royal Soc. London 241B*, 101–111.

Raich, J.W. & W.H. Schlesinger. (1992). The global carbon dioxide flux in soil respiration and its relationship to vegetation and climate. *Tellus 44B*, 81–99.

Ram, A.S.P., Nair, S. & Chandramohan, D. (2003). Seasonal shift in net ecosystem production in a tropical estuary. *Limnol. Oceanog. 48*, 1601–1607.

Ray, S., (1987). *Ecology of littoral larval dipterans (Arthropoda: Insecta) of the mangrove ecosystem of Sundarbans, India.* PhD Thesis, Calcutta University.

Ray, S. & **Straskraba,** M. (2001). The impact of detritivorous fishes on a mangrove estuarine system. *Ecol. Model. 140,* 207–218.

Ribes, M., Coma, R. & Gili, J.M. (1999). Seasonal variation of particulate organic carbon dissolved organic carbon and contribution of microbial communities to the live particulate organic carbon in a shallow near-bottom ecosystem at the Northwestern Mediterranean Sea. *J. Plankton Res. 21,* 1077–1100.

Robertson, A.I. (1988). Decomposition of mangrove leaf litter in tropical Australia. *J. Exp. Mar. Biol. Ecol. 116,* 235–247.

Saha, S. & Choudhury, A. (1995). Vegetation analysis of restored and natural mangrove forest in Sagar Island, Sundarbans, east coast of India. *Indian J. Mar. Sci. 24,* 133–136.

Sebastian, R. & Chacko, J. (2006). Distribution of organic carbon in tropical mangrove sediments (Cochin, India). *Int. J. Environ. Studies 63,* 303–311.

Shiels, A.B. (2006). Leaf litter decomposition and substrate chemistry of early successional species on landslides in Puerto Rico. *Biotropica 38,* 348–353.

Singh, J.S. & Gupta, S.R. (1977). Plant decomposition and soil respiration in terrestrial ecosystems. *Botanical Rev. 43,* 449–526.

Smith, S.V. & Hollibaugh, J.T. (1993). Coastal metabolism and the oceanic organic carbon cycle. *Rev. Geophysics 31,* 75–89.

Smith, S.V. & Hollibaugh, J.T. (1997). Annual cycle and interannual variability of ecosystem metabolism in a temperate climate embayment. *Ecol. Monogr. 67,* 509–533.

Steinke, T.D., Holland, A.J. & Singh, Y. (1993a). Leaching losses during decomposition of mangrove leaf litter. *S. African J. Botany 59,* 21–25.

Steinke, T.D., Rajh, A. & Holland, A.J. (1993b). The feeding behaviour of the red mangrove crab *Sesarma meinerti* De Man, 1887 (Crustacea: Decapoda: Grapsidae) and its effect on the degradation of mangrove leaf litter. *S. African J. Mar. Sci. 13,* 151–160.

Strickland, J.D.H. & Parsons, T.R. (1972). *A Practical Handbook of Seawater Analysis* (2nd edition). Bulletin of the Fisheries Research Board of Canada.

Sugiyama, Y., Anegawa, A., Kumagai, T., Harita, Y., Hori, T. & Sugiyama, M. (2004). Distribution of dissolved organic carbon in lakes of different trophic types. *Limnol. 5,* 165–176.

Takahashi, T., Olafsson, J., Goddard, J.G., Chipman, D.W. & Sutherland, S.C. (1993). Seasonal variations of CO_2 and nutrients in the high latitude surface oceans: a comparative study. *Global Biogeochem. Cycles 7,* 843–878.

Tortell, P.D. & Morel, F.M.M. (2002). Sources of inorganic carbon for phytoplankton in the eastern subtropical and equatorial Pacific Ocean. *Limnol. Oceanogr. 47,* 1012–1022.

Tranvik, L.J. (19920. Allochthonous dissolved organic matter as an energy source for pelagic bacteria and the concept of the microbial loop. *Hydrobiol. 229,* 107–114.

Twilley, R.R. (1985). The exchange of organic carbon in basin mangrove forests in a southwest Florida estuary. *Estuar. Coast. Shelf Sci. 20,* 543–557.

Wang, X. & J. Veizer. (2000). Respiration–photosynthesis balance of terrestrial aquatic ecosystems, Ottawa area, Canada. *Geochim. Cosmochim. Acta 64,* 3775–3786.

Wetzel R.G. (2001). *Limnology:Lake and River Ecosystems* (3rd edition). Academic Press, San Diego, California, USA.

Winkler, J.P., Cherry, R.S. & Schlesinger, W.H. (1996). The Q_{10} relationship of microbial respiration in a temperate forest soil. *Soil Biol. Biochem. 28,* 1067–1072.

Zweifel, U.L., Wikner, J., Hagstrom, A., Lundberg, E. & Norrman, B. (1995). Dynamics of dissolved organic carbon in a coastal ecosystem. *Limnol. Oceanogr. 40,* 299–305.

Chapter 7

LONG TERM NITRATE TRANSPORT THROUGH WATERSHEDS TO ESTUARIES ON CAPE COD: CLIMATE AND TRANSIENT SCALES

George A. Seaver

Cataumet, Massachusetts, US

ABSTRACT

Nutrients in precipitation, in the groundwater of two subwatersheds and in two adjacent estuaries on Cape Cod, Massachusetts were measured regularly year-round from 1988 to the present. This 22-year data set captured low frequency (climate scale) equilibrium changes in nutrient levels, as well as transient changes brought about by an unexpected NO_3 increase to one of the subwatersheds-estuary systems. Both estuaries exhibit low background levels of NO_3 (< 0.2 mg/L or ~3 μM), which increase 6-fold over 3–5 year cycles. Dissolved oxygen concentration remains at 70–90% saturation, summer to winter. The annual average NO_3 concentration in one subwatershed-estuary system was correlated strongly with regional driving forces (low frequency precipitation and water table height), but was not correlated with anthropomorphic sources. Nitrate in the second subwatershed-estuary system exhibited the same correlation with regional driving forces until 2004, when a large, unexpected (20-fold) increase in groundwater NO_3 concentration occurred in this estuary, apparently introduced by an event in the up-gradient watershed. The correlation with regional driving forces disappeared and, after a year's delay, high frequency (3–4 month) NO_3 variability appeared during the winter collapse of the biological NO_3 demand; this cycle lasted for four years. The estuary then returned to its pre-2004 NO_3 level, even with the groundwater NO_3 input remaining high. The transient intervening high frequency behavior is consistent with "regime shift" concepts in their pre-shift stage; if so, the groundwater NO_3 input conditions for a full eutrophic regime shift remain to be found. The generalized case would be the following: in a temperate estuary seasonal NO_3 dynamics are statistically independent of long-term trends and cycles, much like weather is distinct from climate or turbulence from mean flow. The seasonal time scale then can be separated from long- term climate events. However, sudden events can destroy this scale separation and its simplification, with summer algae dynamics then leading to winter NO_3 variability after the collapse of the seasonal biological NO_3 demand. Estuary dynamics can bring about a return to pre-transient conditions, even with continued high NO_3 input, but this latter process is not yet well understood.

INTRODUCTION

Nutrient enrichment in estuaries, most notably by NO_3, has become a matter of public interest over the past two decades, both from a scientific and a political perspective. This situation is particularly evident in the heavily-populated areas along the Atlantic coast from Chesapeake Bay to Boston Harbor. Consequently, reliable, long term nitrate data, as well as the responses of the local estuaries to both slow and abrupt changes in these quantities, is needed. The present estuarine area of study is south of Boston Harbor, is neither pristine nor severely degraded, and therefore makes a good laboratory in which to consider indicators of ecological transitions. The ability of an estuary to process incoming nitrate in a biologically useful manner—termed an equilibrium state in this study—varies over both time and space in the estuaries of Buzzards Bay, Vineyard Sound and the Cape Cod Bay region. A few estuaries are severely affected by nutrient loads, much like Chesapeake Bay, but most are not. Some estuaries temporarily move into a "nutrient loaded" condition, and then recover. Estuaries in the latter category, on the eastern side of Buzzards Bay and the western boundary of Cape Cod, Massachusetts, are the subjects of the present study.

A unique measurement program, extending over the past 22 years on Cape Cod, Massachusetts, by sampling nutrient concentrations in two estuaries, transport from up-gradient groundwater, and from watershed precipitation, has provided insight into long term NO_3 cycles and correlations. Fortuitously, short term, transient NO_3 events in one estuary were observed during the study, so that the other estuary served as a natural control. Low frequency correlations reveal regional equilibrium with near-climatic scale driving forces, and the transient responses reveal that an estuary can adjust to a new, much higher level of NO_3 input, albeit from a low starting point. Phosphate, nitrite and ammonia time series also were collected and examined, but did not exhibit the correlations and coherence observed in the NO_3 results.

Year round, regularly sampled measurements, beginning in 1988, were made of nutrients in precipitation, in the groundwater of two subwatersheds, and in two adjacent estuaries located on upper Buzzards Bay, Massachusetts. The measurement series separated into two distinct periods: 1988–2003, when annual NO_3 averages had a high correlation with NO_3 load, regional low frequency water table variation, and precipitation changes; and 2004–2009 when an abrupt, large increase in NO_3 input to one of the estuaries created a high frequency, transient response in that estuary, thereby overwhelming the low frequency correlations. This chapter considers first the low frequency, regional equilibrium period, 1988–2003; then discusses the high frequency, transient period 2004–2009; and finally, offers an empirical explanation from numerical estuary models. Equilibrium in estuary dynamics occurs when the average annual NO_3 change is linearly related to the change in average annual NO_3 load. This implies a balance (considered in the third section of this chapter) between biological NO_3 uptake and regeneration.

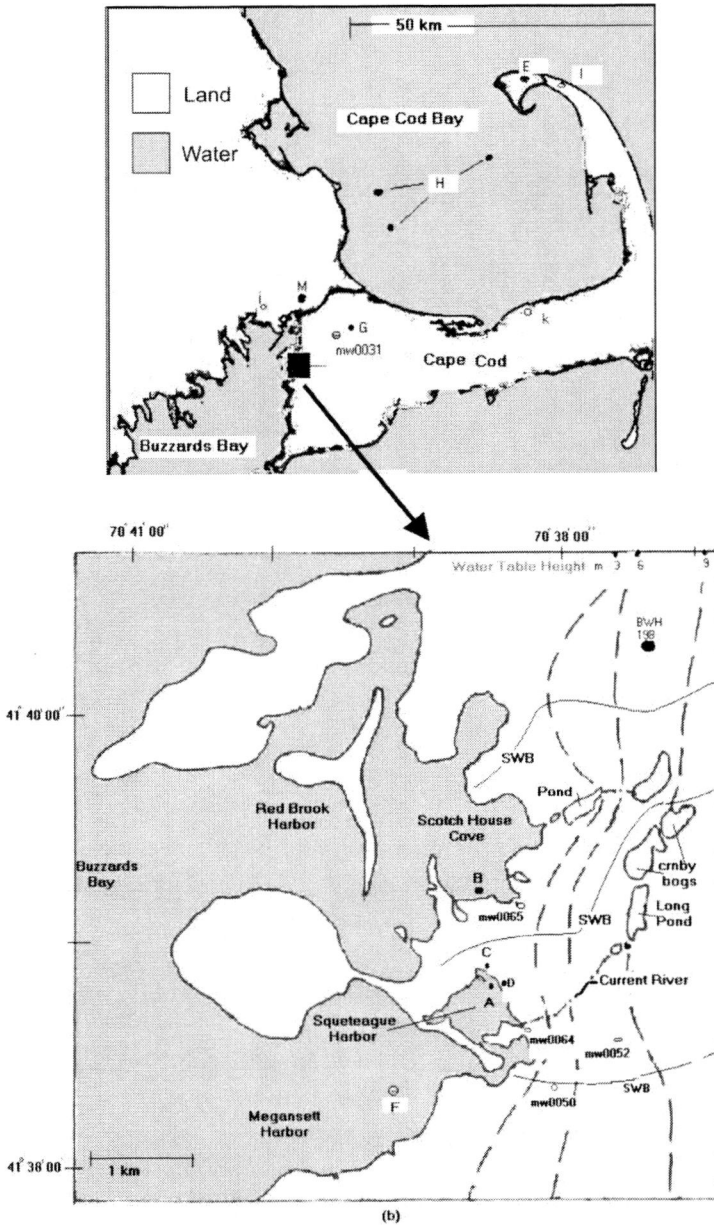

Figure 1. The watershed, water table and estuaries adjoining Buzzards Bay, Massachusetts. The main map shows the east coast of Massachusetts and Cape Code. The Provincetown NADP precipitation sampling station is point E, and the East Wareham NWS precipitation station is point M. The outer Megansett Harbor point is F, the top of the Sagamore lens is point G and the Cape Cod Bay sampling points are H. Inset: Buzzards Bay and the Squeteague Harbor sampling location (point A), and 0.8 km north the Scotch House Cove sampling location (point B) in Red Brook Harbor. The coastal groundwater sampling locations are points C and D. The water table contours (m above msl) are dashed lines. The thin solid lines labeled SWB indicate the subwatershed boundaries that separate the ground water feeding Scotch House Cove from Squeteague Harbor and from Megansett Harbor. The 5 AFCEE groundwater wells are also shown as mw0031, 0065, 0064, 0052, and 0050. The USGS water table measurement wells are points j, k, l and BWH 198. BWH 198 is 3.6 km north northeast of the Squeteague Harbor sampling location (point A) and 2.8 km northeast of the Scotch House Cove sampling location (point B) in Red Brook Harbor.

The experimental area is shown in Figures 1a and 1b. The estuaries and groundwater of this study are in the terminal moraine of Buzzards Bay, are shallow (1–2 m at mean low water), well mixed and are in a lightly populated area (~ 1.25 house per hectare). From Bowen et al. (2007), for population densities greater than 2.5 people per hectare, dissolved inorganic nitrogen (DIN) is correlated with population increase, and below this value it is uncorrelated. The region of this study is at the boundary of this uncorrelated category. However, these estuaries are also subject to large, periodic multi-year increases in NO_3 concentrations from causes other than population, thus we can observe an estuary's response to a slowly varying change in groundwater NO_3 input. Dissolved oxygen concentrations over this period remained at 70–90% of saturation, summer to winter. Squeteague Harbor (point A in figure 1b), Scotch House Cove (point B) and the groundwater input (point C) have been monitored quarterly since 1988 for many parameters, including precipitation, water table height, and DIN. These results were reported in Seaver & Kuzirian (2007) and in Seaver (2010).

Estuary in Equilibrium, 1988–2003

For the period 1988 to 2003, strong correlations were observed between annually averaged NO_3, water table height and winter precipitation. No NO_3 trend was observed over this 16-year period; that is, on an annual basis the inputs and losses were equal and the estuary was in equilibrium with its surroundings. For the 1988–2003 period the NO_3 average for both estuaries was 0.22 mg/L (3.5 µM) and for the groundwater it was 0.59 mg/L (10 µM). Large multi-year excursions from these values were observed, for example between 1996 and 1999 the annual average Squeteague Harbor NO_3 concentration increased by a factor of 3.5 above the 16-year average.

Figure 2 presents the full, un-averaged results of the NO_3 measurements taken at approximately uniform intervals over 22 years in Squeteague Harbor (SQT, station A), Scotch House Cove (SHC, station B) between 1988 and 2010, and at the ground water discharge point at Squeteague Harbor (station C, Figure 1b) between 1991 and 2003. First, in order to see if there were long-term trends in NO_3, 2–3 year NO_3 averages of the baseline were calculated at the beginning and at the end of the two periods. The average value of NO_3 in Squeteague and Scotch House Harbors (combined) for the 2½ years from March 1988 to August 1990 was 0.114 mg/L, for 2000–2002 it was 0.115 mg/L, and for 2008–2010 it was 0.092 mg/L. No trend in NO_3 concentration for the 14-year or 18-year period was observed. Human population increase in the Cataumet area (Figure 1b) was monotonic: 26% for year-round and about 30% for summer residents over this period. If the above analysis had been done after 10 years (in 1997), a misleading rising trend of 0.6 mg/L would have been reported (see Figure 3). The record length must exceed 10 years to include the low frequency variability.

Figure 3 presents the annually averaged NO_3 concentration for Squeteague and Scotch House Cove Harbors (combined), the ground water NO_3 at Squeteague (station C), and the inland water table height over the 16 years of the first period. The results were quite interesting with respect to multi-year, periodic variability, showing a 1997 NO_3 concentration average of 1.4 mg/L in ground water, some 14 times the 1992 average. Natural cycles of variable period and amplitude in NO_3 concentration, much like the variability seen in other climate scale phenomena, will be discussed later. The NO_3 variability shown in Figure 3 occurred in both the two harbors and the ground water, with the ground water showing the largest periodic amplitude. The groundwater NO_3 absolute value was also generally greater than the harbor values and all three were strongly correlated.

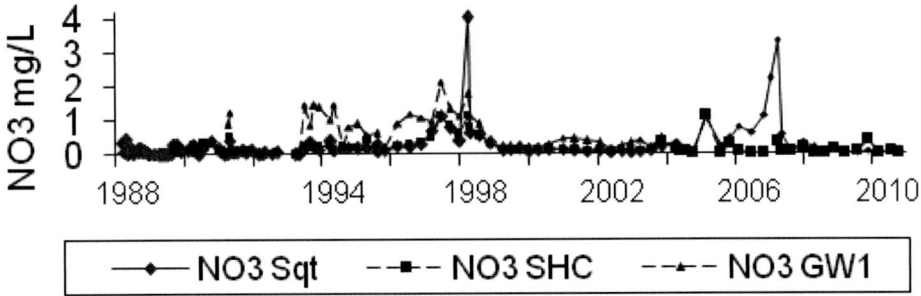

Figure 2. Dissolved NO_3 concentrations in mg/L for Squeteague Harbor (SQT), Scotch House Cove (SHC) and the ground water station C (GW1) plotted against time from March 1988 to November 2010. The Scotch House Cove monitoring began in September 1989, and groundwater in 1991. The frequency of sampling after 1993 was approximately every 3 months with a total of 118 measurements for Squeteague Harbor, 85 for Scotch House Cove and 35 for groundwater. SQT and SHC diverge in 2005 and come back together in 2008.

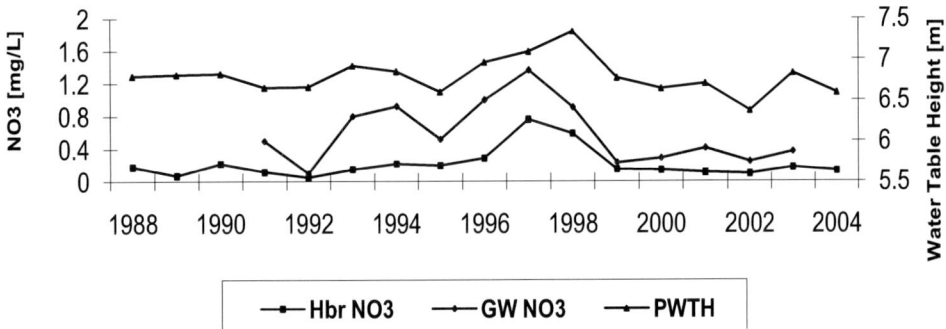

Figure 3. Annually averaged NO_3 concentration time series for Squeteague Harbor and Scotch House Cove (combined as Hbr), for the ground water station C (GW), and for the annually averaged water table height at BWH 198 (PWT) for the years 1988 to 2004. The error bars for the water table height are ±0.61 cm (not shown).

The annually averaged (low frequency) nitrate load to the estuaries exhibits the familiar relationship with the estuary nitrate concentration found in the literature. Valiela et al. (2004) showed that *"predictions of mean annual concentrations of DIN in the Waquoit Bay estuaries consistently increased as ... modeled annual land-derived N load entering the different estuaries increased."* Their graph of the N load and the estuary DIN concentration has a linear correlation of 0.97. This result is supported by the earlier work of McClelland et al. (1997), McClelland & Valiela (1998) and Valiela et al. (2000). Valiela et al. (1997), defined the nitrogen loading to estuaries as *"the sum of atmospheric, fertilizer, and wastewater nitrogen that enters the mosaic of land covers, and survives the complex gauntlet of loss that occurs on the watershed surface, the vadose zone, the aquifer, and in freshwater bodies, to emerge at the seepage face on the estuary shore"*. In the present case the N load is taken as the nitrate load and is approximated at the seepage face by the water table height annual changes above a specified reference level multiplied by the groundwater nitrate concentration using Darcy's Law for groundwater velocity.

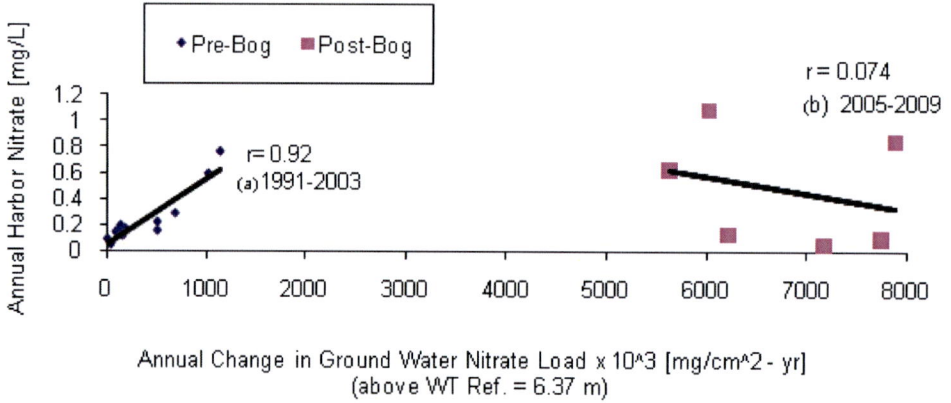

Figure 4. Annually averaged combined Squeteague and Scotch House Cove (Harbor) NO_3 concentration plotted against the input ground water NO_3 flow rate change: (a) between 1991 to 2003 (the equilibrium period), and (b) between 2005 and 2009 (the inequilibrium period). The input flow rate was calculated using the annually averaged water table height (above 6.37 meters) and the ground water NO_3 concentration. The correlation coefficient for (a) was 0.92 and for (b) was 0.074.

The regression curves shown in Figure 4 are an important result. For the first period (1991–2003), the curve is remarkably linear with ($r = 0.92$; 95% confidence limits 0.75–0.98). Thus, at climatic scales and under equilibrium conditions, the harbor NO_3 concentration responds linearly to the input of the ground water NO_3 load (flow rate). The results of Figure 4 are supported by the findings in the literature, in particular, of Valiela et al. (2004). However, the slope and zero load intercept of Figure 4 differ from those reported by Valiela et al. (2004), which included NH_4, the labile portion of dissolved organic nitrogen, and direct atmospheric deposition. Also, the Valiela et al. (2004) input was normalized over the estuary surface area, rather than over the groundwater input cross section, as was done here. The important agreement is that for annual averages in an equilibrium system there is a linear relationship between nitrogen load input and estuary DIN concentration; the relationship is statistically independent of the seasonal dynamics. With negligible or slowly changing long-term trends, the variables are in equilibrium with their driving forces. For the second period of Figure 4 (2005–2009) the variables are not in equilibrium, and the correlation is low. This will be considered in the second section.

CROSS-WATERSHED CORRELATIONS

The correlation coefficient of the full (un-averaged) NO_3 concentration data from Figure 2 between Squeteague Harbor and Scotch House Cove for the 1988 to 2003 period was 0.92, with each estuary being in a separate subwatershed. The correlation was 0.93 in the case of the annual average NO_3 data. Consequently, in subsequent graphs these two series appear as one curve labeled "harbor nitrate". However, as each of these harbors was fed from different subwatersheds, their separate results represent two independent statements. Figure 3 includes the annually averaged water table height with the harbors and groundwater NO_3 for 1988–2003, and a visual inspection indicates that they are strongly correlated. The correlation coefficient between the groundwater annually averaged NO_3 concentrations and that of both Squeteague and Scotch House Cove Harbors was 0.81 (p = 0.0001).

The annually averaged NO_3 concentration in these two harbors was also strongly correlated with the averaged up-gradient water table height. The monthly water table height (Figure 5) was measured at the USGS monitoring well shown in Figure 1 as BWH 198. This station was located 3.2 km north and 1.6 km inland from the Squeteague Harbor monitoring station (A) and north of the subwatersheds of stations A and B. Figure 5 shows monthly, seasonal and weak inter-annual variations, and, just as for the raw harbor NO_3 data of Figure 2, the Figure 5 shows how the low frequency signal is obscured by seasonal fluctuations. Thus, annual averaging was necessary in order to look at the low frequency phenomenon (Figure 3). The correlation coefficient between the annually averaged water table height and the annually averaged ground water NO_3 concentration was 0.74 (p < 0.001). An interesting result here was that, at interannual time scales, increasing recharge water table height was associated with increased NO_3 concentrations, contrary to what some previous investigators found at seasonal time scales.

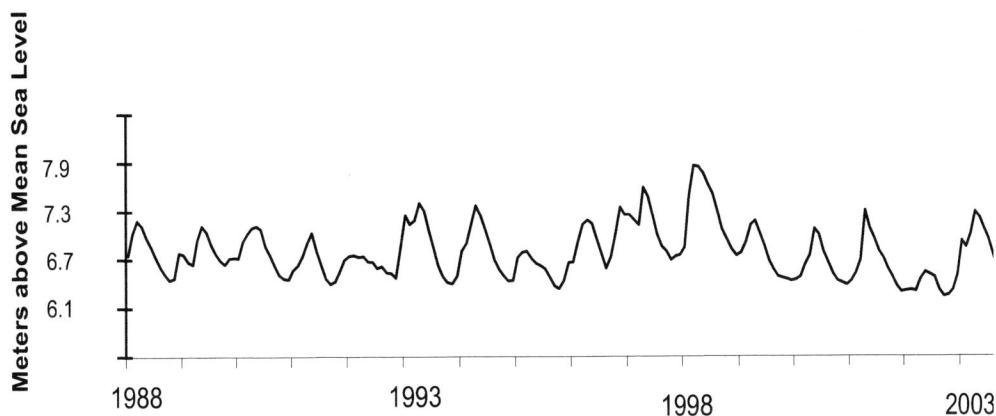

Figure 5. Unaveraged water table height at USGS well BWH 198 1988–2004. There were 204 monthly measurements.

The linear correlation coefficient between the annual water table height (WTH) and the down-gradient annual average NO_3 in the combined estuaries was 0.76 (p = 0.0001). These correlation coefficients and significance levels were strong for marine processes. However, at shorter averaging times these correlations decreased. That is, the climate scale changes were decoupled from the chemically reactive NO_3 changes occurring at seasonal time scales. This is an important result that will be discussed in detail next.

SHORTER TIME SCALES – DECOUPLING OF CLIMATE AND SEASON

Correlations of NO_3 with WTH decreased dramatically at shorter time scales. This result suggests a time scale separation and independence between the climatic scale of interest in this study and the biologically-reactive seasonal time scale. To explore this further, the seasonal-scale variations of NO_3 and WTH were factored out by subtracting the annual averages from the raw monthly real or interpolated data. As stated previously, the annual WTH and ground water NO_3 correlation was 0.74, but when only the seasonal scale was considered this coefficient dropped to an insignificant 0.06. In the case of the more local NO_3 variations in the coastal ground water, Squeteague Harbor and Scotch House Cove, the correlations declined from 0.8 to the 0.6 range. A parallel can be drawn to the separation of time scales found in weather and climate or turbulence and mean flow. For example, if one were to look at the velocity at two points in a turbulent flow that were separated by greater than the turbulent (eddy) length scale, at short times they would not be correlated. But if the velocity was averaged for a time greater than the time scale of the turbulence, then you would begin to see mean velocities and the 2-point correlation would grow. A change in the driving force would show up at both points, but would not be noticed in the turbulent velocity correlations. Perhaps this construct could be applied to the NO_3 transport at climatic scales, with the seasonal scale then becoming a parameter to the mean values of the longer-term process, as is done with climate prediction models. At any rate, if this scale separation holds, one does not have to solve one scale to understand the other.

INDEPENDENT UP-GRADIENT AND SEAWARD CORRELATIONS

In summary, it has been established that locally the low frequency NO_3 concentrations in Squeteague Harbor and, independently, in Scotch House Cove were strongly correlated with the NO_3 in the ground water at the coast and with the up-gradient water table height. An important question is how far up-gradient in the groundwater and seaward toward the bays do these NO_3 correlations persist. Because of a plume coming from the landfill on a military reservation 5.6 km up-gradient, there are about 100 monitoring wells extending from the coast easterly beyond the landfill to a distance of 6.4 km inland, nearly to the top of the ground water lens (point G of Figure 1a). These wells were monitored for NO_3 over the period 1997–2002 by the Air Force Center for Engineering and the Environment (AFCEE) on the military reservation. To assess the inland penetration of the observed coastal NO_3 periodicity, I compared the time series at the coast (Figure 3) with that of the AFCEE well 4.2 km inland from the coast (well mw 0031 in Figure 1a). The time series of 4.5 years was considerably shorter than the 16 years of the coastal ground water series, and trend information was not available. It was found that the period of NO_3 variability was the same 4.0 km inland as it was at the coast, the periodic amplitude was less (at 40% as shown in Figure 6), and the correlation with the coast was 0.47 ($p = 0.08$). The likely reason this correlation was less than for the coastal and harbor data is that the AFCEE time series were much shorter.

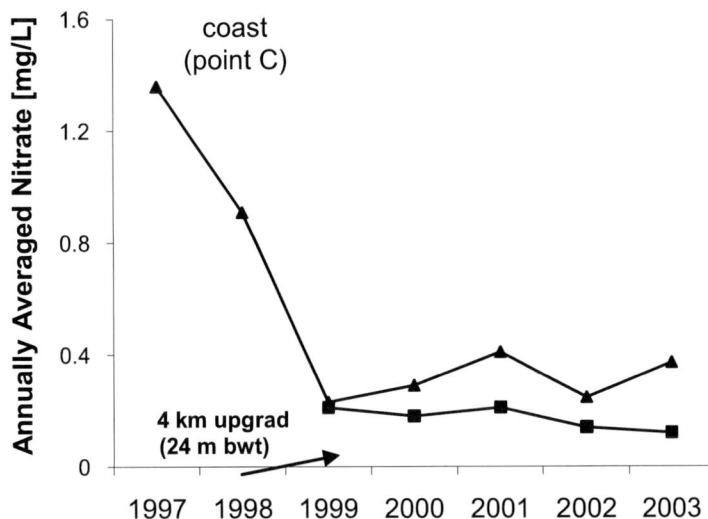

Figure 6. Ground water NO_3 concentrations at the coast (point C) and 4.2 km inland. The inland monitoring well (mw0031) was sampled by the AFCEE. The inland sample was taken at a well screen depth of 24 meters below the water table. The correlation between the coast and inland curves for this short period (4.5 years) was $r = 0.5$.

These correlations are complemented by another independent data set from Cape Cod Bay (CCB, 3 stations H in Figure 1a), taken by the MWRA (M. Mickelson, personal communication 2010) between 1992 and 2009 as part of the Boston Harbor outfall pipe-monitoring project. These results are shown in Figure 7, which compares the annual averages of stations H and the combined harbor NO_3 from stations A and B. The low frequency signals exhibited a multi-year cycle, and the correlation between this NO_3 time series and that of Squeteague Harbor and Scotch House Cove in 1992–2003 was 0.8 (p = 0.002). After 2003 there was a slightly increasing NO_3 trend, a likely result of the relocation of the Boston sewer outfall pipe out to Massachusetts Bay. Also shown in Figure 7 from the Buzzards Bay side of Squeteague is the annually averaged NO_3 concentration from station F in Megansett Outer Harbor (Figure 1b) for the period 1987–1998, and for SHC for the later period, 2005–2010, of relevance to the later inequilibrium section. Turner et al. (2000) investigated 8 stations in Buzzards Bay over the period 1987 to 1998 to determine what effect a sewer outfall pipe in New Bedford had on the bay proper. The station F data set was part of that work, and showed that its average NO_3 concentration was lower than Squeteague station A, at 0.05 vs. 0.115 mg/L, as would be expected from a harbor open to Buzzards Bay, and also showed no NO_3 trend for 1987–1998. The low frequency nature of the results in Figure 7 is what is of interest here, confirming the periodic behavior of NO_3 concentration. The station F Megansett data had a correlation of –0.74 with Squeteague and Scotch House Cove harbors (p = 0.002), but was not in phase with them (Figure 7). As Figure 1 indicates, the subwatershed for Megansett Harbor was separate from those of Squeteague and Scotch House Cove Harbors, but the reason for Megansett's 180° phase lag remains to be determined. The correlated, low frequency behavior of NO_3 presented to this point now extends over two separate bays, two estuaries and a significant portion of the groundwater between them.

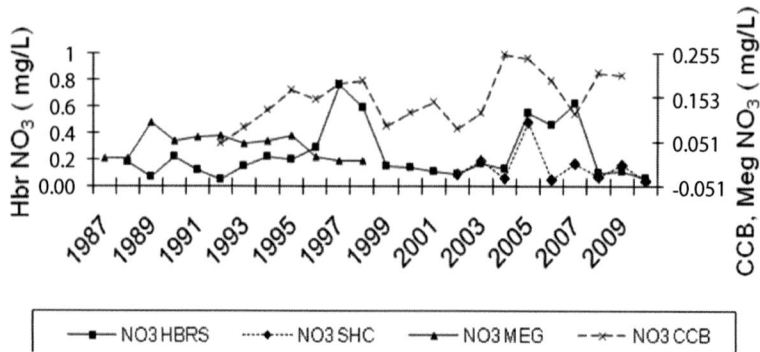

Figure 7. Annually averaged NO_3 concentration time series for Scotch House Cove, for Squeteague Harbor and Scotch House Cove combined as Hbr, for Cape Cod Bay (stations H) and for outer Megansett Harbor (station F), 1987–2010.

REGIONAL CORRELATIONS

How far these correlations persist beyond the local groundwater lens and bays is the regional question. The answer requires regional WTH and precipitation data. The most complete source of precipitation data for this area is from NADP at the National Seashore in Truro, MA (Portnoy 2002). This site is 64 km NE of BWH 198 (point E in Figure 1a). The NADP results showed that NO_3 concentrations in precipitation ranged from 0.7 to 1.3 mg/L annually and 0.6 to 2.1 mg/L in winter between 1988 and 2003. The correlation of the averaged winter precipitation NO_3 concentration with the Squeteague Harbor NO_3 of Figure 3 was –0.5; the negative sign meant that when the NO_3 concentration of the infiltration precipitation increased the NO_3 concentration in the harbors decreased, as a dilution effect of precipitation (Jordan et al. 2000).

A continuity investigation of the annual NO_3 load of precipitation, known by NADP as the annual deposition rate (kg ha^{-1} yr^{-1}), further supported this conclusion. Figure 4 showed that on the system output side, the harbor NO_3 concentration was strongly correlated ($r = 0.92$) with the groundwater NO_3 flow rate (deposition rate equivalent) for the 1988–2003 period. But on the input side this did not prove to be the case. The correlation between the winter NO_3 deposition of the precipitation and the annual groundwater NO_3 flow rate was 0.14 ($p = 0.35$) suggesting independence of these variables. It is clear that NO_3 deposition by precipitation (at Truro) was not the source of the ground water NO_3 load described earlier and in Figure 4.

Given the preceding arguments, a correlation should be seen between the precipitation *amount,* WTH, and NO_3 concentrations in Squeteague Harbor and Scotch House Cove. The nearest monitoring station for precipitation amount was the National Weather Service (NWS) Observation Station at the University of Massachusetts Experimental Station in East Wareham. This location was 10.0 km NNW of the WTH well BWH 198 on the other side of the Cape Cod Canal (point M in Figure 1a). Although Wareham was not an NADP station and did not analyze for NO_3 concentrations in precipitation, it was upwind from BWH 198 for the prevailing winter winds, and winter was the most important period for recharge to the ground water. Figure 8 shows the 16-year time series of the winter precipitation at both East Wareham and Truro, along with the annually averaged WTH. Winter was defined as January-March, precipitation included both liquid and solid forms, and the Truro station was point E, described earlier. The precipitation series at the two stations were very similar with a correlation of $r = 0.8$. The correlation between the precipitation and the water table height in figure 8 is obvious to the eye, with $r = 0.5$ and 0.6, respectively, for the two stations. Winter precipitation was used because it infiltrates to the water table, as a result of low evapotranspiration rates. The precipitation - water table correlation declined to $r = 0.1$ if the annual precipitation was used; that is, only 1% of WTH variability was explained by annual precipitation.

The regional nature of these low frequency correlations was further confirmed by comparing annually averaged WTH between Pocasset and Barnstable (mid Cape Cod, station k), Wareham (mainland side of the Cape Cod Canal, station j) and Truro (far Cape, station l; USGS 2003). These WTH are shown in Figure 9; well locations are shown in Figure 1a. The correlation of Pocasset with Barnstable was 0.92 ($p < 10^{-6}$); with Wareham $r = 0.62$ ($p = 0.003$), and with Truro $r = 0.73$ ($p < 0.0001$). Table 1 provides a summary of all the correlations cited between the NO_3 concentrations in the 4 harbors, the ground water, the WTH, and winter precipitation and infiltration

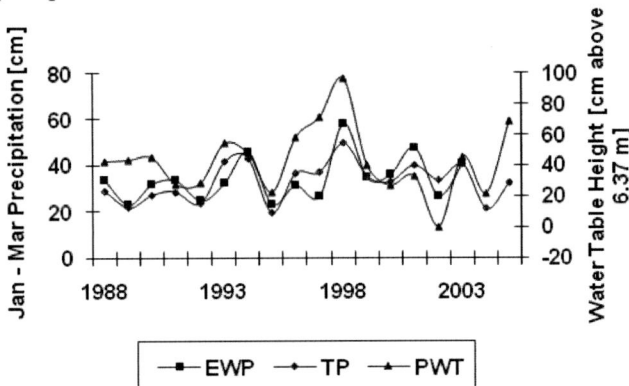

Figure 8. Comparison of the annually averaged time series of precipitation at East Wareham (EWP) and Truro (TP), along with the water table height at BWH 198 (PWT). There is a strong correlation visually between these 3 time series, even though the precipitation stations are about 64.3 km apart, as shown in Figure 1. Correlations are given in the text.

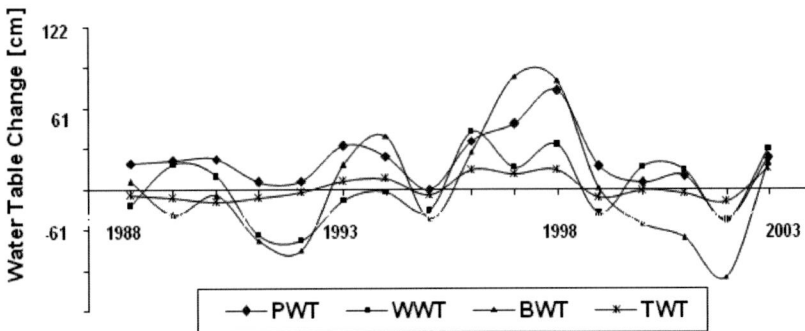

Figure 9. Comparison of the annually averaged water table heights time series throughout the Cape peninsula between 1988 and 2003. The locations of the Pocasset (PWT= BWH 198), Wareham (WWT, point j), Barnstable (BWT, point k) and Truro (TWT, point l) monitoring wells are shown in Figure 1. The correlations of these time series were between 0.6 and 0.9 and demonstrated the regional nature of the low frequency results discussed in this chapter.

Table 1. Correlations between low frequency (annual averaging) NO$_3$ concentrations in 4 harbors, ground water, water table heights and winter precipitation/ infiltration across the Cape Cod region

	SQT	SHC	CGW	CGWf	CCB	MEG	MMR$_1$	MMR$_2$	PWT	WWT	BWT	TWT	EWP	TP	TD
SQT	1.0	0.93	0.81	0.92	0.70	-0.74			0.76				0.68	0.66	
SHC		1.0	0.81	0.92	0.70	-0.74			0.76				0.68	0.66	
CGW			1.0			-0.81	0.47*	0.54	0.74						
CGWf				1.0											0.14**
CCB				0.62	1.0				0.54*						
MEG						1.0			-0.69						
MMR$_1$							1.0								
MMR$_2$								1.0							
PWT									1.0	0.62	0.92	0.73	0.50*	0.58	
WWT										1.0					
BWT											1.0				
TWT												1.0			
EWP													1.0	0.80	
TP														1.0	
TD															1.0

* marginally significant; ** not significant. All the other correlation values are significant. The MEG/SQT/SHC correlations are negative because the values in Megansett are out of phase with those in SQT/SHC.

Letters in parentheses refer to Figure 1. SQT = Squeteague Harbor (A); SHC = Scotch House Cove (B); CGW = Coastal ground water (C); CGWf = Coastal ground water NO$_3$ flux (C); CCB = Cape Cod Bay (H); MEG = Megansett Harbor (F); MMR$_1$ = Ground water 4 km inland (G); MMR$_2$ = Ground water 6.4 km inland (G); PWT = Pocasset water table height (BWH198); WWT = Wareham water table height (j); BWT = Barnstable water table height (k); TWT = Truro water table height (I); EWP = East Wareham winter precipitation (E); TP = Truro winter precipitation (M); TD = Truro winter NO$_3$ deposition (E).

The observable variable of interest is the harbor NO_3 concentration, not the WTH. The correlation of the Truro winter rain amount with the harbor NO_3 was $r = 0.66$ ($p = 0.002$). The annual winter rainfall amount 64 km NE of Buzzards Bay was a fairly good predictor for the annually averaged NO_3 concentrations in each of these two harbors. This is an important result at the regional scale. Assuming that a dynamic relationship exists between these variables, more specific comments can be made about a possible climatic driving mechanism. The low frequency period of variation in harbor NO_3 concentration is determined by the low frequency variability of the winter hydrological cycle, involving evaporation, transport and condensation. Jenkins et al. (2003) suggest that, in the open ocean case, nutrient regulation also occurs at climatic time scales involving the general circulation and upwelling (the "nutrient spiral"), which in turn may regulate climate change and, as in this study, is statistically independent from seasonal processes. In the coastal nutrient case, low frequency phenomena, such as the North Atlantic Oscillation (NAO), have been investigated for correlations with WTH and harbor NO_3 that might explain the low frequency NO_3 variability (cycles) presented in this paper. It is known that the NAO climate index has periods in the 2–10 year range, affects precipitation and operates primarily in the winter. It also is known to have significant environmental consequences on northeast North America (Hurrell 2004). That investigation is ongoing, but is beyond the scope of this chapter.

ESTUARIES IN INEQUILIBRIUM: TRANSIENT EVENTS 2004–2010

With the previously described long-term equilibrium condition of the SQT/SHC harbor-groundwater system as a starting point, this section examines an estuary's transient response to an abrupt, large increase in NO_3 input. This is done both by subtracting out the annual averages and through a comparison with an adjacent estuary as a natural control (SHC; Seaver 2010). A second consideration is the dynamics behind these events and whether they represent an estuary's capability to "digest" large increases in NO_3 input by adjusting the manner in which it processes nutrients through the food web. A third question is whether there is a limit to this capacity to process NO_3, beyond which a regime shift occurs, as suggested in the literature under certain conditions.

In 1998, dredging of a pond in the up-gradient subwatershed of Squeteague Harbor disturbed its bottom sediments and caused the release of NO_3 (but not other nutrients) into the groundwater. This effect was measured almost immediately at station A in Squeteague Harbor through the surface runoff of the dredged NO_3 via Current River (see Figure 1b). This NO_3 spike was observed at the March 23, 1998 measurement point (Figure 2); it had disappeared 30 days later. Six years later, in March 2004, a large (20-fold) increase in groundwater NO_3 concentration appeared at station C at Squeteague Harbor, and only at Squeteague Harbor (Figure 10). Both station C and, after October 2005, station D reflected this large increase, whereas monitoring wells north and south of the Squeteague watershed did not (see stations mw 065 and 050 in Figures 1b and 10). The subsequent 6 years of data resolved the amplitude and phase of the estuary's response to this large and sudden (< 2months) increase in the NO_3 input. The high groundwater NO_3 level continued for 2004–2010, and, after a one-year delay, resulted in a 3-fold increase in Squeteague Harbor's annual NO_3 concentration over the previously-cited long term average. It also caused a 20-fold increase in transient NO_3 variability. This phenomenon lasted for 4 years, at which time SQT station A NO_3 returned to background level. The average NO_3 concentration for 2009–2010 was 0.092 mg/L, as compared to 0.114 mg/L for 1988–1990. Thus, more complex biological dynamics were at work than simple groundwater mixing could explain.

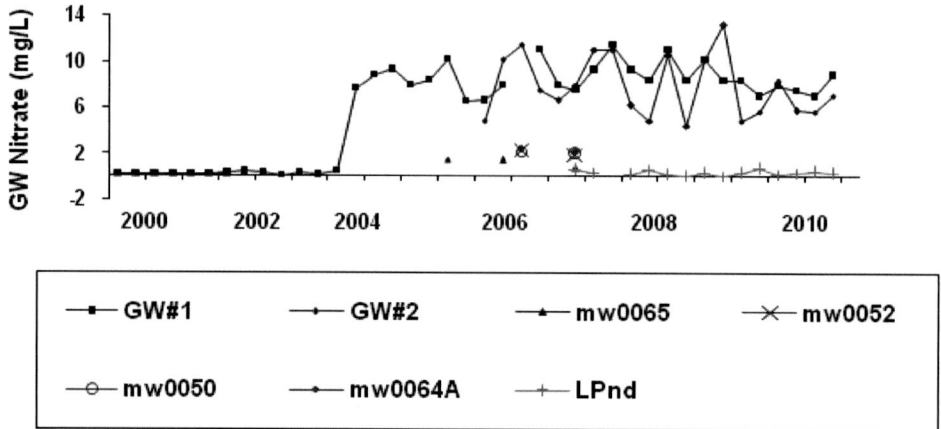

Figure 10. Dissolved NO_3 concentration from 7 groundwater wells at the coast plotted against time, 1999–2010. The GW#1 ground water monitoring at station C began intermittently in April 1991 and continuously in February 1993. GW#2 at station D began in October 2005. The frequency of sampling for stations C and D was every three to four months with the total number of measurements: 70 for GW#1; 19 for GW#2; 14 for Long Pond (station E). The other four wells (mw0050, mw0052, mw0064A, and mw0065) were sampled intermittently and were between Long Pond and the coast with mw0050 and mw0065 in the Megansett and SHC watersheds, respectively.

The twice daily tidal exchange in SQT is 1–2 times SQT's volume at mean low water (mlw), depending upon the moon's phase, and the daily input of groundwater into SQT is about one-sixth of its volume at mlw. Walter and Whealan's (2005) numerical modeling, "MODFLOW-2000," of the surface and groundwater flow in this area (the Sagamore lens) gives an average input into SQT of 14.8×10^6 liters per day. If none of the incoming groundwater was diluted by the tidal inflow, SQT NO_3 at low tide would increase steadily to 8 mg/L in about 6 days. On the other extreme, if only one tenth of the daily groundwater input were to remain with the resident SQT mlw water, in 2 months half of the original SQT water would have been replaced and the NO_3 concentration in SQT would steadily increase to about 4 mg/L. These considerations are the tidal prism calculations discussed in the literature (Valiela et al. 2004). Under any mixing process, SQT would experience a significant, steady increase in NO_3 concentration after March 2004; however, Figure 2 shows that for over a year after that date SQT showed little increase in NO_3 over the SHC reference. Thus, the process in SQT is more than one of physical mixing. The NO_3 utilization in an estuary is a dynamic process involving uptake, respiration, denitrification and sedimentation of NO_3, principally through biological processes.

A study conducted by the Marine Biological Laboratory investigated field-scale nutrient uptake by macroalgae (Teichberg et al. 2007). They spiked estuaries with NO_3, and found that complete uptake by algae took 10 days. The likely process in SQT is a dynamic one where, during the first year, there is complete uptake of the additional groundwater NO_3 by the resident algae. A proposed mechanism for the subsequent SQT response compared to SHC involves these algae blooms, their respiration, denitrification, sedimentation, and decomposition, until recycling occurs in the following years.

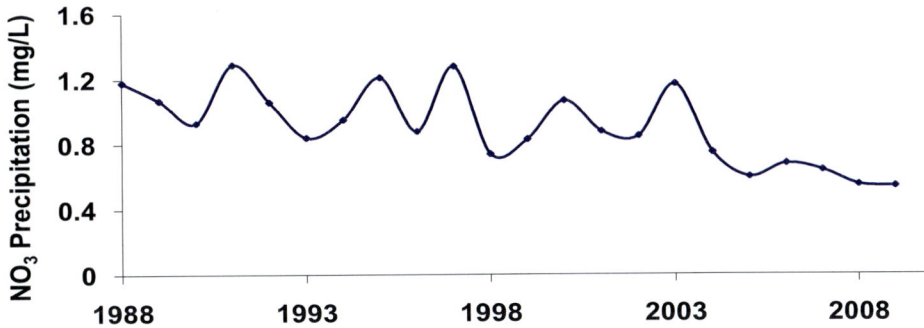

Figure 11. NADP annually averaged NO_3 concentration in precipitation at Provincetown, MA, 1988–2009. Provincetown is 60-km northeast of Squeteague Harbor. The isopleth chart at the NADP website shows that the 50% drop since 1988 applies across the Cape Cod region.

It is known that direct atmospheric deposition of NO_3 to an estuary can be important in NO_3 dynamics, particularly for more lightly developed areas; Valiela (2002, 2004) provides values of 4–20% of total NO_3 input. However, neither NO_3 inputs from the atmosphere, nor from the tidal effect of Buzzards Bay, could explain the above-described changes in Squeteague Harbor. Figure 11 shows the annual precipitation at the Provincetown station (#MA02) of the NADP, 64 km north east of SQT (NADP, 2009); the annual NO_3 concentration for precipitation in this area is now about 0.6 mg/L, and has declined by 50% over the past two decades. The isopleth map at the NADP web site indicates that this decline is across the entire Cape Cod region, and is in response to the regulatory reduction in allowable nitrogen emissions across the nation. Also, because of the construction of secondary treatment at a sewage treatment plant in New Bedford across Buzzards Bay in 1996, NO_3 concentrations in the Bay, and in the inlet to Squeteague Harbor in particular, have declined by an order of magnitude since 1996 (Turner et al. 2000), and are well below those of SQT; this data set was discussed in the first section.

By July 2005 NO_3 concentrations in SQT began to rise and diverge from those of SHC (Figures 2 and 7). As described in the first section of this chapter and in Seaver & Kuzirian (2007) the low frequency (annual average) NO_3 concentration in SQT and SHC for 1987–2003 was highly correlated (r = 0.92) with the low frequency groundwater NO_3 load input, implying non-transient dynamics. Consequently, these annual averages have been subtracted from the raw data of Figure 2 in order to consider only the remaining high frequency (hf) transient estuary response to the sudden groundwater NO_3 increase that was observed in March 2004. Figure 12 shows the resulting hf NO_3 variation in SQT harbor for 1988–2009. This procedure is similar to that used by Carpenter & Brock (2006) to simulate nutrient-induced regime shifts in lakes with dynamic linear modeling, where annual phosphate means, comparable to our annual NO_3 averages, were updated each year. The residual variability is largely that of recycling from the sediments, which, their model finds, is an indicator of impending regime shift. The Carpenter & Brock (2006) regime shift model will be considered later to investigate the possibility of an irreversible threshold to SQT NO_3 increase. Also included in Figure 12 is the difference between the NO_3 concentrations in Squeteague Harbor and that of Scotch House Cove, another means of assessing the transient components. This difference exhibits similar hf variability, reinforcing the assumption that the Squeteague watershed and groundwater inputs are the source of the SQT variability. The daily and monthly (hf) NO_3 concentrations for precipitation at the previously mentioned Provincetown station (NADP 2009) were investigated to see if they could offer a partial explanation for the hf NO_3 variability seen in Figure 12. The correlation between the values at Provincetown and those of Figure 12 was 0.03, which would rule out precipitation as the cause of this SQT variability.

As mentioned above, the daily groundwater input into SQT is about 15% of its volume at mlw, and after 1.3 years of an order of magnitude increase in the groundwater NO_3 input, SQT had not diverged from the reference harbor, SHC. However, after July 2005 there was a clear change in the SQT response (Figure 2). The SQT NO_3 not only diverges from that of SHC, but more importantly, the periodic response shown in Figure 12 (Figure 2 minus annual average) becomes significantly larger. The March 1998 NO_3 spike to 4.1 mg/L (Figure 2) is misleading in that it represents a one-time river discharge from Long Pond. As a result, the hf NO_3 fluctuation of 1998 (Figure 12) is also misleading, as it simply reflects the effects of the 4.1 mg/L spike on the 1998 average. Further, the February 16, 2005 spike (Figure 12) is also misleading, and does not appear in the (SQT-SHC) result. Ice covered much of both harbors on that date, thus preventing full mixing of the groundwater input with the resident estuary water.

The SQT annual average NO_3 concentration began to rise above that of SHC in 2005 (Figure 7), one year after the arrival of the large increase in groundwater NO_3, reaching a level 4 times that of SHC in 2006 and 2007. During this same period, SQT NO_3 peaks occurred in January 2006, March 2007 and January 2008. The March 2007 NO_3 peak was 10 times greater than the pre-2004 variability, and all peaks lasted for 2–3 months during the winter. The January to March peaking in NO_3 is not an unexpected result, is customarily caused by seasonal algae decay and subsequent nitrification, and has been documented, for example, by the Massachusetts Water Resources Authority (2007) in Boston Harbor, Massachusetts Bay, and Cape Cod Bay for 15 years. The unique result here is that these winter peaks are 3 times greater than those of Boston Harbor, and they disappear after 3 years. By 2010 these large transients were gone and the SQT annual average had returned to the background level of SHC.

A POSSIBLE REGIME SHIFT

From the results presented in Figure 12, particularly after 2008, it is clear that a regime shift, as defined in the literature (Contamin & Ellison 2009, Biggs et al. 2008, Carpenter & Brock 2006), had not persisted. However, the characteristics of the mean and transient NO_3 changes during 2005–2008 suggest that the beginning stages, at least, of such an event might have occurred. This warrants further consideration of the properties of regime shifts, particularly in their early stages.

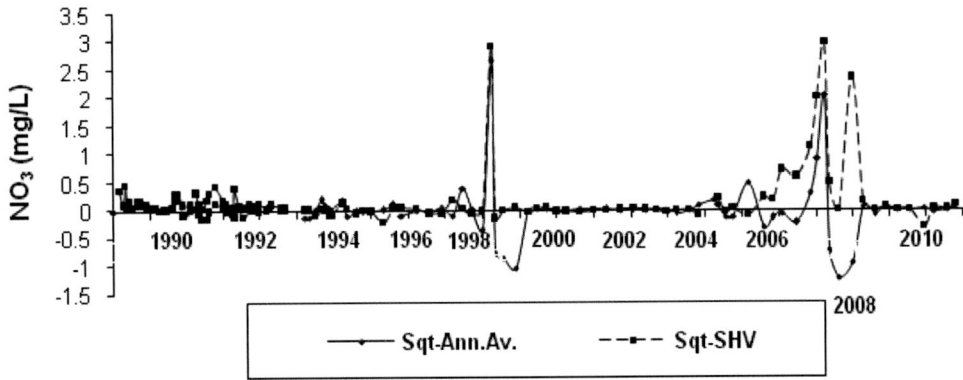

Figure 12. High frequency NO_3 variability at station A in Squeteague Harbor, calculated by subtracting the annual averages from the raw data (Figure 2). The spike of February 1998 reflects the spike seen in Figure 2, caused by Long Pond river runoff. The NO_3 variability increases significantly a year after the arrival of the groundwater NO_3 pulse in March 2004. Also shown is the difference between SQT and SHC, which supports the post-2005 increase in the high frequency NO_3 variability. The hf variability disappears in 2009.

The SQT hf response was analyzed in view of the regime shift models of Carpenter & Brock (2006), Biggs et al. (2008) and Contamin & Ellison (2009), by extending their analysis of phosphate and fish in fresh water to NO_3 in estuaries. From Figures 2 and 12 we know that a change occurred in SQT after 2005 when the groundwater NO_3 input (Figure 10) exceeded ≈ 8 mg/L. Figure 12 is the estuary equivalent of the Contamin & Ellison (2009) graph describing the rising phosphate variability that occurred after recycling from the sediments had begun, but prior to a regime shift, even in the presence of significant input variability. This type of analysis is effective even when "only incomplete information about system drivers and processes is available" (Contamin & Ellison 2009). Thus, knowledge about uptake, respiration, sedimentation, denitrification and recycling is not required.

The record of dissolved oxygen and water transparency in SQT does not support the conclusion that a regime shift occurred in SQT in 2006, 2007 and 2008, as defined by the eutrophication state. Figure 13 shows the 21-year record of dissolved oxygen in both SQT and SHC and does not indicate a decline after 2004, nor a divergence of SQT from SHC after 2004, as we observed for NO_3. Also, the percentage of oxygen saturation ranged (winter to summer) from 104% to 61 % with no trend. The relative minimum water transparency in SQT for the 21-year period, determined with an instrument similar to a secchi disk, usually had a minimum in August. These data do not show a post-2004 change for SQT. In this paper a regime shift is defined as an increase in algae growth and a decline in oxygen, whereas in the Carpenter & Brock (2006) and Contamin & Ellison (2009) studies it was a shift in their model results from an oligotrophic to a eutrophic state signified by the onset of recycling from the sediments.

Contamin & Ellison's (2009) model states that the log of the ratio of the spectral density of variability before and after recycling from the sediments has begun is the most sensitive predictor of future regime shift. From Figure 12 for SQT:

$$\log_{10} [SPEC_{after}/SPEC_{before}] \approx 1.0;$$

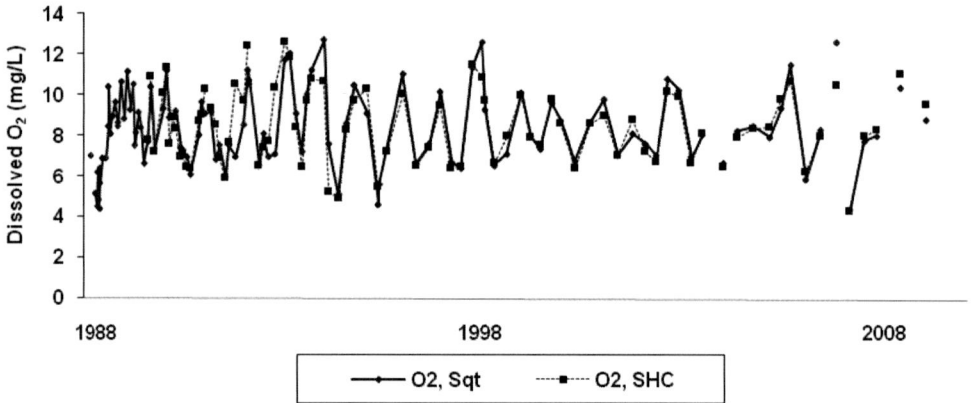

Figure 13. Oxygen concentration (mg/L) in SQT (A) and SHC (B), 1988–2009. Percentage of saturation ranged from 60% to 109% (summer to winter). There was no trend over the 21 years of this record, and SQT did not diverge from SHC.

where $SPEC_{after}$ = hf NO_3 intensity for 2007 and $SPEC_{before}$ = hf NO_3 intensity for 1997; SPEC intensity is defined as the maximum spike determined by the maximum difference of NO_3 calculated at adjacent measurement points within each year's variability. Contamin & Ellison (2009) found in their model runs that this was a more sensitive indicator of regime shift than traditional variance.

This value is comparable to what Contamin & Ellison (2009) found to be capable of predicting a regime shift (1.0–2.5), and suggests that recycling from the sediments in SQT greatly accelerated soon after the arrival of the dramatic groundwater increase. A difference in this study from Contamin & Ellison (2009) is that their modeled system started in an oligotrophic state, and so recycling actually did begin at some point in their model run, whereas in the present study, recycling was part of the long term seasonal cycle and (harmful) eutrophication, defined as impairment in DO concentrations, did not occur.

Biggs et al. (2008) state that an increase in variance provides no indication of how close a regime shift is, or even whether the sustainable NO_3 level has been exceeded for the long term. To illuminate this part of the problem, the model of Biggs et al. (2008) predicts a shift in the spectral density from higher to lower frequency processes as the sediment recycling begins. When the ration of the amplitudes of low to high frequencies in the indicator variable exceeds 1 in their formulation, a warning of an impending regime shift is provided. Furthermore, Biggs et al. (2008) found that this model predictor performed well when the driving variables (the ground water NO_3 input in our case) were "noisy." From Figure 12 of the SQT hf NO_3 variability, we can calculate the following spectral density ratios:

    a.   1993–1996, fluctuations ≅ 5 x 10^{-3} cycles/day (200 day period), amplitude ≅ 0.14 mg/L (before input jump);

    b.   2004–2005, fluctuations ≅ 3 x 10^{-3} cycles/day (360 day period), amplitude ≅ 0.5 mg/L, spectral *ratio* ≅ 3.5 (just after input jump);

    c.   2006–2007, fluctuations ≅ 2 x 10^{-3} cycles/day (440 day period), amplitude ≅ 1.0 mg/L, spectral *ratio* ≅ 7 (2.5 years after input jump).

These spectral ratios are comparable to those of Biggs et al. (0.5–10) and predict a regime shift at some point in the future. However, from Figure 12 for the last half of 2008 and all of 2009 and 2010 the fluctuations decline in amplitude. Our data are from a shorter time series than in Biggs et al. (2008); Carpenter & Brock (2006) and Contamin & Ellison (2009) caution that the time required to develop reliable indicators is greater than the 5 years observed so far in the SQT experiment. In summary it appears that, even though the groundwater-estuary system's response to an abrupt, large NO_3 increase exhibits the characteristics of a pre-regime shift, it does not cross the line, and actually recovers back to an equilibrium condition, even as the higher input levels continue.

ESTUARY INEQUILIBRIUM DYNAMICS

The equilibrium state of the groundwater-estuary system came to an abrupt end in 2005 with the abrupt increase in groundwater NO_3 (Figure 10); the correlation between the NO_3 load from the groundwater and the annual estuary NO_3 concentration went from r = 0.92 for 1988–2003 to 0.07 for 2005–2009 (Figure 4). In 2004 Ivan Valiela and colleagues at the Marine Biological Laboratory in Woods Hole, Massachusetts published the results of their lengthy and detailed investigation of NO_3 transport through groundwater and into estuaries in the form of an Estuary Loading Model (ELM; Valiela et al. 2004) which had 92 terms and 16 input coefficients. In it they quantify the estuary's external NO_3 sources (groundwater, direct atmospheric deposition, nitrogen fixation, DON regeneration) and sinks (denitrification, sediment burial, algae uptake, export to the sea) as the NO_3 traversed from the atmosphere and groundwater, through the estuary, and into the bay or ocean.

Burial, *regeneration,* and water turnover time are the components of the ELM that are the most relevant to explaining the present results, and will be investigated in order to understand the transient dynamics, their phase delay and their ultimate decay during 2005–2010. The regeneration component includes regeneration in both subtidal sediments and the water column; in the equilibrium state assumed by the Valiela ELM model, plant uptake and regeneration of DIN were assumed equal. In the present transient case this is no longer true. The working hypothesis is that the increased groundwater NO_3 input was taken up entirely by PON (algae) and then later, in the winter, was regenerated as DIN, with a residual then leading to increasing annual average NO_3. The question as to why the transients did not appear in the first year and why they died out after three years also has to be explored. The question asked is whether the system moves to a different equilibrium state between the ELM sources and losses. That is, can this be seen as a reordering of the losses enumerated in the ELM: of denitrification, of burial in sediments, of regeneration in sediments and water column, and of the flushing time and exchange with the Bay. Because of the control offered by the second harbor, this is a credible laboratory-like setup of an important estuary response mechanism. The NO_3 conditions in SHC were a control for the effects of the abrupt increase in NO_3 on SQT, as SHC was in a different watershed and had an NO_3 correlation of 0.93 with that of SQT harbor prior to the onset of this abrupt event. The experimental results suggest that an estuary can assimilate an increase in input, at least up to a point. In the ongoing work the ELM model will be configured to look at this phenomenon on a seasonal scale.

CONCLUSION

Sixteen years of NO_3 data from two adjacent, independent estuaries and the groundwater input into them, along with WTH and regional precipitation, have been analyzed for low frequency signals. Annually averaged values of these variables have revealed an estuary in equilibrium with climatic-scale driving forces that modulate NO_3 concentrations through the groundwater system to the estuaries, as shown by the high correlations between these series and the low frequency infiltration of precipitation. Over 50% of the low frequency harbor NO_3 variability is explained by this infiltration and WTH variation. The analysis of groundwater NO_3 concentrations up-gradient from the coast and in harbors to the south and to the north supports this conclusion.

Subsequently, an abrupt, large increase in groundwater NO_3 input overwhelmed the equilibrium relationship and the estuary was able to absorb the sudden increase in NO_3 for a year without an increase in water column NO_3 concentration. After a year, both the mean and hf variability in NO_3 concentrations in the estuary began to increase, in comparison to an adjoining estuary. Dissolved oxygen and water transparency (as indicators of algal production) did not indicate any change in the estuary's eutrophication state. The estuary initially was in a low nutrient condition.

After 4 years of continued high NO_3 input, both the mean and hf variability of NO_3 returned to pre-increase levels. Appealing to the literature, an analysis was done based upon the change in NO_3 variability spectral density, which lent credence to an early regime shift condition during this phase, and an analysis of an estuary loading model provided insight into the internal dynamics of the hf transient phenomenon.

REFERENCES

AFCEE (Air Force Center for Engineering and the Environment). 2003. Landfill-1 2002 annual system performance and ecological impact report. Jacobs Engineering, Document no. ENR-J23-35Z15609-M31-0003. January 2003.

Biggs, R., Carpenter, S. & Brock, W. 2009. Turning back from the brink: detecting an impending regime shift in time to avert it. *PNAS 106*, 826–831.

Carpenter, S. & Brock, W. 2006: Rising variances: a leading indicator of ecological transition. *Ecol. Letters 9*, 311–318.

Contamin, R & Ellison, A. 2009. Indicators of regime shift in ecological systems: What do we need to know and when do we need to know it? *Ecol. Appl. 19*, 799–816.

Hurrell, J.W. & Dickson, R.R. 2004. Climate variability over the North Atlantic. In: N.C. Stenseth, G. Ottersen, J.W. Hurrell & A Belgrano (Eds.) *Marine Ecosystems and Climate Variation: The North Atlantic: A Comparative Perspective*. Oxford University Press, 264 pp.

Jenkins, W. & Doney, S. 2003. The subtropical nutrient spiral. *Global Biogeochem. Cycles 17*, 21.

Jordan, C. & Talbot, R.W. 2000. Direct atmospheric deposition of water-soluble nitrogen to the Gulf of Maine. *Global Biogeochem. Cycles 14*, 1315–1329.

McClelland, J. W., Valiela, I. & Michener, R. H. (1997). Nitrogen-stable isotope signatures in estuarine food webs: a record of increasing urbanization in coastal watersheds. *Limnol. Oceanogr. 42*, 930–937.

McClelland, J. W. & Valiela, I. (1998). Linking nitrogen in estuarine producers to land-derived sources. *Limnol. Oceanogr. 43*, 577–585.

MWRA (Massachusetts Water Resource Authority). (2009). Water column monitoring in Massachusetts Bay: 1992 – 2006. MWRA Environmental Quality Department Report ENQUAD 2007-11.

NADP (National Atmospheric Deposition Program). (2009). Annual precipitation. http://nadp.sws.uiuc.edu/nadpdata Accessed March 25, 2011.

National Weather Service. Monthly precipitation data 1988–2001. University of Massachusetts Observation Station, 1 State Bog Rd., East Wareham, MA. 02538.

Portnoy, J. 2002. NADP rainfall data for site MA01, Provincetown. National Park Service, Provincetown, MA.

Seaver, G. & Kuzirian, A. (2007). Nitrate migration through groundwater, estuaries and bays at climatic frequencies and scales. *J. Coastal Res. 23*, 1000–1009.

Seaver, G. (2010). Estuary Response to an abrupt, large increase in groundwater nitrate input. *Appl. Geochem. 25*, 1453–1460.

Teichberg, M., Fox, S.E, Aguila, C., Olsen, Y.S. & Valiela, I. (2008). Microalgal response to experimental nutrient enrichment and grazing in Waquoit Bay, MA. *Mar. Ecol. Prog. Ser. 368*, 117–135.

Turner, J., Lincoln, J., Borkman, D., Gauthier, D., Kieser, J. & Dunn, C. (2000). *Nutrients, eutrophication and harmful algal blooms in Buzzards Bay, Massachusetts.* Report, Center for Marine Science and Technology, University of Massachusetts.

USGS (U.S. Geological Survey). (2003). Ground water data for Massachusetts and Rhode Island. http://waterdata.usgs.gov/ma/nwis/ Accessed March 25, 2011.

Valiela, I., Collins, G., Kramer, J., Lajtha, K., Geist, M., Seely, B., Brawley, J. & Shame, C.H. (1997). Nitrogen loading from coastal watersheds to receiving estuaries: new method and application. *Ecol. Appl. 7*, 358–380.

Valiela, I., Bowen, J. & Kroger, K. (2002). Assessment of models for estimation of land-derived nitrogen loads to shallow estuaries. *Appl. Geochem. 17*, 935–953.

Valiela, I., Mazzilli, S. Bowen, J. Kroeger, K. Cole, M., Tomasky, G. & Isaji, T. (2004). ELM, an estuary nitrogen loading model: formulation and verification of predicted concentrations of dissolved inorganic nitrogen. *Water, Air Soil Poll. 157*, 365–391.

Valiela, I., Tomasky, G., Hauxwell, J., Cole, M.L., Cebrian, J. & Kroeger, K.D. (2000). Operationalizing sustainability: management and risk assessment for land-derived nitrogen loads to estuaries. *Ecol. Appl. 10*, 1006–1023.

Walter, D. & Whealan, A. (2005). *Simulated water sources and effects of pumping on surface and ground water, Sagamore and Monomoy lenses,* Cape Cod, Massachusetts. U.S. Department of the Interior, USGS. Report No. 2004–5181.

In: Estuaries: Classification, Ecology and Human Impacts   ISBN: 978-1-61942-083-0
Editor: Steve Jordan   © 2012 Nova Science Publishers, Inc.

Chapter 8

CONTAMINANTS IN ESTUARIES IN RELATION TO HUMAN ACTIVITIES

Luis Bartolomé[1], Nestor Etxebarria[2], and Juan Carlos Raposo[1]

[1]General Analysis Service, [2]Department of Analytical Chemistry, Faculty of Sciences & Technology, Universidad del País Vasco, Euskal Herriko Unibertsitatea, Bilbao, Basque Country, Spain

ABSTRACT

The European Water Framework Directive (WFD) considers water management from a wide perspective, looking for the prevention of any future deterioration of water bodies, as well as the protection and improvement of the state of marine ecosystems, in order to obtain "a good status" of water bodies. According to the WFD, good status of water bodies is obtained when concentrations of the priority substances in water, sediment and biota are below the established Environmental Quality Standards (EQSs). All the European Union member states must implement management plans in their river basins, including monitoring programs. In this chapter we compare two estuaries (Bilboa and Urdaibai in northern Spain) that have experienced very different human impacts. They are evaluated by reporting information on (1) concentrations and spatial distribution of trace metals and organic contaminants in sediments along both estuaries since 2002; (2) tools for finding potential sources of micro-contaminants (PCB, PAH) in some of the estuary compartments; (3) relationships between contaminant concentrations and particle size parameters or organic matter content; and, (4) the relevance of total organic carbon (TOC) distribution in the sediment as the principal pathway of transport or persistence of metals between inorganic compartments.

INTRODUCTION

Estuaries are rich ecological environments with numerous biotic communities and very high biological production of organic matter coupled to complex biogeochemical cycles of natural substances and contaminants. Additionally, along the shores of estuaries there are heavily populated urban areas with intensive industrial and leisure activities.

In the Cantabrian Coast (North of Spain), from Galicia to the Basque Country, we can find many short estuaries flowing from the Cantabrian Mountains to the sea. Following the reference work of Valencia et al. (2004), in the Basque Country there are 12 estuaries of different sizes and hydro-morphologies. Among them, the estuaries of Bilbao (Nervión-Ibaizabal Rivers) and Urdaibai (Oka River) are two of the most distinctive. On the one hand, the estuary of Bilbao has the most extensive estuarine area of the Cantabrian Coast, and as one of the most polluted areas of northern Spain, its natural features have been deeply modified by anthropogenic activities such as industry, human settlements and harbor development (Azkona et al. 1984, Cearreta et al. 2000). On the other hand, the estuary of Urdaibai has been included among the natural UNESCO Biosphere Reserves since 1984. In this estuary industrial, urban and agricultural activities coexist with the ecological richness of the estuarine habitat (Dept. of Environment and Regional Planning, Basque Government[1]).

As a consequence of the contrasting needs for restoration of a degraded estuary and the protection of an ecologically rich estuary, the estuaries of Bilbao and Urdaibai have become benchmarks for study and understanding of the fates and the effects of many pollutants. In the case of the estuary of Bilbao, many of those studies are focused on evaluating the effects of past pollution events and following the implementation of pollution abatement policies and strategies intended to restore the ecological status (Leorri et al. 2008, Garcia-Barcina et al. 2006, Cotano & Villate 2006), and in the case of the estuary of Urdaibai, most of the studies are focused on monitoring anthropogenic stresses on various biotic communities (Cortazar et al. 2008, Puy-Azurmendi et al. 2010a,b).

Hydrological descriptions of both estuaries are given in the literature and summarized in Table 1. Briefly, the estuary of Urdaibai (Oka river) is 12 km long; the most populated area is the town of Gernika (20,000 inhabitants), at the limit of tidal influence (Irabien & Velasco 1999, Liria et al. 2009, Monge-Ganuzas et al. 2008), as shown in Figure 1. The estuary of Bilbao is 15 km long and drains a watershed of about 1700 km^2. Bilbao and its metropolitan area (1 million inhabitants) are located along the main channel of the estuary (Cearreta et al. 2000, Leorri et al. 2008, Grifoll et al. 2009).

Many estuaries in Europe are grossly polluted (Hawkins et al. 2002) and the European Union (EU) has led several key actions to establish policies for the protection of the environment. Among them, the Water Framework Directive (WFD, Directive 2000/60/EC[2]) provides to national and regional bodies the legislative basis to protect and enhance aquatic environments, and to ensure a progressive reduction of polluting discharges, as recently reviewed by Borja (2005). This legislation requires systematic monitoring of quantitative ecological, biological and chemical parameters, which means the application of quality standards to all the measurements (traceability, comparability, standardization, etc.) as well as the need to use referenced validated methods (Quevauviller et al. 2007, Lepom et al. 2009) in order to determine the level of compliance with the EQS and to determine temporal trends.

---

[1] http://www.ingurumena.ejgv.euskadi.net/r49-12892/en/ (last accessed on March 2011).
[2] Directive, 2000/60/EC of the European Parliament and the Council of 23.10.2000. A framework for community action in the field of water policy. Official Journal of the European Communities 22.12.2000, p. 72.

Table 1. Main geomorphological and hydrological characteristics of the estuaries of Bilbao and Urdaibai (data obtained from Valencia et al. 2004)

Estuary	Basin area (km^2)	River flow (m^3 s^{-1})	Estuary length (km)	Estuary depth (m)	Estuary volume (m^3 10^6)	Estuary volume/river flow (days)	Current surface (km^2)	% of subtidal surface	% of intertidal surface
Bilbao	1755	36	22.0	30	200	65.0	24	100	0
Urdaibai	178	3.6	12.5	10	3.3	10.6	7.6	30	70

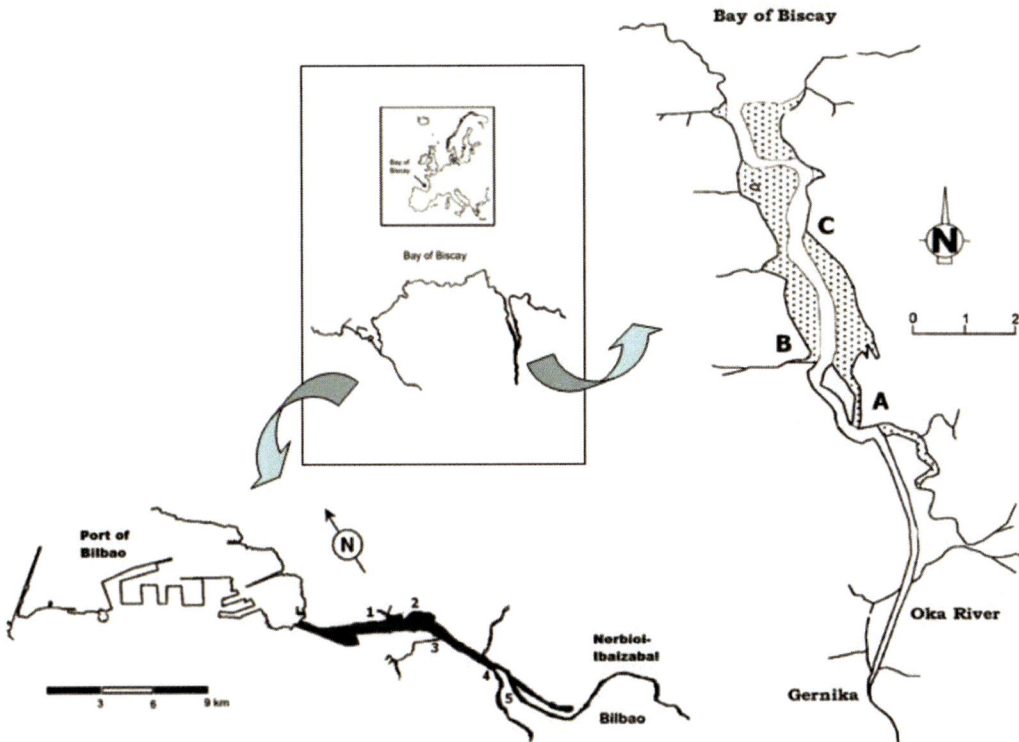

Figure 1. Study areas of the estuaries of Bilbao (samples areas, 1. Gobela, 2. Udondo, 3. Galindo, 4. Zorroza, 5. Kadagua) and Urdaibai (samples areas, A: Arteaga, B: Murueta, C: Kanala and Gernika), in the Basque Coast (northern Spain).

The EU chemical EQS values are supported by the priority list of pollutants (Directive 2008/105/EC[3]) but they are limited to the water column and do not apply to sediments. For the latter, the Sediment Quality Guidelines (SQG) provide a set of suitable reference values, in the sense that they were intended to be either protective or predictive of adverse effects on biological resources (Long et al. 1995, 1998, 2006). The SQGs typically have been developed

---

[3] Directive, 2008/105/EC of the European Parliament and the Council of 16 December 2008 on environmental quality standards in the field of water policy, amending and subsequently repealing Directives 82ç7172/EEC, 83/513/EEC, 84/156/EEC, 84/491/EEC and 86/280/EEC and amending Directive 2000/60/EC.

using large databases with matching measures of sediment chemistry and toxicity from field-collected samples. As a consequence of those relationships, a large variety of algorithms have been used to define specific concentrations associated with particular levels of effect or no-effect such as ERLs (effects range-low), ERMs (effects range-median), PELs (probable effects levels) and TEL (theshold effect levels)[4], as well as the average quotient (ERM-Q) calculated from the combined ratios of individual pollutants (Hyland et al. 1999).

The fact that the Basque Country was heavily industrialized, especially between the 1960s and late 1970s, increased public concern about the environmental status of many streams and estuaries. Undoubtedly the main attention was focused on the estuary of Bilbao, where the density of population and heavy industry had increased in the previous decades, and the estuary had become a tidal channel to dispose of industrial and urban wastewaters (Garcia-Barcina et al. 2006, Saiz-Salinas & Gómez-Oreja 2000). Consequently, the local water council pushed a wastewater treatment scheme for Metropolitan Bilbao to collect and treat most of the urban wastewaters. The central wastewater treatment plant (WWTP) of Galindo receives the wastes of all the sewer network and works with a maximum throughput of 12 m^3/s in biological treatment. As seen in Figure 2, since the functioning of this treatment plant (1990) the population that has been connected has been continuously increasing.

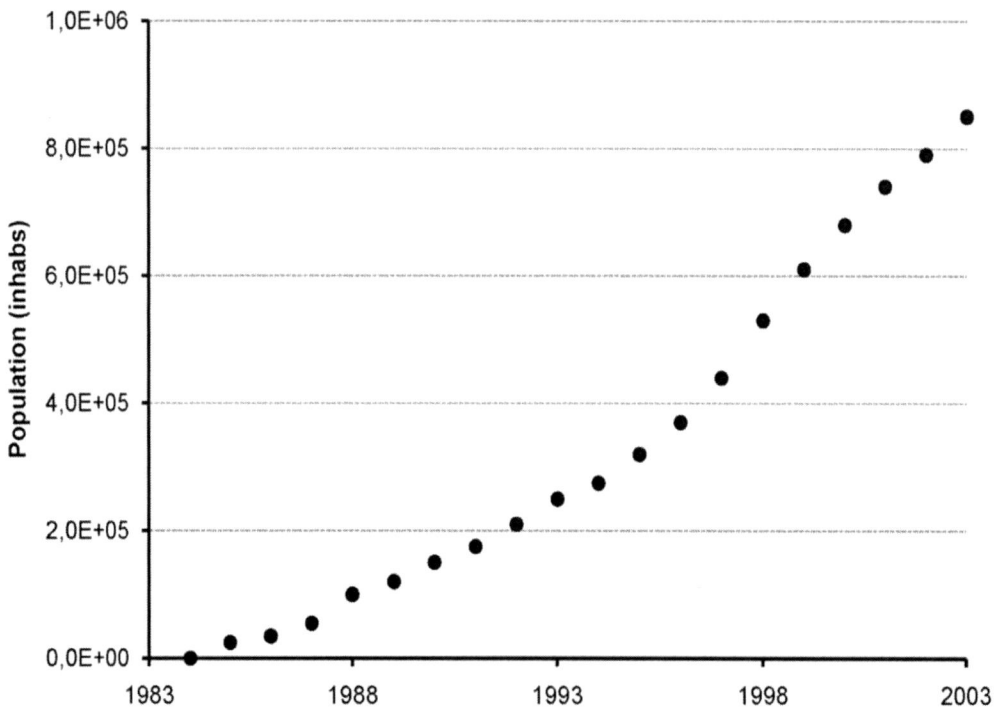

Figure 2. Evolution of the population served by the WWTP of Galindo in the estuary of Bilbao.

---

[4] One source of these reference values is the NOAA Screening Quick Reference Table, available at http://response.restoration.noaa.gov/book_shelf/122_NEW-SQuiRTs.pdf (last accessed on April 2011).

Although the role of WWTPs is outside the scope of this work we are aware of a continuous debate. On the one hand, the discharges of a large treatment plant are beyond the maximum allowable concentrations fixed by the laws, and probably we should consider that the steady disposal of contaminants may behave as a chronic source of pollution. On the other hand, recent literature points out that the effluents of wastewater-treatment plants (WWTPs) are the main sources of pharmaceuticals and personal care products (PPCP) in the environment, because conventional water-treatment processes are highly inefficient in removing PPCP from sewage (Miege et al. 2009, Kasprzyk-Hordern et al. 2009 Ellis et al. 2006).

In this chapter we will focus on chemical monitoring studies of the estuaries of Bilbao and Urdaibai over the past ten years, and in addition we will take into account some previous work to show the long term variation in the distribution of contaminants. Basically, most of the included works refer to the analysis of sediments, not only because they are considered the usual sink for many contaminants, but also because this way we can offer a comprehensive description of both estuaries. The main issue of this chapter is to show the impact of anthropogenic activities and the risk to environmental health status associated with them.

Although there are some recent works that refer to the presence of contaminants in biota (essentially mussels and oysters) and in the water column, this chapter will focus on the distribution of metals and organic micro-pollutants in the sediments.

DISTRIBUTION OF TRACE METALS IN SEDIMENTS
IN THE ESTUARY OF BILBAO

The loading and the fate of metals in estuarine environments has been discussed extensively in the literature. It is well documented that trace metals can be delivered to intertidal zones from the catchment via fluvial transport, atmospheric deposition, and/or local wastewater discharge. Therefore, to assess the environmental quality of estuaries, it is essential to evaluate the trace metal concentrations in intertidal zone sediments (Williams et al. 1994, Ahn et al. 1995, Cochran et al. 1998, Yang et al. 2001, Zhang et al. 2001, Feng et al. 2004, Censi et al. 2006, Qiao et al. 2007).

The evaluation of metals in sediments can be justified from various viewpoints: (1) to assess the environmental and toxicological effects of seasonal variations in trace metal contamination (Birch et al. 2001, Caccia et al. 2003), (2) to investigate chronological inputs (French 1993) and (3) to identify sources and sinks (Emmerson et al. 1997), so as to understand the mobility, accumulation, retention and loss in estuarine systems (Comber et al. 1995).

Many estuaries share common biogeochemical processes but the final outcomes are very sensitive to subtle variations in environmental features such as hydrodynamic residence times, mixing patterns and transport processes. In fact, many of the estuaries of the Basque coast may be differentiated considering their geomorphological and hydrological features, as shown by Valencia et al. (2004), and consequently the patterns of trace metal distribution may differ from one site to another.

As a consequence, there is no a general pattern of trace metal behavior in estuarine sediments. To give an example, if we consider trace metal partitioning between dissolved and particulate phases, it depends on a number of factors including pH, salinity, temperature, redox-conditions, dissolved organic carbon, and composition of the suspended particulate matter (SPM). In particular, there is a considerable body of literature documenting the accumulation of different pollutants in sediments from the Bilbao estuary: organic matter (Cotano & Villate 2006), arsenic (Raposo et al. 2006) and trace metals (Cearreta et al. 2002, Landajo et al. 2004).

The general features of the sediment in the estuary of Bilbao are summarized in Table 2. The particle size distribution of the surface sediments clearly depends on the geochemical composition of the substrate. Broadly speaking, sediments of the estuary of Bilbao are 55% fine-grained. This fact is due to accumulation zones of fine material in the estuary in the vicinity of docks, dykes and the inner part of meanders. In addition, the central part of the channel is permanently submerged, where particles from the water column can accumulate in the sediments depending on the environmental conditions.

The redox potentials show mostly negative values in the sediments together with high contents of fine particles and organic matter, indicating a generally anoxic condition. This fact is also supported by the remarkable stratification of the water column (redox potential, salinity, dissolved oxygen, etc.) (Belzunce et al. 1998).

Table 2. General characteristics of the estuary of Bilbao (collected from Solaun et al 2009)

Estuary of Bilbao	Gravel (%)	Sand (%)	Mud (%)	Redox (mV)	OM (%)	C/N
Mean	6.8	38.1	55.1	−44	8.7	42
Max	85.6	99.2	98.5	230	28	173
Min	0	1.3	0	−254	1.3	16
SD	14.6	34.4	37.6	147	5.2	40

Natural Compounds Found in the Estuary of Bilbao

The most common deposits of minerals and rocks are calcareous minerals, quartz sand, micaceous silicates and oxides (essentially aluminum and iron mixtures), and conglomerates (clay, gravel, mud, till, silt and any cementing material such as silica, calcium carbonate and some organic materials). Apart from this natural distribution of sedimentary compounds, it is necessary to emphasize the high iron concentration in the basin of estuary of Bilbao, with pyrites, cassiterites and iron oxides the most common natural mineral phases, which can also be found in those particular sediments, as a consequence of natural transport processes in the basin.

Based on Raman spectroscopic analysis, Villanueva et al. (2008) designated the following as natural compounds: quartz, mica, aragonite, calcite and pyrite, as shown in Figure 3. Quartz and silicate-type micas, which are natural compounds of silicate oxides, are compounds spread in metamorphic and sedimentary rocks like those present in the basins considered in this work. Additionally, calcium carbonate is also a natural compound, probably coming from isolated carbonate-rich rock masses at the source of some rivers and transported across the whole river until their mouth. Finally, haematite is considered as a natural phase of iron (III) oxide. The basin of the estuary of Bilbao estuary is well known for its iron mines, and therefore high concentrations of iron species are expected along the estuary.

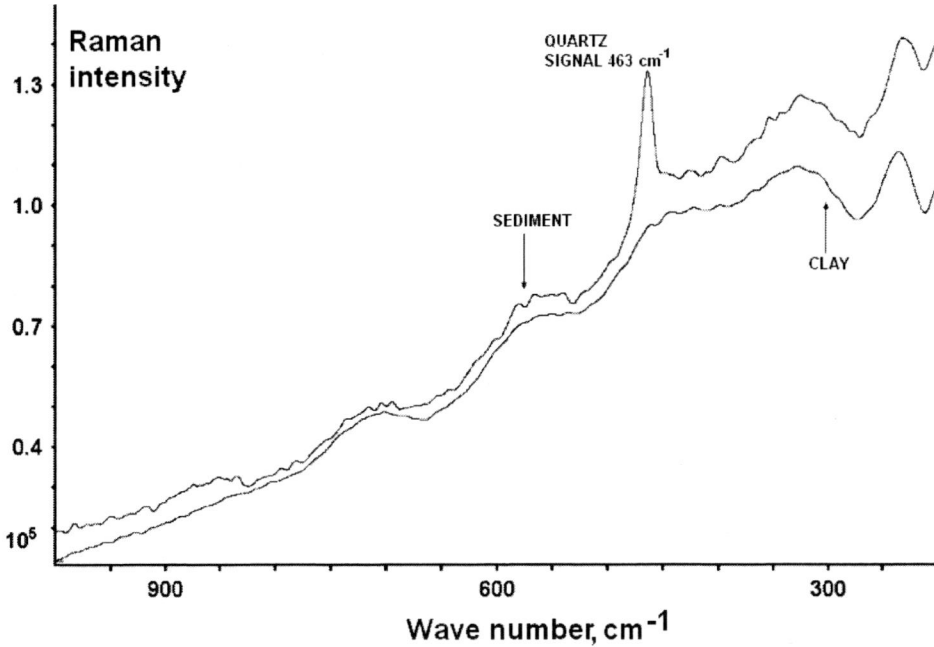

Figure 3. Raman spectra of quartz and clay in a sediment from the estuary of Bilbao.

Evolution of the Environmental State of the Estuary of Bilbao

The distribution of trace metals in estuarine sediment is generally established from the fine-grained fraction (< 63 μm). Though the analysis of metal can be carried out by a variety of methods, we have considered mostly data obtained by mineral acid leaching (HCl:HNO₃ mixtures) and atomic spectrometric analysis (AAS, ICP-OES, ICP-MS).

As an example, Figures 4a and 4b show the variation of the total organic carbon (TOC) (as loss of weight at 400 C) and the concentrations of Cd, Co, Cr and Ni in the sediments collected from one particular station in the estuary of Bilbao (Ruiz & Salinas 2000). This plot reflects the trace metal concentrations in surficial sediments in the highly polluted intertidal flats up to the early 1990s, when less than 25% of the urban and industrial discharges were treated in the WWTP.

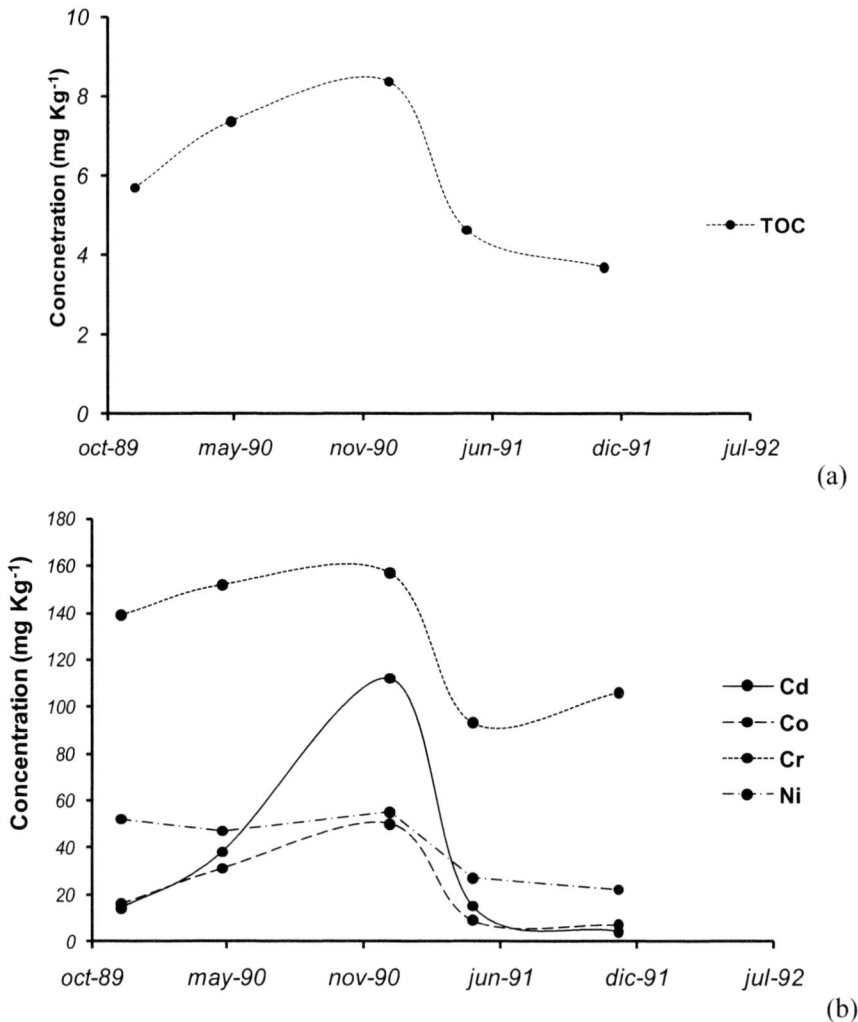

Figure 4. (a) Mean concentration of total organic carbon (TOC) and (b) some trace metals in the sediments of the estuary of Bilbao.

In general, it can be concluded that the trace metal concentration levels were equally distributed following a seasonal pattern. Highest concentrations were measured in winter (around January) and the lowest in summer (Figure 4). Though the role of organic matter in natural systems is complex, the variation of trace metal concentrations seems to be closely linked to the variation of organic matter. Therefore, an increasing amount of organic matter in sediments might induce the complexation of trace metals so that the sediments would act as sinks of contamination. This general pattern was extended for the other trace metals considered in the experimental study (Ruiz & Salinas 2000).

Another fundamental issue is to evaluate the environmental quality status of a given estuary or environmental compartment. In this case, as has been pointed out in the WFD, we need quality standards or references to compare with. In the case of sediments, surface sediments are useful for monitoring changes in current environmental tendencies, and core sediments allow reconstructing the historical input of pollutants and their impact in the ecosystem. Both surface and core samples are widely used in order to define the contamination factor (CF) of a sampling site. This CF can be calculated from the following equation (Tomlinson et al. 1980):

$$CF = \frac{C_n}{B_n} \; ;$$ (1)

where C_n is the trace metal concentration, and B_n is the background level for each trace metal. The reference values of these parameters were proposed by Müller (1979), as shown in Table 3.

Table 3. Scale of pollution levels proposed by Müller, 1979, obtained by the ratio of superficial sediment and background values

CF	Pollution level
> 48	EP: extreme pollution
12–48	HP: high pollution
3–12	P: pollution
1–3	LP: low pollution
<1	NP: no pollution

CF, Contamination factor defined in equation 1.

The use of palaeo-environmental reference conditions to evaluate these results was established for the Bilbao estuary by Cearreta et al. (2002). Further studies in other localities support this methodology (Andersen et al. 2004). The use of the correct reference conditions is even more important if we consider that in the Basque coast pristine or slightly polluted estuarine areas do not exist (Borja 2005). Although comparisons among different trace metal studies are difficult because of analytical differences and statistical approaches, mean background values obtained for this study are similar to those presented in other studies from northern Spain (see Rodriguez et al. 2006 for a summary of regional data).

A pollution classification of the estuary of Bilbao could be done from data in the literature, giving a critical measure of "environmental health" for this estuarine area. Mean concentration levels for both surface (Solaun et al. 2009) and core sediments (Borja et al. 1998) from 1998 to 2001 were selected as one set of reference conditions. Surface and core sediments from 1997 to 2006 also were considered (Cearreta et al. 2002, Leorri et al. 2008); this was the period when significant reductions in contaminant sources and improvements in waste treatment were implemented. Relative pollution levels were calculated by using Equation 1 (Table 4).

The highly variable distribution of the different elements responded to local sources of industrial pollution. Although these values were extremely high in superficial sediments, previous work had documented elevated levels of metals and As in cored samples from the middle estuary, in general even higher than those reported for superficial materials (Cearreta et al. 2002). These results indicate that the channel was heavily polluted. The improvement of environmental conditions in the Bilbao estuary began with the economic recession in industrial activities and closure of the mineral washeries. This improvement has been more noticeable since the construction of the Galindo central WWTP, which started physicochemical treatment in 1991. Ten years later (2001–2002) biological treatment became operative (Figure 2). Table 4 confirms the experimental evidence obtained by Landajo et al. (2004) where it was estimated that WWTP releases to the estuary amounted to almost 50% of the Cr, Pb, Cd, and Ni and up to 77% of Mn that the plant received from local domestic sewage (which is an important source of metal contamination).

In addition to these monitoring and speciation studies of trace metals, study of the presence of these trace metals and their differentiation from the compounds that belong to the sediments as natural compounds is fundamental in order to understand the mobility and retention processes of the trace metals, and thus, their bioavailability. For example, speciation studies of solid phase materials downstream from the WWTP can supply the information required to understand the effects of this type of facility.

Table 4. Pollution levels defined for the estuary of Bilbao according to the values obtained in eq. 1

Trace metal	Pollution level 1998–2001	Pollution level 1997–2006
As	P	NP
Cd	P	LP
Cr	P	LP
Cu	P	LP
Fe	LP	LP
Mn	LP	LP
Ni	P	LP
Pb	P	NP
Zn	P	LP

Villanueva et al. (2008) performed an exhaustive study of anthropogenic compounds, characterizing them by means of Raman spectroscopy. In that study, gypsum ($CaSO_4 \cdot 2H_2O$) was found in a sampling location of the estuary of Bilbao in front of the sewage treatment plant of the main iron production plant of the Basque Country, the ACB factory. This compound probably had been precipitated in situ by reaction of the soluble sulfate and calcium ionic species.

Calcium arsenate was always found in sediments downstream of iron and steel production factories. Arsenic is usually associated with iron minerals and is found in higher concentrations near iron-producing factories and sulfuric production plants using pyrites. If dissolved arsenate reaches a flow with high calcium content, then solid calcium arsenate is formed at the pH of estuarine and river waters.

Iron hydroxyoxide (FeOOH) was found in several locations of the estuary of Bilbao. Iron hydroxyoxide was also detected in the sediments in front of the main steel production plant in the estuary of Bilbao, where this compound has been found associated with iron oxide. The FeOOH is thought to have been formed anthropogenically in the WWTP of the steel factory or by chemical reaction in the seawater of the estuary where the pH of the river water changes from ~7.2 to ~8.3. Probably in both cases, the iron is bound to the surface of natural mineral particles, iron oxides or pyrite (FeS$_2$), suggesting the in situ formation of the FeOOH compound (Figure 5).

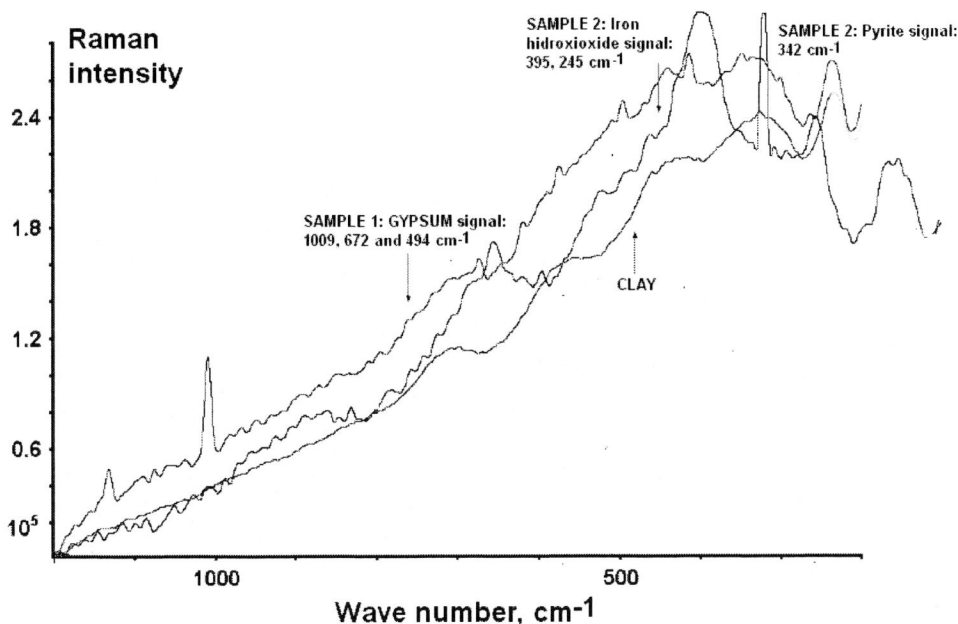

Figure 5. Raman spectra of gypsum, iron hydroxyoxide, pyrite and clay in a sediment from the estuary of Bilbao.

Finally, Villanueva et al. (2008) pointed out the extensive presence of pyrite not only on a clay background as a natural compound but also together with an anthropogenic species of iron. This fact was confirmed by means of SEM-EDS analysis (scanning electron image - X-ray microanalysis). In Figure 6a, a typical electron image for sediment from the estuary of Bilbao (Galindo) is shown. From the five EDS spectra, two of them (spectra 2, 3 and 4) belong to the background sediment as seen by the main components of clays (Al, Si, K, Fe; Figure 6b-top). Additionally, the other two spectra are associated with PbS and FeS species as shown in Figure 6b-bottom. These compounds were not detected by Raman spectroscopy, surely because of the small size and the dispersed presence of these species. The presence of sulfides of lead and iron can be attributed to anthropogenic origins. This fact has been also confirmed by the image shown in Figure 7 (a, b) where a spherical ball of iron oxide was captured by the microscope.

ORGANIC MICRO-CONTAMINANTS

Levels and Toxicity

Polycyclic aromatic hydrocarbons (PAHs), included by the U.S. Environmental Protection Agency (EPA) among the priority pollutants (Keith & Telliard 1979), make up a group of compounds, which is the largest known class of chemical carcinogens and mutagens.

The 16 compounds included by the EPA or the eight included by the WFD are used to define environmental quality standards in many compartments (EC 2008).

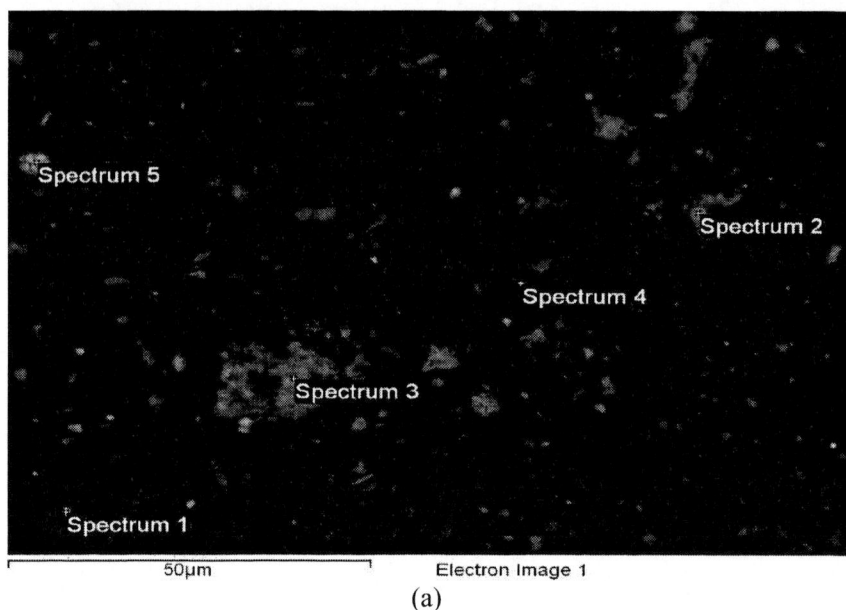

(a)

Figure 6. (Continued)

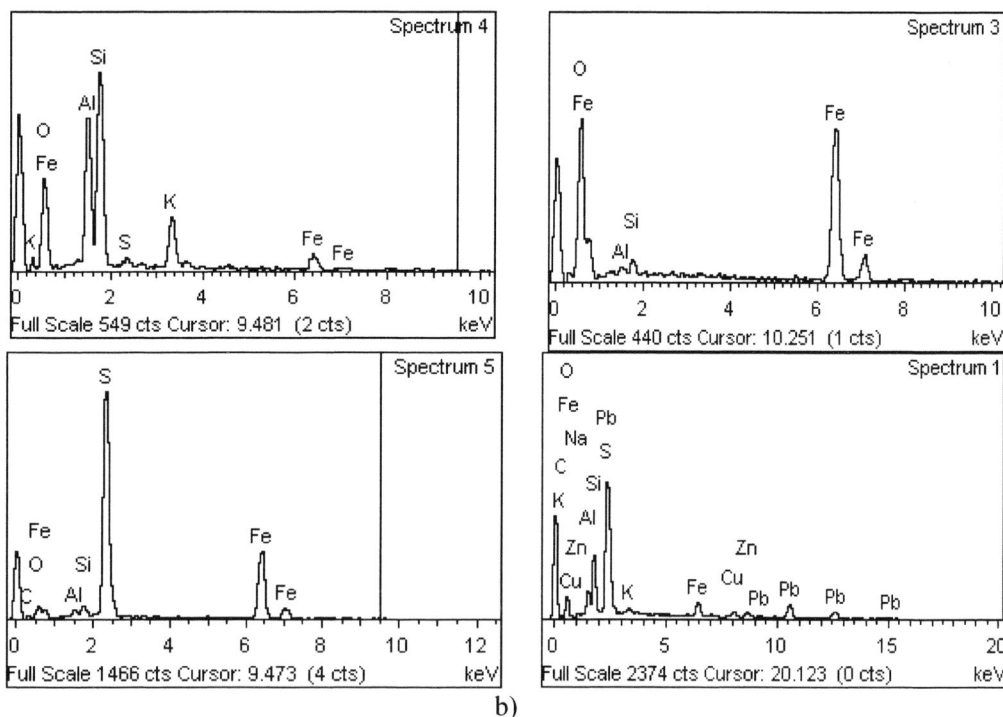

b)

Figure 6. (a) SEM-EDS image of a sediment sample obtained in the station of Galindo and (b) the EDS spectra of four different shots (see text for further explanations).

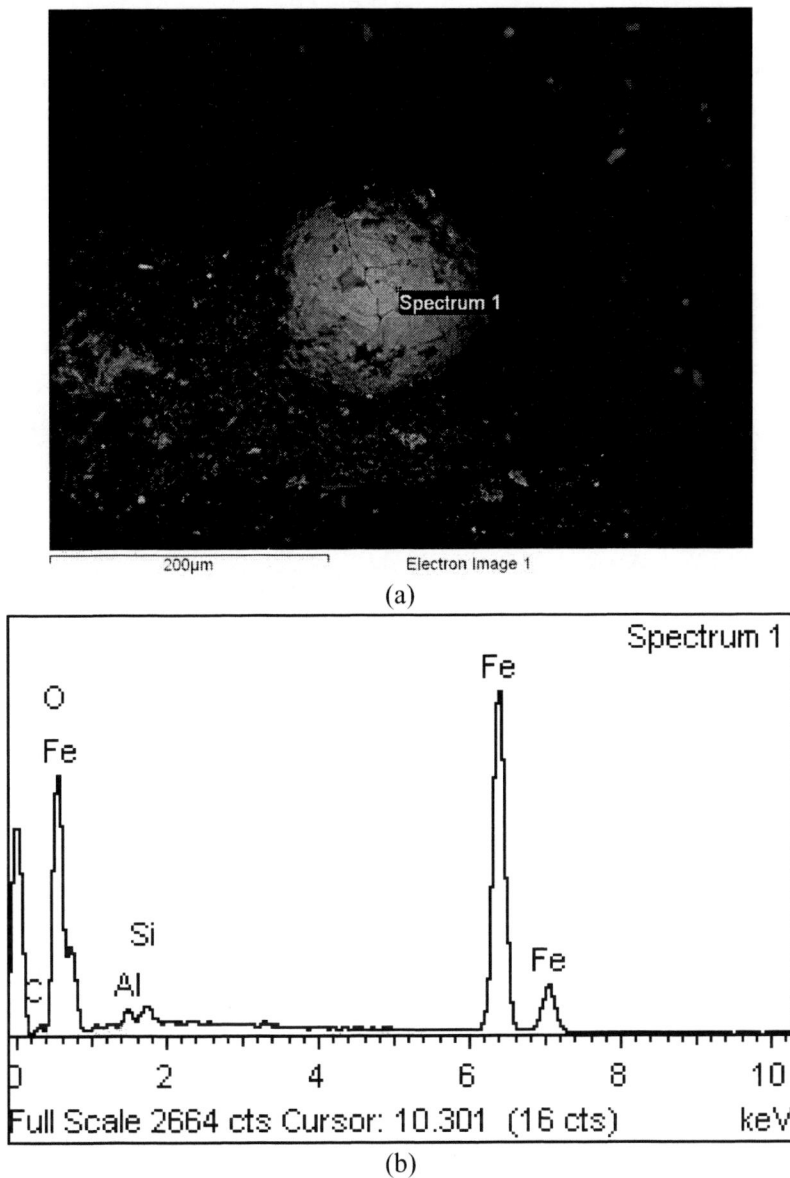

Figure 7. (a) SEM-EDS image of a sediment sample obtained in the station of Galindo and (b) the EDS spectra (b).

Although incomplete combustion in high temperature processes is the main anthropogenic source of PAHs, not all the sources of these compounds are linked to human activities. Natural sources such as fires, volcanic eruptions or other natural phenomena may contribute to the presence of these compounds in almost every environmental compartment.

In contrast, polychlorinated biphenyls (PCBs), found in various environments all over the world, are a consequence of anthropogenic activities. In fact, PCBs constitute a class of 209 compounds (called congeners) that vary in biological activity and toxicity, as a result of differences in the number and position of chlorine atoms in the molecular structure.

(a)

(b)

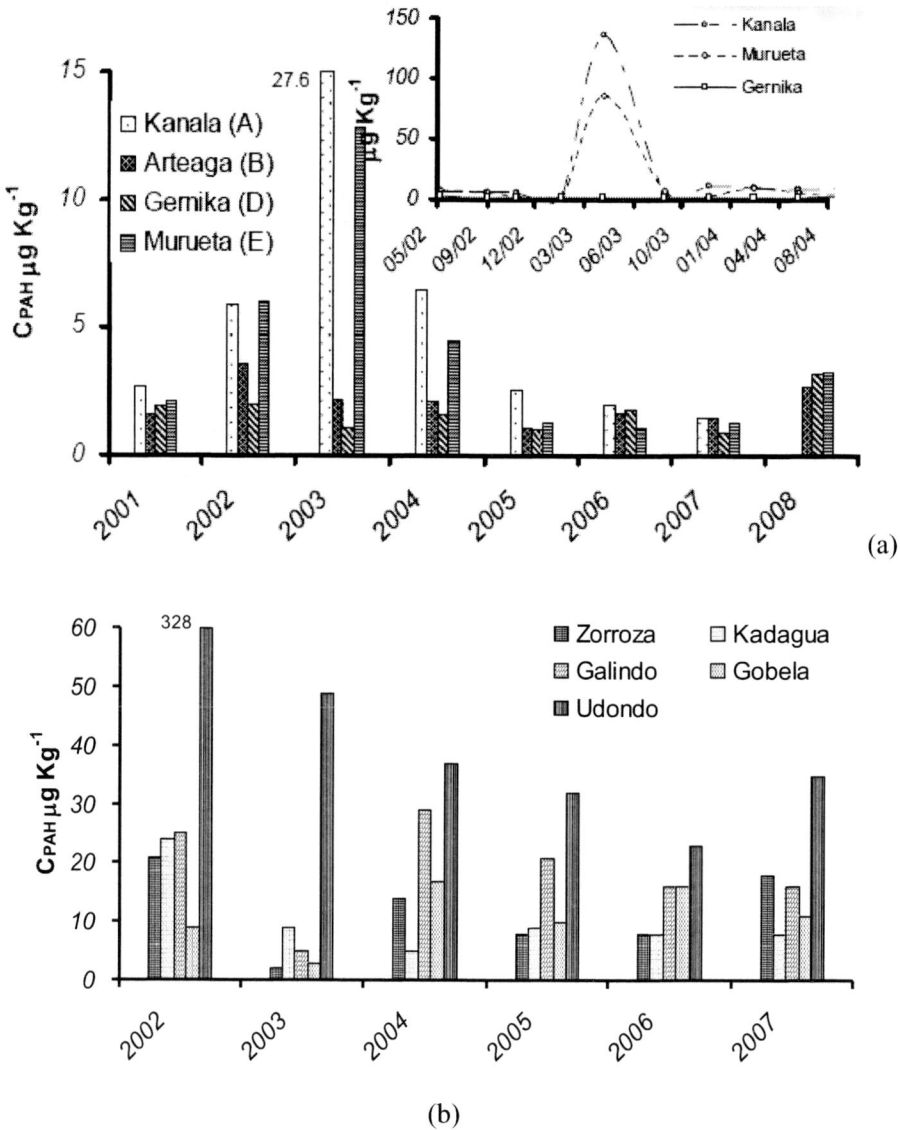

Figure 8. (a) Total mean concentration (in mg kg^{-1}) for PAHs of four sampling points in Urdaibai, 2001–2008. Inset: PAH concentration profile for Kanala, Murueta and Gernika from February 2002 to August 2004. (b) Total mean concentration (in mg kg^{-1}) for PAHs of five sampling points in the Estuary of Bilbao, 2002–2007. In both cases the annual mean values were calculated with different numbers of seasonal samples.

The PAHs and PCBs are released into estuaries through a variety of ways: from coke and petroleum refining industries, accidental oil spills and leakages (Franco et al. 2006), aerial fallout (Lim et al. 2007, Freells et al. 2007), rainwater runoff (Bozlaker et al. 2008), from forest and prairie fires, vehicle traffic and domestic heating (Gaga & Ari 2011), and wastewater effluents (Masclet et al. 1986, Leung 2007). It is generally accepted that human and industrial activities constitute the main source of PAHs, but in environmental studies it is difficult to discriminate between pyrolytic products and petroleum derived ones, because of the complexity governing PAH distributions in the environment and the possible coexistence of various contamination sources.

PAH and PCB concentrations in sediments are often several orders of magnitude greater than aqueous phase levels, because their high K_{ow} and their long persistence due to their slow degradation rates. PAH and PCB tend to accumulate in fat-rich tissues and therefore they are found in fish and in edible shellfish (Arias et al. 2010), thus their presence in the water column and sediments may constitute a human and environmental health hazard. Consequently, surveillance of environmentally sensitive areas includes monitoring of PAH and PCB, among many other contaminants, and the identification of the fate and distribution of these pollutants.

The analysis of organic micro-contaminants in sediments includes sampling the upper sediments (most recent), stored in pre-cleaned flasks with 1 mL 4% (v/v) formaldehyde added to avoid bacterial growth, and transport in cooled boxes to the laboratory. Once in the lab, sediment samples were pre-treated (frozen, freeze-dried and, in some cases, sieved in different fractions) and were kept in the refrigerator at 4 C until analysis. There are multiple analytical methods for PAH and PCB determination in the literature but most of them consist of an assisted solvent extraction (microwave, ultrasound, pressurized solvents, etc.) and subsequent detection by gas or liquid chromatography after a clean-up step. Most of the results obtained in this chapter have been determined through a simultaneous method for both families of compounds (PAH and PCB) and other organic pollutants of interest (Bartolomé et al. 2005).

Figure 8 shows the total average concentrations for 16 PAH in the sediments at the Urdaibai and Bilbao estuaries from 2001-2008. The concentration at each sampling point is the average of 3 or 4 seasonal sampling campaigns taking into account only the fine sediment fraction (< 63 μm).

Between 2002 and 2004, the most remarkable observations in the estuary of Urdaibai were the high concentrations measured in the downstream reaches of the estuary (Kanala and Murueta). At both sampling points, the average concentration of PAH was more than 10 mg kg^{-1} with the highest average concentration in 2003 at Kanala (27.6 mg kg^{-1}). If the original data were considered, a high concentration outlier could have been observed at mid-year 2003 in both sampling stations (Figure 8a).

The fact that the average concentrations in the innermost sampling points remained constant is another remarkable fact. One of the sampling points, Gernika, is located very close to the most populated village within the Biosphere Reserve (see Sampling Area section). At first glance, Gernika might have been considered an increasing focus of pollution due to human impacts, but according to the results, the measured levels remained constant during the sampling period (2001-2008) at this site.

Because Urdaibai is an unstressed area, mean organic pollutants concentrations remain almost constant over long periods of time and, for this reason, contamination events are especially noticeable. In this regard, massive oil spills from the beginning of 2002 until mid-2003 as a consequence of the shipwreck of the *Prestige* tanker have been cited to explain why these high concentrations were found in the outermost points of the estuary, while the innermost areas (Arteaga & Gernika), have remained virtually unchanged (Figure 8a).

In the case of the estuary of Bilbao, the highest concentrations for PAH were always found in Udondo, with mean concentrations of 328 mg kg^{-1} in 2003 (Figure 8b). Although this station was located in the seaward part of the estuary of Bilbao and a tidal dilution effect could be expected, this station always shows the highest values (Bartolomé et al., 2005) probably due to the presence of multiple industrial and urban effluents around in the sampling site. Moreover, the station is at a dock that is totally flooded only when the tide is high, so contaminated sediments from the upper part of the estuary tend to accumulate at the structure, forming a sink for upstream contaminants.

In addition to these findings, a significant downward trend in the concentration of PAH was observed over the sampling period. One of the reasons could be the dredging carried out by the authorities of the harbor to keep the channel navigable and to avoid sediment accumulation and flooding risk. A common practice is to dump the dredged sediments in the outer harbor to aid the reclamation processes. Another reason for this downward trend could be new sediment deposition, whereby the earlier contaminated sediments are covered by less contaminated sediments.

The rest of the stations could be classified into two different groups according to the mean concentration of PAH: the first group including Galindo and Udondo with a mean concentration about 19 mg kg^{-1} and a second group including Gobela, Kadagua and Zorroza averaging about 10 mg kg^{-1}. Galindo and Gobela are two tributaries of the main estuary channel and the sampling points are close to the junction. On the contrary, Udondo is a dock.

The Galindo station is downstream from the city sewage treatment plant (which discharges ~400,000 m^3 per day). This fact makes this station especially remarkable since it shows the highest values in many parameters as shown in the following descriptions.

This difference between the sampling sites along the estuary of Bilbao is also seen in the case of the concentration of PCB. As can be seen in Figure 9b, the stations Galindo (2866 μg kg^{-1}), Udondo (932 μg kg^{-1}) and Gobela (367 μg kg^{-1}) are the sampling sites where the highest concentrations were observed, whereas the average concentrations in Kadagua (~80 μg kg^{-1}) and Zorroza (~200 μg kg^{-1}) remained more or less constant over a range much lower than the above three points. As discussed above, Udondo is a dock where pollutants tend to accumulate, and Galindo is close to the sewage treatment plant.

Generally, the average concentration in the estuary of Urdaibai is 10–100 times lower than in the estuary of Bilbao, but as shown in Figure 9a, the concentration of those pollutants has increased steadily at all stations.

Since sediment chemistry data alone do not provide enough clues for assessing the environmental risks to aquatic organisms, interpretive tools such as the Sediment Quality Guidelines (SQGs) may be an interesting and valuable tool to fill the gap between the concentrations found in sediments and the expected effects observed in organisms.

Based on empirical data two sediment quality assessment values for each compound were developed. The ERL (effects range low) was calculated as the lower 10th percentile of effects concentrations and the ERM (effects range medium) as the 50th percentile of effects concentration (Long & Morgan 1990).

The results of this comparison should be studied carefully since the threshold values were for total concentrations in sediments, whereas the concentrations in our study were from mixed sediment fractions (<63μm and 63<x<250μm). As can be seen in the pie charts (Figure 10), from a total of 146 samples collected from 2001–2008 in the Urdaibai estuary the total concentration of PAHs or PCBs exceeded the ERM limit in only 3% of the cases. On the contrary, from the 226 samples collected in the estuary of Bilbao, 18% of samples exceeded the ERM for PAHs and 62% for PCBs.

(a)

(b)

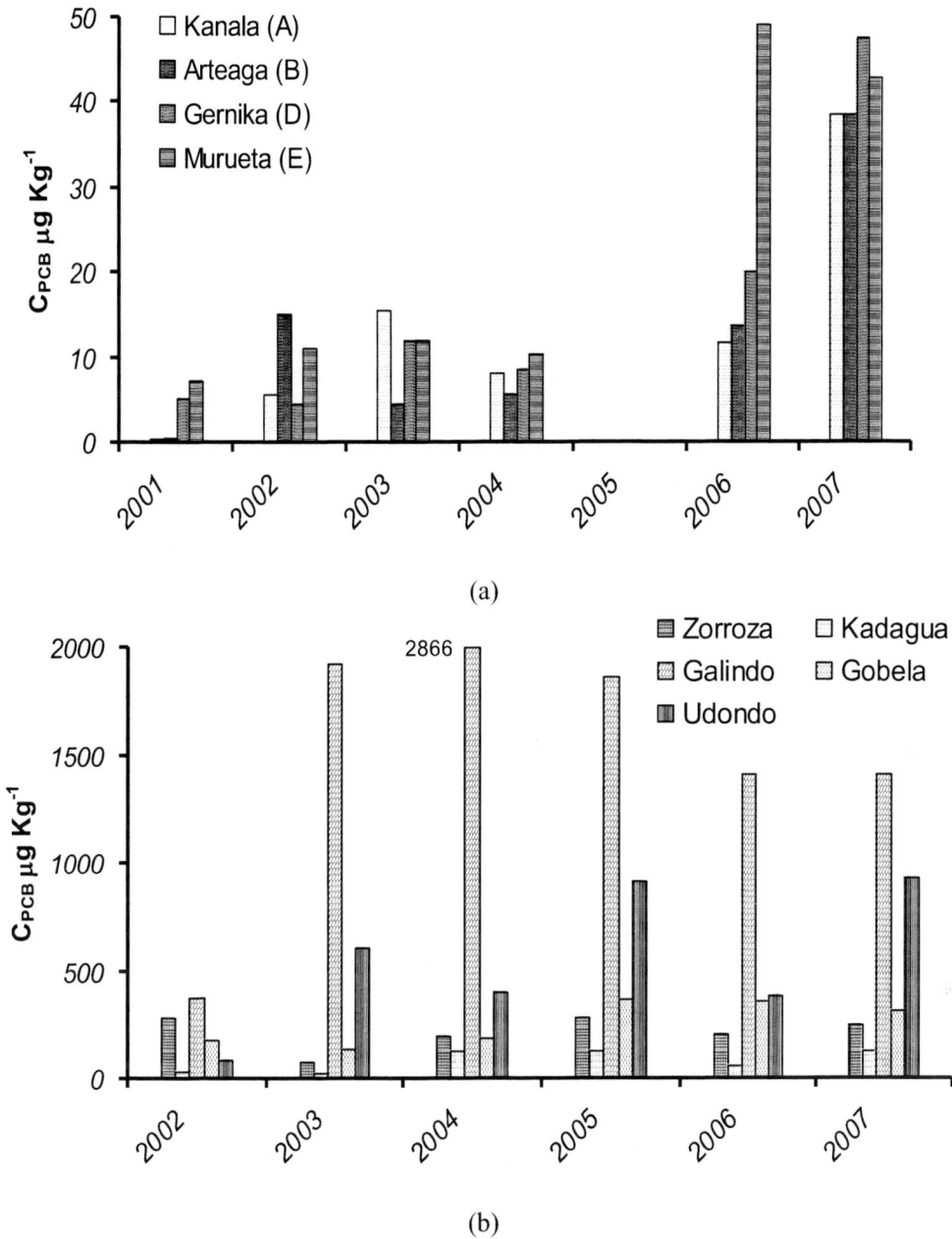

Figure 9. (a) Total mean concentration (in μg kg^{-1}) for PCB at four sampling points in Urdaibai, 2001–2008. (b) Total mean concentration (in μg kg^{-1}) for PCB at five sampling points in the Estuary of Bilbao, 2002–2007. In both cases annual mean values were calculated with different numbers of seasonal samples.

The results obtained showed low to moderate contamination levels in Urdaibai estuary and high contamination levels for PAH and, in particular, PCB in the estuary of Bilbao. This was especially remarkable at sampling points such as Galindo, where virtually none of the sample concentrations were below the ERL limit for PCB.

Toxicity in Urdaibai (PAHs)

Toxicity in Urdaibai (PCBs)

Toxicity in estuary of Bilbao (PAHs)

Toxicity in estuary of Bilbao (PCBs)

□ CPAHs<ERL   ⊠ ERL<CPAHs<ERM   ⊞ CPAHs>ERM   □ CPCBs<ERL   ⊠ ERL<CPCBs<ERM   ⊠ CPCBs>ERM

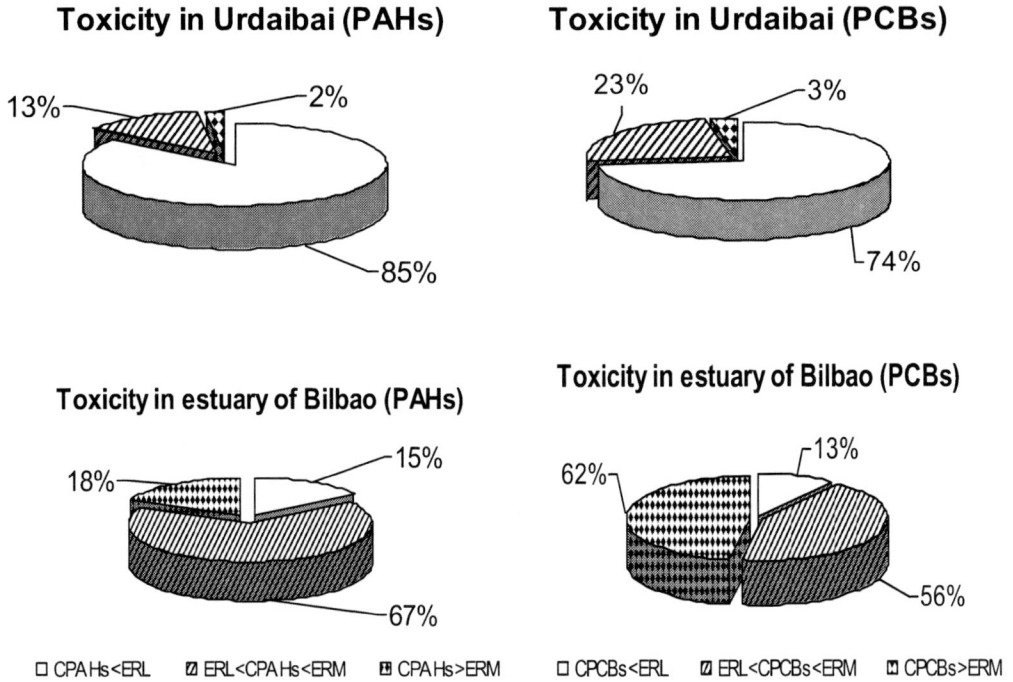

Figure 10. Sediment quality guidelines (SQG) values for PAH and PCB and percentages of samples within categories of SQG toxicity for Urdaibai and Bilbao estuaries. ERL for ΣPAH = 4022 µg kg^{-1}; ERM for ΣPAH = 44,792 µg kg^{-1}; ERL for ΣPCB = 22.7 µg kg^{-1} ; ERM for ΣPCB = 180 µg kg^{-1}.

ORGANIC CONTAMINANTS DISTRIBUTION: PARTICLE SIZE AND ORGANIC MATTER

Non-polar organic contaminants, namely those with log K_{ow}, (octanol/water partition coefficients) values of 5 or higher, such as PAH and PCB, are strongly absorbed to organic matter or adsorbed to particles and tend to accumulate in sediments. Consequently, bed sediments play an important role as accumulators for PCBs and PAHs in the aquatic environment.

Although the partition coefficient increases with increasing hydrophobicity of the xenobiotics and the amount of organic matter in sediment, the role of organic matter associated with the sediment particles has not been thoroughly evaluated. In the case of dissolved organic matter, changes in the structure can alter the observed partition coefficient even on a carbon normalized basis (Kukkonen & Oikari 1991).

As a consequence of high K_{ow} values, the dissolved fraction of non-polar pollutants is very small and the fate and bioavailability of hydrophobic chemicals in aquatic ecosystems are affected by the presence of dissolved and particulate organic matter (Schwarzenbach et al. 2003). This is the case when benthic organisms living in contaminated sediment accumulate high body burdens and transfer them to other components of the food chain. Accumulation of sediment-associated contaminants may occur either via the aqueous phase, i.e., passively through direct contact with pore water, or through ingestion of contaminated sediment particles. The importance of these routes to contaminant accumulation depends on the ecology and feeding behavior of the organism and characteristics of the sediment and chemicals.

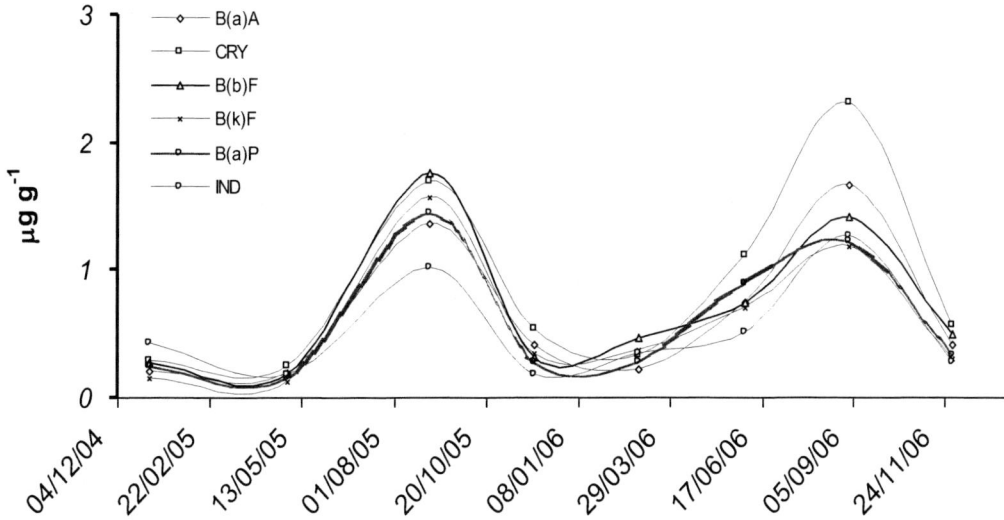

(a)

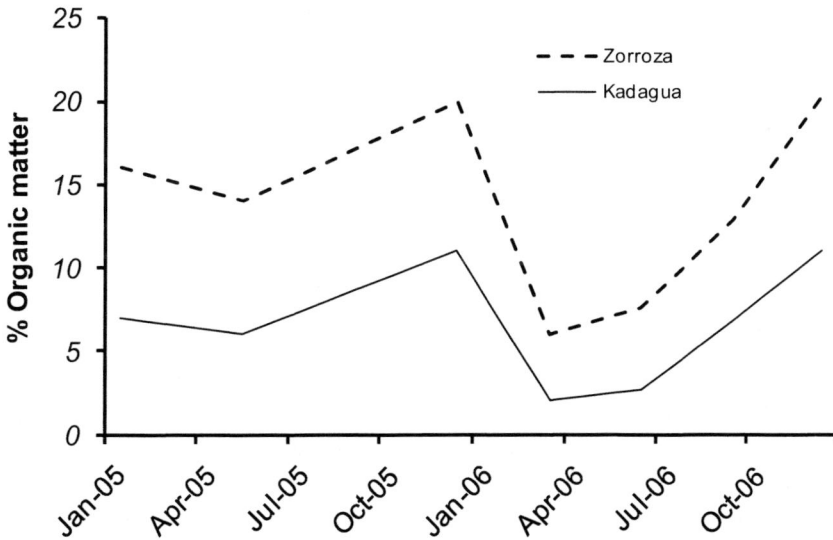

(b)

Figure 11. Seasonal variation for heavy PAH studied at (a) Zorroza sampling point (estuary of Bilbao). (b) Variation of organic content in sediments in two locations: Zorroza and Kadagua. (Abbreviations: Benzo[a]anthracene (B[a]A), Chrysene (Chr), Benzo[b]fluoranthene (B[b]F), Benzo[k]fluoranthene (B[k]F), Benzo[a]pyrene (B[a]P), Indene[1,2,3-cd]pyrene (Ind)).

In order to evaluate the importance of particle ingestion in the accumulation process it is necessary to determine the partitioning of the xenobiotics to different particle size fractions. Since 2003, the sediments collected in the sampling campaigns along both estuaries were fractionated into two particle sizes: x<63 µm and 63 µm <x<250 µm. The relationship between concentration and particle size showed a clear dependence in the case of Bilbao estuary for both contaminant families (Bilbao estuary $0.6<R^2_{PCB}<0.8$, $0.7<R^2_{PAH}<0.8$). In the case of Urdaibai estuary this relationship is not so clear Urdaibai ($0.4<R^2_{PCB}<0.5$, $0.1<R^2_{PAH}<0.4$), but it was stronger in those sites with the highest contaminant levels (Gernika and Murueta). Another factor that may influence the relationship between the concentrations of pollutants in the different fractions of sediment could be the percentage of organic matter. Organic matter content in sediment fractions from Bilbao estuary was almost always higher than in Urdaibai estuary.

Seasonal variations in organic matter may explain this pattern. An example was found in Bilbao estuary during 2005–2007 sediment sampling. As shown in Figure 11a, a seasonal pattern was detected in Kadagua for heavy PAH. At this station the highest concentrations were found in summer (i.e. chrysene Sept. 2005, 2694.6 ng g^{-1}; chrysene Sept. 2006, 2680.9 ng g^{-1}) and the lowest ones in winter (i.e., chrysene May 2005, 350.2 ng g^{-1}; chrysene March 2006 747.0 ng g^{-1}). This profile was also observed in Zorroza but with slight differences in the period of maximum and minimum concentrations. Additionally, these two stations showed the lowest levels of xenobiotics (see Figure 11b). Therefore, this pattern could be related to geochemical cycles in the sediments, probably associated with variations in organic matter and its potential behavior as scavenger of organic micro-contaminants. Variations of organic matter content in sediments probably were attributable to a significant terrestrial plant contribution (up to 35%) in the fine-grained sediments nearer the inflowing rivers (Volkman et al. 2008) and during the autumn. Therefore, an increasing amount of organic matter in the sediment increased the absorption of PAH and PCB, and they were buried in the sediment. In spring-summer, when the contribution of natural organic matter decreased, either the contaminants are released from the sediment, or fresh sediments with lower organic matter content are deposited on top of the older, polluted sediments.

Sources of PAH and PCB Contaminants

PAH contamination can be generated from three main processes: (1) combustion at a very high temperature of coal, wood, fuel, etc.; (2) release of petroleum into the environment; or (3) diagenetic processes (degradation of organic matter). The first two processes are usually anthropogenic; in the third case the source is natural. The molecular PAH patterns generated by each source are like fingerprints and it is possible to determine the processes that generate PAHs when studying sediment PAH distributions (Yunker 2002).

Pyrolytic aromatic hydrocarbons are characterized by the occurrence of PAH with a wide range of molecular weights, whereas petroleum hydrocarbons are dominated by the lowest molecular weight PAH, with penta- and hexa-aromatics present only at trace levels (Yunker et al. 2002).

Several factors, such as chemical composition of the organic matter or temperature, affect yields and distribution of PAH formed during incomplete combustion of organic matter or during its thermal maturation. Thermal PAH formation can occur over a wide range of temperatures. At low temperatures, the compound distribution is governed by thermal stability and the most stable isomers are formed, while at high temperatures, PAH of higher formation enthalpy can be formed.

Pyrolytic aromatic hydrocarbon profiles are dominated by unsubstituted PAH with a wide molecular weight range, up to the hexa-aromatics, and even heavier compounds. On the contrary, in naturally formed PAH, such as those obtained during the slow maturation of petroleum, lighter aromatic hydrocarbons predominate, with a complete suite of alkylated PAH.

Therefore, the study of the distribution of alkylated derivatives and the parent PAH according to their molecular weight can provide information on the process from which these compounds were generated. For example, the PAH/C1-PAH (concentration of the parent compound relative to the sum of the concentrations of the methyl compounds) and high/low (concentration of the high molecular weight compounds relative to the concentration of the low molecular weight compounds) indices can discriminate different sources. These values are higher for PAH generated by pyrolysis of organic matter at high and very high temperatures than for those derived from petroleum.

Based on thermodynamic and empirical data Yunker et al. (2002) proposed a set of indices comparing the ratios of several isomers (anthracene and phenanthrene, fluoranthene and pyrene, benzo-a-anthracene and chrysene, etc.) that can be useful in identifying the most likely sources of PAH.

A first estimate of the origin of pollution in the estuary of Bilbao sediments in 2006–2008 was obtained by plotting the values of the molecular index anthracene/(anthracene + phenanthrene), abbrev. Anth/(Anth + Phen) (Figure 12a). The zone defined by >0.1 ratios is characteristic of pyrolytic origin, whereas the one defined by ratios <0.1 is characteristic of petroleum-derived PAH. The major observation is that for all the sediments analyzed, the contaminants originate mainly from pyrolytic combustion. If another ratio, indene/(indene + benzo(ghi)perylene), abbrev. ind/(ind + B(ghi)P), is presented for the same data, pyrolytic PAH sources could be further divided into two groups where fuel combustion was the main source (Figure 12b). This source in this estuary could be related to industries, wastewaters, traffic and different types of heating during the cold months. From the analysis of frequencies of the four ratios in Urdaibai estuary, the main source in the sediments of the whole estuary is combustion (>90%) (Figure 13a), although it is difficult to distinguish the different raw materials (petroleum, coal, biomass, etc.). In any case, observations in Urdaibai estuary are consistent with a higher contribution from typical rural contamination sources linked to coal and wood combustion than in the Bilbao estuary. However, the time profile of some of those four ratios gives added information about the behavior of the distribution of PAH in those sediments. In fact, as can be seen in Figure 13b for Kanala, the ratios benzo(a)anthracene/(benzo(a)anthracene + chrysene), abbrev. BaA/(BaA + Chy) and Indene/(Indene + Benzo(ghi)perylene), abbrev. Ind/(Ind + BghiP) show a depression between May 2003 and March 2004, which was generally observed in all the sites and for both particle-size fractions. From the ranges described by Yunker (2002), the petroleum and combustion sources of those two ratios are 0.20–0.35 and 0.20–0.50, respectively, and values higher than those mean a net combustion source. From two other ratios—Anth/(Anth + Phen) and fluoranthene/(fluoranthene + pyrene), abbrev. Flu/(Flu + Pyr)—the combustion of petroleum (<30% frequency) and other combustion (>60% frequency) are inferred, but without any time pattern.

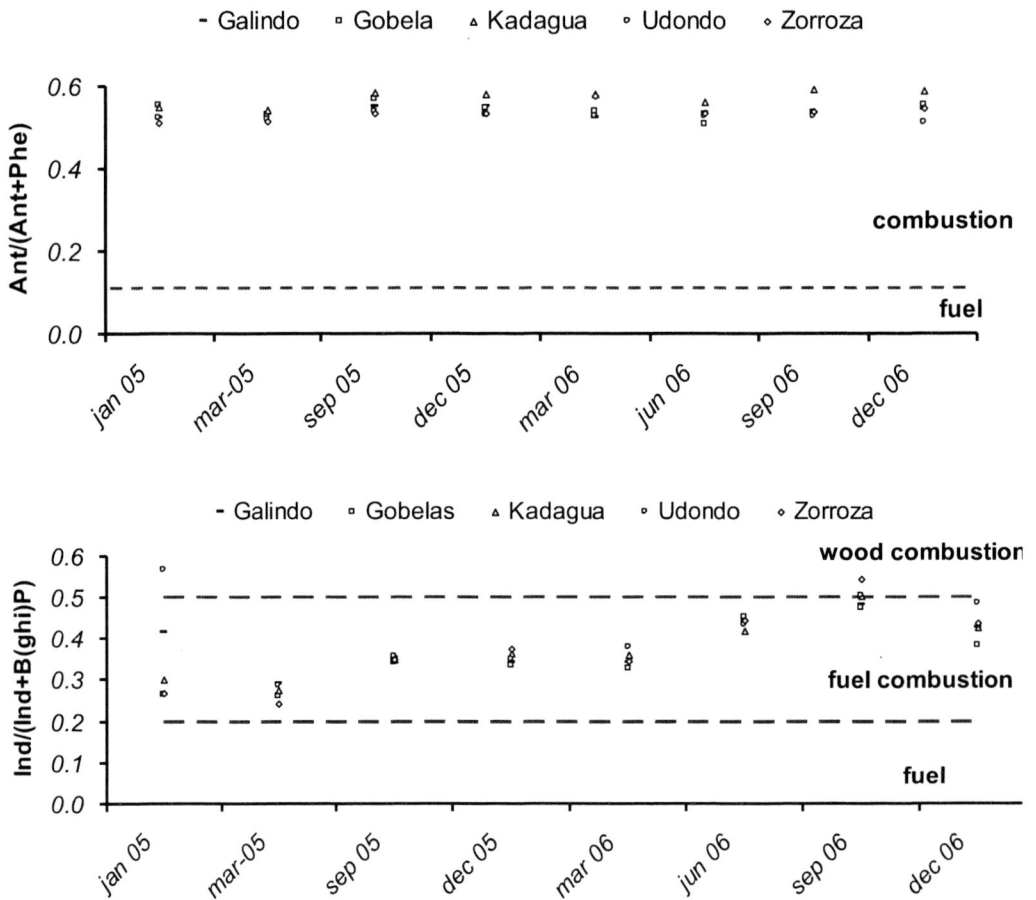

Figure 12. Ant/(Ant+Phe) and Ind/(Ind + BghiP) ratios in sediments from five sampling points from estuary of Bilbao from 2005–2007 (Abbreviations: *Anthracene*, Ant; *phenanthrene*, Phe; *benzo[g,h,i]perylene* B(ghi)P, *indene[1,2,3-cd]pyrene*, Ind.

From the ratios shown in Figures 13a and b, the sudden and steady alteration introduced in the estuary in that period of time was explained either by direct petroleum contamination or the combustion of petroleum sources. It should be recalled that this change was not observed for total concentrations (only the sudden increase in Murueta and Kanala in May 2003, Figure 8a), but these ratios changed at all of the stations and the original state was not recovered for almost two years (2005–2006). This and many other signs pointed to the oil spill of the *Prestige* shipwreck as the most probable source of the observed hydrocarbon pollution (Cortazar 2008, Puy-Azurmendi 2010).

All of the PCB compounds are synthetic, but it is possible to determine specific sources. As mentioned above, PCB have been used extensively as commercial products and in various industrial applications (Platonow et al. 1971, Hutzinger et al. 1974, Costabeber & Emanuelli 2003). Industrial use of PCB started in the early 1930s, and since then over 2 million tons of PCB have been produced as commercial mixtures of 60–90 congeners such as Aroclor in United States and Great Britain, Kaneclor in Japan, Pyralene in France and Solvol in Russia and the former Soviet states (Kim et al. 2004). It is estimated that about 40% of this worldwide production entered the environment. Although most countries have prohibited PCB production, a considerable amount of these compounds is still in use, mainly in electronic equipment (Penteado & Moreira Vaz 2001). Due to its many possible sources, it is an interesting problem to determine the specific source of pollution in an estuary, whether a direct use of banned products or the accumulation of diffuse sources.

As an example of identifying the direct sources, a simple study was carried out along Urdaibai estuary with sediment samples collected from 2002 to 2004 in both size fractions. Although all existing congeners were not measured, it can be estimated roughly which kind of Arochlor could be the source of PCB contamination in sediments. In Table 5 we have collected the observed mean ratio values for different congeners measured (CB28, CB52, CB153, CB180) versus CB 138 in various commercially available Arochlors (according to Frame et al. 1996) and the experimental values obtained from the studied sediments. These theoretical ratios could be compared with the environmental ratios we observed in estuarine sediments.

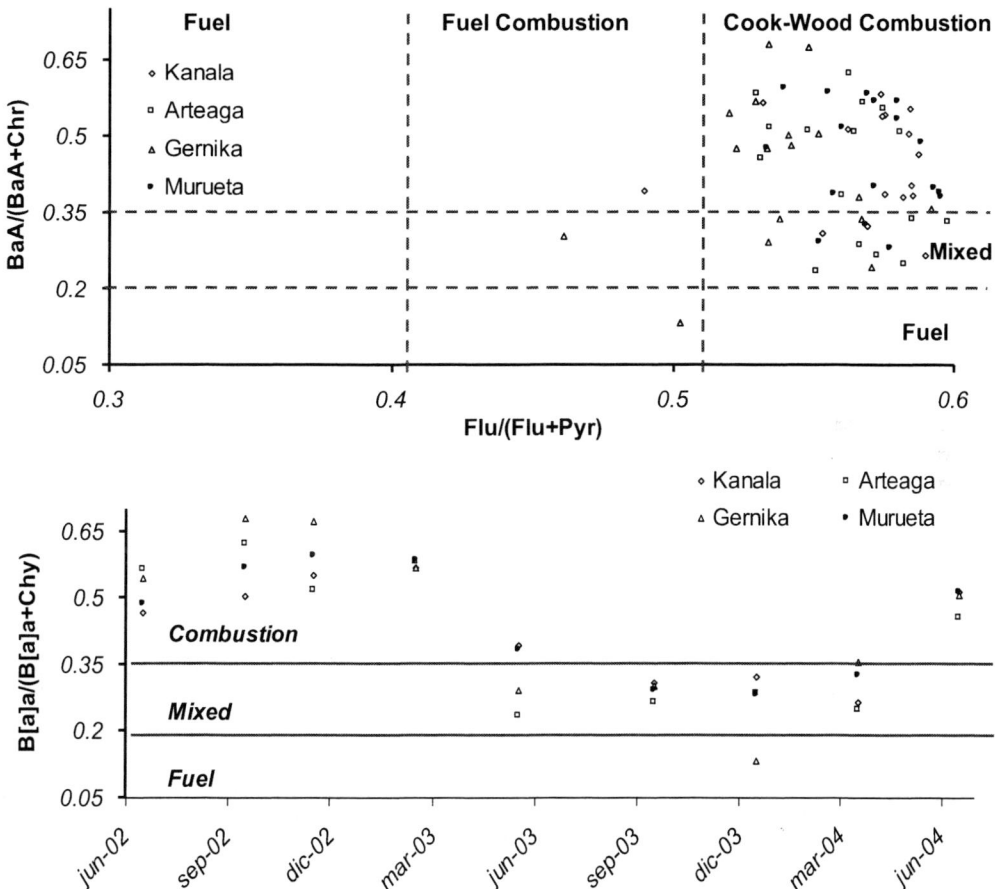

Figure 13. BaA/(BaA + Chy) against Flu/(Flu+Pyr) and B[a]A/(B[a]A+Chy) ratios in sediments from Urdaibai from June 2002 to June 2004.

Table 5. PCBs ratios measured in different available Arochlors and sediment samples from Urdaibai estuary from 2002–2004

Arochlor	CB28/CB138	CB52/CB138	CB153/CB138	CB180/CB138
1248 A3,5	9.58	18.48	0.6	0.04
1248 G3,5	13.47	13.49	1.03	0.5
1254 A4	0.01	0.14	0.55	0.07
1254 G4	0.03	0.92	0.65	0.12
1260	0.005	0.04	1.44	1.74
63µm<sediment<250µm	0.21 ± 0.18	0.16 ± 0.21	1.02 ± 0.52	0.84 ± 0.33
Sediment <63µm	0.23 ± 0.26	0.34 ± 0.36	0.74 ± 0.37	0.83 ± 0.39

According to the results, pollution from uses of various types of Arochlor 1248 can be discarded because the expected ratios with these Arochlors would be higher (for CB28 and CB52) or lower (CB153 and CB180) than measured in our sediments. For the same reason a single source from Arochlor 1260 could be also excluded. Therefore, the most probable Arochlor source for PCB congeners along the Urdaibai estuary would be Arochlor 1254, probably mixed with other potential sources of individual PCB congeners or other Arochlor contributions.

In the identification of diffuse sources we can use the data obtained during 2006–2007 surveys in Urdaibai. The total concentration of PCB ranged between 58 and 220 µg kg^{-1}. Moreover, 4–6 chlorine-atom PCB were the most abundant, with a predominance of CB-194 isomer (this congener was not measured in previous surveys). The profile of the selected PCB was studied over the sampling period at each sampling location.

As can be seen in Figure 14, while a singular pattern was found for each sampling location during 2006, the same profile was found in all the sediments along the estuary of Urdaibai in the two last sampling campaigns, which suggests a common source of PCB during this period of time (2007). Besides, this common profile, found in the last two sampling campaigns, had been observed previously during 2006 from a new sampling station at Gernika, downstream from the WWTP. This fact suggests that this treatment plant could concentrate diffuse sources of PCB from all of the urban and industrial sewage discharges, so that this profile became apparent only downstream from the WWTP effluent.

Statistical Analysis in Environmental Studies

In addition to the sources or temporal and spatial distributions of pollutants, statistical analysis of large environmental data sets can provide information that could have been hidden in a preliminary analysis.

One of the most used statistical tools is principal components analysis (PCA) (Einax et al. 1997). It is an exploratory, multivariate, statistical technique that can be used to examine data variability. Based on multivariate techniques we can consider a variety of factors affecting the data variability (Boruvka et al. 2005) and therefore offer significant advantages over univariate techniques.

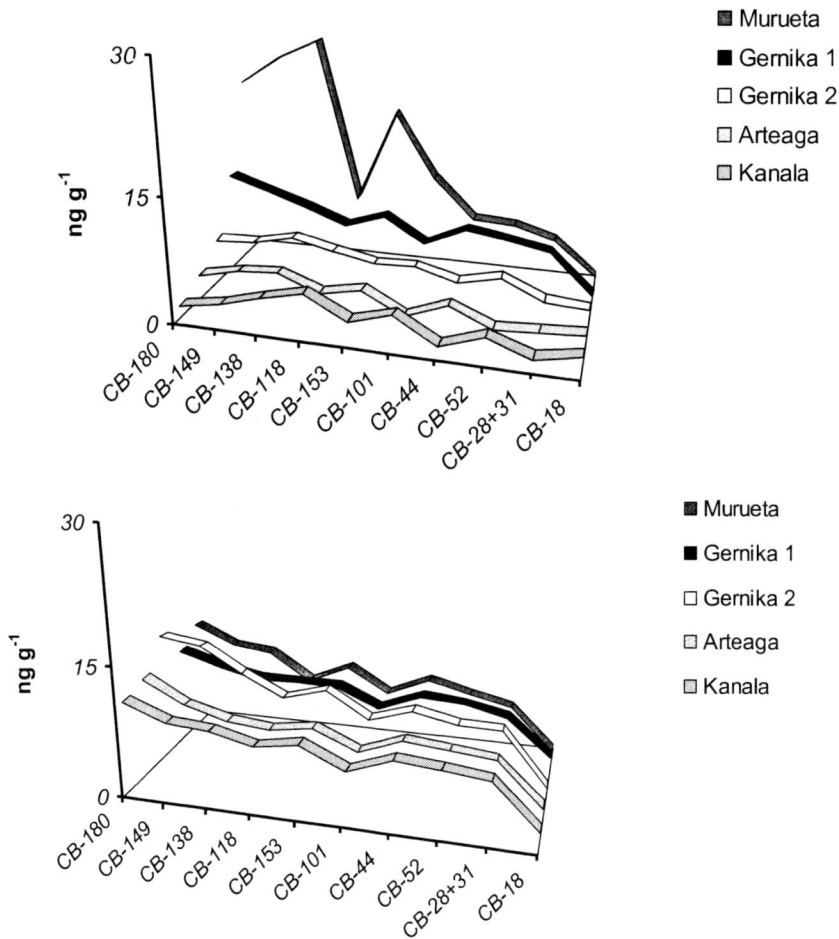

Figure 14. PCB profiles in sediments from all the sites sampled in September 2006 (a) and June 2007 (b). The CB 194 was omitted in the figure for clarity.

Environmental applications of PCA are diverse and widespread. This technique has been applied to various estuary compartments such as surface and ground waters (Chen et al. 2007), sediments (Spencer 2002) and biota (Bartolomé et al. 2010). Typically, PCA has been used to examine the spatial variability of contaminants. However, it has also been used to discriminate between contaminant sources and to identify key variables for environmental monitoring purposes (Carlon et al. 2001).

A principal component analysis (PCA) of the analytical data was carried out in order to find latent structures in the Bilbao estuary for the 2005–2007 campaigns. To avoid missing values in the statistical study, PCB were grouped, taking into account the number of chlorine atoms per molecule. Concentrations were logarithmically transformed to assure a normal distribution of the data. All PC models were built by full cross-validation.

PCA is widely used in unsupervised pattern recognition to identify principal components (PC) which best characterize a data set. The information or the structure hidden in a data table (samples x variables) can be revealed by a systematic decomposition of the original data in PCs which are represented by two sub-matrixes: loadings and scores. The loadings are the projection of the PCs on the original variables (showing the inner structure of the variables) and the scores are the projection of the samples on the PCs (showing the clustering of the samples) (Livingstone, 2009).

As can be seen in Figure 16, where scores are plotted, the samples can be roughly grouped according to the source, with Galindo (GA) in the left most extreme of the first PC and Kadagua (KA marked with + symbol) opposite. This pattern is explained by the content of the different contaminants measured in those samples, as can be seen in the loadings plot (bold annotations on Figure 15). If we focus on the distribution of the measured contaminants, on the one hand we can see that the overall concentrations of PAH, PCB and other contaminants are behind the first PC. Essentially, this first PC responds to the total concentration of organic micro-contaminants and metal content. On the other hand, the PCB are behind the first and third PCs. This distribution of the samples can be interpreted, as it has been pointed out before, in terms of relative contamination levels: the samples from Kadagua and Zorrotza are those with the lowest overall concentration of all the compounds, while Galindo, Udondo and Gobela are the most contaminated ones. This PCA model can be the basis of a classification model of the sample sites according to the specific distribution features of the contaminants along the estuary of Bilbao: PCBs in Galindo where the sewage treatment plant is located, PAHs in Udondo (a dock) and other contaminants (organotins and metals) in Gobela (industrial and wastewater effluents area).

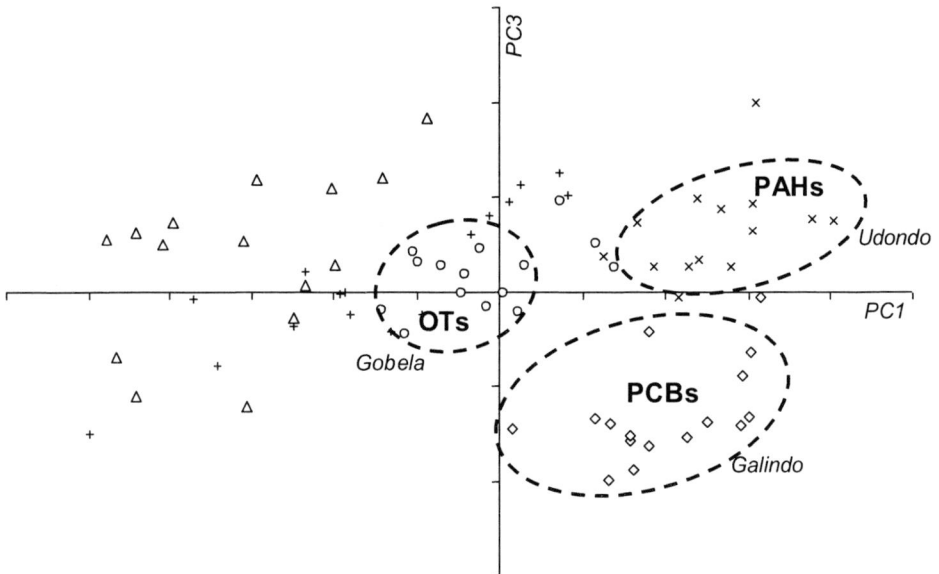

Figure 15. Scores and loadings (in bold) plot for PCA. OTs: organotins.

CONCLUSIONS

As a result of the all the work dealing with the pollution in the estuaries of Bilbao and Urdaibai, we have highlighted the most remarkable data obtained from the sediments. We are aware that sediments represent only one of the media in most environmental assessments, but from them, we can gain historical information about the fate of pollutants in estuarine areas.

In the case of the estuary of Bilbao, it was clear from the beginning that this was a severely degraded area that is undergoing a long term restoration process. The data shown in this chapter reveal the remnants of past continuous pollution events, primarily in the patterns of metal distribution but also in the distribution of key organic micro-pollutants, and the consequences of the pollution abatement measures to minimize the impact of urban and industrial activities in the estuary. On the other hand, the estuary of Urdaibai is an example of the risks that a presumed pristine area may suffer when the development of urban, industrial, or even leisure activities is not conducted with the highest environmental quality standards.

The recorded pollution in the sediment shows not only the effects of a continuous intense human activity but, as a paradox, illustrates the risks associated with the introduction of remedies such as WWTP. Finally, the scenario depicted in this chapter can be easily extended to most of the estuaries of the Bay of Biscay.

ACKNOWLEDGMENTS

This work has been partially financed by the Sediport project (CTM2006-13880-C03-02/MAR) of the Spanish Ministry of Education. The authors gratefully acknowledge G.A. López for the SEM images and EDS spectra.

REFERENCES

Ahn, I.Y., Kang, Y.C. & Choi, J.W. (1995). The influence of industrial effluents on intertidal benthic communities in Panweol, Kyeonggi Bay (Yellow Sea) on the west coast of Korea. *Mar. Poll. Bull. 30*, 200–206.

Andersen, J.H., Conley, D.J. & Hedal, S. (2004). Palaeoecology, reference conditions and classification of ecological status: the EU Water Framework Directive in practice. *Mar. Poll. Bull. 49*, 283–90.

Arias, A.H., Spetter, C.V., Freije, R.H. & Marchoveccio, J.E., (2010). Polycyclic aromatic hydrocarbons in water, mussels (Brachidontes sp., Tagelus sp.) and fish (Odontesthes sp.) from Bahía Blanca Estuary, Argentina. *Estuar. Coast. Shelf S. 85*, 67–81.

Azkona A., Henkins S.H. & Roberts H.M.G. (1984). Sources of contamination of the estuary of the River Nervión, Spain – a case study. *Water Sci. Technol. 16*, 95–125.

Bartolomé, L., Cortazar, E., Raposo, J.C., Usobiaga, A., Zuloaga, O., Etxebarria, N. & Fernández, L.A. (2005). Simultaneous microwave-assisted extraction of polycyclic aromatic hydrocarbons, polychlorinated biphenyls, phthalate esters and nonylphenols in sediments. *J. Chromatogr A. 1068*, 229–236.

Bartolomé, L., Navarro P., Raposo, J.C., Arana, G., Zuloaga, O., Etxebarria, N. & Soto, M., (2010). Occurrence and distribution of metals in mussels from the Cantabrian Coast. *Arch. Environ. Contam. Toxicol. 59*, 235–243.

Belzunce, M.J., A. Borja, V. Valencia, & J. Franco, (1998). *Estudio de la contaminación por metales pesados y compuestos orgánicos en los sedimentos de los estuarios del Nerbioi, Barbadun y Lea*. Informe final, para el Departamento de Ordenación del Territorio, Vivienda y Medio Ambiente, Gobierno Vasco. UTE AZTI-LABEIN. 103 pp.

Birch, G., Taylor, S. & Matthai, C. (2001). Small-scale spatial and temporal variance in the concentration of heavy metals in aquatic sediments: a review and some new concepts. *Environ. Pollut. 113*, 357–372.

Borja A. (2005). The European water framework directive: A challenge for nearshore, coastal and continental shelf research. *Cont. Shelf Res. 25*, 1768–1783.

Borja, A., Franco J., Valencia V., Uriarte A., & Castro R. (1998). *Red de vigilancia y control de la calidad de las aguas del País Vasco: otoño 1996-verano 1997.* Informe final, para el Departamento de Ordenación del Territorio, Vivienda y Medio Ambiente, Gobierno Vasco. UTE AZTI-LABEIN, 333 pp + anexos. Inédito.

Boruvka, L., Vecek, O. & Jenlika, (2005). Principal component analysis as a tool to indicate the origin of potentially toxic elements in soil. *Geoderma 128*, 289–300.

Bozlaker, A., Muezzinoglu, A. & Odabasi, M. (2008). Atmospheric concentrations, dry deposition and air–soil exchange of polycyclic aromatic hydrocarbons (PAHs) in an industrial region in Turkey. *J. Hazard. Mater. 3*, 1093–1102.

Caccia, V.G., Millero, F.J. & Palanques, A. (2003). The distribution of trace metals in Florida Bay sediments. *Mar. Pollut. Bull. 46*, 1420–1433.

Carlon, C., Critto, A., Marcomini, A. & Nathanail, P., (2001*).* Risk based characterization of contaminated industrial site using multivariate and geostatistical tools. *Environ. Pollut. 111*, 417–427.

Cearreta, A., Irabien, M.J., Leorri, E., Yusta, I., Quintanilla, A., Zabaleta, A. (2002) Environmental transformation of the Bilbao estuary, N. Spain: microfaunal and geochemical proxies in the recent sedimentary record. *Mar. Pollut. Bull. 44*, 487–503.

Cearreta, A., Irabien, M.J., Leorri, E., Yusta, I., Croudace, I.W. & Cundy, A.B. (2000). Recent anthropogenic impacts on the Bilbao Estuary, Northern Spain: Geochemical and microfaunal evidence. *Estuar. Coast. Shelf S. 50*, 571–592

Censi, P., Spoto, S.E., Saino, F., Sprovieri, M., Mazzola, S., Nardone, G., Di Geronimo, S.I., Punturo, R. & Ottonello, D. (2006). Heavy metals in coastal water systems: A case study from the northewstern Gulf of Thailand. *Chemosphere 64*, 1167–1176.

Chen, K., Jiao, J.J., Huang, J. & Huang, R., (2007). Multivariate statistical evaluation of trace elements in groundwater in a coastal area in Shenzhen, China. *Environ. Pollut. 147*, 771–780.

Cochran, J.K., Hirschberg, D.J., Wang, J. & Dere, C., (1998). Atmospheric deposition of metals to coastal waters (Long Island Sound, New York, USA): Evidence from salt marsh deposits. *Estuar. Coast. Shelf S. 46*, 503–522.

Comber, S.D.W., Gunn, A.M. & Whaley, C. (1995). Comparison of the partitioning of trace metals in the Humber and Mersey Estuaries. *Mar. Pollut. Bull. 30*, 851–860.

Cortazar E., Bartolomé L., Arrasate S., Usobiaga A. Raposo J.C., Zuloaga O. & Etxebarria N. (2008). Distribution and bioaccumulation of PAHs in the UNESCO protected natural reserve of Urdaibai, Bay of Biscay. *Chemosphere 72,* 1467–1474.

Costabeber, I. & Emanuelli, T., (2003*).* Influence of alimentary habits, age and occupation on polychlorinated biphenyls levels in adipose tissue. *Food Chem. Toxicol. 41*, 73–80.

Cotano U. & Villate F. (2006). Anthropogenic influence on the organic fraction of sediments in two contrasting estuaries: a biochemical approach. *Mar. Pollut. Bull. 52*, 404–414.

EC. (2008). Directive 2008/105/EC of the European Parliament and of the Council of 16 December 2008 on environmental quality standards in the field of water policy, amending and subsequently repealing Council Directives 82/176/EEC, 83/513/EEC,

84/156/EEC, 84/491/EEC, 86/280/EEC and amending Directive 2000/60/EC of the European Parliament and of the Council. In: Commun, O.J.E. (Ed.). *European Parliament and Council, Brussels*, pp. 84–97.

Einax J.W., Zwanziger H.W. & Geib S. (1997). *Chemometrics in Environmental Analysis*, VCH, Weinheim, Germany.

Ellis J.B. (2006). Pharmaceutical and personal care products (PPCPs) in urban receiving waters, *Environ. Pollut. 144*, 184–189.

Emmerson, R.H.C., O'Reilly-Weisse, S.B., MacLeod, C.L. & Lester, J.N. (1997). A multivariate assessment of metal distribution in inter-tidal sediments of the Blackwater Estuary, UK. *Mar. Poll. Bull. 34*, 960–968.

Feng, H., Han, X., Zhang, W. & Yu, L. (2004). A preliminary study of heavy metal contamination in Yangtze River intertidal zone due to urbanization. *Mar. Pollut. Bull. 49*, 910–915.

Fernández-Ortiz de Vallejuelo S., Arana G., de Diego A. & Madariaga J.M. (2010). Risk assessment of trace elements in sediments: The case of the estuary of the Nerbioi–Ibaizabal River (Basque Country) *J. Hazard Mater. 181*, 565–573.

Field, A. (2000). *Discovering Statistics Using SPSS for Windows*. Sage Publications Ltd., London.

Frame, G.M., Cochran, J.W. & Boewadt, S.S., (1996). Complete PCB congener distributions for 17 aroclhor mixtures determined by 3 HRGC systems optimized for comprehensive, quantitative, congener-specific analysis. *J. Sci. High Chrom. 12*, 657–668.

Franco, M.A., Viñas, L., Soriano, J.A., de Armas, D., González, J.J., Beiras, R., Salas, N., Bayona, J.M. & Albaige's, J. (2006). Spatial distribution and ecotoxicity of petroleum hydrocarbons in sediments from the Galicia continental shelf (NW Spain) after the Prestige oil spill. *Mar. Pollut. Bulletin 53*, 260–271.

Freels, S., Chary, L.K., Turyk, M., Piorkowski, J., Mallin, K. & Dimos J. (2007). Congener profiles of occupational PCB exposure versus PCB exposure from fish consumption. *Chemosphere 69*, 435–43.

French, P.W., (1993). Seasonal and inter-annual variation of selected pollutants in modern inter-tidal sediments, Aust Cliff, Severn Estuary. *Estuar. Coast. Shelf S. 37*, 213–219.

Gaga, E.O. & Ari, A. (2011). Gas–particle partitioning of polycyclic aromatic hydrocarbons (PAHs) in an urban traffic site in Eskisehir, Turkey. *Atmos. Res. 99*, 2 207–216.

García-Barcina J.M., González-Oreja J.A. & De la Sota A. 2006. Assessing the improvement of the Bilbao estuary water quality in response to pollution abatement measures. *Water Res. 40*, 951–960.

Grifoll M., Fontán A., Ferrer L., Mader J., González M. & Espino M. (2009). 3D hydrodynamic characterization of a meso-tidal harbour: The case of Bilbao (northern Spain). *Coastal Eng. 56*, 907–918

Hawkins S.J., Gibbs P.E., Pope N.D., Burt G.R., Chesman B.S., Bray S., Proud S.V., Spence S.K., Southward A.J. & Langston J.W. (2002). Recovery of polluted ecosystems: the case of long-term studies. *Mar. Environ. Res. 54*, 215–222.

Hyland, J.L., Van Dolah R.F. & Snoots, T.R. (1999). Predicting stress in benthic communities of southeastern U.S. estuaries in relation to chemical contamination of sediments. *Environ. Toxicol. Chem. 18*, 2557–2564

Hutzinger, O., Safe, S. & Zitko, V. (1974). *The Chemistry of PCBs*. CRC press, Cleveland, OH.

Irabien M.J. & Velasco F. (1999). Heavy metals in Ola river sediments (Urdaibai National Bisophere Reserve, northern Spain): lithogenic and anthropogenic effects. *Environ. Geol. 37*, 54–63.

Kasprzyk-Hordern B., Dinsdale R.M. & Guwy A.J. (2009). The removal of pharmaceuticals, personal care products, endocrine disruptors and illicit drugs during wastewater treatment and its impact on the quality of receiving waters. *Water Res. 43*, 363–380.

Keith, L.H. & Telliard, W.A. (1979). Priority pollutants 1 – a perspective view. *Environ. Sci. Technol. 13*, 416–423.

Kim, M., Kim, S., Yun, S., Lee, M., Cho, B., Park, J., Son, S. & Kim, O. (2004). Comparison of seven indicator PCBs and three coplanar PCBs in beef, pork and chicken fat. *Chemosphere 54*, 1533–1538.

Kukkonen J., & Oikari A. (1991). Bioavailability of organic pollutants in boreal waters with varying levels of dissolved organic material. *Water Res. 25*, 455–463.

Landajo, A., Arana, G., de Diego, A., Etxebarria, N., Zuloaga, O. & Amoroux, D. (2004). Analysis of heavy metals distribution in surficial estuarine sediments (estuary of Bilbao, Basque Country) by open-focused microwave-assisted extraction and ICP-OES. *Chemosphere 56*, 1033–41.

Leorri, E., Cearreta, A., Irabien, M.J. & Yusta, I. (2008). Geochemical and microfaunal proxies to assess environmental quality conditions during the recovery process of a heavily polluted estuary: The Bilbao estuary case (N. Spain). *Sci. Total Environ. 396*, 12–27.

Lepom, P., Brown, B., Hanke, G., Loos, R., Quevauviller, Ph. & Wollgast, J. (2009). Needs for reliable analytical methods for monitoring chemical pollutants in surface water under the European Water Framework Directive. *J. Chromatogr. A. 1216*, 302–315.

Leung, A.O.W., Luksemburg, W.J., Wong A.S. & Wong, M.H. (2007). Spatial distribution of polybrominated diphenyl ethers and polychlorinated dibenzo-p-dioxins and dibenzofurans in soil and combusted residue at Guiyu, an electronic waste recycling site in Southeast China. *Environ Sci. Technol. 41*, 2730–2737.

Lim, L., Wurl, O., Karuppiah, S. & Obbard. J.P. (2007). Atmospheric wet deposition of PAHs to the sea-surface microlayer. *Mar. Pollut. Bull. 54*, 8, 1212–1219.

Livingstone D. (2009). A Practical Guide to Scientific Data Analysis, Wiley, UK.

Liria, P., Garel, E. & Uriarte, A. (2009). The effects of dredging operations on the hydrodynamics of an ebb tidal delta: Oka estuary, northern Spain. *Cont. Shelf Res. 29*, 1983–1994.

Long, E.R., Ingersoll, C.G. & MacDonald, D.D. (2006). Calculation and uses of mean sediment. *Environ. Sci. Technol. 40*, 1726–1736.

Long, E.R. & MacDonald, D.D. (1998) Recommended uses of empirically derived, sediment. *Hum. Ecol. Risk Assess. 4*, 1019–1039.

Long, E.R., MacDonald, D.D., Smith, S.L. & Calder, F.D. (1995). Incidence of adverse biological effects within ranges of chemical . *Environ. Manage. 19*, 81–97.

Masclet, P., Mouvier, G. & Nikolaou, K. (1986). Relative decay index and sources of polycyclic aromatic hydrocarbons. *Atmos. Environ. 20*, 439–446.

Miège, C., Choubert, J.M., Ribeiro, L., Eusebe,M. & Coquery, M. (2009). Fate of pharmaceuticals and personal care products in wastewater treatment plants – Conception of a database and first results. *Environ. Poll. 157*, 1721–1726.

Monge-Ganuzas, M., Cearreta, A. & Iriarte E. (2008). Consequences of estuarine sand dredging and dumping on the Urdaibai Reserve of the Biosphere (Bay of Biscay): the case of the "Mundaka left wave". *J. Iberian Geol. 34*, 215–234.

Müller, G., (1979). Schwermetalle in den Sedimenten des Rheins. Veränderungen seit 1971. *Umschau 79*, 78–783.

Penteado, J.C.P. & Moreira Vaz, J. (2001). O legado das bifenilas policloradas (PCBs). *Química Nova 24*, 390–398.

Phillips, D.J.H. & Rainbow, P.S. (1993). *Biomonitoring of trace aquatic contaminants.* Elsevier Science Publishers Ltd, London.

Platonow, N.S., Sachenbrecker, P.W. & Funnell, H.S. (1971). Residues of polychlorinated biphenyls in cattle. *Canadian Vet. J. 12*, 115–118.

Porte, C., Janer, G., Lorusso, L.C., Ortiz-Zarragoitia, M. & Cajaraville, M.P. (2006). *Endocrine disruptors in marine organisms: approaches and perspectives. Comp. Biochem. Physiol. C 143*, 303–315.

Puy-Azurmendi E., Navarro A., Olivares A., Fernandes D., Martínez E., López de Alda M., Porte C., Cajaraville M.P., Barceló D. & Piña B. (2010b). Origin and distribution of polycyclic aromatic hydrocarbon pollution in sediment and fish from the biosphere reserve of Urdaibai (Bay of Biscay, Basque Country, Spain). *Mar. Environ. Res. 70*, 142–149.

Puy-Azurmendi E., Ortiz-Zarragoitia M., Kuster M., Martínez E., Guillamón M., Domínguez C. Serrano T., Barbero M.C., López de Alda M., Bayona J.M., Barceló D. & Cajaraville M., (2010a). An integrated study of endocrine disruptors in sediments and reproduction-related parameters in bivalve molluscs from the Biosphere's Reserve of Urdaibai (Bay of Biscay). *Mar. Environ. Res. 69*, S63–S66.

Qiao, S., Yang, Z., Pan, Y. & Zuo, Z., (2007). Metals in suspended sediments from the Changjiang (Yangtze River) and Huanghe (Yellow River) to the sea and their comparison. *Estuar. Coast. Shelf S. 74*, 539–548.

Quevauviller Ph., Borchers U. & Gawlik B.M. (2007). Coordinating links among research, standardisation and policy in support of water framework directive chemical monitoring requirements. *J. Environ. Monitor. 9*, 915–923.

Raposo, J.C., Zuloaga, O., Sanz, J., Villanueva, U., Crea, P., Etxebarria, N., Olazábal, M.A., Madariaga, J.M. (2006). Analytical and thermodynamical approach to understand the mobility/retention of arsenic species from the river to the estuary. The Bilbao case study. *Mar. Chem. 99*, 42–51.

Rodrígez, J.G., Tueros, I., Borja, A., Belzunce, M.J., Franco, J., Solaun, O., Valencia, V., Zuazo, A. (2006). Maximum likelihood mixture estimation to determine metal background values in estuarine and coastal sediments within the European Water Framework Directive. *Sci. Total Environ. 370*, 278–93.

Ruiz, J.M., Saiz Salinas, J.I., (2000). Extreme variation in the concentration of trace metals in sediments and bivalves from the Bilbao estuary (Spain) caused by the 1989–90 drought. *Mar. Environ. Res. 49*, 307–317.

Saiz-Salinas, J.I., González-Oreja, J.A. (2000). Stress in estuarine communities: lessons from the highly impacted Bilbao estuary (Spain). *J. Aquatic Ecosystem Stress Recovery 7*, 43–55.

Schwarzenbach R.P. & Westall J. (1981). Transport of nonpolar organic compounds from surface water to ground water. Laboratory sorption studies. *Environ. Sci. Technol. 15*, 1360–1367.

Solaun, O., Belzunce, M.J., Franco, J., Valencia, V. & Borja, A. (2009). Estudio de la contaminación en los sedimentos de los estuarios del País Vasco (1998–2001). *Revista de Investigación Marina 10*, 47–53.

Spencer, K.L., (2002). Spatial variability of metals in the inter-tidal sediments of the Medway Estuary, Kent, UK. *Mar. Pollut. Bull. 44*, 933–944.

Tomlinson, D.L, Wilson J.G., Marris C.R. & Jeffrey, D.W. (1980). Problems in the assessment of heavy metal levels in estuaries and the formation of pollution index. *Helgolander Meeresuntersuchugen 33*, 566–575.

Valencia, V., Franco, J., Borja, A. & Fontán, A. (2004). Hydrography of the southeastern Bay of Biscay. In: Borja, A., Collins M. (Eds) *Oceanography of the Basque Country*. Elsevier, Amsterdam, pp. 159–194.

Villanueva, U., Raposo, J.C., Castro, K., de Diego, A., Arana, G., Madariaga, J.M. (2008). Raman spectroscopy speciation of natural and anthropogenic solid phases in river and estuarine sediments with appreciable amount of clay and organic matter. *J. Raman Spectrosc. 39*, 1195–1203.

Volkman, J.K., Revill, A.T., Holdsworth, D.G. & Fredericks, D. (2008). Organic matter sources in an enclosed coastal inlet assessed using lipid biomarkers and stable isotopes. *Org. Geochem. 39*, 689–710.

Williams, T.P., Bubb, J.M. & Lester, J.N. (1994). Metal accumulation within salt marsh environments: a review. *Mar. Pollut. Bull. 28*, 277–290.

Yang, S.L., Ding, P.X. & Chen, S.L. (2001). Changes in progradation rate of the tidal flats at the mouth of the Changjiang (Yangtze) River, China. *Geomorphology 38*, 167–180.

Yunker, M.B., Macdonald, R.W., Vingarzan, R., Mitchell, R.H., Goyette, D. & Sylvestre, S. (2002). PAHs in the Fraser River basin: a critical appraisal of PAH ratios as indicators of PAH source and composition. *Org. Geochem. 33*, 489–515.

Zhang, W., Yu, L., Hutchinson, S.M., Xu, S., Chen, Z. & Gao, X. (2001). China's Yangtze Estuary: I. Geomorphic influence on heavy metal accumulation in intertidal sediments. *Geomorphology 41*, 195–205.

In: Estuaries: Classification, Ecology and Human Impacts
Editor: Steve Jordan
ISBN: 978-1-61942-083-0
© 2012 Nova Science Publishers, Inc.

Chapter 9

LONG-TERM MONITORING OF TRACE ELEMENTS IN ESTUARINE WATERS, A CASE STUDY: THE ESTUARY OF THE NERBIOI-IBAIZABAL RIVER (BILBAO, BASQUE COUNTRY)

Ainara Gredilla, Silvia Fdez-Ortiz de Vallejuelo, Gorka Arana, Alberto de Diego and Juan Manuel Madariaga

Department of Analytical Chemistry, University of the Basque Country, P. O. Box 644, Bilbao, Basque Country, Spain

ABSTRACT

Estuaries are usually highly affected by pollutant discharges of both urban and industrial origin. Environmental risk assessment studies in estuaries most often include trace elements, bacause of their high toxicity and regular presence in these environments. The fate of trace elements in an estuary is clearly influenced by changes in physico-chemical properties of the estuary as mediated by tidal effects. Rigorous monitoring programs are necessary for the adequate study of the spatial distribution and evolution in time of trace elements in estuarine waters. These programs should include factors such as salinity, sampling depth, proximity to pollution inputs and sampling season. In this work we describe a monitoring program carried out in the estuary of the Nerbioi-Ibaizabal River (metropolitan area of Bilbao, Basque Country) to investigate the fate of several trace elements in water. This estuary is located in one of the most populated areas (~1 million inhabitants) on the Bay of Biscay, with numerous industrial, urban and recreational activities. The exploitation of local iron at the end of the 19th century and ship-building in the bay also have had clear influence on the water quality of the estuary. Surface and deep (in contact with the sediment) water samples were collected approximately every three months from January 2005 to October 2010 (22 sampling campaigns) at eight different points of the estuary (four in tributary rivers, one in an enclosed dock area, two in the main channel, and one in the mouth of the estuary). The sites were sampled at low and high tides. Physico-chemical parameters such as temperature, redox potential, dissolved oxygen (DO) and conductivity were measured in situ by means of a multi-parameter probe. The values found were typical from a highly stratified estuary; the average dissolved oxygen concentration was ~65% of saturation. Concentrations of Al, As, Cr, Cu, Fe, Mn, Ni and Zn, measured simultaneously in the samples by inductively coupled plasma/mass spectrometry (ICP/MS), were comparable to those of similar estuaries in Europe, with the exception of As, Cu and Fe, which presented significantly higher values. Concentrations of most of the metals at most of the sampling sites significantly increased with time over the period investigated, a trend that was especially noticeable in samples collected at high tide at the bottom. Decreasing trends were found systematically in time series of pH, whatever the sampling site and the characteristics of the samples (deep, surface, high tide and low tide). Possible sources of each metal to the estuary were also investigated.

Keywords: Trace elements, estuaries, Nerbioi-Ibaizabal, Bilbao, estuarine waters, monitoring programs, risk assessment

INTRODUCTION

Various definitions have been proposed in the literature to describe estuaries adequately. Most of them agree that estuaries are semi-enclosed transitional water systems with a permanent tidal effect, in which sea water is diluted with fresh water from rivers (Fairbridge 1980, Liu et al. 2003, Caruso et al. 2010; Jordan, Chapter 1).

Because of their complex hydrodynamics (Zwolsman & Van Eck, 1999), coupled with shallow depth and tidal influence, estuaries are highly variable water bodies, with strong spatial and temporal variations (Batlle-Aguilar et al. 2007). These variations affect the physico-chemical properties of estuaries, producing vertical and lateral gradients in salinity, pH, suspended particulate matter, DO and organic and inorganic nutrients (Nolting et al. 1999, Abowei 2009, Schäfer et al. 2009). Salinity gradients, for example, play a fundamental role in the biogeochemical, physical and biological processes of estuarine systems (Zwolsman et al. 1997, Chaudry & Zwolsman, 2008, Fernández et al. 2008). The pH also has direct influence in chemical reactions associated with the formation and dissolution of minerals, and depends on the presence of organic acids and the extent of biological and physical processes (Inland Waters Directorate 1979).

With close connections to land and open sea, estuaries are favorable environments for propagation and growth of many organisms (Bierman et al. 2011). Apart from their ecological importance, estuaries throughout history have been notable economic resources, as centers of industrial and urban activities, among other uses. Moreover, estuaries provide convenient routes for transportation and conduits for disposal of wastewaters (Spencer 2002). For all of these reasons they are considered the most productive coastal environments, even more than the open sea.

Although the evaluation of the ecological status of coastal marine ecosystems has grown significantly since the 1990s, the Water Framework Directive (WFD 2000)—for restoring the quality of surface and ground waters in aquatic systems within the European Union—was not introduced until October 2000. Because information about background levels of dissolved metals in waters is scarce (Tueros et al. 2008, Zaldívar et al. 2008), water quality indexes (WQI), flow rates, tidal currents, and degradation coefficients have been used as surrogates to assess water environmental capacity[1] (Sánchez et al. 2007, Periáñez 2009, Li et al. 2010). Monitoring programs are, however, the most popular tools to fulfill this task. Regular monitoring programs provide useful, representative and specific information about water quality (Devlin et al. 2007, Sánchez et al. 2007, Summers et al. Chapter 11).

A variety of parameters typically is monitored in estuaries. Physico-chemical variables, such as the above-mentioned DO, conductivity and pH, are commonly used as pollution indicators (Micheletti et al. 2001, Raj and Azeez 2009). Apart from these, other parameters related to anthropogenic inputs are often measured, because these inputs can damage severely the environmental capacity of transitional waters. Concentrations of persistent organic pollutants and inorganic substances (trace elements principally) are commonly quantified (Grant & Middleton 1990, Williams & Benson 2009, Grifoll et al. 2010). Specifically, anthropogenic inputs of metals come principally from tributary rivers, direct discharges associated mainly with urban wastes, mining and industrial activities, and atmospheric deposition (Deepti et al. 2009). All of these inputs, as free metal ions in solution, are easily accumulated by aquatic organisms (Vicente-Martorell et al. 2009, Wallner-Kersanach et al. 2009) or adsorbed on suspended matter, and finally removed to the sediments (Spencer 2002, Ip et al. 2007, Fdez-Ortiz de Vallejuelo et al. 2010). Although sediments are potentially more hazardous, as the result of their capacity for accumulating trace elements, water analyses provide data for metal concentrations in a specific area at a particular moment.

Following the recommendations of the WFD, in this work we have carried out surveillance monitoring of metal pollution in waters of the estuary of the Nerbioi-Ibazabal River. This estuary is located in one of the most populated areas on the Cantabrian coast (~1 million people) and has suffered throughout its history abundant industrial, mining and urban activities, which significantly degraded its environmental quality (Belzunce et al. 2001, Cearreta et al. 2002, Fernández et al. 2008, Tueros et al. 2008, Fdez-Ortiz de Vallejuelo et al. 2009). The extent of metal pollution in sediments of the estuary has been investigated rather thoroughly (Cearreta et al. 2002, Sanz-Landaluze et al. 2004, Bartolomé et al. 2006, Raposo et al., 2006, Moros et al. 2009, Fdez-Ortiz de Vallejuelo et al. 2010, Fdez-Ortiz de Vallejuelo et al. 2011, Moros et al. 2010), but analogous studies of its waters are scarce (Borja et al. 2004, Borja et al. 2006, Fernández et al. 2008).

Table 1. Experimental conditions used in the analysis by ICP/MS

Nebulisation flow	0.94 L min^{-1}
Plasma flow	15 L min^{-1}
Auxiliary flow	1.2 L min^{-1}
Sample flow	1 mL min^{-1}
Measured isotopes	^{27}Al, ^{75}As, ^{52}Cr, ^{63}Cu, ^{57}Fe, ^{55}Mn, ^{60}Ni, ^{66}Zn
Radiofrequency power	1000W
Integration time	1000 ms
Replicates	3-4
Cones material	Ni
Internal Standards	^{9}Be, ^{45}Sc, ^{115}In and ^{209}Bi

---

[1] Environmental capacity is defined as "a property of the environment and its ability to accommodate a particular activity or rate of an activity, such as the discharge of contaminants, without unacceptable impact" (GESAMP 1986).

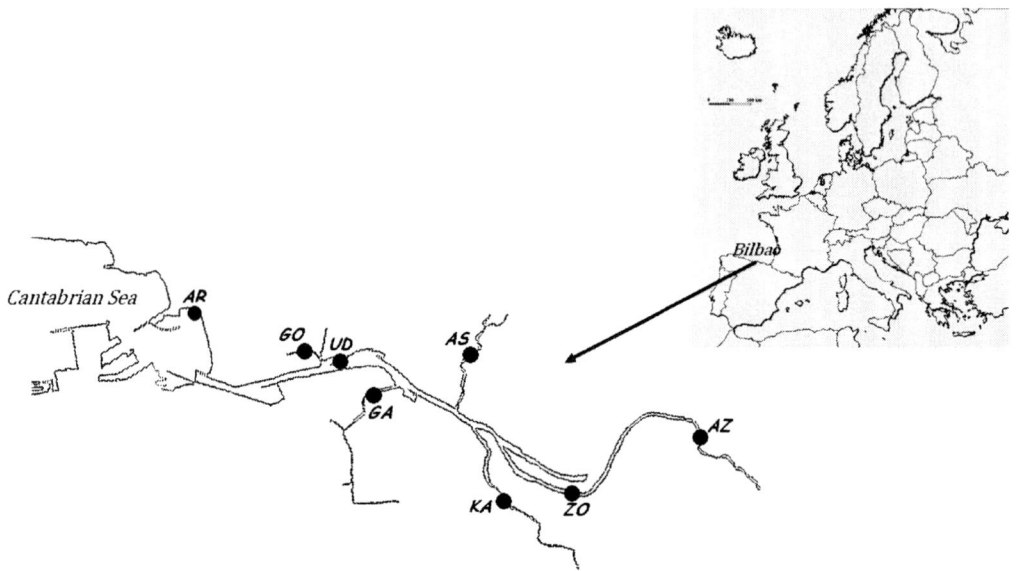

Figure 1. Estuary of the *Nerbioi-Ibaizabal* River and location of the eight sampling stations: Arriluze (*AR*), Gobela (*GO*), Udondo (*UD*), Galindo (*GA*), Asua (*AS*), Kadagua (*KA*), Zorrotza (*ZO*) and Alde Zaharra (*AZ*).

An intensive monitoring program carried out from January 2005 to October 2010 is described in this work. Deep and surface water samples were collected—at high and low tides—at eight points in the estuary every three months. Physico-chemical parameters, i.e. temperature, conductivity, pH, redox potential and dissolved oxygen were measured in situ. Concentrations of Al, As, Cr, Cu, Fe, Mn, Ni, and Zn were measured also. The results allowed us to investigate spatial and temporal trends in the dataset, and to draw some conclusions about possible pollution sources in the estuary.

MATERIALS AND METHODS

Study Area

The estuary of the Nerbioi-Ibaizabal River is located on the continental shelf of the Cantabrian Sea in the southeastern corner of the Bay of Biscay. It is 15 km long, and although its principal freshwater input comes from the Nerbioi and Ibaizabal Rivers, it also receives contributions from four tributaries: Kadagua (27%), Galindo (4%), Asua (0.7%), and Gobela (0.3%) (Leorri et al. 2008). The average flow of the estuary is about 30 $m^3 \cdot s^{-1}$, discharging into Abra Bay, which is 30 m deep and 3.5 km wide (Belzunce et al. 2001). According to the annual average salinity, the estuary is classified as oligohaline at low tide and as euhaline at high tide (WFD, 2000).

As a consequence of mining and industrial activities starting in the last decades of the 19th century, the estuary was extremely affected by uncontrolled releases of organic and inorganic pollutants (Saiz-Salinas 1997, Belzunce et al. 2001, Cearreta et al. 2002). Moreover, to make navigation easier, the original channel was significantly modified. However, the condition of the estuary improved significantly at the end of the 20th century because most of the industries were closed and, starting in 1984, an ambitious plan for the environmental recovery of the estuary was implemented (Barreiro and Aguirre 2005, Fdez-Ortiz de Vallejuelo et al. 2010)

Sampling and Analysis

Surface (top 10 cm) and deep (in contact with the sediment) water samples were collected at eight different locations along the estuary (Figure 1), both at low and high tides. Two of the sampling sites were located in the main channel (*ZO* and *AZ*), one in the mouth of the estuary (*AR*), another in a semi-enclosed dock area (*UD*), and the rest were located in the tidal part of the principal tributaries (*GO*, *AS*, *KA* and *GA*). The sites were sampled every three months from January 2005 to October 2010, a total of 22 sampling campaigns. Deep water at high tide was collected only at one site (*ZO*) in the last seven sampling campaigns. It had been observed in previous studies (Fernández 2008) that the characteristics of deep water at high tide were very similar all over the estuary.

Trace element analysis in natural waters requires careful procedures to avoid sample contamination. All the material in contact with the samples was first cleaned with tap water and soap, thoroughly washed with Ellix water (κ <0.2 μS· cm^{-1}, Millipore, Billerica, MA, USA)[2], then immersed in a 10% HNO_3 bath for 24 h, and finally soaked with Milli-Q water (κ <0.05 μS·cm^{-1}, Millipore, Billerica, MA, USA). When necessary, the pre-cleaned material was dried in an oven at 50°C.

Water was collected with a 2 L Van Dorn type all-plastic water sampler (KD Denmark, Research Equipment, Silkeborg, Denmark), specifically designed for trace metal sampling. After homogenizing and filling a 0.5 L Pyrex bottle with the sample, it was protected against light using aluminium foil and transported to the lab in a cool box. Procedural blank samples with Milli-Q water were handled in the same way. Temperature (C ± 0.1), redox potential (mV ± 0.1), DO (% of saturation ± 0.1) and electrical conductivity (mS·cm^{-1} ± 0.01) were measured in situ using a YSI 556 multi-parameter probe (YSI Environmental, Yellow Springs, OH). Salinity and total dissolved solids (TDS) were estimated from the conductivity measurement.

Once in the laboratory, the water samples were filtered through 0.45 μm cellulose filters (Whatman) in a pre-cleaned standard borosilicate vacuum system and acidified with HNO_3 to an approximate concentration of 1%. The acidified samples were stored in 50 mL polyethylene vials at 4°C and protected against light until analysis. Concentrations of Al, As, Cr, Cu, Fe, Mn, Ni, and Zn were measured simultaneously by ICP/MS (Elan 9000, Perkin Elmer, Ontario, Canada), using the external calibration method with internal standard correction (Be, Sc, In and Bi) in a Class 100 clean room. The calibrants and sample dilutions were prepared on a mass basis. The operating conditions used in the ICP/MS measurements are summarised in Table 1. Further details on the experimental procedure can be found elsewhere (Fernández et al. 2008). The accuracy of the method was checked by triplicate analysis of the NIST 1640 certified reference material (natural water, National Institute of Standards and Technology) with satisfactory results (Figure 2). The detection limits (in

---

[2] κ = specific conductance, or conductivity.

$\mu g \cdot kg^{-1}$) estimated for the analytes (3 times the standard deviation of 8 replicates of the blank) were as follows: Al, 2.7; As, 0.6; Cr, 0.8; Cu, 1.2, Fe, 18; Mn, 0.1; Ni, 0.6; and Zn, 2.1.

RESULTS AND DISCUSSION

Physico-chemical Parameters – Spatial Distribution

The physico-chemical variables were all measured in deep and surface water at low and high tide. Average, maximum and minimum values detected in each campaign are summarized in Tables 2–5.

As expected, water temperature exhibited seasonal variation (4.0–26.2 °C), but there was no significant difference among sites within each sampling period. The lowest values usually were observed in January at high tide in the bottom water. As a consequence of the huge wastewater treatment plant 100 m upstream from the *GA* sampling point, slightly higher temperature was recorded routinely at this point, especially in winter and in surface water samples.

The pH of an aquatic system is determined by several factors, such as dissolution of carbonate rocks, atmospheric CO_2 exchange, and respiration of aquatic organisms (Mosley et al. 2010). Values of pH from 6.0 to 8.6 are typical for estuarine waters. No significant difference was observed among the pH values measured in samples collected at different points, but the highest were observed regularly in surface waters. The range of pH was 6.1–8.5, within that defined for unpolluted estuarine waters (ANZECC/ARMCANZ 2000, Spiteri et al. 2008, Abowei 2009, Williams and Benson 2009).

Salinity and total dissolved solids (TDS) were calculated from conductivity measurements, and the trends observed for conductivity were extrapolated to the other two variables. As expected, samples collected at high tide always exhibited significantly higher conductivity than those collected at low tide. Furthermore, because denser marine water flows in the deeper portions of the channel, deep water collected at high tide showed the highest conductivity, with maximum values close to 40 $mS \cdot cm^{-1}$ in most of the areas sampled. No significant differences were observed among the conductivities measured in deep waters at high tide along the estuary. Conversely, important spatial variability was clearly distinguished in the conductivity of surface samples, especially at high tide (Figure 3a). In the obvious pattern, the highest conductivity values were found at the mouth of the estuary (*AR*), decreasing from the outermost stations toward the head of the estuary. Similar differences (but less marked) were also observed in deep waters collected at low tide. In this way, it was possible to know the real influence of the tides. The marked differences observed between surface and deep waters confirmed the strong stratification of the water column (low salinity water on the surface and high salinity water at the bottom) previously described (García-Barcina et al. 2006).

Redox potential largely influences the chemical speciation of metals in water and, hence, their fate in estuaries. The values we recorded were typical of estuarine waters, with an average value of 262 mV, minimum of 47 mV and maximum of 453 mV. No significant spatial variation was observed between low and high tides, surface and deep waters, or among sampling sites. It is worth noting, however, the high values measured between July 2009 and April 2010, especially in surface waters. In this period redox potentials between 350 and 450 mV were observed along the entire estuary.

The concentration of DO in water is another physico-chemical parameter commonly used in quality assessment of environmental areas (Simões et al. 2008). The DO ranged from 3.8 to 118% of saturation, with an average value of 66.5%. Slightly higher values were found in surface than in deep waters. No significant difference was detected, however, between waters collected at low and high tide. The lowest DO values were in dry seasons, probably due to a decrease in water flow and to decomposition of organic matter from aquatic plants (Whitall et al. 2010). In deep water samples collected at high tide a clear trend was observed, with higher average DO in the outermost sites (*AR*, 75.0%) and lower average values in the innermost sites (*AZ*, 41.3%). As shown in Figure 3b, a similar trend was detected in deep water at low tide (*AR*, 85.2%; *AZ*, 43.7%). In other combinations (surface-high tide and surface-low tide) no remarkable differences in DO among sampling campaigns were identified.

Physico-Chemical Parameters – Temporal Distribution

To detect increasing or decreasing trends in the time series of the physico-chemical parameters, the Mann-Kendall test was applied separately to deep or surface samples collected at low or high tide. All possible combinations were taken into account. The Mann-Kendall test is a non-parametric method which is able to detect increasing or decreasing trends in time series with a given confidence level. The calculations were made by the Excel template MAKESENS (Mann-Kendall test for trend and Sen´s slope estimates on annual data, Version 1.0, 2002). The statistical basis of this macro is described elsewhere (Ravichandran 2003, Fdez-Ortiz de Vallejuelo et al. 2010). The results are shown in Table 6.

The Mann-Kendall test was not applied to the time series corresponding to the deep waters from *GO, UD* and *ZO* at low tide. The weak flow of the river at low tide at those sites did not allow us to distinguish between deep and surface water. Thus, those samples were considered to be surface waters. Similarly, only one sample per campaign was collected at *AR* at low tide and treated as deep water.

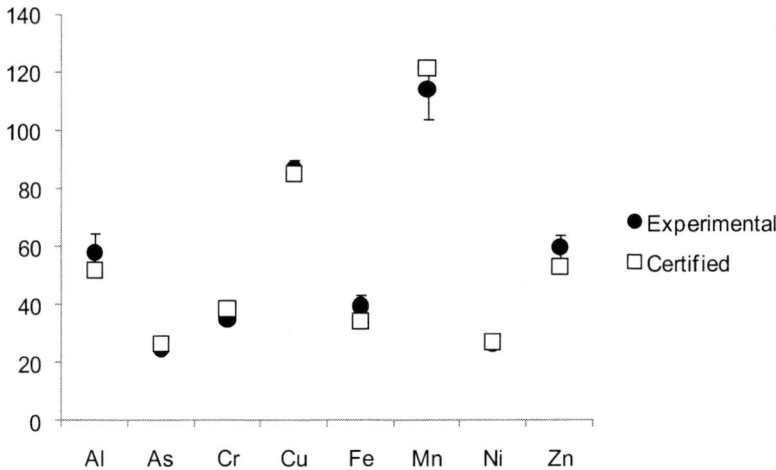

Figure 2. Trace element concentrations ($\mu g \cdot Kg^{-1}$) found after triplicate analysis (n=3) of the NIST 1640 reference material (black circles), together with the corresponding certified values (white squares).

Table 2. Average, maximum and minimum temperature (T), pH, electrical conductivity (K), dissolved oxygen (DO), redox potential (ORP), total dissolved solids (TDS) and salinity (SAL) found in each sampling campaign at high tide in deep water. NA = not analyzed

Campaign	T(°C)	TDS (g L⁻¹)	SAL (g L⁻¹)	DO (% of saturation)	κ (mS·cm⁻¹)	pH	ORP (mV)
Jan-05	NA	NA	NA	NA	NA	NA	NA
May-05	16.2 (17.2 - 15.1)	24.13 (27.27 - 15.67)	23.6 (27.0 - 14.7)	56.1 (75.5 - 22.6)	37.12 (41.94 - 24.11)	7.84 (7.97 - 7.54)	161.5 (231.0 - 101.3)
Sep-05	22.0 (24.6 - 20.8)	31.83 (33.21 - 30.43)	32.0 (33.6 - 30.4)	49.3 (96.9 - 5.4)	48.94 (51.10 - 46.72)	7.87 (8.04 - 7.64)	200.4 (285.3 - 67.5)
Dec-05	10.7 (12.1 - 9.4)	18.40 (32.44 - 0.89)	17.9 (32.6 - 0.7)	58.8 (71.1 - 32.6)	20.93 (37.61 - 0.97)	7.97 (8.53 - 7.64)	405.0 (449.6 - 314.9)
Mar-06	11.9 (14.9 - 10.9)	21.78 (33.70 - 11.06)	21.2 (34.0 - 10.0)	41.3 (47.6 - 31.8)	25.14 (38.57 - 12.64)	7.69 (7.90 - 7.49)	NA
Jul-06	21.9 (22.6 - 21.6)	34.11 (36.48 - 29.42)	34.6 (37.3 - 29.3)	59.5 (75.7 - 33.1)	49.39 (53.09 - 43.21)	NA	NA
Oct-06	20.4 (21.2 - 20.1)	35.01 (36.25 - 34.18)	35.7 (37.1 - 34.7)	59.5 (78.4 - 22.4)	49.33 (50.92 - 47.68)	7.83 (8.03 - 7.39)	270.3 (318.0 - 184.0)
Jan-07	14.1 (14.6 - 13.9)	35.24 (36.11 - 34.51)	35.8 (36.8 - 35.0)	59.4 (81.0 - 29.9)	42.96 (43.94 - 41.98)	7.97 (8.10 - 7.68)	253.5 (283.0 - 150.0)
Apr-07	16.1 (17.5 - 15.1)	29.39 (36.15 - 13.87)	28.6 (33.9 - 12.9)	73.4 (107.2 - 52.5)	34.15 (43.99 - 18.30)	7.86 (8.11 - 7.45)	267.4 (292.0 - 231.0)
Jul-07	20.8 (21.6 - 19.9)	35.27 (37.91 - 30.47)	34.8 (37.3 - 30.3)	68.7 (90.7 - 35.1)	47.81 (54.98 - 37.26)	7.75 8.037.34	243.3277.0 206.0
Oct-07	15.3 (16.1 - 14.9)	30.27 (34.39 - 24.18)	28.6 (34.9 - 23.4)	52.3 (70.0 - 16.7)	36.52 (42.62 - 28.68)	7.79 (7.94 - 7.43)	253.4 (258.0 - 243.0)
Jan-08	12.2 (12.8 - 11.7)	29.25 (31.09 - 25.77)	29.1 (31.1 - 25.3)	61.2 (78.6 - 33.4)	33.98 (36.68 - 29.84)	7.60 (7.73 - 7.45)	244.9 (276.0 - 213.0)
Apr-08	13.7 (15.2 - 13.1)	14.96 (29.07 - 0.70)	14.4 (28.9 - 0.5)	65.7 (87.4 - 7.8)	18.00 (34.90 - 0.83)	7.66 (8.04 - 7.31)	240.1 (253.0 - 232.0)

Campaign	T(°C)	TDS (g L^{-1})	SAL (g L^{-1})	DO (% of saturation)	κ (mS·cm^{-1})	pH	ORP (mV)
Jul-08	19.3 (22.0 - 13.5)	28.14 (31.51 - 14.97)	28.0 (31.7 - 13.9)	76.8 (88.4 - 68.7)	38.43 (43.11 - 21.70)	7.71 (7.77 - 7.51)	198.6 (254.0 - 107.0)
Oct-08	15.4 (18.5 - 10.6)	23.17 (31.03 - 15.77)	22.7 (31.2 - 14.7)	45.2 (70.7 - 23.6)	29.73 (41.66 - 17.59)	7.48 (7.68 - 7.19)	244.1 (271.0 - 216.0)
Jan-09	9.3	NA	NA	77.1	NA	7.53	268.0
Apr-09	13.7	13.77	12.7	67.6	16.59	7.10	284.0
Jul-09	20.2	NA	26.6	42.2	36.73	7.03	439.0
Oct-09	15.4	24.11	23.5	37.2	30.31	7.24	347.0
Jan-10	8.9	NA	NA	81.0	NA	7.40	183.0
Apr-10	NA	26.84	26.5	75.9	33.19	7.63	384.5
Oct-10	16.8	26.55	26.2	NA	34.43	7.76	192.0

Table 3. Average, maximum and minimum temperature (T), pH, electrical conductivity (κ), dissolved oxygen (%DO), redox potential (ORP), total dissolved solids (TDS) and salinity (SAL) found in each sampling campaign at high tide in surface water. NA = not analyzed

Campaign	T(°C)	TDS (g L^{-1})	SAL (g L^{-1})	DO (% of saturation)	κ (mS·cm^{-1})	pH	ORP (mV)
Jan-05	5.9 (7.0 – 4.0)	NA	NA	NA	NA	NA	NA
May-05	NA	NA	NA	NA	NA	NA	NA
Sep-05	NA	NA	NA	NA	NA	NA	NA
Dec-05	NA	NA	NA	NA	NA	NA	NA
Mar-06	11.9 (14.0 - 9.8)	6.23 (20.08 - 0.78)	5.7 (19.2 - 0.6)	47.0 (53.2 - 43.6)	7.36 (24.39 - 0.89)	8.06 (8.51 - 7.69)	NA
Jul-06	23.8 (26.2 - 22.6)	25.27 (35.81 - 10.50)	25.0 (36.6 - 9.5)	78.4 (106.8 - 62.1)	37.70 (52.79 - 16.00)	NA	NA
Oct-06	13.9 (22.3 - 19.4)	26.57 (33.62 - 3.19)	26.6 (34.1 - 2.6)	79.6 (80.4 - 60.0)	32.24 (47.15 - 4.39)	7.97 (8.19 - 7.61)	268.0 (320.0 - 192.0)
Jan-07	13.9 (16.7 - 11.3)	26.57 (34.84 - 1.69)	26.6 (35.4 - 1.4)	79.6 (95.9 - 63.3)	32.24 (42.70 - 2.18)	7.97 (8.25 - 7.52)	268.0 (283.0 - 250.0)

Table 3. (Continued)

Apr-07	17.1 (18.4 - 15.1)	15.78 (32.24 - 2.46)	15.2 (31.2 - 2.0)	74.5 (92.6 - 58.3)	19.01 (36.05 - 3.22)	8.03 (8.35 - 7.72)	268.9 (298.0 - 241.0)
Jul-07	22.8 (23.7 - 21.7)	26.03 (37.92 - 4.57)	25.6 (37.4 - 3.9)	88.8 (108.9 - 65.2)	38.27 (58.49 - 6.84)	7.85 (8.43 - 7.17)	248.9 (273.0 - 218.0)
Oct-07	15.1 (17.8 - 13.7)	18.95 (31.71 - 3.98)	17.7 (28.1 - 3.4)	66.7 (90.5 - 56.3)	23.36 (37.71 - 5.28)	7.77 (8.01 - 7.13)	249.3 (256.0 - 233.0)
Jan-08	10.9 (12.0 - 9.1)	10.24 (27.09 - 0.79)	9.5 (26.7 - 0.6)	74.9 (84.8 - 65.2)	11.70 (31.30 - 0.87)	7.77 (8.14 - 7.51)	235.0 (274.0 - 190.0)
Apr-08	13.9 (16.7 - 13.2)	6.51 (23.61 - 0.37)	6.0 (23.0 - 0.3)	75.4 (85.2 - 61.2)	7.84 (28.34 - 0.45)	7.87 (8.19 - 7.41)	233.0 (245.0 - 208.0)
Jul-08	20.9 (23.5 - 19.3)	17.69 (29.88 - 0.99)	17.3 (29.9 - 0.8)	83.9 (105.9 - 72.6)	24.58 (41.65 - 1.46)	7.93 (8.55 - 7.72)	197.7 (259.0 - 103.0)
Oct-08	16.8 (19.8 - 13.1)	15.19 (24.73 - 7.55)	14.3 (24.2 - 6.6)	40.9 (58.3 - 33.6)	19.64 (33.63 - 10.30)	7.49 (7.73 - 6.85)	240.3 (271.0 - 198.0)
Jan-09	10.1 (11.6 - 8.7)	NA	NA	80.5 (109.0 - 67.2)	NA	7.53 (7.84 - 7.25)	272.5 (286.0 - 252.0)
Apr-09	16.5 (19.1 - 14.3)	5.65 (18.12 - 0.34)	5.1 (17.2 - 0.3)	100.0 (111.4 - 88.8)	7.31 (23.85 - 0.43)	7.40 (7.87 - 6.70)	256.6 (273.0 - 208.6)
Jul-09	21.4 (22.3 - 20.2)	NA	12.4 (25.7 - 1.6)	82.9 (103.2 - 66.4)	18.52 (36.00 - 2.81)	7.42 (7.86 - 7.21)	402.3 (453.0 - 353.0)
Oct-09	14.9 (16.8 - 13.8)	14.67 (29.94 - 3.08)	14.0 (29.9 - 2.6)	44.3 (69.1 - 33.0)	18.21 (37.21 - 3.75)	7.39 (7.72 - 7.11)	336.6 (346.0 - 328.0)
Jan-10	9.5 (10.9 - 8.4)	NA	NA	79.8 (100.3 - 48.8)	NA	7.41 (7.60 - 7.06)	276.9 (351.0 - 166.0)
Apr-10	NA	16.69 (28.03 - 3.30)	15.9 (27.8 - 2.8)	79.8 (85.1 - 72.0)	20.81 (36.96 - 4.06)	7.67 (8.03 - 7.44)	358.5 (428.2 - 215.9)
Oct-10	14.2 (16.2 - 11.9)	11.76 (29.89 - 1.14)	11.1 (29.9 - 0.9)	89.9 (97.8 - 78.9)	14.65 (38.24 - 1.31)	7.79 (8.32 - 7.23)	181.4 (213.2 - 129.2)

Table 4. Average, maximum and minimum temperature (T), pH, electrical conductivity (κ), dissolved oxygen (%DO), redox potential (ORP), total dissolved solids (TDS) and salinity (SAL) found in each sampling campaign at low tide in deep water. NA = not analyzed

Campaign	T(°C)	TDS (g L^{-1})	SAL (g L^{-1})	DO (% of saturation)	κ (mS cm^{-1})	pH	ORP (mV)
Jan-05	NA	NA	NA	NA	NA	NA	NA
May-05	17.6 (18.0 - 16.9)	17.69 (24.17 - 9.17)	16.9 (23.6 - 8.3)	60.0 (85.8 - 37.4)	27.33 (37.16 - 14.47)	7.76 (8.04 - 7.29)	241.9 (265.0 - 230.0)
Sep-05	21.5 (21.9 - 21.0)	28.28 (30.64 - 23.58)	28.0 (30.0 - 23.0)	35.5 (70.1 - 3.8)	44.37 (51.79 - 36.27)	7.67 (8.09 - 7.21)	95.9 (184.3 - 60.1)
Dec-05	9.9 (10.4 - 9.4)	9.47 (10.86 - 8.08)	8.4 (9.6 - 7.1)	65.5 (66.7 - 64.3)	10.29 (11.59 - 8.98)	7.84 (8.04 - 7.64)	345.5 (448.9 - 242.1)
Mar-06	11.2 (11.5 - 11.0)	12.95 (21.18 - 4.73)	12.1 (20.2 - 4.0)	46.7 (50.3 - 43.0)	14.62 (23.84 - 5.39)	7.79 (7.82 - 7.75)	NA
Jul-06	23.8 (24.9 - 22.3)	15.59 (36.31 - 2.72)	15.2 (37.1 - 2.3)	55.6 (107.7 - 34.7)	23.48 (55.69 - 3.35)	NA	NA
Oct-06	21.8 (25.1 - 20.3)	14.55 (37.71 - 3.20)	14.1 (38.7 - 2.7)	59.6 (82.3 - 26.5)	21.25 (58.15 - 4.07)	7.47 (7.88 - 6.90)	163.2 (225.0 - 47.0)
Jan-07	13.5 (15.4 - 12.5)	19.91 (36.22 - 3.13)	19.6 (36.9 - 2.6)	56.4 (84.1 - 22.8)	23.78 (42.45 - 3.93)	7.83 (8.06 - 7.64)	268.4 (280.0 - 253.0)
Apr-07	16.9 (19.7 - 15.1)	5.86 (24.55 - 0.92)	5.6 (24.0 - 0.8)	60.7 (104.6 - 42.2)	7.52 (33.96 - 0.18)	7.86 (8.13 - 7.70)	258.8 (288.0 - 238.0)
Jul-07	20.1 (20.8 - 18.8)	7.39 (21.65 - 0.69)	7.7 (25.5 - 0.4)	58.7 (101.0 - 24.7)	10.46 (30.06 - 0.73)	7.60 (8.03 - 7.32)	200.8 (253.0 - 114.0)
Oct-07	15.6 (16.9 - 14.2)	13.40 (29.74 - 4.53)	12.1 (25.6 - 3.9)	59.9 (87.7 - 25.4)	16.52 (35.27 - 5.07)	7.63 (7.93 - 7.23)	260.3 (312.0 - 222.0)
Jan-08	11.0 (14.4. - 8.8)	11.04 (25.91 - 0.53)	10.4 (25.4 - 0.4)	62.0 (82.8 - 37.1)	12.37 (29.28 - 0.56)	7.51 (7.76 - 7.11)	222.4 (264.0 - 199.0)

Table 4. (Continued)

Apr-08	13.4 (13.9 - 13.1)	18.79 (29.01 - 0.42)	18.4 (28.8 - 0.3)	70.9 (83.7 - 56.2)	22.56 (34.74 - 0.50)	7.27 (8.08 - 6.16)	266.0 (297.0 - 203.0)
Jul-08	19.6 (20.0 - 19.3)	27.09 (31.80 - 17.68)	26.9 (32.0 - 16.7)	77.6 (107.0 - 46.3)	37.40 (43.56 - 24.44)	7.58 (7.61 - 7.55)	242.0 (284.0 - 188.0)
Oct-08	18.3 (18.4 - 18.1)	27.72 (32.01 - 20.87)	27.5 (32.3 - 20.1)	33.6 (47.8 - 26.0)	37.15 (42.76 - 28.01)	6.79 (7.36 - 6.12)	244.3 (301.0 - 203.0)
Jan-09	NA	NA	NA	113.0	NA	7.60 8.01 7.14	251.0 272.5 239.6
Apr-09	13.4 14.3 12.5	18.94 27.22 1.88	18.4 26.9 1.5	86.5 108.5 61.1	22.78 32.79 2.21	6.80 7.15 6.38	249.4 329.0 189.0
Jul-09	20.1 20.3 19.9	NA	26.7 29.125.0	50.8 89.8 32.0	36.81 38.73 34.93	6.98 7.37 6.59	384.5 426.0 350.0
Oct-09	15.4 16.4 14.4	22.77 27.91 16.57	22.2 27.7 15.6	42.2 48.6 37.9	28.61 34.57 21.22	7.15 7.32 6.88	359.6 383.0 320.0
Jan-10	9.5	NA	NA	76.3	NA	6.79	366.0
Apr-10	NA	23.21 28.89 15.90	23.2 29.9 14.9	74.7 83.8 58.8	29.87 38.84 19.88	7.44 7.51 7.35	290.6 376.9 160.8
Oct-10	17.2 18.1 16.5	27.54 29.74 26.38	27.3 29.7 26.0	64.4 102.4 27.2	36.07 39.72 34.20	7.62 7.91 7.28	202.0 210.9 186.9

Table 5. Average, maximum and minimum temperature (T), pH, electrical conductivity (κ), dissolved oxygen (% DO), redox potential (ORP), total dissolved solids (TDS) and salinity (SAL) found in each sampling campaign at low tide in surface water. NA = not analyzed

Campaign	T(°C)	TDS (g L^{-1})	SAL (g L^{-1})	DO (% of saturation)	κ (mS cm^{-1})	pH	ORP (mV)
Jan-05	6.9 (11.0 - 5.0)	NA	NA	NA	1.60 (5.99 - 0.38)	7.91 (8.10 - 7.52)	NA
May-05	18.3 (20.3 - 16.5)	7.59 (25.54 - 0.54)	7.2 (25.1 - 0.4)	72.5 (97.5 - 31.5)	11.15 (39.30 - 0.00)	7.84 (8.55 - 7.23)	237.8 (406.4 - 142.6)
Sep-05	21.0 (21.7 - 19.6)	19.58 (30.40 - 2.87)	19.3 (30.5 - 2.4)	44.8 (77.5 - 14.6)	30.13 (46.77 - 4.44)	7.79 (8.11 - 7.55)	153.2 (260.0 - 65.4)
Dec-05	10.2 (13.9 - 8.9)	2.16 (5.52 - 0.38)	1.8 (4.7 - 0.3)	61.1 (68.6 - 55.7)	2.35 (6.04 - 0.41)	7.72 (8.12 - 7.25)	316.3 (360.8 - 190.0)
Mar-06	11.2 (15.0 - 9.7)	2.19 (7.33 - 0.27)	1.9 (6.4 - 0.2)	46.4 (50.1 - 42.0)	2.47 (8.32 - 0.29)	7.65 (7.81 - 7.41)	NA
Jul-06	23.1 (24.3 - 21.9)	4.71 (11.48 - 1.24)	4.2 (10.4 - 1.0)	70.9 (101.6 - 41).9	6.66 (16.95 - 1.49)	NA	NA
Oct-06	20.5 (22.1 - 19.2)	7.34 (13.52 - 2.10)	6.6 (12.9 - 1.7)	66.68 (5.4 - 51.9)	8.68 (15.64 - 2.87)	7.54 (7.95 - 6.82)	179.1 (277.0 - 122.0)
Jan-07	12.4 (15.6 - 10.7)	12.58 (27.28 - 2.16)	12.0 (26.9 - 1.8)	65.1 (71.1 - 55.0)	14.34 (31.39 - 2.73)	7.59 (8.15 - 6.62)	267.0 (292.0 - 227.0)
Apr-07	19.0 (21.0 - 17.2)	3.26 (12.15 - 0.66)	2.9 (11.1 - 0.5)	77.9 (99.0 - 69.6)	4.18 (17.20 - 0.15)	7.94 (8.31 - 7.14)	254.4 (277.0 - 233.0)
Jul-07	20.1 (22.1 - 18.4)	4.38 (11.60 - 0.60)	3.9 (10.6 - 0.4)	64.5 (79.5 - 44.7)	6.11 (14.74 - 0.46)	7.72 (7.99 - 7.21)	199.4 (246.0 - 143.0)
Oct-07	15.7 (20.2 - 13.8)	8.08 (14.41 - 3.05)	7.4 (13.4 - 2.5)	61.9 (76.8 - 53.4)	9.94 (17.74 - 3.68)	7.61 (8.00 - 7.19)	223.6 (286.0 - 132.0)
Jan-08	10.4 (14.4 - 8.8)	5.59 (15.07 - 0.50)	5.0 (14.0 - 0.4)	66.7 (79.5 - 57.6)	6.19 (16.85 - 0.53)	7.61 (8.19 - 7.01)	212.3 (276.0 - 148.0)
Apr-08	13.4 (16.3 - 12.6)	3.68 (8.50 - 0.37)	3.2 (7.6 - 0.3)	69.8 (77.2 - 62.8)	4.15 (10.12 - 0.44)	7.57 (7.79 - 7.19)	250.5 (270.0 - 215.0)

Table 5. (Continued)

Campaign	T(°C)	TDS (g L^{-1})	SAL (g L^{-1})	DO (% of saturation)	κ (mS cm^{-1})	pH	ORP (mV)
Jul-08	20.5 (21.5 - 19.7)	7.31 (19.05 - 1.47)	6.6 (18.2 - 1.2)	81.5 (96.4 - 66.7)	10.27 (26.63 - 2.04)	7.75 (8.13 - 7.53)	217.5 (271.0 - 116.0)
Oct-08	16.5 (18.0 - 13.2)	9.10 (14.83 - 0.12)	8.3 (13.8 - 0.1)	41.3 (44.3 - 39.8)	12.03 (19.78 - 0.15)	7.45 (7.56 - 7.29)	238.5 (266.0 - 213.0)
Jan-09	NA	NA	NA	108.4 (118.0 - 101.9)	NA	7.19 (7.50 - 6.94)	264.2 (275.0 - 250.7)
Apr-09	13.7 (16.9 - 12.3)	5.25 (13.33 - 0.52)	4.7 (12.3 - 0.4)	98.1 (111.2 - 71.9)	6.25 (15.88 - 0.63)	7.11 (7.53 - 6.56)	209.6 (311.0 - 105.4)
Jul-09	20.9 (21.8 - 18.3)	NA	5.7 (18.4 - 1.0)	43.9 (89.4 - 18.8)	8.55 (26.67 - 1.62)	7.09 (7.35 - 6.70)	396.5 (428.0 - 351.0)
Oct-09	15.3 (18.6 - 13.9)	8.59 (22.40 - 2.12)	7.9 (21.7 - 1.7)	56.0 (105.0 - 41.6)	10.68 (28.03 - 2.87)	7.40 (7.77 - 6.62)	350.1 (415.0 - 283.0)
Jan-10	9.3 (11.5 - 8.2)	NA	NA	83.9 (96.1 - 71.4)	NA	7.04 (7.26 - 6.66)	343.2 (363.0 - 335.0)
Apr-10	NA	6.29 (13.83 - 0.95)	5.6 (12.8 - 0.7)	77.7 (91.2 - 66.6)	8.00 (17.95 - 1.23)	7.68 (8.20 - 7.22)	295.7 (382.2 - 196.9)
Oct-10	14.4 (18.1 - 12.4)	6.28 (19.11 - 1.03)	5.7 (18.27 - 0.8)	86.1 (94.8 - 79.6)	7.80 (24.36 - 1.25)	7.74 (8.27 - 7.04)	186.3 (208.3 - 153.3)

Table 6. Increasing (+) and decreasing (–) trends detected in each sampling site after the application of the Mann-Kendall test in time series of physico-chemical parameters measured in deep or surface waters collected at high or low tide. 90, 95, 99 and 99.9 (%) express the confidence level at which the trend was detected. "–" means that the test was not performed in those conditions

Sampling sites	DEEP WATER								SURFACE WATER							
	Low tide				High tide				Low tide				High tide			
	pH	DO	ORP	κ	pH	DO	ORP	κ	pH	DO	ORP	κ	pH	DO	ORP	κ
AR	-99	+90	+90		-99				-	-	-	-	-99.9			
GO	-	-	-	-	-99				-95				-99	+95		
UD	-	-	-	-	-99.9				-99	+95		+90	-99.9			
GA	-99				-99				-90				-90			
AS	-90	+90		+90	-99.9			-99				+95	-95			
KA	-90				-90	+90		-90				+95				
ZO	-	-	-	-	-99				-95	+95			-99			
AZ						+99			-90				-95			

Table 7. Average metal concentrations reported for different water bodies of the world. The concentrations are in µg L⁻¹ unless otherwise stated

Locality	Al	As	Cr	Cu	Fe	Mn	Ni	Zn	Ref.
Amvrakiko Gulf (Greece)	-	-	-	0.62	-	-	1.77	3.34	(Scoullos et al., 1994)
Maliakos Gulf (Greece)	-	-	-	0.76	-	-	2.85	1.62	(Scoullos et al., 1987)
Tinto River (Spain)	-	2.9	-	54.7	-	-	-	381.6	(Elbaz-Poulichet et al., 1999)
Garonne estuary (France)	-	1.5	-	0.8	8	-	0.36	1177	(Krapiel et al., 1997)
Tagus estuary (Portugal)	-	51	-	1.64	12	-	1	15.2	(Elbaz-Poulichet et al., 2001)

Table 7. (Continued)

Locality	Al	As	Cr	Cu	Fe	Mn	Ni	Zn	Ref.
Litheos River (Greece)	-	-	-	5.4	-	-	10	1.7	(Mueller and Foerstner, 1975)
Elbe estuary (German)	-	-	2.5–15.5	3.5–18	-	-	3.8–18.6	9–194	(Morillo et al., 2005)
Huelva estuary (Spain)	-	7.0–40	-	9–240	31–257	8–670	1–23	38–1125	(Morillo et al., 2005)
Thames estuary (England)	-	-	-	10.7	-	-	6.3	29.1	(Power et al., 1999)
Ebro River (Spain)	-	-	-	0.35–5.5	-	-	-	8.2–325	(Ramos et al., 1999)
Mersey estuary (England)	-	-	-	3.17	-	-	-	0.89	(Martino et al., 2002)
Tay estuary (England)	-	-	-	0.05	-	-	-	0.98	(Owens and Balls, 1997)
Gironde estuary (France)	-	-	-	1.35	-	-	-	5.9	(Masson et al., 2006)
Flanders estuary (Belgium)	-	-	-	0.3–14.5	-	-	-	8–445	(Bervoets et al., 1994)
Grande de Xubia River (Spain)	-	-	-	2.16	-	-	-	4.64	(Cobelo-Garcia and Prego, 2004)
Lawrence estuary (France)	<1.99	-	-	6.62	4.02	1.54	0.66	0.3	(Rondeau et al., 2005)
Ennore estuary (India)	-	-	-	3330–1090	-	-	-	-	(Padmini and Geetha, 2007)
Seine River (France)	-	-	-	2	2.28	-	4.2	13.2	(Chiffoleau et al., 1994)
Mhlathuze estuary (KwaZulu)	160–44412	-	14–226	14–76	-	360–23500	54–266	52–260	(Mzimela et al., 2003)
Axios River (Greece)	-	-	-	7	-	-	-	67.3	(Dassenakis et al., 1997)
Odiel River	-	4.7	-	20.38	-	-	-	107.65	(Vicente-Martorell et

Locality	Al	As	Cr	Cu	Fe	Mn	Ni	Zn	Ref.
Padre Santo Channel (France)	-	3.5	-	11.0	-	-	-	52.4	(Vicente-Martorell et al., 2009)
Po River (France)	6.75	-	-	1197	-	-	-	-	(Alberti et al., 2008)
Gao-ping River (Taiwan)	-	0.63-4.33	0.47-338	0.9-34.3	-	-	3.64-35.8	5.14-354	(Doong et al., 2008)
Vigo Ria (Spain)	-	-	-	0.63	-	-	0.36	1.32	(Santos-Echeandia et al., 2009)
Ferrol Ria (Spain)	-	-	-	0.42-0.6	-	-	-	1.12-1.6	(Cobelo-Garcia et al., 2005)
Nerbioi-Ibaizabal (in $\mu g \cdot kg^{-1}$)	13.5	22.4	5.4	263	486	37.4	7.0	20.6	This work

Table 8. Increasing (+) and decreasing (-) trends detected in each sampling site after the application of the Mann-Kendal's test in time series of trace element concentrations measured in deep or surface waters collected at high or low tide. 90, 95, 99 and 99.9 express the confidence level at which the trend was detected. "-" means that the test was not performed in those conditions

DEEP WATER	Low tide								High tide							
	AR	GO	UD	GA	AS	KA	ZO	AZ	AR	GO	UD	GA	AS	KA	ZO	AZ
Al	+95	-	-						+90	+95	+99		+95	+90	+99	
As	+99	-	-		+95		-	+95	+95	+99	+95	+99	+99	+90	+95	+99
Cr	+99	-	-	+90		+95	-	+99	+90	+95	+99	+99.9	+99	+95	+99	+99
Cu	+99	-	-		+99		-	+95	+90	+95	+99	+95	+99	+95	+99	+95
Fe	+99	-	-		+95		-		+99	+95	+95	+90	+90			+95
Mn	+95	-	-				-			+90	+95	+90	+95		+95	
Ni	+90	-	-		+90	+90	-	+95	+95	+99	+95	+99	+99	+99	+95	+99
Zn	+99	-	-		+90	+90	-		+95	+99.9	+95	+99.9	+99.9	+99	+99.9	+99.9

Table 8. (Continued)

	SURFACE WATER															
	Low tide								High tide							
	AR	GO	UD	GA	AS	KA	ZO	AZ	AR	GO	UD	GA	AS	KA	ZO	AZ
Al	-								+95							
As	-			+90					+99.9	+95	+99					
Cr	-		+90	+99		+90	+99.9	+95	+95	+90	+99	+99.9	+95	+95	+95	+95
Cu	-		+99			+90			+99.9		+95					
Fe	-		+95	+99			+90		+95	+90	+90					
Mn	-			+90				+90							+99	
Ni	-		+95			+90			+99		+90	+90	+90			
Zn	-		+99			+95		+95	+99		+99	+90	+95	+95	+95	

Table 9. Correlation matrix obtained using: a) all the data; b) those samples collected at high tide and c) those samples collected in the surface. Correlations significantly different from zero at a 99% confidence level are marked in bold, and the most significant ones (a) >0.46; b) and c) >0.54) are in italic

a)

	Al	As	Cr	Cu	Fe	Mn	Ni	Zn	T	DO	κ	pH	ORP
Al	1.00												
As	**-0.23**	1.00											
Cr	**0.39**	0.04	1.00										
Cu	**-0.18**	*0.90*	-0.04	1.00									
Fe	-0.10	*0.85*	-0.02	*0.90*	1.00								
Mn	0.06	0.07	**0.34**	0.03	0.07	1.00							
Ni	**0.17**	*0.47*	**0.26**	*0.50*	*0.50*	**0.34**	1.00						
Zn	**0.44**	*0.10*	**0.42**	**0.15**	**0.16**	**0.42**	*0.58*	1.00					
T	**-0.18**	**0.23**	**-0.24**	**0.27**	**0.27**	0.07	**0.18**	-0.07	1.00				
DO	**0.17**	**-0.21**	0.08	**-0.25**	**-0.16**	**-0.15**	0.01	**0.15**	-0.05	1.00			
κ	**-0.25**	*0.51*	**-0.39**	*0.60*	*0.55*	**-0.28**	0.11	**-0.14**	**0.35**	**-0.16**	1.00		
pH	-0.09	**-0.23**	-0.10	**-0.18**	**-0.18**	**-0.41**	**-0.37**	**-0.42**	-0.01	**0.22**	-0.05	1.00	
ORP	0.06	**0.15**	0.10	**0.16**	**0.12**	-0.11	0.05	0.04	**-0.13**	-0.10	0.03	**-0.21**	1.00

b)

	Al	As	Cr	Cu	Fe	Mn	Ni	Zn	T	DO	к	H	ORP
Al	1.00												
As	-0.22	1.00											
Cr	0.4	0.17	1.00										
Cu	-0.18	0.93	0.10	1.00									
Fe	-0.11	0.86	0.07	0.92	1.00								
Mn	0.07	0.12	0.43	0.11	0.14	1.00							
Ni	0.13	0.59	0.35	0.59	0.59	0.32	1.00						
Zn	0.44	0.25	0.49	0.28	0.27	0.51	0.58	1.00					
T	-0.28	0.28	-0.24	0.35	0.35	-0.03	0.20	-0.07	1.00				
DO	0.26	-0.24	0.12	-0.25	-0.18	-0.07	0.07	0.22	0.01	1.00			
к	-0.32	0.32	-0.47	0.40	0.41	-0.24	0.01	-0.14	0.49	-0.21	1.00		
pH	-0.10	-0.34	-0.27	-0.25	-0.24	-0.47	-0.33	-0.44	0.06	0.20	0.01	1.00	
ORP	0.02	0.26	0.09	0.22	0.19	-0.04	0.18	0.07	0.00	-0.16	-0.08	-0.27	1.00

c)

	Al	As	Cr	Cu	Fe	Mn	Ni	Zn	T	DO	к	pH	ORP
Al	1.00												
As	-0.32	1.00											
Cr	0.29	0.03	1.00										
Cu	-0.17	0.82	-0.12	1.00									
Fe	-0.08	0.79	-0.02	0.82	1.00								
Mn	-0.11	0.01	0.30	-0.07	-0.03	1.00							
Ni	0.11	0.26	0.11	0.38	0.36	0.28	1.00						
Zn	0.40	-0.05	0.28	0.03	0.09	0.35	0.52	1.00					
T	-0.21	0.29	-0.15	0.27	0.29	0.12	0.18	-0.08	1.00				
DO	0.24	-0.14	0.03	-0.21	-0.09	-0.09	0.02	0.13	0.06	1.00			
к	-0.15	0.60	-0.33	0.73	0.70	-0.27	0.24	-0.06	0.35	-0.07	1.00		
pH	-0.05	-0.13	-0.15	-0.16	-0.17	-0.33	-0.34	-0.47	-0.07	0.15	-0.14	1.00	
ORP	0.10	0.01	0.06	0.08	0.06	-0.20	-0.06	0.10	-0.09	-0.08	0.11	-0.23	1.00

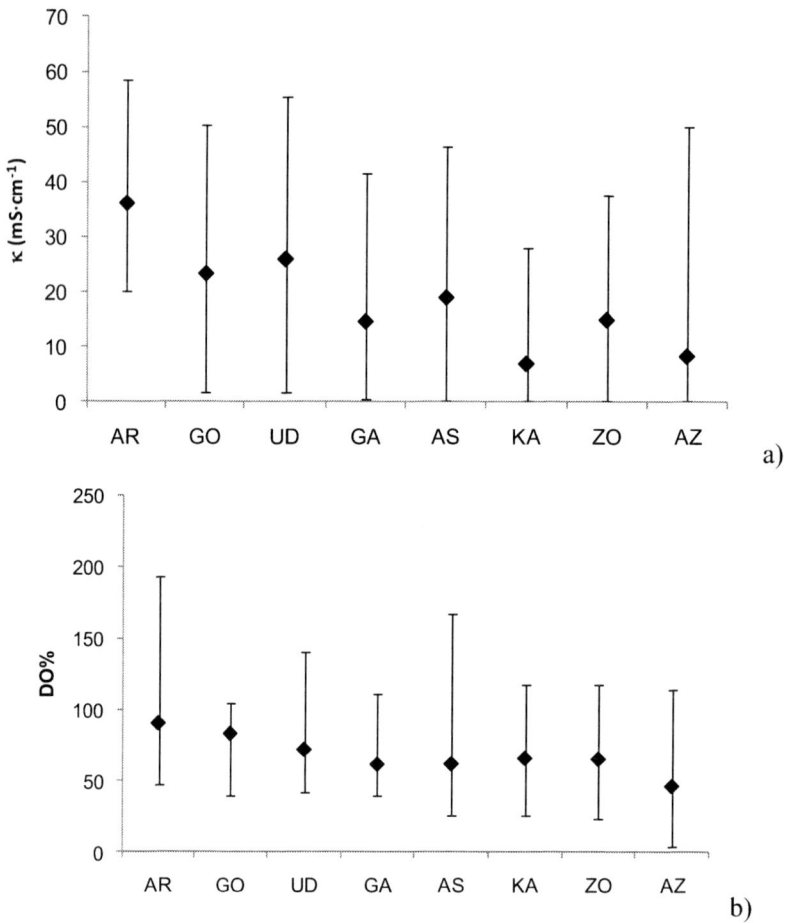

Figure 3. a) Average (black point), maximum and minimum conductivity values (mS·cm⁻¹) measured in surface water samples collected at high tide in each sampling site (22 campaigns are considered, from January 2005 to October 2010); b) Average values of the dissolved oxygen concentration (DO, expressed as % of saturation) measured in deep waters collected at low tide during the sampling periods between March 2006 and April 2010.

The most striking result was a clear decreasing trend in pH in the earlier part of the time series. After April–July 2009 the decreasing trend ended and pH showed an increasing trend thereafter. To illustrate, the time series of pH in deep water from *UD* collected at high tide is shown in Figure 4a.

When significant trends were identified in the time series of DO, they were always positive, indicating overall improvement in water quality of the estuary. The constant increase in water oxygenation (see Figure 4b), had been recognised previously for the period 1993–2003 (García-Barcina et al. 2006).

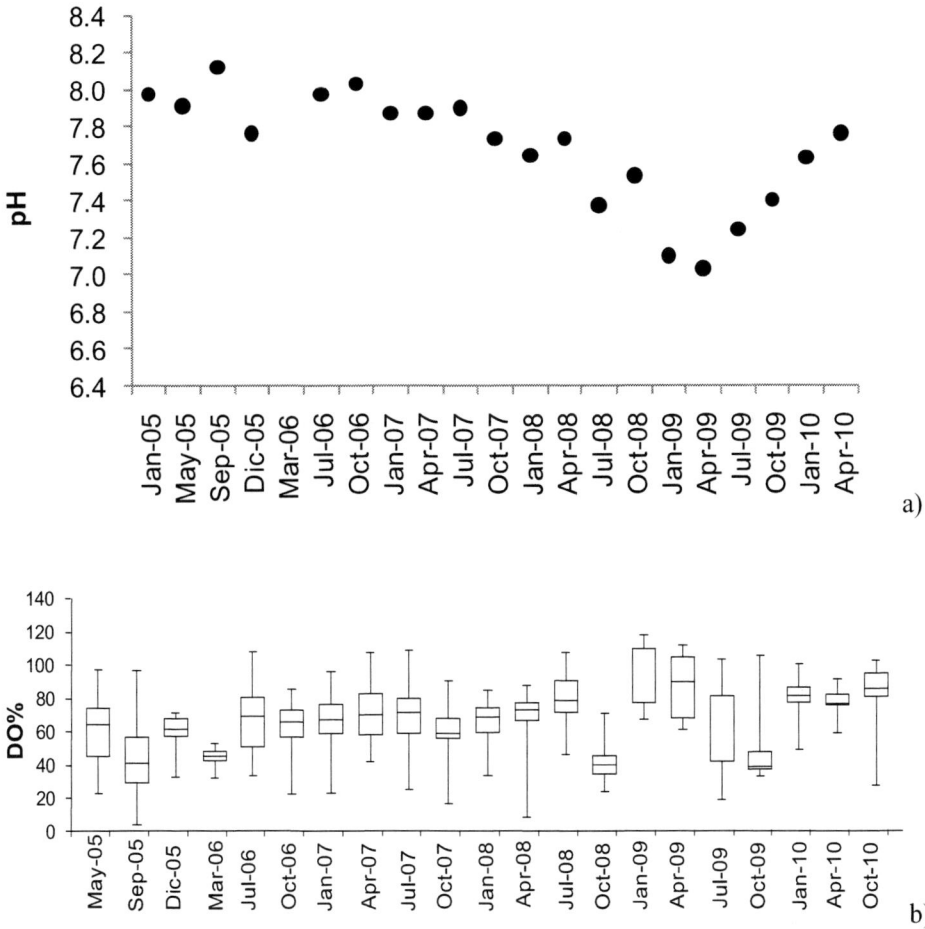

Figure 4. a) pH values measured in deep waters collected in Udondo (*UD*) at high tide between January 2005 and April 2010; b) Box-Whisker plots of the dissolved oxygen concentration (DO, expressed as % of saturation) measured in each sampling campaign. The box shows the 25th percentile and the 75th percentile, and the whiskers represent the lowest and the highest concentrations, while the line inside the box expresses the average value.

For conductivity two different trends were distinguished: increasing conductivity over time in waters collected at low tide, and decreasing trends in samples collected at high tide. In neither case was the number of sampling sites with significant trends sufficient for drawing general conclusions

Trace Metal Concentrations – Comparison with other Estuaries

As can be observed in Table 7, the range of trace element concentrations in waters of the estuary of the Nerbioi-Ibaizabal River cannot be considered alarming in comparison with other natural water bodies. In most of the cases, concentrations are comparable to those found in other estuaries of the world. Concentrations of As, Fe and Cu are the exceptions, with moderately higher average values (22, 486 and 162.3 $\mu g \cdot kg^{-1}$ respectively).

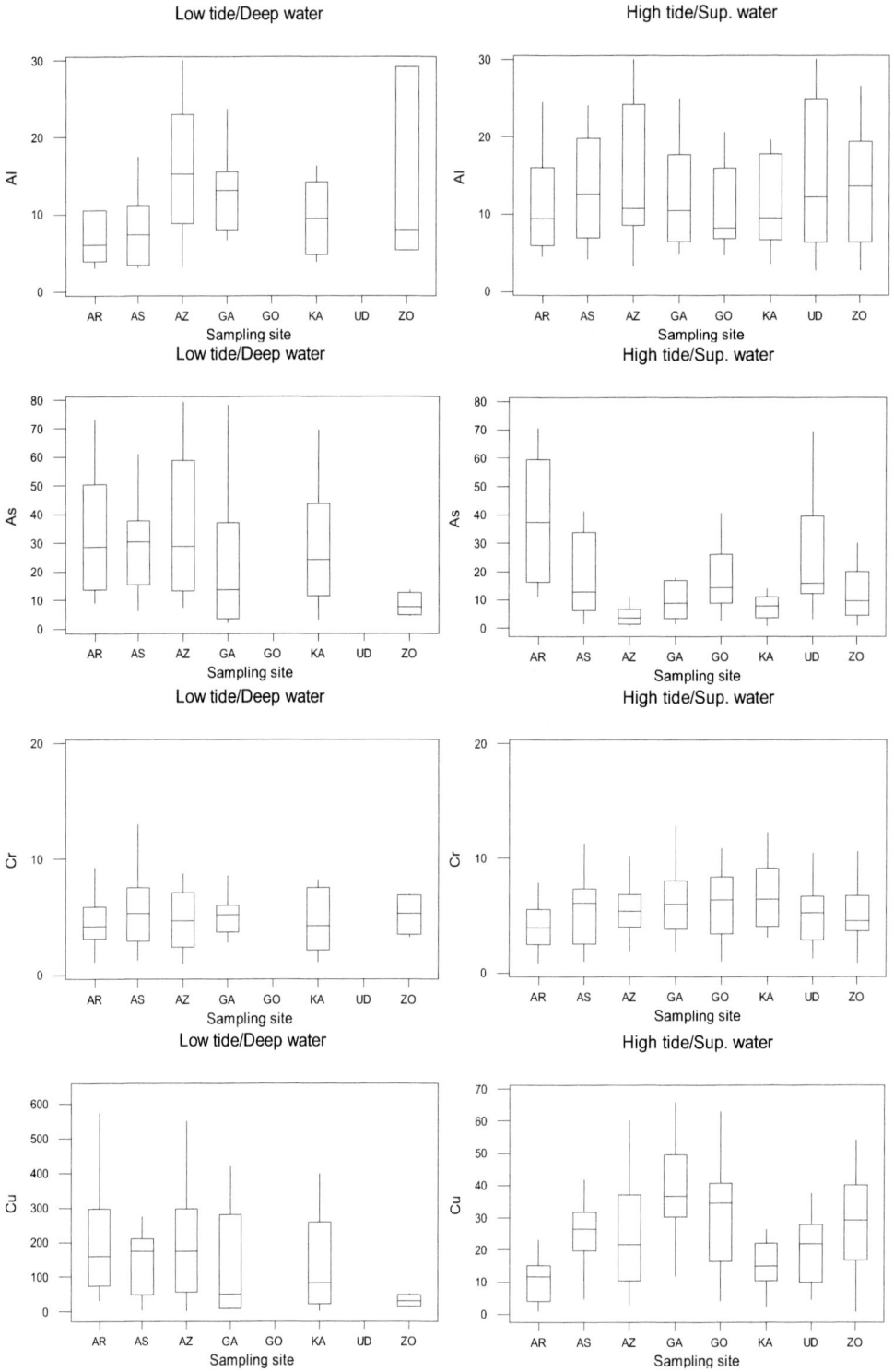

Figure 5. (Continued)

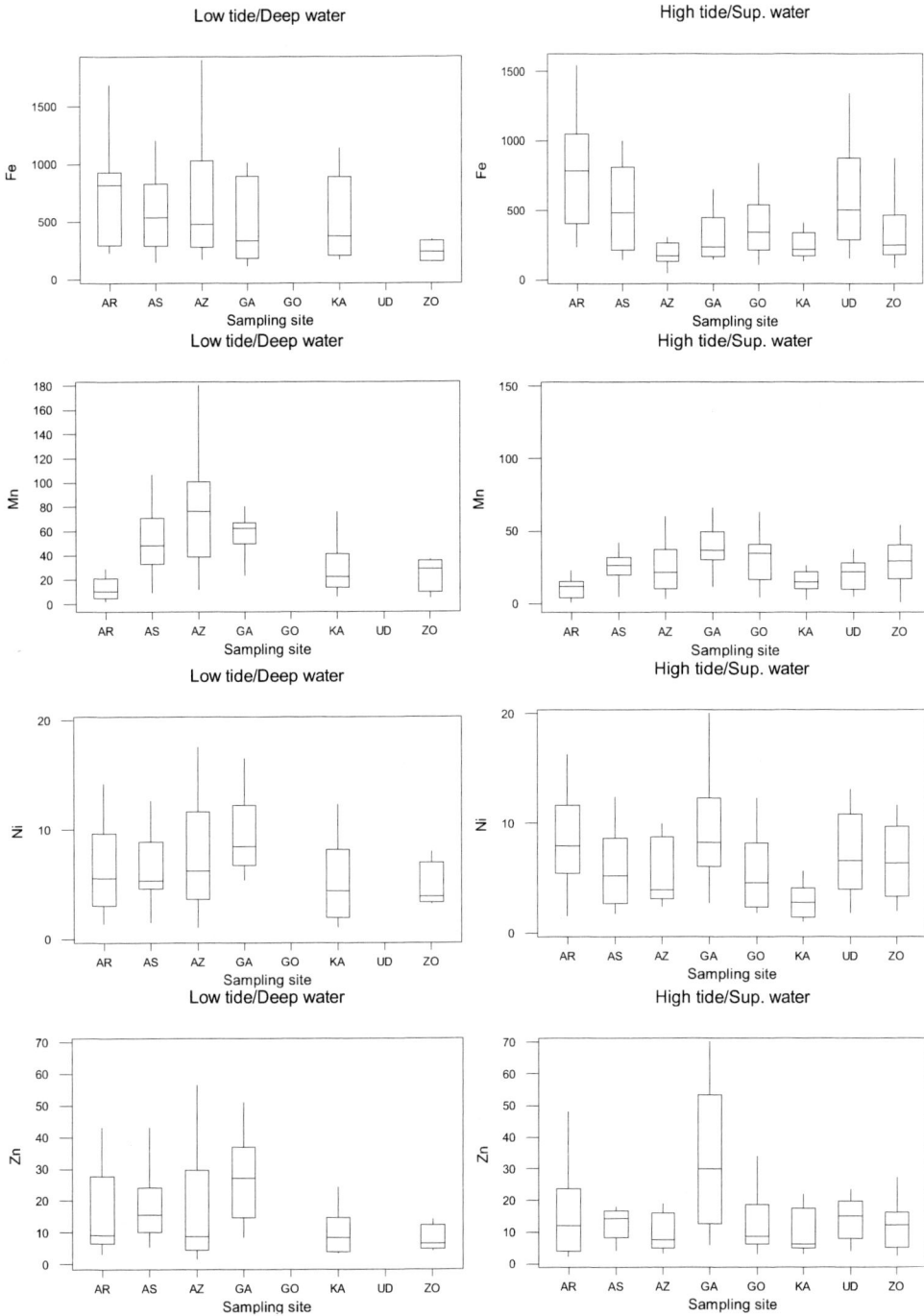

Figure 5. Box-Whisker plots of the trace element concentrations ($\mu g \cdot kg^{-1}$) found in deep waters collected at low tide vs. those ones found in surface waters collected at high tide in the eight sampling sites from May 2005 to October 2010. The box shows the 25th percentile and the 75th percentile, and the whiskers represent the lowest and the highest concentrations, while the line inside the box expresses the average value.

Trace Metal Concentrations – Spatial Distribution and Pollution Sources

Basic statistics (average, maximum, minimum and percentiles) of the trace element concentrations measured in surface and deep waters at low and high tide are summarized in Figures 5 and 6. In general, higher concentrations were observed for As, Cu and Fe in deep samples at high tide, and for Cr in surface waters at low tide. The rest of the metal concentrations did not present clear trends. As to geographical distribution, when all data were taken into account, no remarkable difference among sampling sites was found in the case of Al, Cr and Ni. Zn presented slightly higher concentrations in water samples from *GA*, as did Mn in samples from *AZ*. As, Cu and Fe exhibited similar trends, with maximum concentrations in the mouth of the estuary (*AR*), intermediate values in the middle of the estuary (*UD*, *AS* and *GO*) and minimum values in the upper part (*GA*, *KA*, *ZO* and *AZ*).

More specifically, concentrations of the trace elements in deep water collected at high tide (characteristic of ocean water) tended to be homogeneous all over the estuary. Conversely, the trace element content of surface samples collected at low tide (characteristic of the tributary river inputs) significantly changed from site to site, and the distribution was characteristic for each element. Comparing deep water at high tide (DH) vs. surface water at low tide (SL) (Figure 6) leads to the following conclusions about metal pollution sources in the estuary:

- *Cr and Mn:* Concentrations of Cr were systematically higher in SL than in DH, indicating a diffuse source of pollution for Cr, entering the downstream reaches of the estuary through all the tributary rivers of the system. The situation is similar for Mn, with the exception of *AZ* (concentration in DH surprisingly much higher than in SL) and *KA* (similar concentrations recorded in DH and SL). Consequently, the Gobela, Galindo and Asua Rivers seem to be local sources of net inputs of Mn to the estuary.
- *As, Cu and Fe:* Concentrations in DH were always higher than in SL, so their sources were not freshwater inputs from the tributaries. It has been reported that the concentration of these elements in sediments of the lower estuary is decreasing with time (Fdez-Ortiz de Vallejuelo et al. 2010) so, probably, the sediments themselves are acting as secondary sources of As, Cu and Fe to the water.
- *Ni, Zn:* The water samples from DH and SL did not have meaningful differences in the concentrations of these metals. The exceptions were the Galindo River for Ni and Zn, and the Gobela and Asua tributaries for Zn. These rivers are net sources of Ni and Zn to the estuary.
- *Al:* This element was present at slightly higher concentrations in SL from *GA, KA, ZO* and *AZ*, suggesting that Al enters the system (not in large amounts) through the Galindo, Kadagua and Nerbioi-Ibaizabal (but not Gobela) Rivers.

Trace metal Concentrations – Temporal Distribution

As a first approach to analysing temporal variations in trace elements, time series of the average metal concentrations were inspected visually. In this way, important trends were detected for As, Cu, Fe and Ni, which presented maximum concentrations in dry seasons (July and October) and minimum values in rainy ones (January and April). Very probably this pattern was a consequence of relatively high temperature and low water flow during dry months. As an example, the average of the Cu concentrations measured in surface waters at high tide is plotted against time in Figure 7, together with the flow data reported for each campaign. The flow was estimated by the sum of the partial flows of the three principal tributaries of the estuary (Ibaizabal, Nerbioi and Kadagua) using data reported by the local administration.[15]

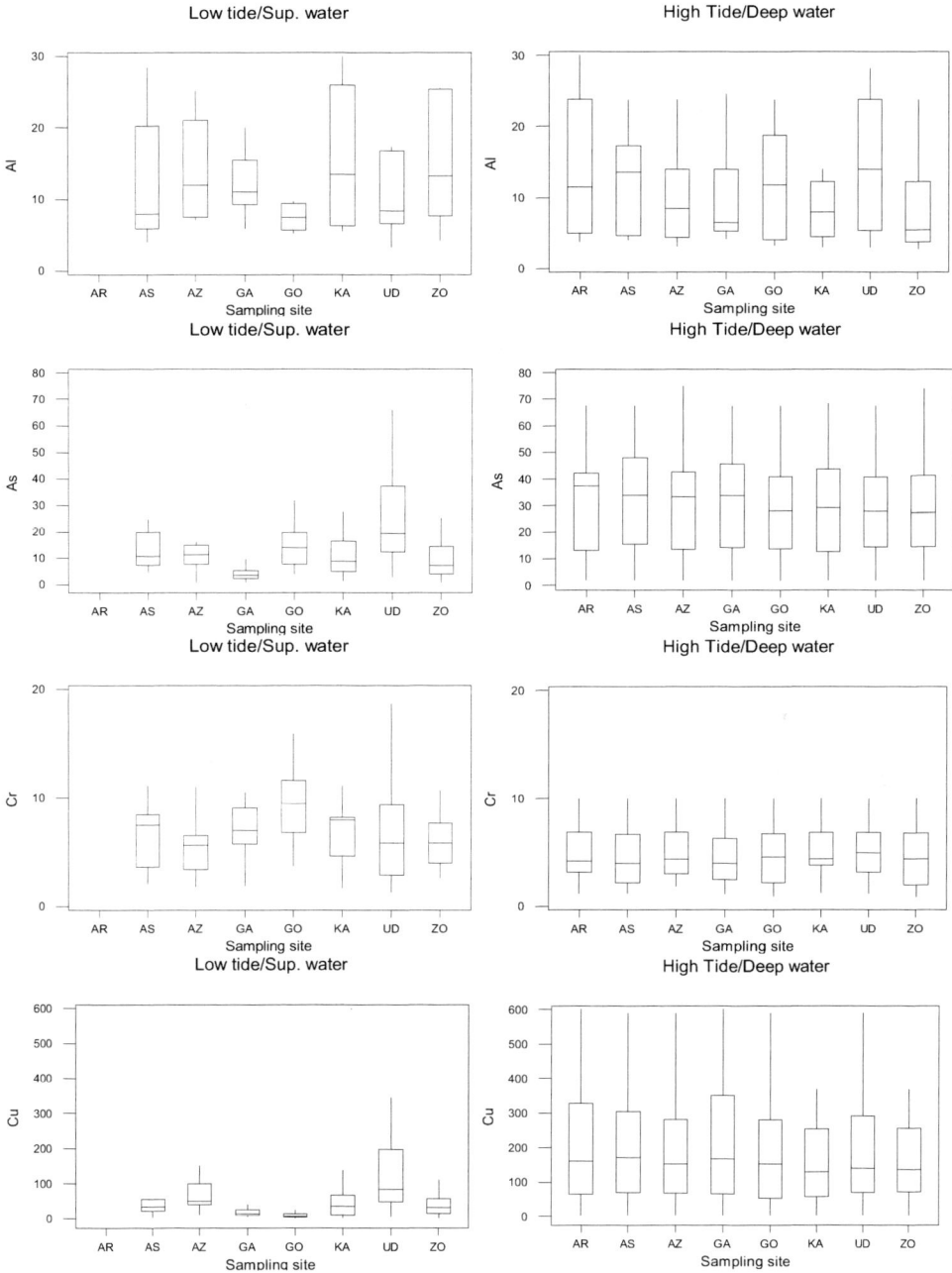

---

[15]   http://www.bizkaia.net/Ingurugiroa_Lurraldea/Hidrologia_Ac/Datos_meteo.asp?Idioma=CA&Tem_Codigo= 2679.

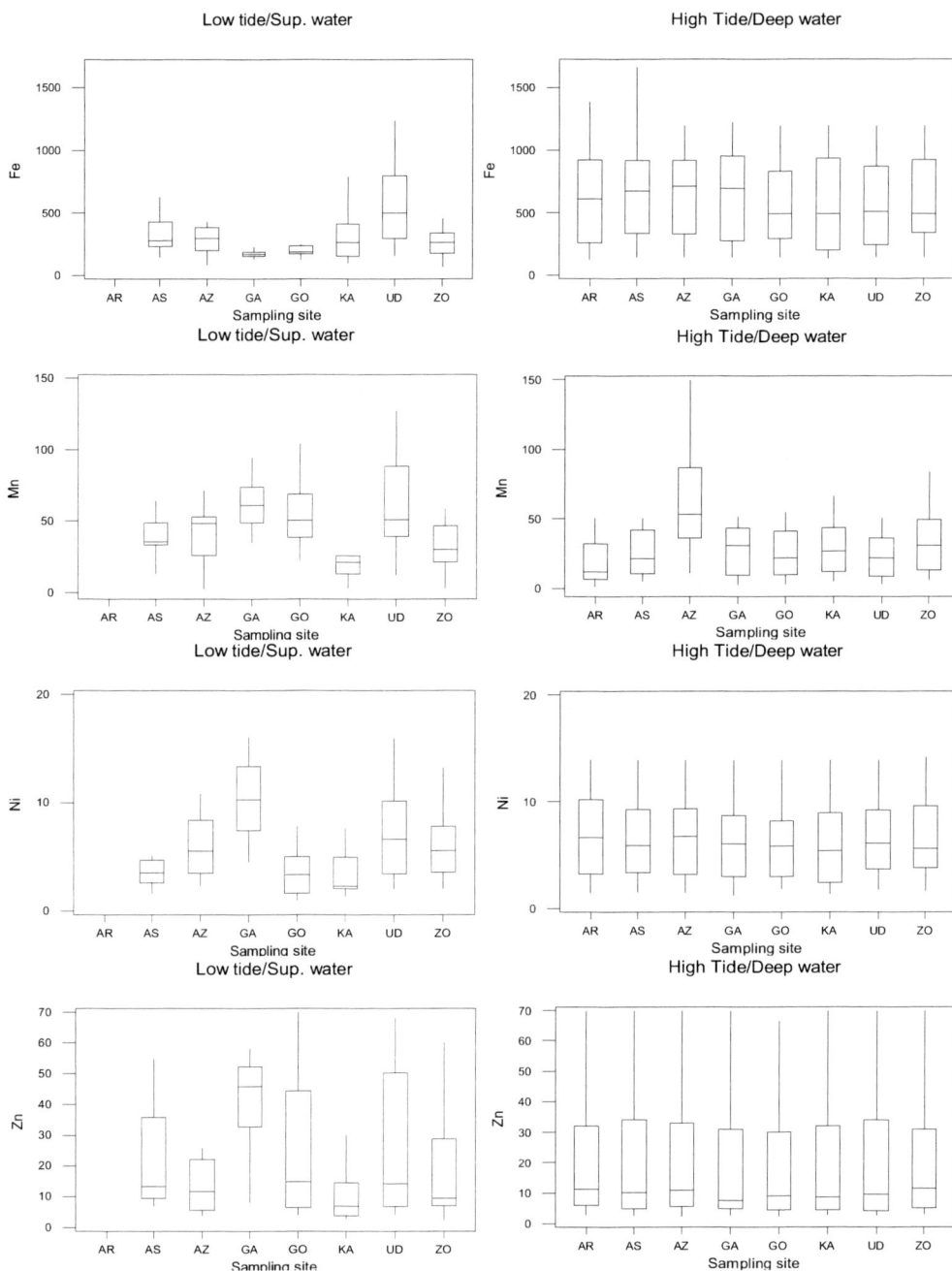

Figure 6. Box-Whisker plots of the trace element concentrations ($\mu g \cdot kg^{-1}$) found in surface waters collected at low tide vs. those ones found in deep waters collected at high tide in the eight sampling sites from May 2005 to October 2010. The box shows the 25th percentile and the 75th percentile, and the whiskers represent the lowest and the highest concentrations, while the line inside the box expresses the average value.

As in the case of physico-chemical parameters, the Mann-Kendall test was used in order to detect significant increasing or decreasing trends in the time series of the concentrations of trace elements (the time series not analysed for physico-chemical parameters also were not analysed for metals, for the reasons given above). The results are summarized in Table 8. It must be emphasized that 1) when a significant trend was observed it was always positive, and 2) the increasing trend was consistent for nearly all the metals (Al was a clear exception) and for sampling sites in deep water at high tide, that is, in water entering the estuary from the ocean (see Figure 8 as an example).

Otherwise, the number of increasing trends decreased in the order: SH>DL>SL, which confirms the likelihood that temporal changes in the characteristics of the ocean water were responsible for the increasing metal concentrations observed in the estuary.

Correlation Analysis between Trace Elements and Physico-chemical Parameters

All correlation analyses were performed by SPSS, PASW statistic (1200) 18.0 software, using Spearman's non-parametric correlation coefficient (r; Gauthier, 2001). First, a correlation analysis with all of the data from the study was carried out. Those cases with "not analysed" or "below detection limit" response in any of the parameters were eliminated from the calculation. The critical value of r (N = 545) is 0.11 at a confidence level of 99%, i.e. r >0.11 or <–0.11 is significantly different from zero. As could be expected with such a large N, almost all the correlation coefficients calculated fulfilled these conditions. Therefore, in order to highlight only the most significant associations, we used a compromise threshold value of r = 0.46. The results are shown in Table 9a. Note that all the coefficients above 0.46 were positive. The strongest relations observed were those ones among As, Cu and Fe, with r = 0.85–0.90. These three trace elements also were correlated with Ni ($r \sim 0.5$) and with conductivity (As, r = 0.51; Cu, r = 0.60 and Fe, r = 0.55). Concentrations of Ni and Zn also were positively correlated (r = 0.58).

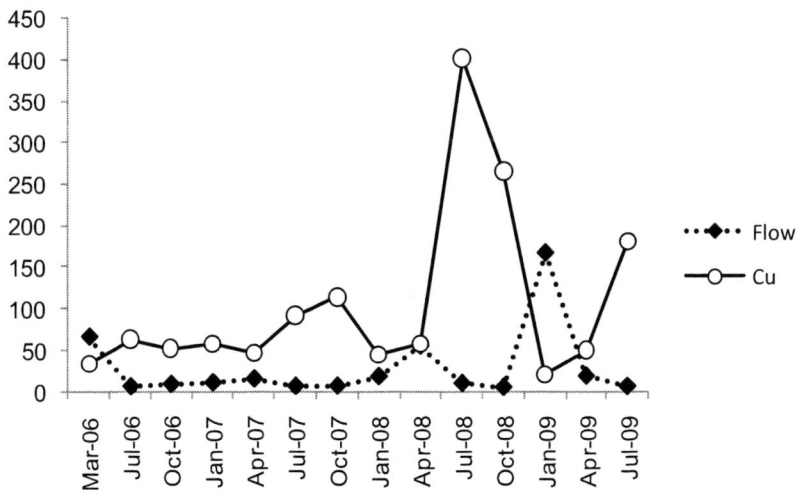

Figure 7. Average concentrations of Cu ($\mu g \cdot kg^{-1}$) measured in surface waters collected at high tide, together with the estimated river flow ($m^3 \cdot cm^{-1}$) in the sampling campaigns from March 2006 to July 2009.

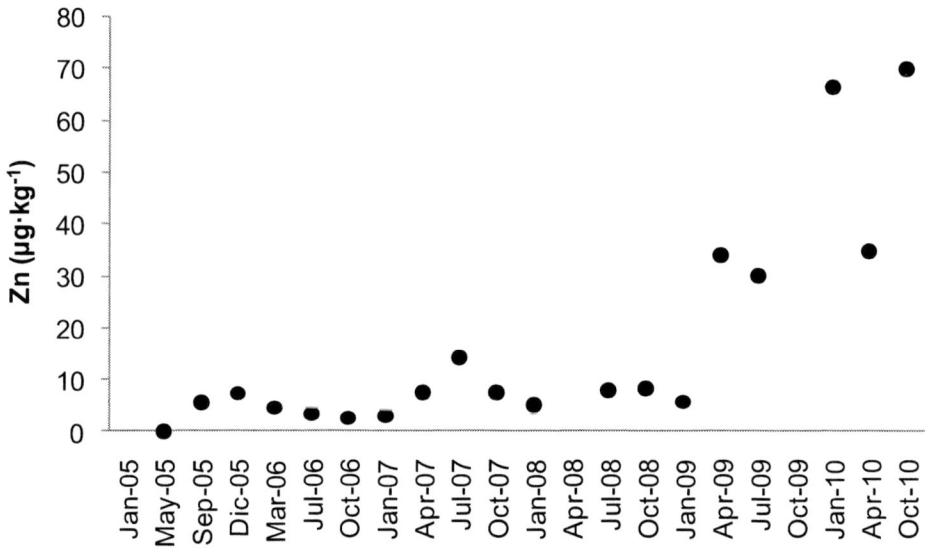

a)

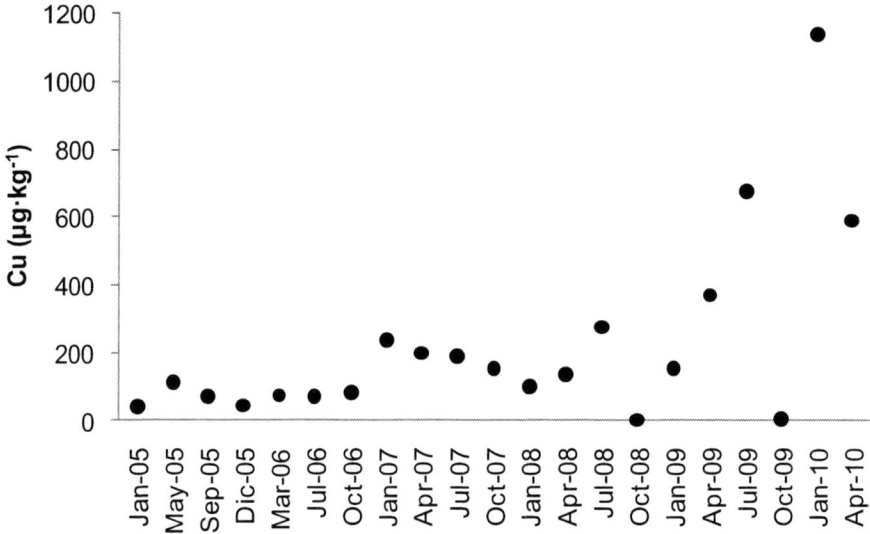

b)

Figure 8. Time series of the a) Zn concentration measured in deep waters of Galindo (GA) collected at high tide between May 2005 and October 2010 and b) Cu concentration measured in deep waters of Zorrotza (ZO) collected at high tide between January 2005 and April 2010.

Second, the samples were divided into four groups, i.e. samples collected at low tide (L), at high tide (H), in surface water (S) and in bottom water (D). The number of samples was approximately half of the initial data (N = 227–318), so that a threshold value of $r = 0.54$ was chosen to investigate the strength of the relationships among variables. Correlations for samples collected at high tide (H) and those from the bottom (D) were similar. In both cases all the significant correlations were positive, and again, the strongest were among As, Cu, Fe and Ni, as one group, and between Ni and Zn, as another. For these samples, no strong association was found between the trace elements and conductivity (Table 9b). The results of the correlation analysis done with the samples collected at low tide (L) and the samples from the surface (S) also were similar (Table 9c). For these samples, apart from the relations observed among As, Cu and Fe (r = 0.85–0.92), there were again significant positive correlations between As, Cu, Fe and conductivity. As an example of the positive correlation, in Figure 9 Fe is plotted against Cu for deep water, low tide samples.

We observe that the correlation patterns obtained for deep and high tide samples, on the one hand, and those for surface and low tide samples, on the other, are similar. Note that deep water at high tide is characteristic of the ocean and surface water at low tide is fluvial. These two groups, SL and DH, were independently analysed. In this case, r = 0.63 was chosen as the threshold value, because of the smaller number of samples (N = 91–169). As expected, the results obtained in SL did not differ from those obtained for S and L independently, and the same finding applies to the DH group.

CONCLUSION

This study confirms the importance of tides and stratification for the distribution of the measured properties. Although the tide was not a limiting factor for DO, the depth at which the sample was collected showed a clear influence on this variable. The evolution in time of DO suggests that the ecological quality of the estuary of the Nerbioi-Ibaizabal River is improving significantly. Data also confirm that the influx of seawater, biological activity, atmospheric deposition, and the tributary rivers should be considered as sources of oxygen in the estuary. Unexpectedly, a significant and general decrease in pH was detected through January 2009, a trend that became positive after that date. In addition, inflow discharge seems to play an important role in the concentrations of some of the trace elements considered. Concentrations of As, Cu, Fe and Ni reflect an important dilution effect, with maximum values in dry seasons and minimum values in wet ones. This fact implies a seasonal trend in the concentrations of these elements.

In general terms, as can be observed in Table 7, the condition of the waters of the estuary with respect to metal pollution is not alarming compared to other similar environments in the world. Three possible metal pollution sources were identified, however, in the estuary: (1) the tributary rivers (except Kadagua); (2) the ocean, and (3) the sediments themselves, which are able to accumulate contaminants for long time periods. In fact, several findings in this work seem to point in the same direction: first, the concentration in water of nearly all trace elements considered increased with time over the period investigated; second, this phenomenon was more pronounced in deep samples collected at high tide, where strong positive correlations were found between As, Cu, Fe, Ni and conductivity; third, a general decreasing trend in water pH was found over most of the study period. Furthermore, we have seen in previous investigations (Fdez-Ortiz de Vallejuelo et al. 2010) that the metal content in sediments of the lower part of the estuary is decreasing with time. As a hypothesis, it could be suggested that sediments are acting as a secondary source of metal pollution (at least As, Fe, Cu and Ni) to the estuary water, with a likely connection to the decreasing trend in pH. Acceptance or rejection of this hypothesis, however, requires further investigation, and we are currently working in that direction.

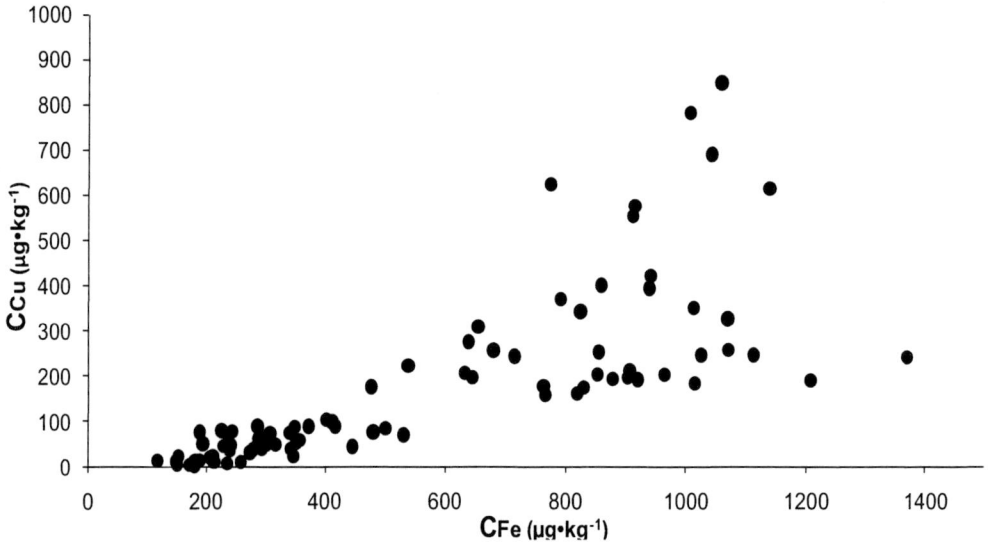

Figure 9. Fe and Cu concentrations in deep water samples collected at low tide from May 2005 to October 2010.

Finally, we want to highlight the design *per se* of the monitoring exercise carried out in this work. On the one hand, collecting deep and surface water at high and low tide is laborious and time consuming, but it has been essential to draw the conclusions compiled in this work. Furthermore, collecting samples every three months allows us to investigate possible seasonal trends, which is not possible when, as most usually occurs, sampling routines are annual. The investigation of seasonal trends, however, requires longer time series if definitive conclusions are to be obtained.

ACKNOWLEDGMENTS

This work has been financially supported by the UNESCO Chair on "Sustainable Development and Environmental Education" of the University of the Basque Country through the UNESCO 09/23 project. Ainara Gredilla is grateful to the University of the Basque Country for her pre-doctoral fellowship.

REFERENCES

Abowei, J.F.N. (2009). Salinity, dissolved oxygen, pH and surface water temperature conditions in Nkoro River, Niger Delta, Nigeria. *Adv. J. Food Sci. Technol. 2*, 36–40.

Alberti, G., Biesuz, R. & Pesavento, M. (2008). Determination of the total concentration and speciation of metal ions in river, estuarine and seawater samples. *Analytical Sci. 24*, 1605–1611.

ANZECC/ARMCANZ. Australian and New Zealand Guidelines for Fresh and Marine Water Quality, 2000.

Barreiro, P. & Aguirre, J.J. (2005). 25 Años del plan integral de Saneamiento de la ría de Bilbao. *Dina Enero-Febrero*, 25–30.

Bartolomé, L., Tueros, I., Cortazar, E; Raposo, J.C; Sanz, J. Zuloaga, O., de Diego, A., Etxebarria, N., Fernández, L.A. & Madariaga, J.M. (2006). Distribution of trace organic contaminants and total mercury in sediments from the Bilbao and Urdaibai Estuaries (Bay of Biscay). *Mar. Pollut. Bull.* 52, 1111–1117.

Batlle-Aguilar, J., Orban, P. & Dassargues, A. (2007). Brouyère, S. Identification of groundwater quality trends in a chalk aquifer threatened by intensive agriculture in Belgium. *Hydrogeol. J. 15*, 1615–1627.

Belzunce, M.J., Solaun, O., Franco, J., Valencia, V. & Borja, A. (2001). Accumulation of Organic Matter, Heavy Metals and Organic Compounds in Surface Sediments along the Nervión Estuary (Northern Spain). *Mar. Pollut. Bull.* 42, 1407–1411.

Bervoets, L., Int Panis, L. & Verheyen, R. (1994). Trace metal levels in water, sediments and Chironomus gr. thumni, from different water courses in Flanders (Belgium). *Chemosphere 29*, 1591–1601.

Bierman, P., Lewis, M., Bertram, O. & Tanner, J. (2011). A review of methods for analysing spatial and temporal patterns in coastal water quality. *Ecol. Indic. 11*, 103–114.

Borja, A., Franco, J., Valencia, V., Bald, J., Muxika, I., Belzunce, M.J. & Solaun, O. (2004). Implementation of the European water framework directive in the Basque Country (northern Spain): a methodological approach. *Mar. Pollut. Bull. 48*, 209–218.

Borja, A., Galparsoro, I., Solaun, O., Muxika, I., Tello, E.M., Uriarte, A. & Valencia, V. (2006). The European Water Framework Directive and the DPSIR, a methodological approach to assess the risk of failing to achieve good ecological status. *Estuar. Coast. Shelf Sci. 66*, 84–96.

Caruso, G., Leonardi, M., Monticelli, L.S., Decembrini, F., Azzaro, F., Crisafi, E., Zappala, G., Bergamasco, A. & Vizzini, S. (2010). Assessment of the ecological status of transitional waters in Sicily (Italy): First characterisation and classification according to a multiparametric approach. *Mar. Pollut. Bull. 60*, 1682–1690.

Cearreta, A., Irabien, M.J., Leorri, E., Yusta, I., Quintanilla, A. & Zabaleta, A. (2002). Environmental transformation of the Bilbao estuary, N. Spain: microfaunal and geochemical proxies in the recent sedimentary record. *Mar. Pollut. Bull. 44*, 487–503.

Chaudry, M. & Zwolsman, J.J.G. (2008). Seasonal dynamics of dissolved trace metals in the Scheldt Estuary: relationship with redox conditions and phytoplankton activity. *Estuar. Coast. 31*, 430–443.

Chiffoleau, J.F., Cossa, D., Auger, D. & Truquet, I. (1994). Trace metal distribution, partition and fluxes in the Seine estuary (France) in low discharge regime. *Mar. Chem. 47*, 145–158.

Cobelo-Garcia, A. & Prego, R. (2004). Behavior of dissolved Cd, Cu, Pb and Zn in the estuarine zone of the Ferrol Ria (Galicia, NW Iberian Peninsula). *Fresen. Environ. Bull. 13*, 753–759.

Cobelo-Garcia, A., Prego, R. & DeCastro, M. (2005). Metal distributions and their fluxes at the coastal boundary of a semi-enclosed ria. *Mar. Chem. 97*, 277–292.

Dassenakis, M., Scoullos, M. & Gaitis, A. (1997). Trace metals transport and behaviour in the Mediterranian Estuary of Achellos River. *Mar. Pollut. Bull. 34*, 103–111.

Deepti, V.G.D., Nayak, G.N. & Basavaiah, N. (2009). Grain size, geochemistry, magnetic susceptibility: Proxies in identifying sources and factors controlling distribution of metals in a tropical estuary, India. *Estuar. Coast. Shelf Sci. 85*, 307–318.

Devlin, M., Best, M. & Haynes, D. (2007). Implementation of the Water Framework Directive in european marine waters. *Mar. Pollut. Bull. 55*, 1–2.

Doong, R., Lee, S., Lee, C., Sun, Y. & Wu, S. (2008). Characterization and composition of heavy metals and persistent organic pollutants in water and estuarine sediments from Gao-ping River, Taiwan. *Mar. Pollut. Bull. 57*, 846–857.

Elbaz-Poulichet, F., Morley, N.H., Cruzado, A., Velasquez, Z., Achterberg, E.P. & Braungardt, C.B. (1999). Trace metal and nutrient distribution in an extremely low pH (2.5) river-estuarine system, the Ria of Huelva (south-west Spain). *Sci. Total Environ. 227*, 73–83.

Elbaz-Poulichet, F., Braungardt, C., Achterberg, E., Morley, N., Cossa, D., Beckers, J.M., Nomerange, P., Cruzado, A. & Leblanc, M. (2001). Metal biogeochemistry in the Tinto-Odiel rivers (Southern Spain) and in the Gulf of Cadiz: a synthesis of the results of TOROS project. *Cont. Shelf Res. 21*, 1961–1973.

Fairbridge, R.W. (1980). The estuary: its definition and geodynamic cycle, in: *Chemistry and Biochemistry of Estuaries*. John Wiley, pp. 1–35.

Fdez-Ortiz de Vallejuelo, S., Barrena, A., Arana, G., de Diego, A. & Madariaga, J.M. (2009). Ultrasound energy focused in a glass probe: An approach to the simultaneous and fast extraction of trace elements from sediments. *Talanta 80*, 434–439.

Fdez-Ortiz de Vallejuelo, S., Arana, G., de Diego, A. & Madariaga, J.M. (2010). Risk assessment of trace elements in sediments: the case of the estuary of the Nerbioi-Ibaizabal River (Basque Country). *J. Hazard. Mater. 181*, 565–573.

Fdez-Ortiz de Vallejuelo, S., Arana, G., de Diego, A. & Madariaga, J.M. (2011). Pattern recognition and classification of sediments according to their metal content using chemometric tools. A case study: the estuary of the Nerbioi-Ibaizabal River (Bilbao, Basque Country). *Submitted to Chemosphere.*

Fernández, S. (2008). Diagnostico, evolución y predicción de la concentración metálica en el estuario del río Nerbioi-Ibaizabal. PhD Thesis, University of the Basque Country.

Fernández, S., Villanueva, U., de Diego, A., Arana, G. & Madariaga, J.M. (2008). Monitoring trace elements (Al, As, Cr, Cu, Fe, Mn, Ni and Zn) in deep and surface waters of the estuary of the Nerbioi-Ibaizabal River (Bay of Biscay, Basque Country). *J. Marine Syst. 72*, 332–341.

García-Barcina, J.M., González-Oreja, J.A. & De la Sota, A. (2006). Assessing the improvement of the Bilbao estuary water quality in response to pollution abatement measures. *Water Res. 40*, 951–960.

Gauthier, T.D. (2001). Detecting trends using Spearman's rank correlation coefficient. *Environ. Forensics 4*, 359–362.

GESAMP (IMO/FAO/Unesco/WMO/WHO/IAEA/UN/UNEP Joint Group of Experts on the Scientific Aspects of Marine Pollution). (1986). Environmental capacity. An approach to marine pollution prevention. *Rep. Stud. GESAMP 30*, 49 pp.

Grant, A. & Middleton, R. (1990). An assessment of total metal contamination in the sediments of the Humber Estuary, UK. *Estuar. Coast. Shelf Sci. 31*, 71–85.

Grifoll, M., Jordà, G., Borja, A. & Espino, M. (2010). A new risk assessment method for water quality degradation in harbour domains using hydrodynamic models. *Mar. Poll. Bull. 60*, 69–78.

Inland Waters Directorate. (1979). Environment Canada Analytical Methods Manual, Chapter III.

Ip, C.C.M., Li, X.D., Zhang, G., Wai, O.W.H. & Li, Y-S. (2007). Trace metal distribution in sediments of the Pearl River Estuary and the surrounding coastal area, South China. *Environ. Pollut. 147*, 311–323.

Krapiel, A., Chiffoleau, J.F., Martin, J.M. & Morel, F. (1997). Geochemistry of trace metals in the Gironde estuary. *Geochim. Cosmochim. Ac. 70*, 1421–1436.

Leorri, E., Cearreta, A., Irabien, M.J. & Yusta, I. (2008). Geochemical and microfaunal proxies to assess environmental quality conditions during the recovery process of a heavily polluted estuary: The Bilbao estuary case (N. Spain). *Sci. Total Environ. 396*, 12–27.

Li, Y., Qiu, R., Yang, Z., Li, C. & Yu, J. (2010). Parameter determination to calculate water environmental capacity in Zhangweinan canal sub-basin in China. *J. Environ. Sci. 22*, 904–907.

Liu, W.X., Li, X.D., Shen, Z.G., Wang, D.C., Wai, O.W,H. & Li, Y.S. (2003). Multivariate statistical study of heavy metal enrichment in sediments of the Pearl River Estuary. *Environ. Pollut. 21*, 377–388.

Martino, M., Turner, A., Nimmo, M. & Millward, G.E. (2002). Resuspension, reactivity and recycling of trace metals in the Mersey Estuary, UK. *Mar. Chem. 77*, 171–186.

Masson, M., Blanc, G. & Schaefer, J. (2006). Geochemical signals and source contributions to heavy metal (Cd, Zn, Pb, Cu) fluxes into the Gironde Estuary via its major tributaries. *Sci. Total Environ. 370*, 133–146.

Micheletti, C., Gottardo, S., Critto, A., Chiarato, S. & Marcomini, A. (2001). Environmental quality of transitional waters: The lagoon of Venice case study. *Environ. Int. 37*, 31–41.

Morillo, J., Usero, J. & Gracia, I. (2005). Biomonitoring of trace metals in a mine-polluted estuarine system (Spain). *Chemosphere 58*, 1421–1430.

Moros, J., Fdez-Ortiz de Vallejuelo, S., Gredilla, A., de Diego, A., Madariaga, J.M., Garrigues, S. & de la Guardia, M. (2009). Use of reflectance infrared spectroscopy for monitoring the metal content of the estuarine sediments of the Nerbioi-Ibaizabal River (Metropolitan Bilbao, Bay of Biscay, Basque Country). *Environ. Sci. Technol. 43*, 9314–9320.

Moros, J., Gredilla, A., Fdez-Ortiz de Vallejuelo, S., de Diego, A., Madariaga, J.M., Garrigues, S. & de la Guardia, M. (2010). Partial least squares X-ray fluorescence determination of trace elements in sediments from the estuary of Nerbioi-Ibaizabal River. *Talanta 82*, 1254–1260.

Mosley, L.M., Peake, B.M. & Hunter, K.A. (2010). Modelling of pH and inorganic carbon speciation in estuaries using the composition of the river and seawater end members. *Environ. Model. Softw. 25*, 1658–1663.

Mueller, G. & Foerstner, U. (1975). Heavy metals in sediments of the Rhine and Elbe estuaries. Mobilization or mixing effect. *Environ. Geol. 1*, 33–39.

Mzimela, H.M., Wepener, V. & Cyrus, D.P. (2003). Seasonal variation of selected metals in sediments, water and tissues of the groovy mullet, Liza dumerelii (Mugilidae) from the Mhlathuze Estuary, South Africa. *Mar. Pollut. Bull. 46*, 659–664.

Nolting, R.F., Helder, W., de Baar, H.J.W. & Gerringa, L.J,A. (1999). Contrasting behaviour of trace metals in the Scheldt estuary in 1978 compared to recent years. *J. Sea Res. 42*, 275–290.

Owens, R.E. & Balls, P.W. (1997). Dissolved trace metals in the Tay Estuary. *Estuar. Coast. Shelf Sci. 44*, 421–434.

Padmini, E. & Geetha, B.V. (2007). A comparative seasonal pollution assessment study on Ennore estuary with respect to metal accumulation in the grey mullet, Mugil cephalus. *Oceanol. Hydrobiol. St. 36*, 91–103.

Periáñez, R. (2009). Environmental modelling in the Gulf of Cadiz: heavy metal distributions in water and sediments. *Sci. Total Environ. 407*, 3392–3406.

Power, M., Attrill, M.J. &Thomas, R.M. (1999). Heavy metal concentration trends in the Thames Estuary. *Water Res. 33*, 1672–1680.

Raj, N. & Azeez, P.A. (2009). Spatial and temporal variation in surface water chemistry of a tropical river, the river Bharathapuzha, India. *Curr. Sci. 96*, 245–251.

Ramos, L., Fernandez, M.A., Gonzalez, M.J. & Hernandez, L.M. (1999). Heavy metal pollution in water, sediments, and earthworms from the Ebro River, Spain. *B. Environ. Contam. Tox. 63*, 305–311.

Raposo, J.C., Zuloaga, O., Sanz, J., Villanueva, U., Crea, P., Etxebarria, N., Olazabal, M.A. & Madariaga, J.M. (2006). Analytical and thermodynamical approach to understand the mobility/retention of arsenic species from the river to the estuary. The Bilbao case study. *Mar. Chem. 99*, 42–51.

Ravichandran, S. (2003). Hydrological influences on the water quality trends in Tamiraparani Basin, South India. *Environ. Monit. Assess. 87*, 293–309.

Rondeau, B., Cossa, D., Gagnon, P., Pham, T.T. & Surette, C. (2005). Hydrological and biogeochemical dynamics of the minor and trace elements in the St. Lawrence River. *Appl. Geochem. 20*, 1391–1408.

Saiz-Salinas, J. (1997). Evaluation of adverse biological effects induced by pollution in the Bilbao Estuary (Spain). *Environ. Pollut. 96*, 351–359.

Sánchez, E., Colmenarejo, M.F., Vicente, J., Rubio, A., García, M.G., Travieso, L. & Borja, R. (2007). Use of the water quality index and dissolved oxygen deficit as simple indicators of watersheds pollution. *Ecol. Indic. 7*, 315–328.

Santos-Echeandía, J., Prego, R. & Cobelo-García, A. (2009). Intra-annual variation and baseline concentrations of dissolved trace metals in the Vigo Ria and adjacent coastal waters (NE Atlantic Coast). *Mar. Pollut. Bull. 58*, 298–303.

Sanz-Landaluze, J., de Diego, A., Raposo, J.C. & Madariaga, J.M. (2004). Methylmercury determination in sediments and fish tissues from the Nerbioi-Ibaizabal estuary (Basque Country, Spain). *Anal. Chim. Acta 508*, 107–117.

Schäfer, J., Norra, S., Klein, D. & Blanc, G. (2009). Mobility of trace metals associated with urban particles exposed to natural waters of various salinities from the Gironde Estuary, France. *J. Soil. Sediment. 9*, 374–392.

Scoullos, M., Dassenakis, M., Zeri, C., Papageorgiou, K. & Rapti, M. (1987). Chemical studies of main estuaries and coastal areas of Greece. Progress report to DG XI, Project GEC-ENV.560 GR.

Scoullos, M., Dassenakis, M. & Pavlidou, A. (1994). A brief account on trace metal levels in Saronikos Gulf, based on the 1986-1993 MED-POL Monitoring Programme. In Proc. of the International Conference on Restoration and Protection of the Environment II, 272–279.

Simões, F.d.S., Moreira, A.B., Bisinoti, M.C., Gimenez, S.M.N. & Yabe, M.J.S. (2008). Water quality index as a simple indicator of aquaculture effects on aquatic bodies. *Ecol. Indic. 8*, 476–484.

Spencer, K.L. (2002). Spatial variability of metals in the inter-tidal sediments of the Medway Estuary, Kent, UK. *Mar. Pollut. Bull. 44*, 933–944.

Spiteri, C., Van Cappellena, P. & Regniera, P. (2008). Surface complexation effects on phosphate adsorption to ferric iron oxyhydroxides along pH and salinity gradients in estuaries and coastal aquifers. *Geochim. Cosmochim. Ac. 72*, 3431–3445.

Tueros, I., Rodríguez, J.G., Borja, A,m Solaun, O,m Valencia, V. & Millán, E. (2008). Dissolved metal background levels in marine waters, for the assessment of the physico-chemical status, within the European Water Framework Directive. *Sci. Total Environ. 407*, 40–52.

Vicente-Martorell, J.J., Galindo-Riaño, M.D., García-Vargas, M. & Granado-Castro, M.D. (2009). Bioavailability of heavy metals monitoring water, sediments and fish species from a polluted estuary. *J. Hazard. Mater. 162*, 823–836.

Wallner-Kersanach, M., de Andrade, C.F., Zhang, H., Milani, M.R. & Niencheski, L.F.H. (2009). In situ measurement of trace metals in estuarine waters of Patos Lagoon using diffusive gradients in thin films (DGT). *J. Brazil. Chem. Soc. 20*, 333–340.

WFD (2000). Directive of the European Parliament and of the Council 2000/60/EC establishing a framework for community action in the field of water policy. Water Framework Directives, European Commission Environment, Luxembourg.

Whitall, D., Hively, W.D., Leight, A.K., Hapeman, C.J., McConnell, L.L., Fisher, T., Rice, C.P., Codling, E., McCarty, G.W., Sadeghi, A.M., Gustafson, A. & Bialek, K. (2010). Pollutant fate and spatio-temporal variability in the Choptank River estuary: Factors influencing water quality. *Sci. Total Environ. 408*, 2096–2108.

Williams, A. & Benson, N. (2009). Interseasonal hydrological characteristics and variabilities in surface water of tropical estuarine ecosystems within Niger Delta, Nigeria. *Environ. Monit. Assess. 165*, 399–406.

Zaldívar, J.M., Cardoso, A.C., Viaroli, P., Newton, A., de Wit, R., Ibañez, C., Reizopoulou, S., Somma, F., Razinkovas, A., Basset, A., Holmer, M. & Murray, N. (2008). Eutrophication in transitional waters: an overview. *Transiti. Waters Monogr. 2*, 1–78.

Zwolsman, J.J.G., Van Eck, B.T.M. & Van Der Weijden, C.H. (1997). Geochemistry of dissolved trace metals (cadmium, copper, zinc) in. the Scheldt estuary, southwestern Netherlands: Impact of seasonal variability. Geochimi. Cosmochim. Ac. 61, 1635–1652.

Zwolsman, J.J.G. & Van Eck, G.T.M. (1999). Geochemistry of major elements and trace metals in suspended matter of the Scheldt estuary, southwest Netherlands. *Mar. Chem. 66*, 91–111.

Chapter 10

MODELING AN ESTUARINE ECOSYSTEM: HOW TIDAL FLATS AND BENTHIC ECOSYSTEMS AFFECT THE EUTROPHIC ESTUARY

Akio Sohma

Environmental, Natural Resources and Energy, Div. Mizuho Info and Research Institute, Tokyo, Japan

ABSTRACT

A newly developed ecosystem model—the first model describing ecological connectivity consisting of both benthic-pelagic and central bay-tidal flat ecosystem coupling while simultaneously describing the vertical micro-scale in the benthic ecosystem—was developed and applied to Tokyo Bay. The model permits prediction and evaluation of the effects of environmental measures, such as tidal flat creation or restoration, sand capping, dredging, and nutrient load reduction from rivers, on the hypoxic estuary from the perspectives of (1) the whole estuary composed of temporal-spatial mutual linkage of benthic-pelagic or central bay-tidal flat ecosystems (holistic approach), and (2) each biochemical and physical process contributing to oxygen production and consumption (elemental approach). The model outputs demonstrated significant ecosystem responses as follows. First, the oxygen consumption in the benthic system during summer was quite low due to low concentrations of dissolved oxygen (DO), i.e. hypoxia, although reduced substances, Mn^{2+}, Fe^{2+}, and S^{2-}, accumulated in the pore water. This result demonstrates the importance of using oxygen consumption rate at high DO concentrations as the index of hypoxia potential. Second, simulations of both tidal flat creation and nutrient load reduction decreased the anoxic water volume and mass of detritus in Tokyo Bay. Simulated creation of tidal flats, however, led to a greater biomass of benthic fauna, whereas nutrient load reduction led to lower biomass of benthic fauna compared to the existing situation. This result clarifies the differences between contrasting goals: (1) a bountiful ocean, non-hypoxic and with rich production of higher trophic level biology, vs. (2) a clear ocean, non-hypoxic and with low levels of particulate organic matter, and also the distinction between a bountiful ecosystem and higher water quality. Lastly, in the simulation, reproducing reclaimed tidal flats, as existed in the earlier Tokyo Bay system, prevented an increase of oxygen consumption potential (hypoxia potential) and a decline of higher trophic production leading to red tides, compared to the existing Tokyo Bay system with reclamation of tidal flats. This result demonstrates higher ecosystem tolerance of the earlier Tokyo Bay to red tide, and the tidal flat's function of keeping an optimum ecological balance resilient to environmental perturbation.

Keywords: Coastal ecosystem, environmental restoration, hypoxia, tidal flat reproduction/restoration, nutrient load reduction

INTRODUCTION

Eutrophication, a serious problem in Japanese estuaries such as Tokyo Bay, has been thought to drive estuarine ecosystems in the direction of red tides, hypoxia, and eventually to decreases in the number of species and biomass of living organisms. In Japan, nutrient load reduction from rivers has been conducted since the 1960s, and in the case of Tokyo Bay, nitrogen and phosphorus have been reduced by 33–50% (Ministry of the Environment 2006). As a result, water quality in terms of chemical oxygen demand (COD), total nitrogen (TN), and total phosphorus (TP) has now recovered as concentrations have dropped below those observed during the worst times. However, the number of species and biomass of living organisms have not recovered. One of the reasons is thought to be the disappearance of shallow waters (tidal flats) as a result of reclamation. Nowadays, a specific vision and direction for how to restore eutrophic estuaries to a more desirable state is required.

As a contribution to our vision and understanding, a new ecosystem model, the Ecological Connectivity Hypoxia Model: ECOHYM ("ZAPPAI"[16] in Japanese; Sohma et al. 2005, 2008) was developed. The features of ECOHYM can be described briefly as (1) modeling both the benthic and pelagic ecosystems and the linkage between them, (2) modeling both tidal-flat and central bay areas and their linkage, and (3) describing a micro-scale vertical spatial resolution of the benthic ecosystem. The vanguard attempt to achieve all three items, (1), (2), and (3) simultaneously, was not only a technological challenge to model development but also a philosophical challenge to our concepts of the estuarine environment.

In this document, firstly, the philosophy and technology of the development of ECOHYM are introduced. Secondly, a detailed description of ECOHYM is provided. Lastly, the extent to which the collaboration of philosophy and technology was achieved and what was revealed from it, are summarized by interweaving several analyses generated by implementing the model for Tokyo Bay.

PHILOSOPHY – HOW WE UNDERSTAND THE ENVIRONMENTAL PROBLEMS OF ESTUARIES

In this section, I provide an overview of significant concepts about the series of eutrophication problems in the estuary. The concepts described in this section are not independent but are linked to each other. Therefore, there may be some repetition. However,

---

[16] "ZAPPAI" is named after a type of Japanese poem born from the general public (commoner) during the samurai period (Edo period) (Miyata, 2003). "ZA" means "miscellaneous" and "PPAI" means "haiku" in Japanese.

all of the visions introduced here are thought to arise from a meaningful philosophy. They were the reasons and motivation for development of a new model, ECOHYM, despite the challenges. It is now possible to estimate and predict quantitative versions of some of the concepts as the result of developing ECOHYM.

Significance of the Ecosystem Chain Response: Interactions between Tidal Flats and the Central Bay Area and between Benthic and Pelagic Systems

The importance of the roles of the tidal-flat ecosystem and benthic ecosystem are explained here with respect to hypoxia in Tokyo Bay. Hypoxia is a serious environmental problem in the semi-closed coastal zone; it is chronic in Tokyo Bay from May to September (Chiba Prefectural Fisheries Research Center 2001–2006, Kanagawa Prefectural Fisheries Research Institute 2005–2006). Hypoxia is related explicitly or implicitly to the phenomena of eutrophication, reclamation of tidal flats, red tide (rapid growth of phytoplankton), blue tide (upwelling of oxygen-depleted bottom water to the sea surface), and decrease of fishery biomass (Suzuki et al. 1998, Imao et al. 2004). Figure 1 is a conceptual diagram of the linkage of hypoxia to accompanying phenomena and processes (Sohma & Sekiguchi 2003). Generally, a tidal flat is an area of high biological productivity where many species of living organisms dominate (Odum 1971). Tidal flats are usually located along the coastal line of the bay and have been targets for reclamation (converted to upland for development) because of the relative convenience and low cost. One of the putative causes for repeated red tides and blue tides is precisely the loss of tidal flats to reclamation (Kikuchi 1993, Ishida & Hara 1996, Aoyama & Suzuki 1997). Disappearance of tidal flats reduces populations of benthic fauna (bivalves, etc.) that once dominated there and thereby reduces the predation pressure on phytoplankton. The low predation pressure induces abrupt increases in the phytoplankton population (red tide) and large amounts of dead phytoplankton settle out on the seafloor. This accumulation of organic matter at the seafloor changes metabolism at the sediment-water interface and increases oxygen consumption at the sea bottom (Furota 1988, Suzumura et al. 2003). The oxygen-depleted bottom water leads to mass mortalities of living organisms not only in deep waters of the central bay area (hypoxic area), but also at the sea surface of the central bay area or in the tidal flat area caused by upwelling and transverse flow. The flow transports hypoxic water from the bottom to the surface or from the central bay to the tidal flats. The mass mortality caused by oxygen depletion promotes further harmful effects in the estuary (Figure 1). In this way, hypoxia has ripple effects across an ecosystem network comprised of mutual interactions between the benthic and pelagic systems and between the hypoxic central bay and tidal flat areas.

Hereinafter, the chain response of the ecosystem in Figure 1, left, beginning with the disappearance of tidal flat areas and leading to environmental deterioration, is what I have defined as "environmental deterioration spiral (negative spiral)" (Sohma & Sekiguchi 2003). The environmental deterioration spiral possibly leads to survival of a few kinds of organisms with high tolerance for low DO. As a result, it is proposed that the negative spiral leads to an estuary characterized by (1) low or depauperate biodiversity, (2) impairment of nutrient flows from lower to higher trophic levels, and (3) low potential for utilization of stored nutrients in the estuary.

● Estuary towards **low biological productivity**
(Environmental deterioration spiral)

Italic letters : Measures spinning out from the environmental deterioration spiral = Traditional method

● Estuary towards **high biological productivity**
(Environmental improvement spiral)

Shallow waters creation: Measures leading to environmental improvement spiral = Future method

[Shallow waters (tidal flat, sea-grass beds)]

Creation (switch)

Reclamation (switch)

Mortality due to lack of oxygen

Decrease of fishery biomass (fish)

Hypoxia (blue tide) generation

Temporal avoidance from increase of nutrient flux and oxygen consumption at seafloor

Increase of oxygen consumption at seafloor

Towards a more deteriorated environmental condition

Sand capping

1. Low/poverty biodiversity.
2. Poor nutrient transition from lower to higher trophic level.
3. Low utilizable potential of stored nutrients in the estuary.

Deterioration of benthic atmosphere

Transport of Carbon, Nitrogen and Phosphorus out of the estuary

Settling of dead shape of phytoplankton to seafloor

Decrease of fishery biomass (Suspension feeders, etc.)

Dredging

High frequency of red tide

Increase of phytoplankton

Load reduction

Transport of Carbon, Nitrogen and Phosphorus out of the estuary

Increase of fishery biomass (Suspension feeders, etc.)

Decrease of phytoplankton

Decrease of red tide

Decrease in the settling of dead shape of phytoplankton to seafloor

1. High/bountiful biodiversity.
2. Smooth nutrient transition from lower to higher trophic level.
3. High utilizable potential of stored nutrients in the estuary.

Manipulation of benthic atmosphere

Toward a more improved environmental condition

Decrease of oxygen consumption at seafloor

hypoxia annihilation

Increase of fishery biomass (fish)

No mortality due to lack of oxygen

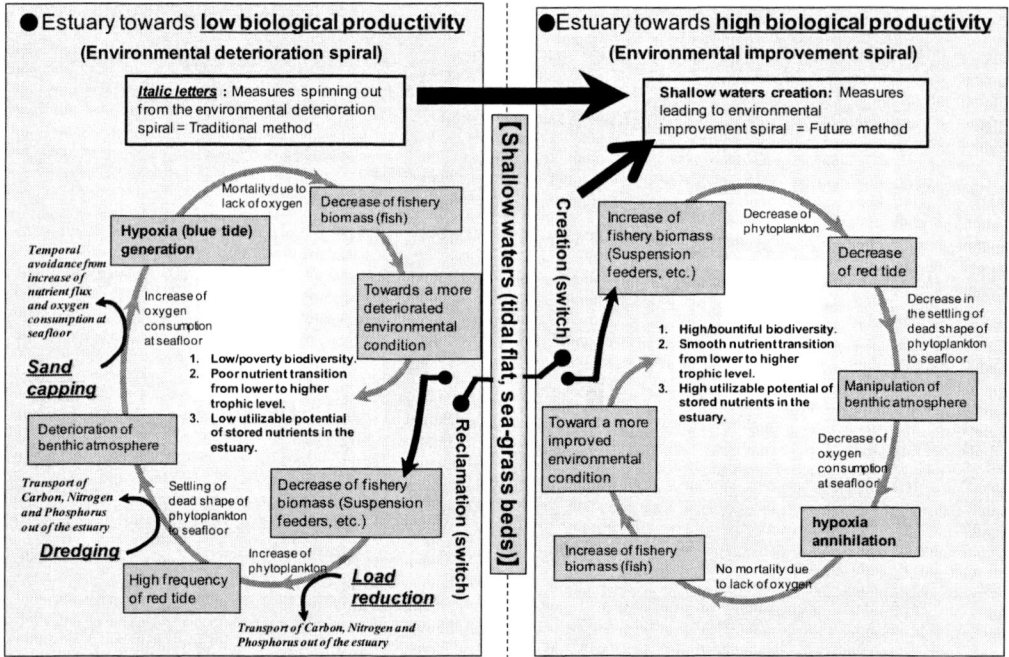

Figure 1. An example of the ecological chain responses in the environmental deterioration spiral and environmental improvement spiral accompanied by tidal flat reclamation and creation. The positions of other environmental measures (load reduction, dredging and sand capping) and the results of spirals are also illustrated.

The Essential Objectives of Environmental Improvement Measures: to Inhibit Eutrophication or to Recover a Bountiful Ecosystem, Considering Positive and Negative Spirals

Considering the objectives of environmental measures such as (1) nutrient load reduction, (2) sand capping, (3) dredging, and (4) tidal flat restoration from the perspective of the environmental deterioration spiral, the first three essentially resemble each other and differ from tidal flat restoration. It means that the objectives of nutrient load reduction, sand capping and dredging are basically to escape from the processes and paths of the environmental deterioration spiral, i.e. (a) increase of phytoplankton, (b) settling of dead phytoplankton to the seafloor, and (c) increase of oxygen consumption at the seafloor (Figure 1, left). The expectation of such measures is to stop and prevent the environmental deterioration spiral. In contrast, tidal flat restoration is based on the idea of generating the driving force of environmental improvement and reversing the spiral from negative to positive (Figure 1, right), rather than stopping or preventing the environmental deterioration spiral. In artificially recreating the tidal flat as a favorable habitat zone for living organisms, the organisms re-establish autonomously, thereby restoring biological and ecological functions. As a result, predation pressure on phytoplankton increases and red tide is prevented. Preventing red tide reduces the sedimentation flux of dead phytoplankton, and the benthic habitat for living organisms recovers. Hypoxia is prevented and the mortality of fish and shellfish from lack of oxygen does not occur. The ecosystem chain response explained above may restore a hypoxic estuary autonomously toward an estuary featuring (1) diverse and bountiful aquatic life, (2) more efficient transfer of nutrients from lower to higher trophic levels, and (3) higher potential utilization of stored nutrients. The reverse of the environmental deterioration spiral mentioned above is what I have defined as the (positive) environmental improvement spiral (Sohma & Sekiguchi 2003). The action inducing the environmental improvement spiral, i.e. tidal flat recreation or restoration, might be characterized as a drastic measure for reversing the environmental deterioration spiral. In contrast, actions intended to stop or prevent the environmental deterioration spiral, such as nutrient load reduction, sand capping, and dredging, are better characterized as stopgap measures.

Considering nutrient load reduction, sand capping, dredging, and tidal flat recreation or restoration from the perspective of nutrient (N and P) cycling, the first three are aimed at transportation of excess nutrients (which cannot be used effectively for higher trophic level production), out of the system (estuary). In contrast, tidal flat recreation or restoration is aimed at increasing the potential for effective nutrient utilization through the recovery of a bountiful ecosystem and assimilation of the nutrients by organisms at higher trophic levels. In simple terms, the primary objective of nutrient load reduction, sand capping, and dredging is inhibition of eutrophication (reduction of excess nutrients), but the primary objective of tidal flat recreation and restoration is recovery of a bountiful ecosystem (increasing the potential for nutrient utilization).

TECHNOLOGY – THE REQUIREMENTS FOR AN UNPRECEDENTED MODEL OF MULTIPLE ECOSYSTEMS AND VERTICAL MICRO-SCALE MECHANISMS OF THE BENTHIC SYSTEM

In order to conduct coastal environmental management strategically, estimates and predictions of the effects of environmental improvement measures (i.e. nutrient load reduction, sand capping, dredging, and tidal flat recreation and restoration) and development (i.e. tidal flat reclamation) on the coastal ecosystem are required. It is best if predictions can demonstrate ecosystem responses (1) at both short and long time scales, (2) quantitatively (the strength of the effect), qualitatively (the direction of ecosystem response), and (3) mechanically to explain the causes of the effect. On the basis of the previous discussion, the ideal method for performing such predictions or estimates is to embrace the whole estuary as a complex of ecosystems with mutual temporal and spatial linkage of each ecosystem between the tidal flat area and the central bay area, and between the benthic system and the pelagic system.

In order to evaluate the ecosystem complex in the context of eutrophication, it is crucial to consider (1) the physical-biochemical processes and (2) their mechanical linkages that are thought to be significant for eutrophication effects on the ecosystem, as well as processes and linkages that occur inside and at the boundary of each ecosystem making up the ecosystem complex (i.e. tidal flat pelagic ecosystem, tidal flat benthic ecosystem, central bay pelagic ecosystem, and central bay benthic ecosystem). The reason is that the ecosystem complex is composed of tangles (interactions) of physical-biochemical processes, and they propagate the effects of environmental measures, development, impact and disturbance. The ecosystem response is the result of spillover effects from the tangles. The positive and negative spirals shown in Figure 1 may be just two examples of results from the tangles.

The ecosystem model describes each physical-biochemical process and the ecological dynamics derived from the mutual interaction (tangles) of physical- biochemical processes. Therefore, it is a powerful tool to reveal both the ecosystem mechanisms and ecosystem responses to seasonal, daily or episodic changes in the external and internal environment of the estuary, such as meteorology, nutrient load from rivers, red tide, sand capping, dredging and recreation and reclamation of tidal flats, etc. Because of these advantages of ecosystem models, many such models have been developed. However, when the development of ECOHYM started, no ecosystem model satisfied all of the following requirements simultaneously, which need to be considered to simulate a hypoxic estuary from the background and perspectives mentioned above.

(1) Requirements for Modeling the Benthic Ecosystem in the Central Bay Area

The main cause of hypoxia is thought to be the consumption of oxygen around the seafloor in the central bay area. This consumption originates from biochemical processes in the sediment, which change precipitously on a micro-scale in the vertical direction (Canfield et al. 1993). Thus, to demonstrate hypoxia dynamics accurately, describing the vertical profiles of biochemical processes in the micro-scale (μm–mm) is required (Rysgaard & Berg 1996). One-dimensional benthic biochemical models with a vertical micro-scale have now reached such a level of sophistication and comprehensiveness that they can accurately reproduce benthic metabolism as well as carbon, nutrients, or oxygen cycling in the sediment or sediment-water interface (Soetaert et al. 1996a, 1996b, 2000, Boudreau 1996, Sohma & Sayama 2002, Dedieu et al. 2007).

(2) Requirements for Modeling the Tidal Flat Ecosystem

A tidal flat, where high potential for hypoxia improvement exists, is an oxygen producing area caused by (a) the photosynthesis of benthic algae, seagrass and seaweed, and (b) the accelerated aeration driven by the emersion/submersion cycle and nearshore waves. Oxygen dynamics in tidal flats are complex and vary considerably over a daily time scale (Kuwae et al. 2003). Furthermore, similar to the central bay area, metabolism in the tidal flat sediment changes precipitously over a micro-scale in the vertical direction (Revsbech et al. 1986, Kuwae et al. 2003). Therefore, in order to evaluate the dynamics of metabolism or oxygen production and consumption mechanisms in a tidal flat, an ecosystem model that simultaneously describes the time course of oxygen changes on a 1-day scale as well as vertical benthic metabolic profiles on a micro-scale is required. Heretofore, several ecosystem models applied to the tidal flat area have been developed (Baretta et al. 1988, Hata & Nakata 1998, Sohma et al. 2000, Nakamura et al. 2004). Some models are able to calculate the dynamics of the boundary between the oxic (aerobic) layer and anoxic (anaerobic) layer in the benthic system, although they do not calculate the details of diagenetic processes along the vertical scale (Baretta et al. 1988, Hata & Nakata 1998, Sohma et al. 2000). Moreover, some models have focused not only on the seasonal dynamics but also dynamics assessed on a daily scale (Sohma et al. 2000, Nakamura et al. 2004). However, to the best of our knowledge, no ecosystem model existed before ECOHYM that simultaneously described both the benthic vertical micro-scale metabolic mechanisms and daily dynamics in a tidal flat.

Requirements for Modeling an Ecosystem Complex

As mentioned above, hypoxia generation and annihilation and the ensuing ripple effects (Figure 1) result from mutual interactions between benthic-pelagic ecosystems and between central bay and tidal flat ecosystems, i.e. an ecosystem complex. Therefore, when evaluating the ecological response of the hypoxic estuary, the model must contain (a) the benthic and pelagic systems in both the central bay and tidal flat areas, and (b) mutual interactions between the benthic and pelagic systems, and between the central bay and tidal flat areas. In recent years, a number of benthic-pelagic coupling models have been developed (Baretta et al. 1988, Baretta et al. 1995, Baretta-Bekker & Baretta 1997, Sohma et al. 2001, Sohma et al. 2004, Luff & Moll, 2004), with several models describing diagenetic (metabolic) processes in detail (Luff & Moll 2004).

Furthermore, the requirements in items (1), (2) and (3) mentioned above are not independent, because vertical micro-scale mechanisms in the benthic system (items 1 and 2), or the daily-scale dynamics in the tidal flat area (item 2) are controlled by the ecosystem network in the ecosystem complex (item 3). Thus, an ecosystem model that treats all the requirements in items (1), (2) and (3) simultaneously is important for evaluating the ecosystem dynamics of a hypoxic estuary.

TOWARD THE COLLABORATION OF PHILOSOPHY AND TECHNOLOGY: OBJECTIVES AND SIGNIFICANCE OF THE RESEARCH; HOLISM AND REDUCTIONISM

On the basis of the philosophical and technological background described above, ECOHYM, the first model to meet all the requirements in items (1), (2) and (3) simultaneously, was developed. With the model, the challenges of clarifying the mechanisms of hypoxia and predicting ecosystem responses and tolerance to environmental measures, development, impact and disturbance were addressed from two perspectives: (1) the whole estuary, composed of temporal-spatial mutual linkage of benthic-pelagic or central bay-tidal flat ecosystems (holism), and (2) each physical-biochemical process contributing to oxygen production and consumption (reductionism).

For ECOHYM to meet these challenges, firstly, selecting how to treat the physical and biochemical processes in the model was significant. The treatment should be based upon confirming what is known about the mechanics of hypoxia and related phenomena. Each physical-biochemical process selected should be formulated based on the latest scientific knowledge as much as possible. This approach is derived from reductionism. The additional requirement for ECOHYM was not to apprehend each physical and biochemical process in fragmentary fashion, nor to describe the ecosystem by superposing (i.e. stacking) them, but rather to describe the mechanical linkages and interactions of each process. Therefore, a numerical construction (Sohma 2005) that described the autonomous responses and feedback effects due to the entanglement of each process, and that could estimate the dynamics of the ecosystem as a whole, had to be applied to ECOHYM. Modeling the internal mechanisms of the benthic and pelagic ecosystems or the tidal flats and central bay ecosystems, and also linking each ecosystem by such a numerical construction enables us to regard the whole estuarine ecosystem as the temporal-spatial mutual linkage of the benthic and pelagic systems across the central bay and tidal flat areas. This approach is derived from holism.

The success of the development of such a model reveals where and how much each modeled physical and biochemical process contributes relatively to oxygen consumption and production, and leads to the clarification of hypoxic mechanisms. In addition, it enables us to predict the response of the whole estuarine ecosystem to environmental measures and perturbations quantitatively and qualitatively, while considering the ripple effects of these changes through the entanglement of physical and biochemical processes. These results are also linked to establishing a foundation for cost-performance evaluation of environmental improvement technology, such as nutrient reduction, dredging, sand capping, and tidal flat restoration. Here, "quantitatively" means the direction and trend of the temporal and spatial dynamics.

In modeling, assumptions often have to be made to account for ambiguous, unknown, or uncertain processes. However, the fact that the model simulates the results of mechanical interactions among many processes suggests that some of the yet-to-be-defined processes may be resolved by inference from the perspective of the whole ecosystem balance (the holistic approach), although they have not yet been resolved by piecemeal research into each process. Such heuristic methods are sometimes effective in increasing knowledge and advancing the frontiers of science.

Because all existing models are condensed and simplified descriptions of real systems, any model includes some approximations and assumptions. Nevertheless, if we want to inform management by reference to the model results, and if the persons engaging in management decisions have sufficient understanding of the approximations and assumptions imposed on the model, the model can be used to aid consensus building in the planning of coastal management actions.

MODEL DESCRIPTION

Construction of the Model

ECOHYM is composed of two models: a hydrodynamics model and an ecological model for the benthic and pelagic systems. The ecological model is generalized to enable its application to both the central bay and tidal flat area. The whole construction of ECOHYM is illustrated in Figure 2. The hydrodynamics model is calculated independently from the ecological model, whereas the ecological model receives input of the flow-temperature field from the hydrodynamics model. Therefore, the physical fields calculated in the hydrodynamics model such as advection, eddy diffusion, and temperature, are not affected by the ecological model variables (plankton, detritus, nutrients, DO, etc.), but model variables of the ecological model are moved passively by the physical field. More specifically, the physical field transports the model variables of the ecological model and changes the biochemical balance among the model variables at any computational grid. Therefore, hydrodynamics affects the biochemical processes, but not the reverse. The interaction between the central bay and tidal flat areas is also driven by the physics (flow field and eddy viscosity) calculated by the hydrodynamics model. The physical processes that drive transport within the sediment or at the sediment-water interface such as molecular diffusion, irrigation, bioturbation, burial, etc. are treated in the ecological model, while not considered in the hydrodynamics model.

Hereinafter, to clarify the definition of the central bay and tidal flat areas in the model, the central bay area is defined as that area where deep water and low transparency hinder benthic algae, seagrass and seaweed photosynthesis. The tidal flat area is defined as that area where high transparency and shallow depth permit photosynthesis by benthic algae, seagrass and seaweed. Areas of submersion and emersion with changes in tidal level are included in the tidal flat area. In eutrophic estuaries in Japan and elsewhere, deep central areas usually have higher potential to become hypoxic than shallow or tidal flat areas.

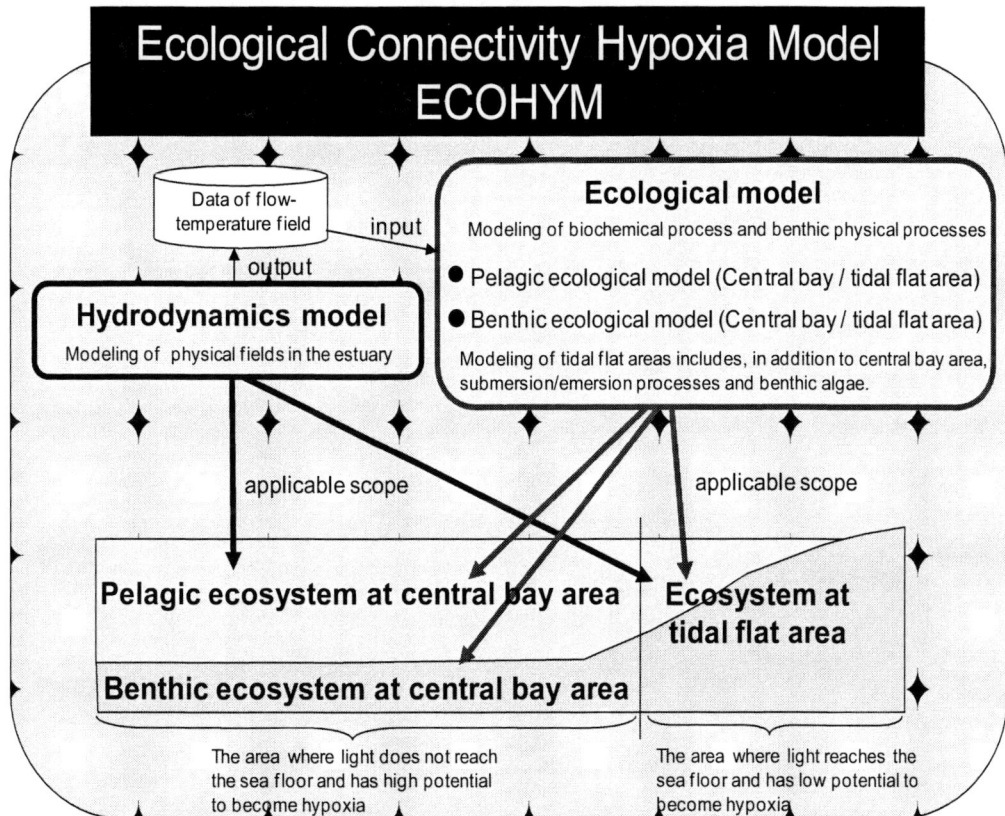

Figure 2. The construction of ECOHYM. The model is composed of two models, the hydrodynamics model and the ecological model for the benthic and pelagic systems. The ecological model is generalized to enable its application to both the central bay and tidal flat areas.

Hydrodynamics Model

The hydrodynamics model simulates the three dimensional physical field in the pelagic system of the estuary and demonstrates the long term variability of flow field, salt and heat transport. The target area of the model is a meso-scale estuary, ($1\sim100$ km^2), a semi-enclosed coastal water body where seawater is exchanged with the ocean and is diluted by the inflow of freshwater from rivers (Pritchard 1967). The model equations and algorithms of the hydrodynamics model are well described by Nakata et al. (1983a, 1983b). Thus, only the outline is described here. The model includes tidal forcing, surface wind and local density gradients with realistic coastal topography and bathymetry described by computational grids/mesh. Under hydrostatic and Boussinesq approximations on a rotating Cartesian coordinate system, the model employs the equations of fluid motion, flow continuity and conservation of heat and salt to determine the local distribution of model variables; i.e., mean velocity components, surface displacement, temperature and salinity.

Ecological Model

The ecological model is a system of equations that establishes the significant components of hypoxia and its related environmental phenomena as the model variables. The model describes the interaction among model variables through the biochemical and physical processes in terms of O_2, C, N, and P cycling. Namely, the ecological model describes the transformations that O_2, C, N, and P undergo as the result of biochemical processes, while considering the physical transport. The dynamics and spatial distribution of the model variables are described by partial differential equations. The equations satisfy the mass conservation of O_2, C, N and P, and are comprised of the production and consumption terms due to biochemical processes and transport terms due to physical processes. Each path of the O-C-N-P coupled cycle caused by biochemical processes is derived from empirical and experimental formulations. The formulations of each biochemical reaction are based on a first kinetic reaction and include (a) several model variables, (b) environmental variables obtained from prescribed functions and data (i.e. temperature, light intensity, etc.), (c) biochemical parameters, and (d) universal constants. The values of the biochemical processes are calculated, changing at each time step. Changes in the biochemical processes affect the dynamics and spatial distribution of the model variables and vice versa. In this way, the ecological model simulates the ecosystem dynamics as the result of entanglement of various interactions. The details of the ecological model are explained below.

Ecological Diagram (Model Variables and Biochemical Processes)

The pelagic and benthic ecological diagrams treated in each grid of the model are shown in Figures 3 and 4. The model variables in the ecological model are phytoplankton, zooplankton, detritus, dissolved organic matter, NH_4-N, NO_3-N, PO_4-P, benthic algae, suspension feeders, deposit feeders (benthic fauna), DO and ODU (oxygen demand units, representing stoichiometric substitute expressions of oxygen demands of Mn^{2+}, Fe^{2+}, and S^{2-}; Soetaert et al. 1996a). In the diagrams, the model variables are illustrated as boxes and biochemical processes as arrows. The model variables and their connecting biochemical processes were selected to describe hypoxic mechanisms and oxygen dynamics based on the lowest trophic levels (primary production and detrivory).

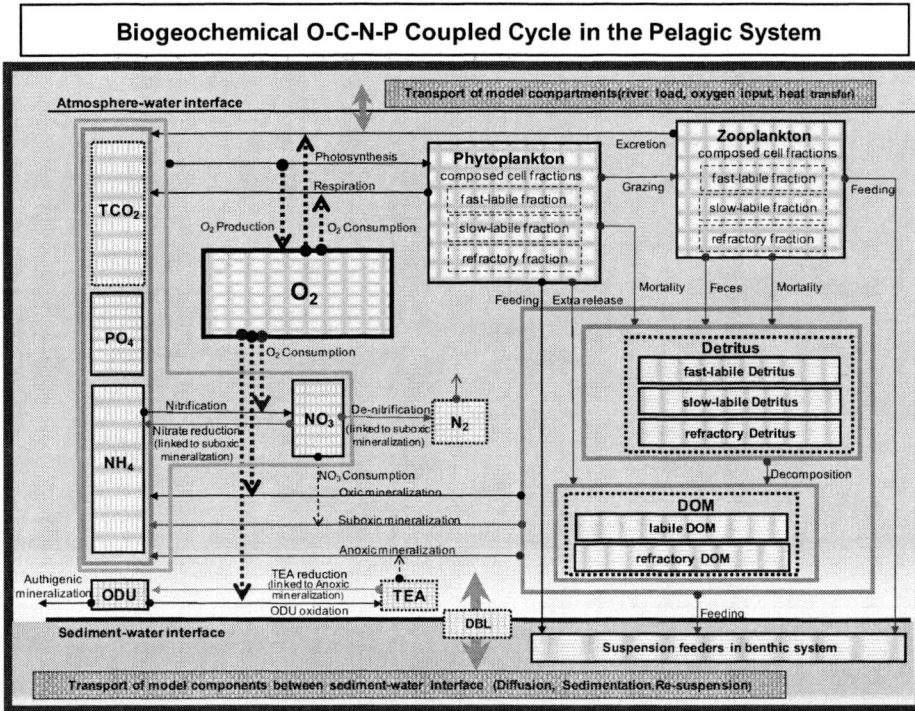

Model variable	Unit	Representation in the above diagram	Notation and [No.] in Appendix
Phytoplankton	mgC/L (μgC/ml)	Phytoplankton	PP [01]
Zooplankton	mgC/L (μgC/ml)	Zooplankton	ZP [02]
Fast labile detritus	mgC/L (μgC/ml)	Fast-labile Detritus	WDE_1 [03,1]
Slow labile detritus	mgC/L (μgC/ml)	Slow-labile Detritus	WDE_2 [03,2]
Refractory detritus	mgC/L (μgC/ml)	Refractory Detritus	WDE_3 [03,3]
Labile dissolved organic matter	mgC/L (μgC/ml)	Labile DOM	WDM_1 [04,1]
Refractory dissolved organic matter	mgC/L (μgC/ml)	Refractory DOM	WDM_2 [04,2]
Ammonium	mgN/L (μgN/ml)	NH_4	WNX [05]
Nitrate	mgN/L (μgN/ml)	NO_3	WNY [06]
Phosphate	mgP/L (μgP/ml)	PO_4	WDP [07]
Reduced substances $(Fe^{2+}, Mn^{2+}, S^{2-})$	mg/L (μg/ml)	ODU	WOU [08]
Dissolved oxygen	mg/L (μg/ml)	O_2	WDO [09]

Figure 3. Model variables and biochemical processes in the pelagic system of the ecological model. Model variables are described by boxes with solid lines while biochemical processes are indicated by arrows with both solid and dotted lines. This ecological diagram is produced from the O-C-N-P coupled cycle. ODU: oxygen demand unit, DOM: dissolved organic matter.

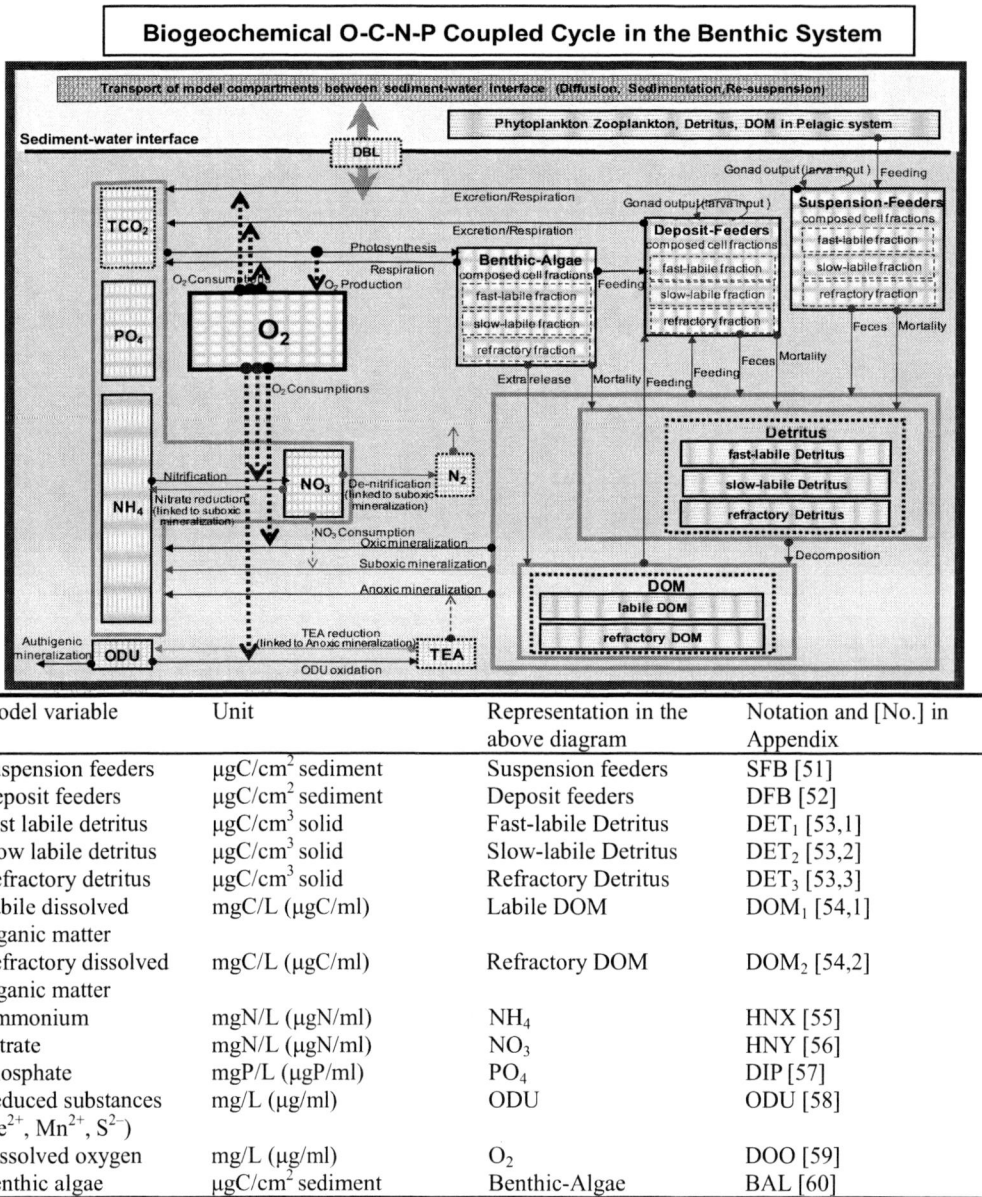

Figure 4. Model variables and biochemical processes in the benthic system of the ecological model. Model variables are described by boxes with solid lines and biochemical processes are indicated by arrows with both solid and dotted lines. This ecological diagram is produced from the O-C-N-P coupled cycle.

Model variable	Unit	Representation in the above diagram	Notation and [No.] in Appendix
Suspension feeders	$\mu gC/cm^2$ sediment	Suspension feeders	SFB [51]
Deposit feeders	$\mu gC/cm^2$ sediment	Deposit feeders	DFB [52]
Fast labile detritus	$\mu gC/cm^3$ solid	Fast-labile Detritus	DET_1 [53,1]
Slow labile detritus	$\mu gC/cm^3$ solid	Slow-labile Detritus	DET_2 [53,2]
Refractory detritus	$\mu gC/cm^3$ solid	Refractory Detritus	DET_3 [53,3]
Labile dissolved organic matter	mgC/L ($\mu gC/ml$)	Labile DOM	DOM_1 [54,1]
Refractory dissolved organic matter	mgC/L ($\mu gC/ml$)	Refractory DOM	DOM_2 [54,2]
Ammonium	mgN/L ($\mu gN/ml$)	NH_4	HNX [55]
Nitrate	mgN/L ($\mu gN/ml$)	NO_3	HNY [56]
Phosphate	mgP/L ($\mu gP/ml$)	PO_4	DIP [57]
Reduced substances (Fe^{2+}, Mn^{2+}, S^{2-})	mg/L ($\mu g/ml$)	ODU	ODU [58]
Dissolved oxygen	mg/L ($\mu g/ml$)	O_2	DOO [59]
Benthic algae	$\mu gC/cm^2$ sediment	Benthic-Algae	BAL [60]

The oxygen-consuming and oxygen-producing biochemical processes described in the ecological model are (1) photosynthesis by phytoplankton and benthic algae, (b) excretion by phytoplankton, zooplankton and benthic fauna (suspension feeders and deposit feeders), (c) oxic mineralization, (d) nitrification, and (e) oxidization of the total reduced substances, Mn^{2+}, Fe^{2+} and S^{2-} (Figure 5).

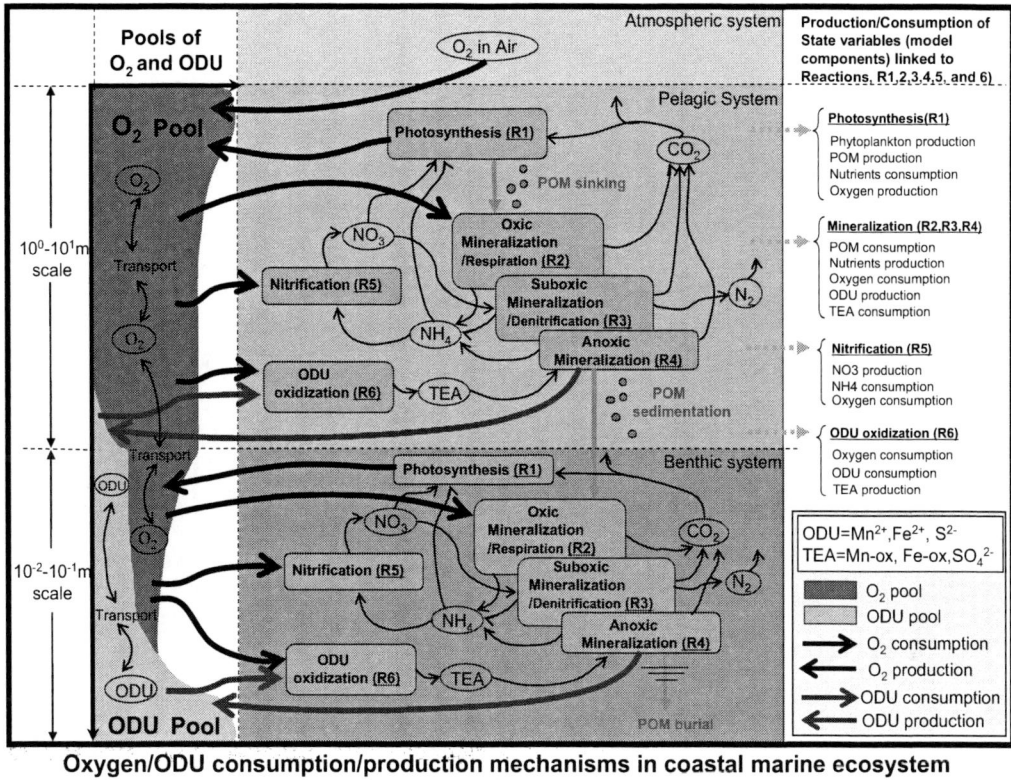

Oxygen/ODU consumption/production mechanisms in coastal marine ecosystem

Figure 5. Modeled oxygen production and consumption mechanisms and associated biochemical processes in the Ecological Connectivity Hypoxia Model, ECOHYM.

The characteristic point of the model is to divide the bacterial mineralization processes into three mechanisms, viz. oxic, suboxic, and anoxic mineralization. All these mineralization processes are formulated as first order kinetics of detritus. Oxic mineralization is limited by oxygen (Michaelis-Menten type kinetics). Suboxic mineralization based on nitrate is inhibited by oxygen (one minus Michaelis-Menten type kinetics) and limited by nitrate (Michaelis-Menten type kinetics). The consumption of oxygen and nitrate as terminal electron acceptors is explicitly modeled. Mineralization processes using other oxidants (manganese oxides, iron oxides, sulfate) are lumped into one process that is inhibited by oxygen and nitrate (one minus Michaelis-Menten type kinetics). Anoxic mineralization produces reduced substances as ODU. Re-oxidation of one mole of ODU requires one mole of O_2 (Soetaert et al. 1996a). ODU re-oxidation and nitrification are formulated to be limited by oxygen. Idealized stoichiometric relationships of each bacterial mineralization process used in our model are shown in Table 1 (Sohma et al. 2001). Modeling the mineralization processes as three pathways with the stoichiometric relationship in Table 1 enables analysis of the mechanisms underlying the relationship between oxygen consumption, nitrate reduction, de-nitrification and ODU production that are caused by mineralization. As shown in Table 1, the stoichiometric relationships of consumption and production by living organisms resulting from respiration and photosynthesis are described in the same way as oxic mineralization by bacteria.

Table 1. Stoichiometric relationships associated with biochemical processes treated in the ecological model

Photosynthesis using NH_4-N

$$m(CO_2) + n(NH_3) + (H_3PO_4) + m(H_2O) \rightarrow (CH_2O)_m(NH_3)_n(H_3PO_4) + m(O_2)$$ (T1.1)

Photosynthesis using NO_3-N

$$m(CO_2) + n(NO_3^-) + (H_3PO_4) + (m+n)(H_2O) + nH^+ \rightarrow (CH_2O)_m(NH_3)_n(H_3PO_4) + (2n+m)(O_2)$$ (T1.2)

Oxic mineralization, Excretion, Respiration

$$(CH_2O)_m(NH_3)_n(H_3PO_4) + m(O_2) \rightarrow m(CO_2) + n(NH_3) + (H_3PO_4) + m(H_2O)$$ (T1.3)

Suboxic mineralization

$$(CH_2O)_m(NH_3)_n(H_3PO_4) + a(HNO_3) \rightarrow m(CO_2) + a(x/2)(N_2) + n(NH_3) + a(1-x)(NH_3) + (H_3PO_4) + b(H_2O)$$ (T1.4)

where, $a = -4m/(3x-8)$, $b = m(3x-4)/(3x-8)$, $0 \leq x \leq 1$, These condition satisfy $a \geq 0$ and $b \geq 0$ at anytime

Anoxic mineralization

$$(CH_2O)_m(NH_3)_n(H_3PO_4) + m(TEA) \rightarrow m(CO_2) + n(NH_3) + (H_3PO_4) + m(ODU) + Q(H_2O)$$ (T1.5)

where, $ODU = 2Mn^{2+}$, $4Fe^{2+}$ and $1/2 S^{2-}$, $TEA = 2 MnO_2$, $2 Fe_2O_3$ and $1/2 SO_4^{2-}$

Nitrification

$$NH_3 + H_2O + 2 O_2 \rightarrow NO_3^- + 2 H_2O + H^+$$ (T1.6)

ODU (Oxygen Demand Unit) oxidization

$$ODU + O_2 \rightarrow TEA$$ (T1.7)

where, $ODU = 2 Mn^{2+}$, $4 Fe^{2+}$ and $1/2 S^{2-}$, $TEA = 2 MnO_2$, $2 Fe_2O_3$ and $1/2 SO_4^{2-}$

(1) m, n denote C, N, P ratio of created or mineralized organic matter, i.e., C:N:P=m:n:1. (2) x denotes ratio of nitrogen reducing to nitrogen gas (N_2) and reducing to ammonium from nitrate by suboxic mineralization. i.e., N_2:NH_3= x:(1-x). (3) a, b are coefficients determined from stoichiometric relation. (4) Nitrogen of N_2 and NH_3 shading and written by italic in the right-hand side are derived from HNO_3 in the left-hand side in the equation "T1.4".

Detritus is divided into three fractions: fast-labile, slow-labile and refractory organic matter. This means that the model is a Multi-G model (Jørgensen 1978). Accordingly, the cells of living organisms (phytoplankton, zooplankton, benthic algae and benthic fauna) are defined as being composed of these three fractions (Sohma et al. 2004). Hence, fluxes between living organisms and detritus (fluxes of uptake, excretion, feces, mortality, etc.) can be modeled while considering these fractions.

In addition, benthic fauna are divided into suspension feeders and deposit feeders in order to describe their functional differences, which have the potential to affect hypoxia dynamics, e.g. the transportation of particulate organic matter (POM = phytoplankton, zooplankton, and detritus) from pelagic to benthic compartments by suspension feeding and mineralization of benthic organic matter as a result of deposit feeder excretion.

Physical Processes

As for the physical processes, (a) water current (flow velocity) and eddy viscosity in the water column, and (b) molecular diffusion, irrigation, bioturbation and deposition in the sediment or sediment-water interface are described in the ecological model equations. In the water column, model variables are transported by water flow (advection) and eddy viscosity (diffusion). As mentioned before, flow velocity and diffusion in the water column are calculated by the hydrodynamics model based on well-established fluid dynamics theory. The time series data of flow velocity and diffusion are entered into the pelagic equation in the ecological model at each time step. In the sediment or at the sediment-water interface, transport of dissolved substances differs from transport of particulate substances. Dissolved substances in the benthic system are transported by molecular diffusion, bioturbation and irrigation due to the activity of benthic fauna, and advection by pore water velocity. Particulate substances in the benthic system are transported by advection (sediment deposition) and bioturbation.

Concerning molecular diffusion, the model uses well-established theory based on statistical physics, and the value of the molecular diffusion coefficient is determined by temperature. Here, the molecular diffusion coefficient in the sediment reflects the effect of tortuosity (Berner 1980) as an empirical function of porosity. The molecular diffusive fluxes of dissolved substances at the sediment-water interface are obtained from the theory of the diffusive boundary layer (DBL) (Boudreau and Jørgensen 2001). Other physical processes in the sediment or sediment-water interface are influenced by the biomass and species composition of benthic fauna and sediment grain size, but no general theory exists for those processes. Therefore, many variations of how to model these processes and determine their parameter values are possible. In our model, bioturbation is modeled as a diffusion-like process. Irrigation is modeled as a diffusion-like process and as a proportional process of the differences in concentration between surface water and each depth of sediment (Berner 1980, Berg et al. 1998). The diffusive coefficients of bioturbation and irrigation are formulated as hyperbolic functions of suspension feeders and deposit feeders (Sohma et al. 2004) and the vertical dependence of bioturbation and irrigation are formulated as exponential functions (Soetaert et al. 1996a).

The advection of solid substances and dissolved substances is obtained from the porosity and sedimentation rate to conserve the volume of liquid and solid phases.

During the period of emersion of the tidal flat area, the fluxes across the sediment-water interface, i.e. diffusion of dissolved substances, sediment deposition and feeding of suspension feeders, are set at zero. The oxygen flux from the atmosphere to pore water due to aeration is solved based on the DBL theory.

Governing General Equations and Assumptions

The following seven equations and four assumptions are applied in the ecological model. These equations, especially for the benthic system, are mostly based on the equations propounded by Berner (1980).

The general equation for the pelagic system is as follows:

$$\frac{\partial C_w}{\partial t} = -(\mathbf{v}_w \bullet \nabla)C_w + \nabla \bullet (\mathbf{K} \bullet \nabla C_w) + \sum R$$

$$= -u_w \frac{\partial C_w}{\partial x} - v_w \frac{\partial C_w}{\partial y} - w_w \frac{\partial C_w}{\partial z} + \frac{\partial}{\partial x}\left(K_x \frac{\partial C_w}{\partial x}\right) + \frac{\partial}{\partial y}\left(K_y \frac{\partial C_w}{\partial y}\right) + \frac{\partial}{\partial z}\left(K_z \frac{\partial C_w}{\partial z}\right) + \sum R$$

(1)

Where C_w = concentration of pelagic substances, i.e. phytoplankton, zooplankton, detritus (fast labile detritus, slow labile detritus, refractory detritus), dissolved organic matter (labile DOM, refractory DOM), NH_4-N, NO_3-N, PO_4-P, DO, and ODU [mass/L^3–liquid], $\mathbf{v}_w = (u_w, v_w, w_w)$ = flow velocity that already has been calculated by the hydrodynamics model [L/T], t = time [T], x, y, z = space coordinates [L], $\sum R$= biochemical reactions and fluxes from outside the system [mass /L^3–liquid/T], $\mathbf{K}$ = eddy diffusion (viscosity) tensor [L^2–liquid/T].

Diagenetic equation for benthic dissolved substances:

$$\frac{\partial(\phi C)}{\partial t} = \frac{\partial\left\{D_B \dfrac{\partial(\phi C)}{\partial z} + \phi(D_S + D_I + D'_B)\dfrac{\partial C}{\partial z}\right\}}{\partial z} + \phi\alpha(C_0 - C)$$

$$- \frac{\partial(\phi v C)}{\partial z} + \phi R_{ads} + \phi\sum R'$$

(2)

Where ϕ = porosity [–], C = concentration of dissolved substances i.e. dissolved organic matter (labile DOM, refractory DOM), NH_4-N, NO_3-N, PO_4-P, DO, and ODU [mass/L^3–liquid], C_0 = concentration of dissolved substances at sediment-water interface [mass/L^3–liquid], D_S = molecular diffusion coefficient in sediment including the effects of tortuosity [L^2–sediment/T], D_B = solid biodiffusion coefficient (intraphase mixing expression) [L^2–sediment/T], D'_B = solid biodiffusion coefficient (interphase mixing expression) [L^2–sediment/T], D_I = irrigation coefficient (diffusion-like expression) [L^2–sediment/T], α = irrigation coefficient2 [1/T], v = velocity of burial of water below the sediment-water interface [L–sediment/T], R_{ads} = reactions of dissolved materials due to equilibrium adsorption or desorption [mass/L^3–liquid/T]. $\sum R'$ = all other slow (irreversible) biochemical reactions [mass/L^3–liquid/T].

Diagenetic equation for benthic particulate substances:

$$\frac{\partial \{(1-\phi)\overline{\rho}_s\overline{C}\}}{\partial t} = \frac{\partial \left[D_B \dfrac{\partial \{(1-\phi)\overline{\rho}_s\overline{C}\}}{\partial z}\right]}{\partial z} + \frac{\partial \left\{D_B'(1-\phi)\overline{\rho}_s \dfrac{\partial \overline{C}}{\partial z}\right\}}{\partial z}$$
$$-\frac{\partial \{(1-\phi)\overline{\rho}_s w\overline{C}\}}{\partial z} + (1-\phi)\overline{\rho}_s\overline{R}_{ads} + (1-\phi)\overline{\rho}_s\sum \overline{R'} \qquad (3)$$

Where $\overline{C}$ = concentration of a particulate substance in terms of mass per unit mass of total solids, i.e. detritus (fast labile detritus, slow labile detritus, refractory detritus), absorbed DOM, absorbed NH_4-N, $\overline{\rho}_s$ = density of total solid phase [mass-solid/ L^3–solid], w = rate of depositional burial of solids [L–sediment/T], $\overline{R}_{ads}$ = reactions of dissolved materials due to equilibrium adsorption or desorption [mass/mass-solid/T], $\sum \overline{R'}$ = all non-equilibrium slow biochemical reactions [mass/mass-solid/T].

Equations for suspension feeders, deposit feeders and benthic algae:

$$\frac{\partial B}{\partial t} = \sum R_B \qquad (4)$$

Where B = biomass, expressed per square of sediment [mass/L^2–sediment], R_B = biochemical reactions [mass/L^2–sediment].

Equation for the relation of adsorption-desorption reaction:

$$\overline{R}_{ads} = \frac{-\phi}{(1-\phi)\overline{\rho}_s} R_{ads} \qquad (5)$$

Equation for mass/volume conservation of benthic solid phase:

$$\frac{\partial \phi}{\partial t} + \frac{\partial (v \cdot \phi)}{\partial z} = \frac{\partial}{\partial z}\left(D_B \frac{\partial \phi}{\partial z}\right) \qquad (6)$$

Equation for mass/volume conservation of benthic liquid phase:

$$\frac{\partial (1-\phi)}{\partial t} + \frac{\partial \{w \cdot (1-\phi)\}}{\partial z} = \frac{\partial}{\partial z}\left(D_B \frac{\partial (1-\phi)}{\partial z}\right) \qquad (7)$$

Note that w and v in the benthic system are calculated to meet the relationship among equations (6) and (7).

The following assumptions are imposed on the equations described above (Berner 1980).

i. Seawater is treated as incompressible liquid: $div\ v_w = 0$

ii. Density of solid does not change with space or time: $\overline{\rho}_s$ is constant

iii. The equilibrium expression for simple linear adsorption: $\overline{C} = K'C$, K' = adsorption coefficient.

iv. Adsorptive property does not change with space or time, in other words, K' is constant.

v. If $\overline{C}$ is adsorbed substances, then there are no slow diagenetic reactions, hence, $\sum \overline{R'} = 0$ in equation (3).

IMPLEMENTATION

In this section, the methodology for applying ECOHYM to Tokyo Bay is discussed.

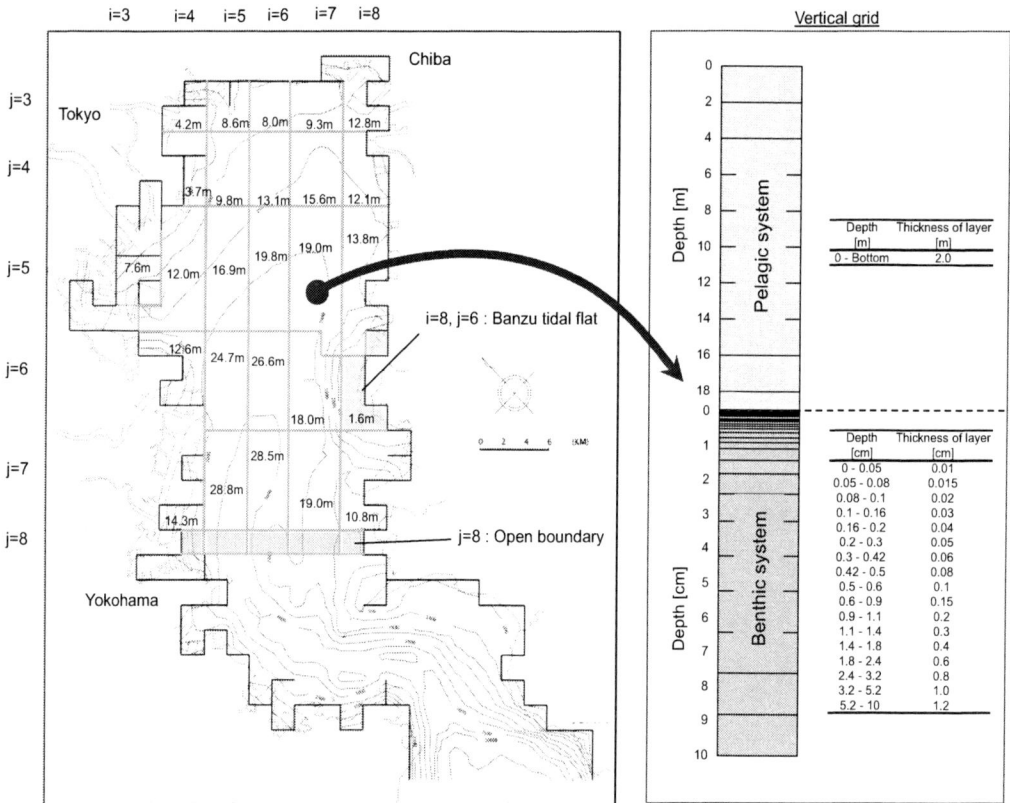

Figure 6. Geographical description of Tokyo Bay (calculated area, spatial resolution both in the vertical and horizontal directions, and coordinates (i, j) on the ecological model application).

Spatial Resolution – Vertical

The vertical and horizontal spatial resolution setting in the ecological model is shown in Figure 6. In the benthic system, as for the biochemical processes, the vertical spatial distributions of processes related to hypoxia—namely, oxic, suboxic and anoxic mineralization, denitrification, nitrification, ODU oxidation and photosynthesis of benthic algae—have a steep gradient within a scale of 1–10 mm (Kuwae et al. 2005, Sayama 2005). As for the transport processes, the advection flow in the sediment (deposition of sediment and advection of pore water) is 0.3 to ~2.0 cm/year in Tokyo Bay (Matsumoto 1983), and the diffusion in the sediment is dominated by molecular diffusion (i.e. small-scale diffusion) as compared to the pelagic system. Thus, the vertical grid interval in the benthic system was set to 0.1–12 mm.

In the pelagic system, because the mass transport due to water currents is larger than that in the benthic system, the vertical profile of pelagic metabolism, which is significant to the hypoxic mechanism, is more moderate than in the benthic profile. In this implementation, the vertical spatial resolution in the pelagic system is set at 1–2 m, which is enough to demonstrate accurately the following hypoxia-related processes: vertical mixing, stratification, sedimentation of particulate substances, oxygen consumption and nutrients flux from the benthic system.

Horizontal Spatial Resolution

Horizontal spatial resolution is different for the hydrodynamic and ecological models. In the hydrodynamic model, a 2 km x 2 km grid was implemented, whereas in the ecological model, the Tokyo Bay area under consideration was divided into 26 zones (boxes). The input data for the ecological model from the hydrodynamics model, i.e. flow velocity, eddy viscosity and temperature etc., were averaged spatially and adjusted to the 26 boxes of the ecological model, while respecting the flow continuity equation of water volume. The reasons for the low horizontal spatial resolution used in the ecological model are as follows: (a) a shorter calculation time is advantageous for iterative model tuning and calibration in the first stage of model development; (b) our focus is on vertical micro-mechanisms, particularly for the benthic and short timescale dynamics in the tidal flat; and (c) setting the ideal vertical and temporal resolution lies at the limit of computational performance, while maintaining a reasonable calculation time. Although the resolution of the 26 boxes is not sufficient to demonstrate details of the horizontal spatial distribution, it still enables us to describe the difference in ecological mechanisms from the central bay to tidal flat area, the difference in benthic-pelagic mutual interaction depending on the water depth (i.e. the achievable mass of pelagic particulate substances to the seafloor), and the difference in the central bay-tidal flat area interaction depending on the distance between each area.

Simulation Period and Time Step

A simulation was carried out to demonstrate the daily and seasonal dynamics of an average year. The prescribed functions (forcing function) of the hydrodynamics and ecological models, i.e. freshwater and nutrient input from rivers, meteorological conditions (light intensity, wind, etc.), and open boundary condition of model variables were set as one-year periodical functions. These functions were created based on the observed data from 1998 to 2002. The convergence state of this simulation describes the dynamics with a one-year period. This state hereafter is called the "annual periodical steady state of the existing Tokyo Bay". The convergence time to achieve the annual periodical steady state was different for (a) the fluid dynamics of the pelagic system, (b) ecological dynamics of the pelagic system in the central bay, (c) the ecological dynamics of the benthic system in the central bay, and (d) the ecological dynamics of tidal flats. The differences mainly result from the differences in the physical or biochemical turnover periods of the modeled substances (model variables) among (a), (b), (c) and (d). Here, the turnover period (residence time) is defined as the inverse of turnover rate, and the turnover rate is defined as shown in Figure 7. As far as the residence time due to physical processes is concerned, the value of the pelagic system, 30–100 days, is the time scale for all existing water in Tokyo Bay to be exchanged once. The value of the benthic system, 30–100 years, is the time scale for the benthic solid phase at the sediment-water interface to achieve the modeled benthic bottom layer depth (10 cm). Meanwhile, in the simulation, the hydrodynamics model achieved the annual periodical steady state after a one year calculation. The ecological model achieved the annual periodical steady state after a 100 year calculation. These results suggest that the convergence time of both the hydrodynamics and ecological models is controlled by their characteristic residence times, which in turn are controlled by physical processes.

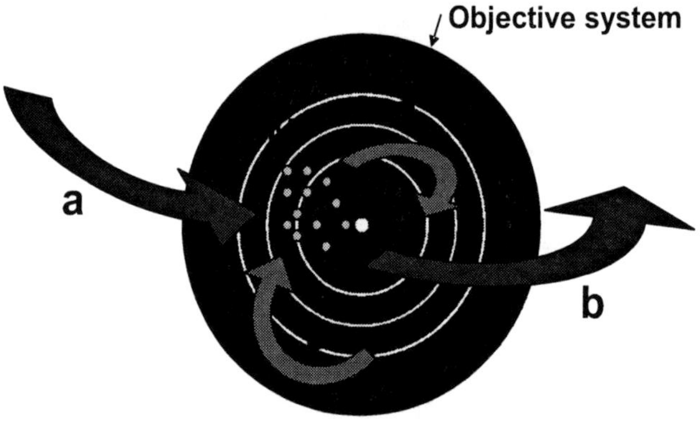

Objective system

●: Target material: material A
a: Summation of influx by transport [g/T]
b: Summation of efflux by transport [g/T]
c: Summation of biochemical production [g/T]
d: Summation of biochemical consumption [g/T]
M: Total mass of target material in objective system [g]

(1) **Turn-over rate of target material due to physical process** $[1/T] = \frac{1}{2} \cdot (a+b) / M$
(2) **Turn-over rate of target material due to biochemical process** $[1/T] = \frac{1}{2} \cdot (c+d) / M$

Figure 7. Concept and formulations of turnover rate due to physical and biochemical processes.

The significant time scale of ecosystem dynamics differs between the benthic and pelagic systems, or between the tidal flat and central bay areas. In the tidal flat, tidal or daily scale dynamics of biochemical processes are dominant because of the shallow depth (e.g., changes in feeding behavior of suspension feeders due to submersion/emersion and changes in the level of photosynthesis of benthic algae due to light intensity at the sediment surface, etc.). These dynamics have the potential to dominate the interaction between the tidal flat and the central bay area. In addition, seasonal dynamics of light and temperature also affect metabolism in the tidal flat. In the central bay area, photosynthesis by phytoplankton, and the transport of substances in the pelagic system change on a tidal and daily scale. Seasonal dynamics of light and temperature also affect metabolism in the central bay area. However, for the benthic system in the central bay where light does not reach the seafloor and there is no submersion/emersion phenomenon, it is suggested that the daily and tidal scale dynamics are not as dominant as in the tidal flat area.

The notable time scale of dynamics in calculations made by ECOHYM ranges from a daily scale to a seasonal time scale. The time step was thus set at 0.2 hrs, a value that enables us to demonstrate both the tidal and seasonal dynamics.

Boundary Conditions and Initial Values

The boundary conditions used in the model are divided into values set at the open sea boundary (open sea boundary conditions) and values set at the atmosphere and inflow points of the river (other boundary conditions). For the "open sea boundary conditions," tidal levels resulting from four component tides (M2, S2, O1 and K1), the concentrations of model variables, salinity, and water temperature were set. For "other boundary conditions," air temperature, surface wind, light, and the quantities of nutrients, detritus, and fresh water discharged from rivers were set. These were prescribed based on the field measurements and were interpolated by the spline method or linear interpolation. Initial values of the model variables in the ecological model were set at the yearly averaged value on April 1, from 1998 to 2002. The dynamics of these boundary conditions become the driving force of the simulation.

Parameter Tuning

Many biochemical parameters are included in the formulation of biochemical processes. We investigated a range of values for these parameters that have been observed or used in other models, and set values within the range or at the same order of magnitude as the investigated value. Values of relevant biochemical parameters established for this study and references for the sources of the parameters are listed in Sohma et al. (2008).

It is very difficult to control ECOHYM because of the inclusion of both benthic and pelagic systems as well as the central bay and tidal flat area. Therefore, in the first stage of the development process, each ecological sub-model, namely, the sub-model of the pelagic system of the central bay area, the sub-model of the benthic system of the central bay area, and the sub-model of the benthic-pelagic-coupled system of the tidal flat area were developed and calculated separately for verification, calibration and validation. In these calculations, the values at the boundary of the focused system of each sub-model (e.g. the benthic boundary condition of pelagic sub-models (the model variables and fluxes from the benthic system), or the pelagic boundary condition of the benthic sub-model (the model variables of bottom water of the pelagic system, etc.) were given by prescribed functions based on the observed data. In the next stage, mutual interactions between sub-models were incorporated by (a) removing the prescribed functions at the boundary between sub-models; (b) assigning the value of the boundary from the calculation of each sub-model at each time step; and (c) making the mutual linkage (dependence) between each sub-model operational. The tuning of the mutually interacting model was implemented based on the specified values of the parameters used in the first stage.

In the early stage of tuning, we initially focused on getting the model to reproduce the observed values of particulate substances (phytoplankton, zooplankton, and detritus in both the sediment and water column). Second, we checked the model's capacity to reproduce the inorganic substance values (dissolved oxygen, NH_4-N, NO_3-N, PO_4-P). This method is based on the assumption that the dynamics of inorganic matter are controlled by the organic matter, given the higher turnover rate of the former, especially in the benthic system. In the case where the model reproduced the observed values of organic matter well but did not reproduce the values of inorganic matter, the fractions of organic matter as fast-labile detritus, slow-labile detritus and refractory detritus, as well as physical parameters such as bioturbation and irrigation, were tuned to reproduce the observed data.

VALIDATION – COMPARISON BETWEEN MODEL OUTPUT AND OBSERVED DATA

Calculated model variables and fluxes were compared with the time series of measurements recorded at monitoring stations in Tokyo Bay, including the tidal flat located at Banzu. The model outputs compared here are the results of calculations in which the values of the parameters were set within the known values, and the mutual interactions between the benthic-pelagic systems or the central bay-tidal flat area were functional (outputs from the final stage calculation mentioned in the tuning scheme discussed previously). Because the temporal and spatial distributions of model variables depend on each other through the ecological network described in the model, it is necessary for the validation to be performed by checking all model variables. The detailed results for the simultaneous reproduction of the observed temporal and spatial dependence of all major model variables were introduced in Sohma et al. (2008) and Sohma (2009). Therefore, I demonstrate here the summary of the result.

Figure 8 shows the comparison between the monitoring data and model outputs regarding seasonal variation of dissoloved oxygen (DO) in the central bay area, and Figure 9 shows the comparison regarding daily variation of DO in the existing tidal flat area, Banzu tidal flat. The monitoring data in Figure 8 was observed from 1998 to 2002, and the data in Figure 9 was observed in 2003. Because of the model demonstrating a one year average condition, model outputs for the comparison, especially in Figure 9, was used at the extracted period of 2-3 days when the phase relationship between the tidal level and the light intensity was almost the same as the monitoring situation in the same month (August). As shown in Figures

8 and 9, the DO dynamics in the model are in good agreement with the monitoring data. For the other model variables, the reproducibility were also the same level as the DO (Sohma et al. (2008) and Sohma (2009)).

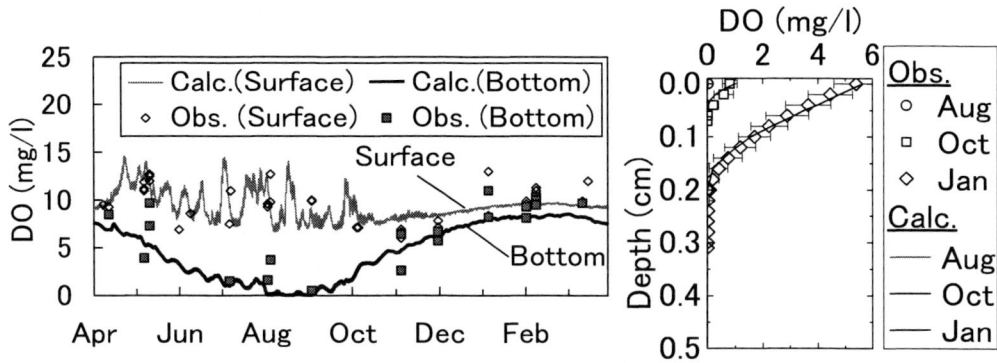

Figure 8. Seasonal variation of dissolved oxygen of the pelagic system (left, (i, j) = (6, 4)) and of the benthic system (right, (i, j) = (5, 4)) in the central bay area.

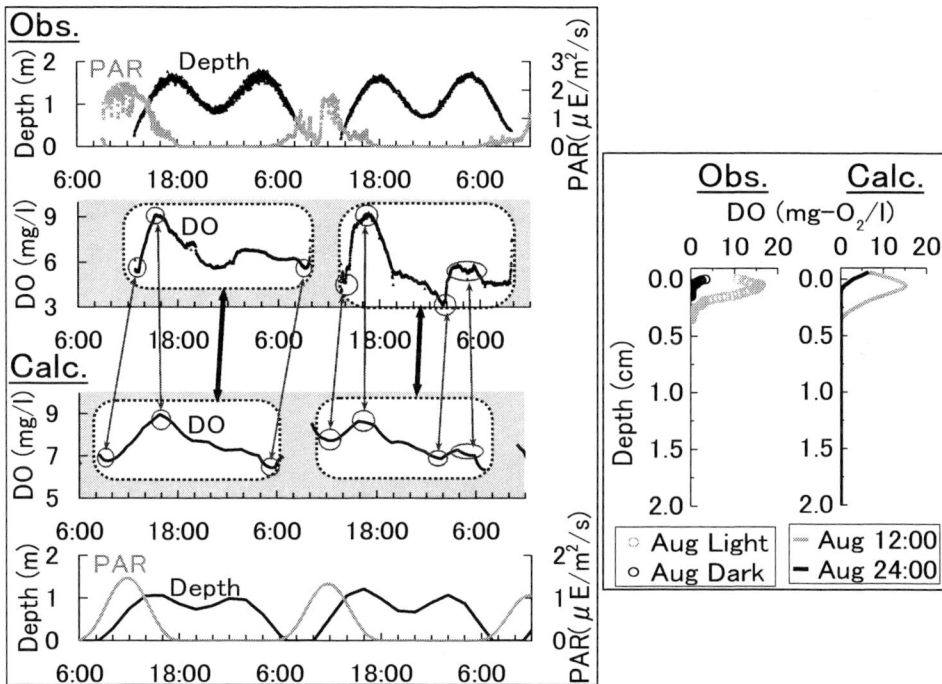

Figure 9. Daily variations in dissolved oxygen of the pelagic system (left, (i, j) = (8, 6)) and of the benthic system (right, (i, j) = (8, 6)) in the tidal flat area.

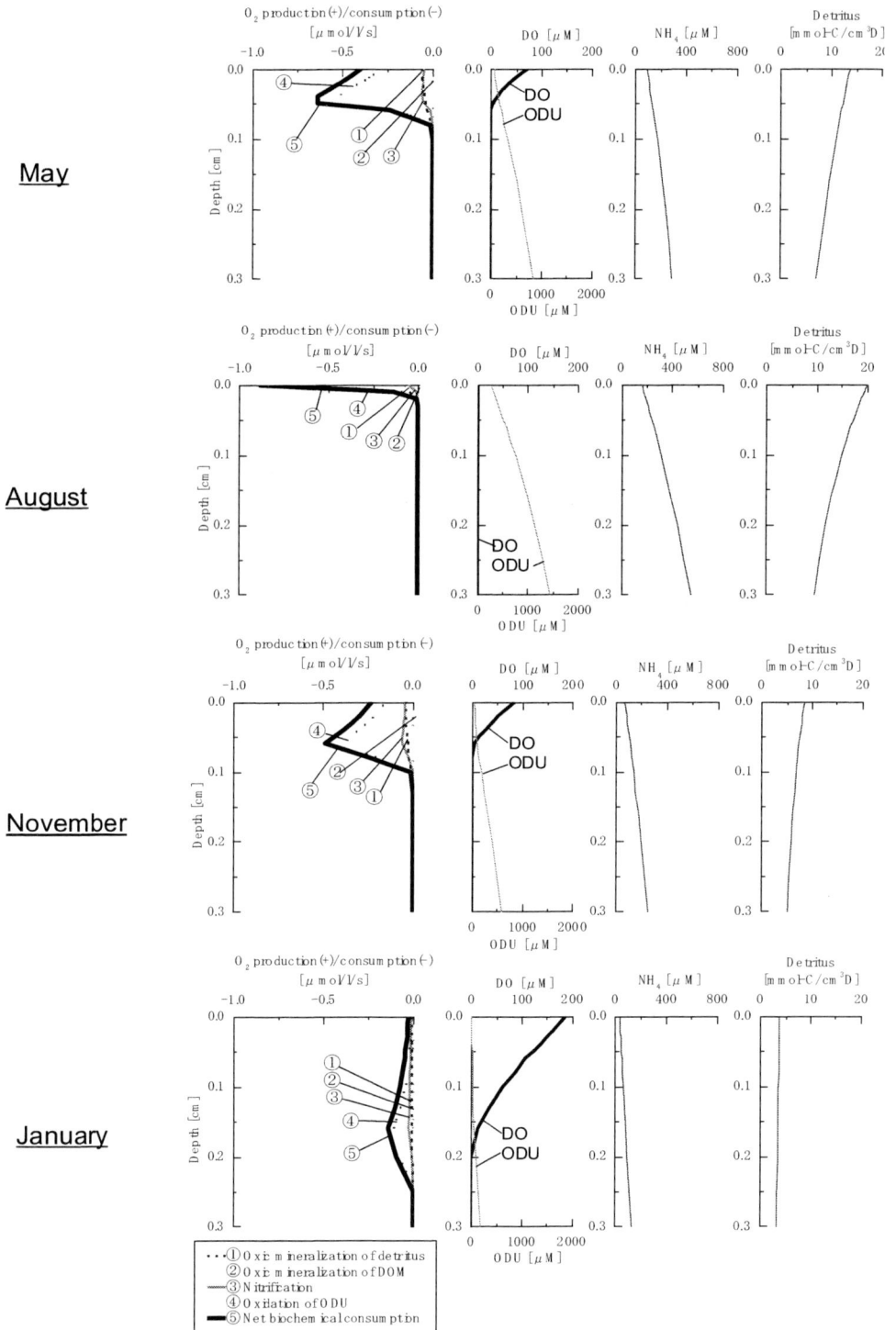

Figure 10. Vertical profiles of oxygen consumed by biochemical processes and model variables (DO, ODU, NH₄-N and detritus) at the benthic system in the central bay area. Grid number (i, j) = (5, 4): spring (May), summer (August), autumn (November) and winter (January).

WHAT WAS REVEALED FROM THE COLLABORATION OF PHILOSOPHY AND TECHNOLOGY?

Several analyses for revealing ecosystem mechanisms and for predicting ecosystem responses have been performed using validated ECOHYM applied to Tokyo Bay. The results help to inspire our interest and idea of how we should understand the environmental problems and the ecosystem in this eutrophic, hypoxic estuary. Some results are introduced in this section.

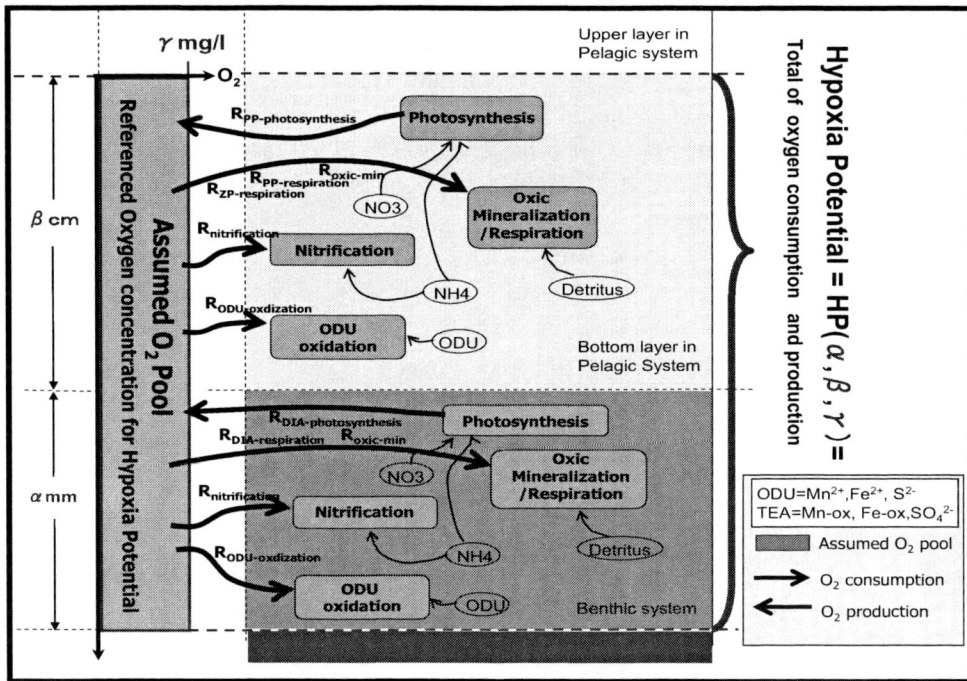

$$HP(\alpha,\beta,\gamma)= \int_0^\alpha R_{oxic-min}(\gamma,POM,T)dz + \int_0^\alpha R_{nitrification}(\gamma,NH_4,T)\,dz$$
$$+ \int_0^\alpha R_{ODU-oxidation}(\gamma,ODU,T)\,dz + \int_0^\alpha R_{DIA-respiration}(DIA,T)dz$$
$$- \int_0^\alpha R_{DIA-photosynthesis}(DIA,I,nutrients,T)dz$$
$$+ \int_0^\beta R_{oxic-min}(\gamma,POM,T)dz + \int_0^\beta R_{nitrification}(\gamma,NH_4,T)\,dz$$
$$+ \int_0^\beta R_{ODU-oxidation}(\gamma,ODU,T)\,dz + \int_0^\beta R_{PP-respiration}(PP,T)dz$$
$$+ \int_0^\beta R_{ZP-respiration}(ZP,T)dz$$
$$- \int_0^\beta R_{PP-photosynthesis}(PP,I,nutrients,T)dz$$

$HP(\alpha,\beta,\gamma)$: Hypoxia potential
α: defined thickness of the benthic system
β: defined thickness of the pelagic system
γ: defined standard concentration of DO
$R_{oxic-min}$: O_2 consumption of oxic mineralization
$R_{nitrification}$: O_2 consumption of nitrification
$R_{ODU-oxidation}$: O_2 consumption of ODU oxdization
$R_{DIA-photosynthesis}$: O_2 production of benthic algae photosynthesis
$R_{DIA-respiration}$: O_2 consumption of benthic algae excretion
$R_{PP-photosynthesis}$: O_2 production of phytoplankton photosynthesis
$R_{PP-respiration}$: O_2 consumption of phytoplankton excretion
$R_{ZP-respiration}$: O_2 consumption of zooplankton excretion
T: Temperature
I: light intensity
POM: detritus
nutrients: NH_4-N, NO_2-N, NO_3-N, and PO_4-P
DIA: benthic algae biomass
PP: phytoplankton concentration
ZP: zooplankton concentration
NH_4-N: ammonium nitrogen concentration
ODU: oxygen demand unit concentration

Figure 11. The concept and formulation of hypoxia potential (HP). Hypoxia potential is the oxygen consumption rate under the defined standard concentration of DO. The right side describes the formulation of hypoxia potential in ECOHYM; 1st to 5th terms of the right hand side are oxygen consumption and production flux in the benthic system; 6th to 11th terms are oxygen consumption and production flux in the pelagic system. Each term is calculated by the biochemical formulation in Tables A5 and A6.

Benthic Oxygen Consumption during Summer Is Low – The Motivation for Focusing on Hypoxia Potential

In Figure 10, the vertical profiles of oxygen producing and oxygen-consuming fluxes and model variables at the benthic system in the central bay area are shown. The values in Figure 10 are model output at 1200 one day of each spring (May), summer (August), autumn (November) and winter (January) in area (i, j) = (5, 4). At the benthic system in the central bay, light does not reach the sea floor. Thus, there is no photosynthesis and there are no oxygen producing processes. The oxygen consuming processes are oxic mineralization, nitrification and ODU oxidization. Respiraiton of the benthic fauna consumes oxygen near the benthic-pelagic boundary. Therefore, we could assume that benthic fauna use oxygen in the benthic, pelagic, or both systems. In this study, the benthic fauna consume oxygen in the bottom layer of the pelagic system, not in the benthic system. What was revealed from the net biochemical oxygen consumption in the benthic system was that (a) ODU oxidization is the largest oxygen consumer among all biochemical processes during any season, and that (b) oxygen consumption in summer is less than in other seasons in terms of the vertically integrated value. The high contribution of ODU oxidization to oxygen consumption, by conjecture, is characteristic of most of the hypoxic estuaries in Japan. Low oxygen consumption during summer is attributed to the lack of oxygen for aerobic metabolism at the seafloor in the presence of hypoxia. In contrast to oxygen consumption, ODU concentration is greatest during summer. This phenomenon results because anoxic mineralization rates are higher in summer than in other seasons, attributable to (a) larger masses of detritus from high phytoplankton production, (b) higher rates of anoxic mineralization than oxic mineralization when oxygen is lacking, and (c) high temperature during summer.

Model results and observations both show that higher levels of detritus, ODU and NH_4-N are accumulated in the benthic system under summer hypoxic conditions. In this situation, even if oxygen is supplied to the seafloor by some kind of vertical disturbance, it is consumed immediately by oxic mineralization, oxidization of ODU, and nitrification. Whether oxygen consumption substrates are highly accumulated or not, oxygen consumption at the seafloor is negligible or absent under hypoxic conditons because there is no oxygen to consume. Therefore, oxygen consumption rate during hypoxia is not suitable for estimating the tenacity of hypoxia or tendency toward hypoxia. The difference in oxygen consumption rates between the state of high accumulation of oxygen-demanding substrates to the state of low accumulation is revealed only when DO is supplied to the seafloor. Based on this background, I have defined the oxygen consumption rate at the defined standard DO concentration at the seafloor as "hypoxia potential" (Sohma 2005, Sohma et al. 2005), and its value is used to estimate the tenacity of hypoxia and the tendency to become hypoxic. The concept of hypoxia potential can be used both for experiments (i.e. measurement of oxygen consumption rate in sediment core samples soaked by seawater in which DO is controlled at a defined standard concentration), and for numerical model analysis.

To estimate hypoxia potential with ECOHYM, hypoxia potential was calculated as the oxygen consumption rate both in the pore water of the benthic system (at the range from the sediment-water interface to α mm below) and in the bottom water of the pelagic system (at the range from sediment-water interface to β cm above) under defined standard DO (γ mg O_2 L^{-1}). The processes contributing to oxygen consumption are oxic mineralization ($R_{oxic-min}$), nitrification ($R_{nitrification}$), ODU oxidization ($R_{ODU-oxidization}$), phytoplankton respiration (R_{PP-}

$_{\text{respiration}}$), zooplankton respiration ($R_{\text{ZP-respiration}}$), and benthic algae respiration ($R_{\text{DIA-respiration}}$). In addition, in order to estimate hypoxia potential in shallow water areas where light can reach the sea bottom, oxygen producing processes, i.e. phytoplankton photosynthesis ($R_{\text{PP-photosynthesis}}$) and benthic algae photosynthesis ($R_{\text{DIA-photosynthesis}}$) are effects that decrease hypoxia potential. Each oxygen consumption and production process is calculated by the formulation described in Tables A5 and A6 in the Appendix. For the calculation, DO is set γ mg O_2 L^{-1} and the other model variables set at the values of pre-calculated results (i.e. the values of the results of regular simulation without fixing the level of DO at γ mg O_2 L^{-1}) at each time step. The conceptual model and calculation method of hypoxia potential, HP (α, β, γ), is described in Figure 11. In this study, the values of α, β and γ were set as: $\alpha = 1$ mm, $\beta = 10$ cm and $\gamma = 2$ mg O_2 L^{-1}. The values of α, β and γ are assumed to be, respectively, the mean thickness of the benthic oxic layer during hypoxia, the range and thickness of the hypoxic water column in the pelagic system, and the lowest DO level for 100% survival of benthic fauna.

A Clean Ocean Is Different from a Bountiful Ocean

A scenario reproducing the early tidal flat in Tokyo Bay that was reclaimed in the past was compared to a scenario of 50% reduction of the nutrient load to Tokyo Bay from rivers. The geography of the tidal flat reproduction case is shown in Figure 12. These two scenarios were calculated as sensitivity analyses based on the model implementation in section 6, i.e. the annual periodical steady state of the existing Tokyo Bay (control case). Both scenarios started from the ecological state of the control case, with initial values of the model variables on 1 April. In the tidal flat reproduction scenario, the tidal flat ecosystems were simulated geographically, with the model variables for the reconstructed tidal flats initiated at the same values as the existing tidal flat (Banzu tidal flat in Figures 6 and 12). Then their states were calculated autonomously in the model subsequently. For the nutrient load reduction case, the nutrient load from rivers was reduced by 50%. Both cases achieved a new annual periodical steady state after a 15 year run.

The comparison between the new annual periodical steady state of the two scenarios, tidal flat reproduction and nutrient load reduction, are shown in Figure 13. Both cases show decreases in particulate organic matter (detritus) compared to the control case (existing system). In addition, both cases demonstrate the same level of hypoxia reduction from the control case. In other words, measures of tidal flat reproduction and nutrient load reduction lead to the "clean ocean" in terms of water quality. However, the benthic fauna responded differently to the tidal flat reproduction scenario, which increased benthic biomass relative to the control, than to the 50% nutrient load reduction scenario, which decreased benthic biomass. Although a 50% decrease in nutrient load leads to recovery from eutrophication, it does not lead to a rich ecosystem with diverse and abundant living organisms but rather leads to a poor ecosystem with relatively few living organisms. In contrast, tidal flat reproduction promotes the effective utilization of excess nutrients and organic matter by assimilating the materials into benthic fauna, leading to a rich ecosystem and bountiful ocean. Although the model output shown here is based upon several assumptions and hypotheses, the results reveal essential differences between restoring tidal flats and reducing nutrient loads – tidal flat restoration recovers a rich ecosystem that assimilates nutrients into biomass and raises the potential for effective nutrient utilization, whereas nutrient load reduction does not have this effect, although it may prevent mortality from hypoxia[17].

Robust and Healthy Balance of the Ecosystem – A Motivation for Focusing on Production by Lower and Quasi-Higher Trophic Level Production

The vision of a smooth transition from a lower trophic level to a higher trophic level is significant. Coastal areas where eutrophication is proceeding and hypoxia is a serious problem have attenuated links from lower to higher trophic levels. As a result, out of balance ecosystems such as those with red tides or other blooms of harmful phytoplankton may develop[18]. In contrast, by accelerating smooth transitions lower to higher trophic levels, a robust ecosystem balance and bountiful ocean with diverse and abundant living organisms may be achieved.

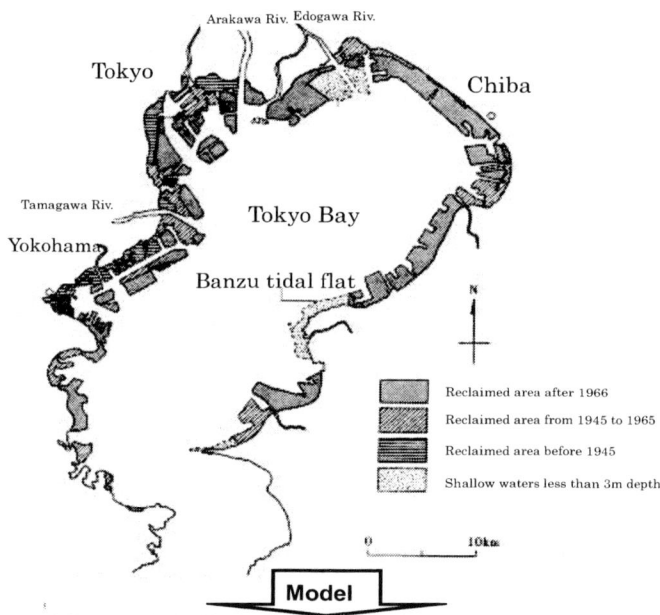

Figure 12. (Continued)

[17] The result demonstrated here does not mean that the reduction of nutrient loads from rivers is not necessary. Excess nutrients should not be discharged into the estuary or the ocean. However, if the bountiful ocean can be recovered by restoring tidal flats, the amount of nutrients incorporated into biomass by living organisms increases. As a result, the amount of nutrients considered to be excess decreases and the required amount of nutrient reduction also decreases.

[18] The discussion here is based on the concept that robust and healthy estuaries are in a state of well–balanced material cycling. Although it is difficult to define well–balanced or unbalanced material cycling directly, the overview explanation from the concept of stocks and flows is as follows. For example, carbon exists as various forms (i.e. "stocks") such as plankton, fish tissue, detritus, and CO_2 in the estuary. The form varies through biological, chemical, and physical processes (i.e. "flows"), the paths of carbon cycling. Red tide is the state in which most of the carbon and nutrients are taken up by phytoplankton, representing the unbalanced state of stocks of carbon, nutrients, and materials. Unbalanced stocks lead to unbalanced flows and vice versa.

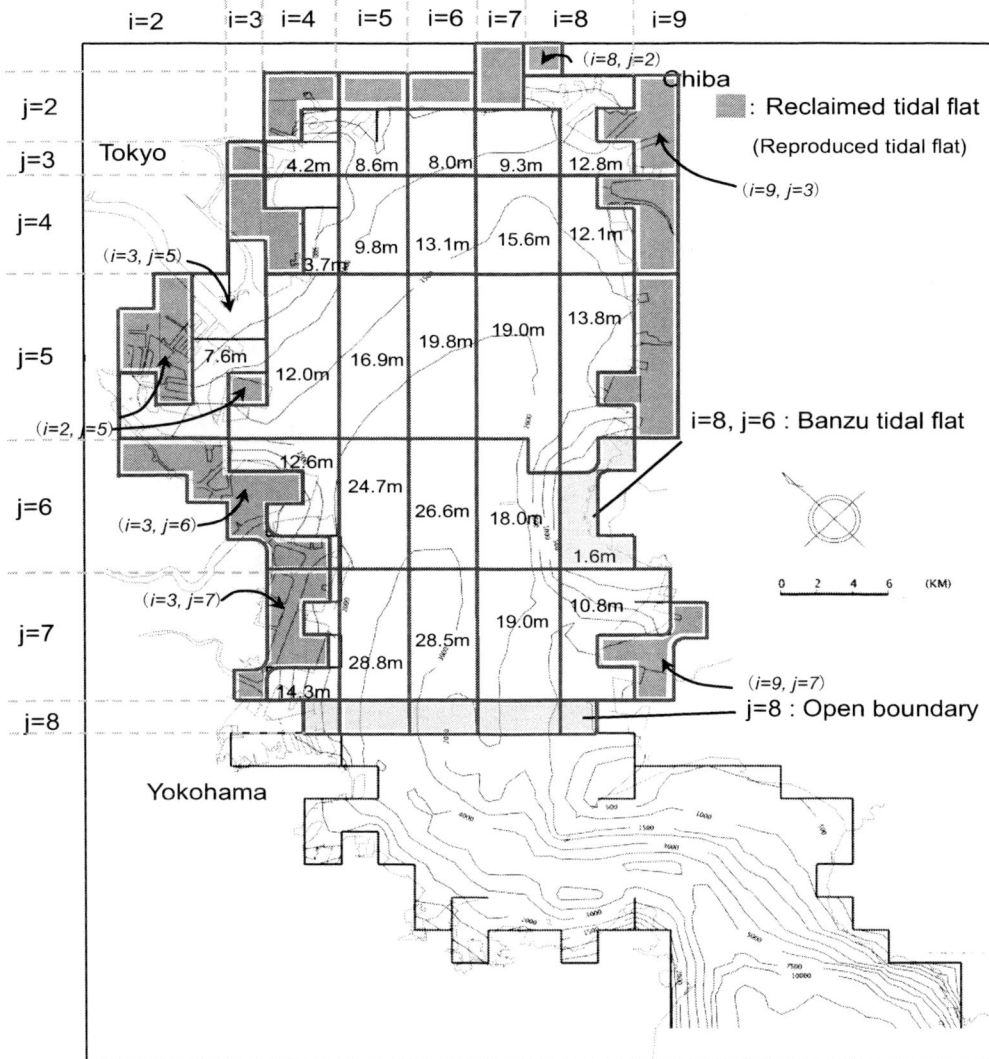

Reference: Koike (2000).

Figure 12. Upper figure; the early Tokyo Bay before reclamation (tidal flat reproduction system) and the existing Tokyo bay after reclamation (existing system/tidal flat disappearance system). Lower figure; the boxes/axis/depth set on the simulation using ECOHYM.

Based on the above background, by demonstrating the fluxes of lower and higher trophic productions quantitatively, (1) the smooth transition from a lower trophic level to a higher trophic level, (2) effective utilization of excess nutrients by higher level living organisms and (3) the recovery of a robust ecosystem balance derived from (1) and (2) were estimated. Lower trophic production was defined as the total flux of photosynthetic phytoplankton production and zooplankton production due to grazing (Figure 3). Fish, representing higher trophic levels, were not modeled in ECOHYM. Therefore, the transition from phytoplankton, zooplankton and detritus to benthic fauna, at a higher trophic level than plankton, was regarded as "quasi-higher trophic level production." Specifically, "quasi-higher trophic production" was defined as the total flux of feeding of benthic fauna (suspension and deposit feeders) in Figure 4. All required fluxes for estimating lower trophic level production and quasi-higher trophic level production are simulated by ECOHYM.

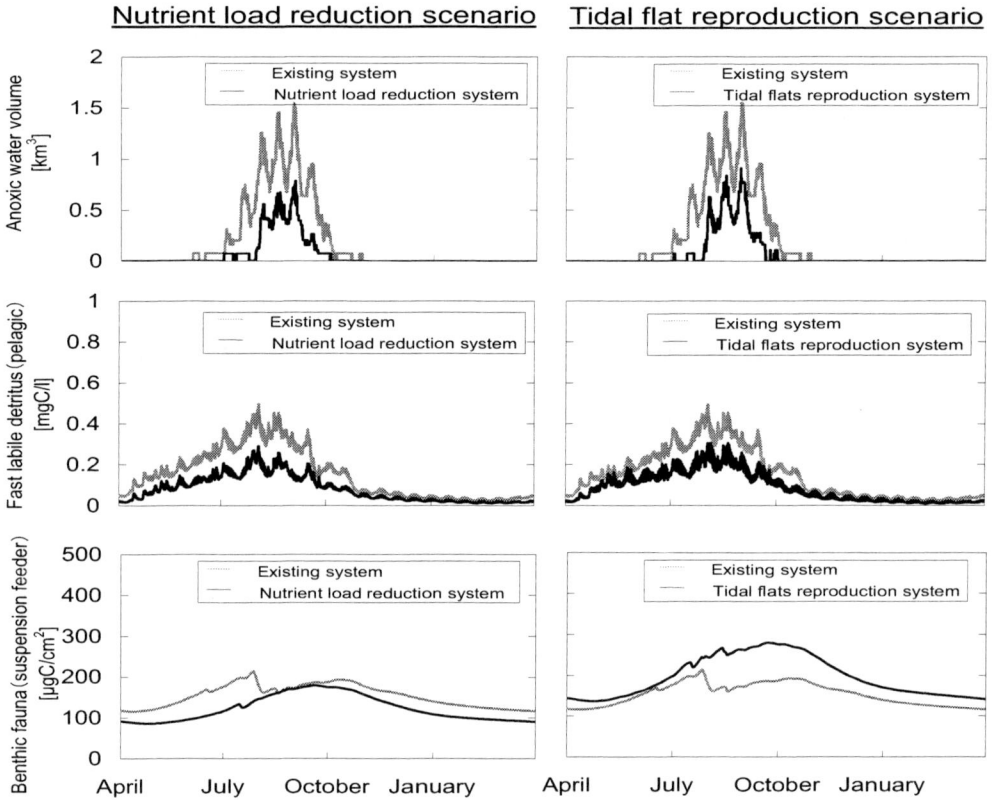

Figure 13. The seasonal variation of Tokyo Bay under the scenario of 50% nutrient load reduction from rivers and the scenario of reproduction (restoration) of early tidal flats reclaimed in the past, compared to the existing Tokyo Bay (existing system). (a) Anoxic water volume (DO < 2 mg O_2 L^{-1}) in Tokyo Bay, (b) concentration of detritus (fast labile detritus) in the pelagic system, and (c) biomass of benthic fauna (suspension feeders) in the benthic system. The values in figures are mean values of the existing Tokyo Bay areas.

It's Possible that the Environmental Recovery Spiral Could Occur

The differences of ecosystem response to the increase and decrease of nutrient load from rivers were estimated in terms of the following indices: (1) hypoxia potential, (2) lower trophic level production and (3) quasi-higher trophic level production, between the historical Tokyo Bay before reclamation of the tidal flats and the present Tokyo Bay after reclamation of most of the tidal flats (existing system with tidal flat disappearance). Figure 14 shows the dependences of these indices on increased and decreased nutrient loads. The values of the indices are annually averaged values integrated over all areas of the existing Tokyo Bay (the indices for the tidal flat reproduction system do not include the value of the reproduced tidal flat areas[19]) and are analyzed on the periodical seasonal steady state in the control case after reducing or increasing nutrient loads from rivers. As shown in Figure 14, hypoxia potential and lower trophic production are lower and the quasi-higher trophic potential is higher for tidal flat reproduction than for tidal flat disappearance at any load input. In addition, hypoxia potential and lower trophic production for tidal flat reproduction and disappearance both increase in accordance with load increase. However, increases of hypoxia potential and lower trophic production with tidal flat disappearance are greater than with tidal flat reproduction. Meanwhile, quasi-higher trophic production increases in accordance with nutrient load increase in the range from 0.5 to 1.5 times the current load. In the range > 1.5 times the current load, quasi-higher trophic production with tidal flat disappearance declines as nutrient load increases. The decline results from the mortality of benthic fauna due to oxygen depletion. These ecosystem responses demonstrated by ECOHYM suggest the possibility of significant effects derived from the tidal flat reproduction. The effects are (1) preventing hypoxia, (2) transferring nutrients from lower trophic to higher trophic levels, and (3) recovering a bountiful ocean with high biological production, i.e. driving the "environmental improvement spiral".

A Bountiful Ocean Has High Tolerance for the Imbalanced Ecosystem State

In order to investigate the differences between tidal flat reproduction and tidal flat disappearance scenarios on ecosystem tolerance to environmental stress, a red tide pulse (temporary high level of phytoplankton) was forcefully introduced to both scenarios. The responses of each system to the red tide were evaluated in terms of hypoxia potential, lower trophic production, and quasi-higher trophic production.

Firstly, a simulated red tide pulse of 3 mg C L^{-1} in the level of phytoplankton, starting from 0:00 on 15 August and continuing 24 hrs, was forced on the tidal flat reproduction and tidal flat disappearance scenarios, and the time series of ecosystem responses was investigated. The red tide pulse was introduced to the entire surface layer of Tokyo Bay (results in Figure 15). The longitudinal axis in Figure 15 represents the differences between (1) the case with the red tide impact and (2) the case without the red tide impact, i.e. (1) minus (2) in the level of hypoxia potential, lower trophic production, and quasi-higher trophic production. For hypoxia potential, the maximum peak of the tidal flat reproduction system is not higher than the tidal flat disappearance system, and increase of hypoxia potential of the tidal flat reproduction system is alleviated promptly compared to the tidal flat disappearance system. Lower trophic production reaches the maximum value immediately after the red tide impact in both scenarios. Then, 5–6 days later, the levels are alleviated to the same value as the case without the red tide impact (zero of longitudinal axis in Figure 15). For the maximum value of lower trophic production on 16 August, the tidal flat reproduction scenarios results in

---

[19] The area of Tokyo Bay's tidal flat reproduction system has expanded compared to the tidal flat disappearance system due to the additional areas of reproduced tidal flats (Figure 12). However, the results shown in Figures 13, 14, 15, 16 and 17 are estimated excluding the reproduced tidal flats area for the estimation of the tidal flat reproduction and disappearance systems to be compared over the same area of spatial integration. In the case of the estimation including the reproduced tidal flat areas, the differences between tidal flat reproduction system and tidal flat disappearance system in the three indexes are greater than the results shown in Figures 13, 14, 15, 16 and 17.

a higher value than the tidal flat disappearance scenario. Levels of quasi-higher trophic production increase in both scenarios, at the moment the red tide impact arises. However, 2–3 days later, in the case of tidal flat disappearance, the value of quasi-higher trophic production with red tide is lower than the case without the red tide impact (negative value of longitudinal axis in Figure 15). In contrast, for case of tidal flat reproduction, the value of quasi-higher trophic production with red tide returns immediately to the same value as without the red tide impact (zero value of longitudinal axis in Figure 15). The model results shown here represent that the tidal flat reproduction system absorbs red tide immediately, prevents the red tide from inducing hypoxia, and alleviates the decrease of quasi-higher trophic production resulting from red tide.

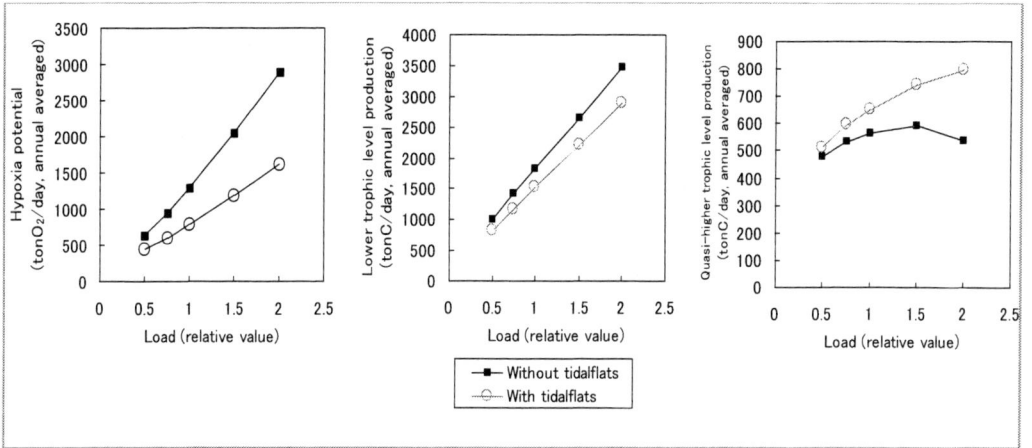

Figure 14. Ecological response of the early Tokyo Bay (tidal flats reproduction system; with tidal flats) and of the existing Tokyo Bay (without tidal flats) to the increase and decrease of nutrient load from rivers. The values in figures are integrated values of the existing Tokyo Bay areas.

Secondly, the dependence of ecosystem tolerance on the concentration of red tide was evaluated for tidal flat reproduction and tidal flat disappearance. Figure 16 shows hypoxia potential, lower trophic production and quasi-higher trophic production resulting from the ecosystem response to various concentrations of phytoplankton (simulated red tide). The red tide was set to start at 0:00 on 15 August and to continue for 24 hr over the entire surface layer of Tokyo Bay. The longitudinal axis in Figure 16 represents, as in Figure 15, the differences between (1) the case with red tide and (2) the case without red tide, i.e. (1) minus (2) in the level of hypoxia potential, lower trophic production, and quasi-higher trophic production. The result shown in Figure 16 represents that as the level in the concentration of red tide increases, hypoxia potential and lower trophic level production increase in both scenarios. Quasi-higher trophic level production with tidal flat disappearance decreases as the concentration of red tide increases. However, with tidal flat reproduction, quasi-higher trophic level production is greater (positive value in Figure 16) with < 4 mg C L^{-1} in the concentration of red tide. The activity becomes weaker than the case without red tide (negative value in Figure 16) when red tide concentration is > 4 mg C L^{-1}.

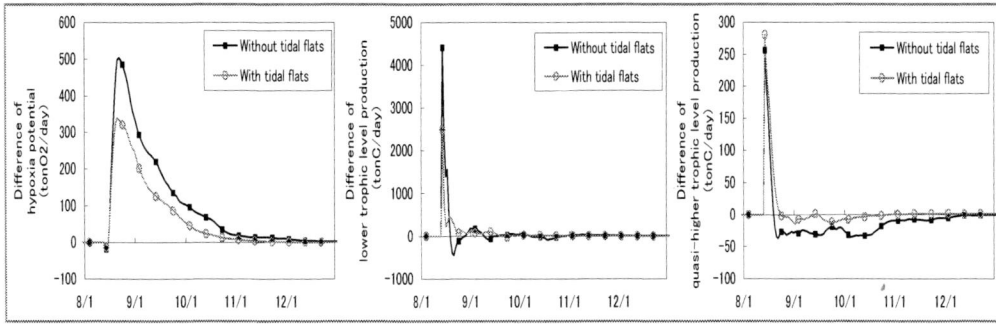

Figure 15. Ecological response of historical Tokyo Bay (tidal flats reproduction; with tidal flats) and of the existing Tokyo Bay (without tidal flats) to a red tide pulse (temporal dynamics after the red tide impact). The values in figures are integrated values of the existing Tokyo Bay areas.

Finally, the dependence of ecosystem tolerance on the duration of the simulated red tide was evaluated for both tidal flat scenarios. Figure 17 shows the results of ecosystem response to the red tide in terms of hypoxia potential, lower trophic production and quasi-higher trophic production. In this study, the forced red tide started at 0:00 on 15 August, with durations from one-half day to one week, extending over the entire surface layer of Tokyo Bay. In all cases, the level of red tide (phytoplankton) was fixed at 3 mg C L^{-1}. The longitudinal axis in Figure 17 represents, as in Figures 15 and 16, the differences from (1) the case with the red tide impact to (2) the case without the red tide impact, i.e. (1) minus (2). From the results of this analysis, hypoxia potential and lower trophic production increase as the duration of the red tide increases in both tidal flat scenarios. However, for tidal flat reproduction, the quasi-higher trophic level production is greater than the non-red tide situation with red tide duration up to 4 days, after which it decreases, with a greater rate of decrease in the tidal flat disappearance scenario. In addition, with red tide duration > 4 days, rates of increase for hypoxia potential and lower trophic production are higher in the tidal flat disappearance scenario than in the tidal flat reproduction scenario.

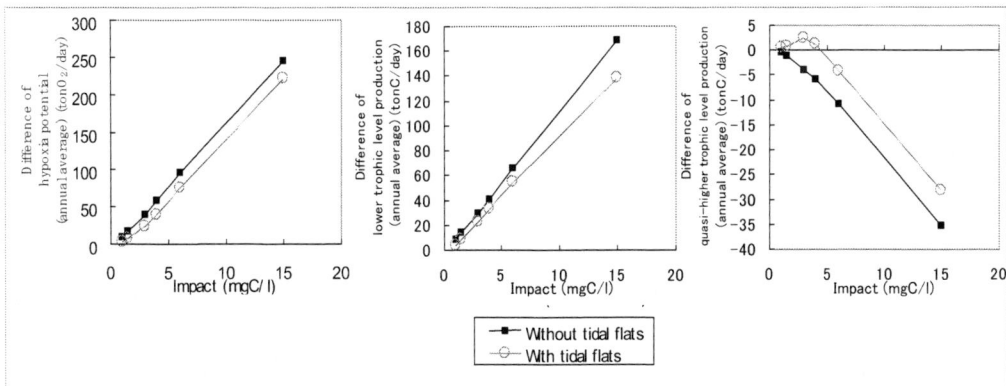

Figure 16. Ecological response of historical Tokyo Bay (tidal flats reproduction; with tidal flats) and of the existing Tokyo Bay (without tidal flats) to a red tide pulse (dependence on the concentration of phytoplankton). The values in the figure are integrated from temporally from 15 August to 1 April the following year, and spatially over the existing area of Tokyo Bay.

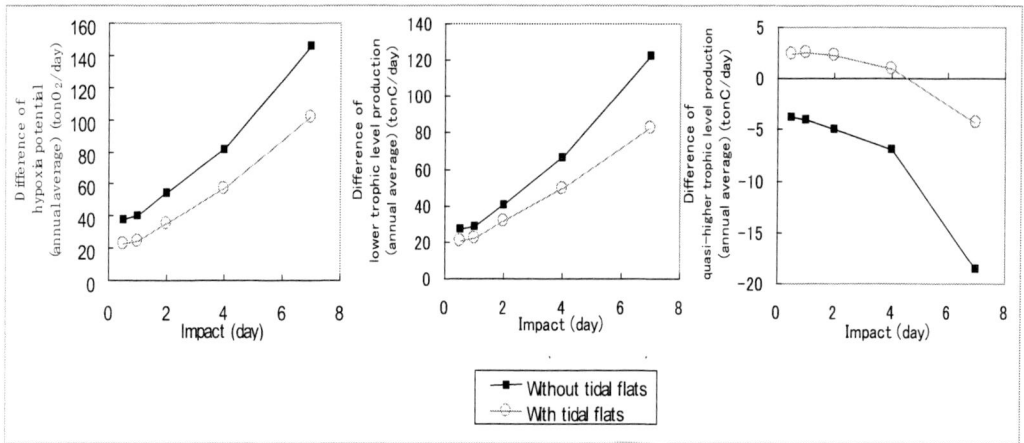

Figure 17. Ecological response of the early Tokyo Bay (tidal flats reproduction system; with tidal flats) and of the existing Tokyo Bay (without tidal flats) to the red tide pulse impact (dependence on the running days of red tide). The values in figures are temporal-spatial integrated values. Integrated period is from 15th August to 1st April the following year and the integrated area is the existing Tokyo Bay areas.

The results described above reveal the higher differences in tolerance to red tide from the tidal flat reproduction scenario to the tidal flat disappearance scenario. In summary, for tidal flat reproduction, the increase of hypoxia potential with red tide generation rapidly reduces after red tide annihilation and the quasi-higher trophic production does not decrease except in the case where a strong red tide impact occurs. Red tide of a few days duration has the potential to increase quasi-higher trophic level production rather than lower it. In contrast, for tidal flat disappearance, hypoxia potential does not reduce rapidly after red tide annihilation and it increases even if a weak red tide impact occurs. With red tide duration of more than several days, the ecosystem gets caught in an environmental deterioration spiral (negative spiral), in which hypoxia generation and lower trophic production are accelerated and quasi-higher trophic production is decreased.

CONCLUDING REMARKS – THE ECOSYSTEM MODEL AS A COMMUNICATION PLATFORM

With the development of ECOHYM, philosophical concepts such as "bountiful ocean," "robust and healthy balance of the ecosystem" and "environmental deterioration and improvement spirals" have been made tangible, while the technology of prediction, explanation, and evaluation of these concepts have been established in quantitative terms. However, the scenarios analyzed in this chapter are bold (i.e. restoration of all formerly reclaimed tidal flats, and 50% reduction of nutrient loads from rivers), so that they may be not realistic to actualize. Nevertheless, these bold scenarios and exciting hypotheses are meaningful to understanding and discussing the direction of the environmental restoration of the estuary. In addition, even if the feasible measures for tidal flat restoration and creation, or for nutrient load reduction from rivers are small in scale, ecosystem responses in the areas where measures are applied will, with high probability, follow the same trends as the ecosystem responses presented in this chapter. Detailed analysis of ecosystem responses to small scale measures is now proceeding by using an advanced version of ECOHYM with higher vertical spatial resolution. If ecosystem responses at small spatial scales are revealed to have the trends described above, from the results of this detailed analysis, we may postulate a more realistic method to manage eutrophic estuaries: (1) select a limited area to model in accordance with the desired ecosystem recovery or the intended use – from the perspectives not only of environmental conservation, but also the development of industry and national land/ocean use plans, (2) to determine a goal for each model area, and (3) to create a strategy for implementing the measures needed to achieve the goals.

Based on ECOHYM, expectations for future enhancements of ecosystem models and their applications should include the following:

Contributions to natural science
- understanding how autonomous ecosystem processes affect responses to environmental change (i.e. chain responses)
- revealing unknown biological and chemical processes
- predicting the phenomena overlooked by observation and monitoring
- designing efficient and effective strategies for observation and monitoring

Contributions to policy
- As environmental management tools
- estimating the costs and performance of environmental improvement measures (including the effects of autonomous ecological responses)
- predicting the effect of environmental development and improvement measures (including the effects of autonomous ecological responses)
- As communication tools for consensus-building
- illuminating the consensus-building process with quantitative explanations of how each ecological process contributes to the problems and solutions of interest

I expect that, with further development, ecosystem models such as ECOHYM and its progeny will become powerful tools for communication and collaboration among physical, biological, and chemical investigations of estuaries, as well as for policy-making.

APPENDIX. FORMULATION OF MAJOR BIOCHEMICAL PROCESSES

The formulations of biochemical processes in ECOHYM are shown in Tables A1– A6.

Table A1. Notation of model variables (pelagic system)

Notation and No. of Model Variables	Unit	Description
PP : [01]	mgC/l	Phytoplankton
ZP : [02]	mgC/l	Zooplankton
WDE_1, WDE_2, WDE_3 : [03,1], [03,2], [03,3]	mgC/l	Detritus (fast-labile, slow-labile, and refractory/very slow-labile detritus)
WDM_1, WDM_2 : [04,1], [04,2]	mgC/l	Dissolved organic matter (labile and refractory DOM)
WNX : [05]	mgN/l	Ammonium
WNY : [06]	mgN/l	Nitrate
WDP : [07]	mgP/l	Phosphate
WOU : [08]	mg/l	Oxygen demand unit (ODU)
WDO : [09]	mg/l	Dissolved oxygen

Table A2. Notation of model variables (benthic system)

Notation of Model Variables	Unit	Description
SFB : [51]	$\mu gC/cm^2$	Suspension feeders
DFB : [52]	$\mu gC/cm^2$	Deposit feeders
DET_1, DET_2, DET_3 : [53,1], [53,2], [53,3]	$\mu gC/cm^3$ solid	Detritus (fast-labile, slow-labile, and refractory/very slow-labile detritus)
DOM_1, DOM_2 : [54,1], [54,2]	$\mu gC/ml$	Dissolved organic matter (labile and refractory DOM)
HNX : [55]	$\mu gN/ml$	Ammonium
HNY : [56]	$\mu gN/ml$	Nitrate
DIP : [57]	$\mu gP/ml$	Phosphate
ODU : [58]	$\mu g/ml$	Oxygen demand unit (ODU)
DOO : [59]	$\mu g/ml$	Dissolved oxygen
BAL : [60]	$\mu gC/cm^2$	Benthic algae

Table A3. General variables and prescribed functions in the model

General Variables and Prescribed Functions	Unit	Description
z	cm	Water depth or sediment depth
Δz	cm	Thickness of layer
dt	h	Calculation time step
ϕ	-	Porosity
$\bar{\rho}_s$	g/cm^3	Density of sediment
TmpW, TmpB	°C	Temperature of sea water and sediment
I_0, I_B	$\mu E/m^2/s$	Light intensity on sea surface and sediment surface

Table A4. Ratio and distribution functions (prescribed function or calculated in the model)

Functions	Description
$R_{FOD51.1}$, $R_{FOD51.2}$, $R_{FOD51.3}$	Composition ratio (ratio of fast-labile, slow-labile and refractory/very slow-labile part) of prey of suspension feeders
$R_{ncFOD51}$, $R_{pcFOD51}$	N/C, P/C ratio of prey of suspension feeders
R_{Zfec51}, R_{Zmor51}	Vertical distribution of feces and mortality of suspension feeders
$R_{FOD52.1}$, $R_{FOD52.2}$, $R_{FOD52.3}$, $R_{FOD52.541}$, $R_{FOD52.542}$	Composition ratio (ratio of fast-labile, slow-labile and refractory/very slow-labile part) of prey of deposit feeders
$R_{ncFOD52}$, $R_{pcFOD52}$	N/C, P/C ratio of prey of deposit feeders
R_{Zfee52}, R_{Zfec52}, R_{Zexc52}, R_{Zmor52}	Vertical distribution of feeding, feces, excretion and mortality of deposit feeders
R_{Zpho60}, R_{Zres60}, R_{Zmor60}	Vertical distribution of photosynthesis, base respiration and mortality of benthic algae
$R_{FOD52.60}$	Ratio of benthic algae to prey of deposit feedeers

Table A5. Formulation of essential biochemical processes (pelagic system)

Biochemical Processes	Formulation [min(a, b) = a (a > b), or b (a < b) ; g(X, a_{half}) = X / (X + a_{half})]	Unit	Parameters [see Table 2, Table 3]
[Phytoplankton : 01]			
Biochemical net production / consumption	$\dot{C}_{pp} = Dpp_{pho} - Dpp_{ext} - Dpp_{res} - Dpp_{mor} - Dzp_{gra} - Dpp_{sff}$	mgC/l/h	
Photosynthesis	$Dpp_{pho} = v_{pho01} \cdot u_{pho01a} \cdot u_{pho01b} \cdot PP$	mgC/l/h	
Maximum growth rate	$v_{pho01} = \alpha_{pho01} \cdot \exp(\beta_{pho01} \cdot TmpW)$	1/h	α_{pho01}, β_{pho01}
Nutrient limitation	$u_{pho01a} = min[\ g(WNX+WNY, Hf_{n,pho01}), g(WDP, Hf_{p,pho01})\ ]$	-	$Hf_{n,pho01}$, $Hf_{p,pho01}$
Light availability	$u_{pho01b} = (I_0e^{-kz} - I_{min01}) / ((I_0e^{-kz} - I_{min01}) + (I_{hf01} - I_{min01}))$	-	I_{hf01}, I_{min01}
Light attenuation	k: Calculated in model depend on PP, ZP, WDE_i	1/cm	
Extra-release	$Dpp_{ext} = Dpp_{pho} \cdot 0.135 \cdot \exp(-0.00201 \cdot R_{chl} \cdot PP \cdot 10^3)$	mgC/l/h	R_{chl}
Respiration	$Dpp_{res} = \alpha_{res01} \cdot \exp(\beta_{res01} \cdot TmpW) \cdot PP$	mgC/l/h	α_{res01}, β_{res01}
Mortality	$Dpp_{mor} = \alpha_{mor01} \cdot \exp(\beta_{mor01} \cdot TmpW) \cdot PP$	mgC/l/h	α_{mor01}, β_{mor01}
Suspension feeder feeding (bottom layer only)	$Dpp_{sff} = \dfrac{PP}{PP + ZP + \sum\limits_{i=1}^{3} WDE_i} \cdot \dfrac{1}{A_{Depq}} \cdot Dsfb_{fee}$	mgC/l/h	A_{Depq}
[Zooplankton : 02]			
Biochemical net production / consumption	$\dot{C}_{zp} = Dzp_{gra} - Dzp_{fec} - Dzp_{exc} - Dzp_{mor} - Dzp_{sff}$	mgC/l/h	
Grazing	$Dzp_{gra} = \alpha_{gra02} \cdot \exp(\beta_{gra02} \cdot TmpW) \cdot (1 - \exp(A_{ivl02} \cdot (A_{kai02} - PP))) \cdot ZP$	mgC/l/h	α_{gra02}, β_{gra02}, A_{ivl02}, A_{kai0}
Feces	$Dzp_{fec} = (1 - R_{ege02}) \cdot Dzp_{gra}$	mgC/l/h	R_{ege02}
Excretion	$Dzp_{exc} = (R_{ege02} - R_{grt02}) \cdot Dzp_{gra}$	mgC/l/h	R_{ege02}, R_{grt02}
Mortality	$Dzp_{mor} = \alpha_{mor02} \cdot \exp(\beta_{mor02} \cdot TmpW) \cdot ZP$	mgC/l/h	α_{mor2}, β_{mor02}
Suspension feeder feeding (bottom layer only)	$Dzp_{sff} = \dfrac{ZP}{PP + ZP + \sum\limits_{i=1}^{3} WDE_i} \cdot \dfrac{1}{A_{Depq}} \cdot Dsfb_{fee}$	mgC/l/h	A_{Depq}
[Detritus : 03,i] *[i = 1 (fast-labile), 2 (slow-labile), 3 (refractory / very slow-labile)]*			
Biochemical net production / consumption	$\dot{C}_{wde,i} = R_{PP,i} \cdot Dpp_{mor} + R_{PP,i} \cdot Dzp_{fec} + R_{ZP,i} \cdot Dzp_{mor}$ $- Dwde_{omi,i} - Dwde_{smi,i} - Dwde_{ami,i} - Dwde_{dec,i}$ $- Dwde_{sff,i} + Dwde_{sfc,i}$	mgC/l/h	$R_{PP,i}$, $R_{ZP,i}$ [$i = 1 - 3$]
Oxic mineralization	$Dwde_{omi,i} = m_{03,i} \cdot g(WDO, Hf_{o2w,omi}) \cdot WDE_i / G$	mgC/l/h	$Hf_{o2w,omi}$
Suboxic mineralization	$Dwde_{smi,i} = m_{03,i} \cdot g(WNY, Hf_{no3w,smi})$ $\cdot (1 - g(WDO, Hf_{o2w,smi})) \cdot WDE_i / G$	mgC/l/h	$Hf_{o2w,smi}$, $Hf_{no3w,smi}$
Anoxic mineralization	$Dwde_{ami,i} = m_{03,i} \cdot (1 - g(WNY, Hf_{no3w,ami}))$ $\cdot (1 - g(WDO, Hf_{o2w,ami})) \cdot WDE_i / G$	mgC/l/h	$Hf_{o2w,ami}$, $Hf_{no3w,ami}$
Mineralization rate	$m_{03,i} = \alpha_{mi03,i} \cdot \exp(\beta_{mi03,i} \cdot TmpW)$ $G = g(WDO, Hf_{o2w,omi})$ $+ g(WNY, Hf_{no3w,smi}) \cdot (1 - g(WDO, Hf_{o2w,smi}))$ $+ (1 - g(WNY, Hf_{no3w,ami})) \cdot (1 - g(WDO, Hf_{o2w,ami}))$	1/h - 	$\alpha_{mi03,i}$, $\beta_{mi03,i}$ [$i = 1 - 3$] $Hf_{o2w,omi}$, $Hf_{o2w,smi}$, $Hf_{no3w,smi}$ $Hf_{o2w,ami}$, $Hf_{no3w,ami}$
Decomposition	$Dwde_{dec,i} = R_{dec03,i} \cdot (Dwde_{omi,i} + Dwde_{smi,i} + Dwde_{ami,i})$	mgC/l/h	$R_{dec03,i}$ [$i = 1 - 3$]
Suspension feeder feeding (bottom layer only)	$Dwde_{sff,i} = \dfrac{WDE_i}{PP + ZP + \sum\limits_{i=1}^{3} WDE_i} \cdot \dfrac{1}{A_{Depq}} \cdot Dsfb_{fee}$	mgC/l/h	A_{Depq}
Suspension feeder feces (bottom layer only)	$Dwde_{sfc,i} = R_{ZWfec51} \cdot R_{FOD51,i} \cdot \dfrac{1}{\Delta z} \cdot Dsfb_{fec}$	mgC/l/h	$R_{ZWfec51}$
[Dissolved organic matter (DOM) : 04,j] *[j = 1 (labile), 2 (refractory)]*			
Biochemical net production / consumption	$\dot{C}_{wdm,j} = R_{ext01,j} \cdot Dpp_{ext} + \sum\limits_{i=1}^{3} R_{DOM,ji} \cdot Dwde_{dec,i}$ $- Dwdm_{omi,j} - Dwdm_{smi,j} - Dwdm_{ami,j}$	mgC/l/h	$R_{ext01,j}$, $R_{DOM,ji}$ [$i = 1 - 3, j = 1 - 2$]
Oxic mineralization	$Dwdm_{omi,j} = m_{04,j} \cdot g(WDO, Hf_{o2w,omi}) \cdot WDM_j / G$	mgC/l/h	$Hf_{o2w,omi}$
Suboxic mineralization	$Dwdm_{smi,j} = m_{04,j} \cdot g(WNY, Hf_{no3w,smi})$ $\cdot (1 - g(WDO, Hf_{o2w,smi})) \cdot WDM_j / G$	mgC/l/h	$Hf_{o2w,smi}$, $Hf_{no3w,smi}$
Anoxic mineralization	$Dwdm_{ami,j} = m_{04,j} \cdot (1 - g(WNY, Hf_{no3w,ami}))$ $\cdot (1 - g(WDO, Hf_{o2w,ami})) \cdot WDM_j / G$	mgC/l/h	$Hf_{o2w,ami}$, $Hf_{no3w,ami}$
Mineralization rate	$m_{04,j} = \alpha_{mi04,j} \cdot \exp(\beta_{mi04,j} \cdot TmpW)$	1/h	$\alpha_{mi04,j}$, $\beta_{mi04,j}$ [$j = 1 - 2$]

Table A5. (Continued)

$[NH_4\text{-}N : 05]$

Biochemical net production / consumption

$$\dot{C}_{wnx} = R_{nc01} \cdot Dpp_{res} + R_{nc01} \cdot Dzp_{exc} + Dwnx_{red}$$

$$+ \sum_{i=1}^{3} R_{nc03,i} \cdot (Dwde_{omi} + Dwde_{ami} + Dwde_{ami})$$

$$+ \sum_{j=1}^{2} R_{nc04,j} \cdot (Dwdm_{omj} + Dwdm_{ami} + Dwdm_{ami})$$

$$- R_{nc01} \cdot Dpp_{pho} \cdot \frac{WNX}{WNX+WNY} - Dwnx_{nit}$$

$$+ R_{ncFOD51} \cdot \frac{1}{\Delta z} \cdot Dsfb_{exc} \qquad \text{(bottom layer only)}$$

mgN/l/h $R_{nc01}, R_{nc03,i}, R_{nc04,j}$ $[i = 1 - 3, j = 1 - 2]$

Nitrification $Dwnx_{nit} = \alpha_{nit05} \cdot \exp(\beta_{nit05} \cdot TmpW) \cdot g(WDO, Hf_{o2,nit05}) \cdot WNX$ mgN/l/h $\alpha_{nit05}, \beta_{nit05}, Hf_{o2,nit05}$

Nitrate reduction
$$Dwnx_{red} = (14/12) \cdot (4/(8-3 \cdot R_{den06})) \cdot (1 - R_{den06})$$
$$\cdot \left(\sum_{i=1}^{3} Dwde_{ami} + \sum_{j=1}^{2} Dwdm_{ami} \right)$$
mgN/l/h R_{den06}

$[NO_3\text{-}N : 06]$

Biochemical net production / consumption

$$\dot{C}_{wny} = Dwnx_{nit} - Dwny_{den} - Dwnx_{red}$$

$$- R_{nc01} \cdot Dpp_{pho} \cdot \frac{WNY}{WNX+WNY}$$

mgN/l/h R_{nc01}

De-nitrification
$$Dwny_{den} = (14/12) \cdot (4/(8-3 \cdot R_{den06})) \cdot R_{den06}$$
$$\cdot \left(\sum_{i=1}^{3} Dwde_{ami} + \sum_{j=1}^{2} Dwdm_{ami} \right)$$
mgN/l/h R_{den06}

$[PO_4\text{-}P : 07]$

Biochemical net production / consumption

$$\dot{C}_{wdp} = R_{pc01} \cdot Dpp_{res} + R_{pc01} \cdot Dzp_{exc}$$

$$+ \sum_{i=1}^{3} R_{pc03,i} \cdot (Dwde_{omi} + Dwde_{ami} + Dwde_{ami})$$

$$+ \sum_{j=1}^{2} R_{pc04,j} \cdot (Dwdm_{omj} + Dwdm_{ami} + Dwdm_{ami})$$

$$- R_{pc01} \cdot Dpp_{pho}$$

$$+ R_{pcFOD51} \cdot \frac{1}{\Delta z} \cdot Dsfb_{exc} \qquad \text{(bottom layer only)}$$

mgP/l/h $R_{pc01}, R_{pc03,i}, R_{pc04,j}$ $[i = 1 - 3, j = 1 - 2]$

$[ODU : 08]$

Biochemical net production / consumption

$$\dot{C}_{wou} = (32/12) \cdot \left(\sum_{i=1}^{3} Dwde_{ami} + \sum_{j=1}^{2} Dwdm_{ami} \right)$$

$$- Dwou_{aut} - Dwou_{oxi}$$

mg/l/h

Oxidation
$$Dwou_{oxi} = \alpha_{oxi08} \cdot \exp(\beta_{oxi08} \cdot TmpW) \cdot g(WDO, Hf_{o2,oxi08}) \cdot WOU$$
$$+ R_{oxi08} \cdot (32/12) \cdot \left(\sum_{i=1}^{3} Dwde_{ami} + \sum_{j=1}^{2} Dwdm_{ami} \right)$$
mg/l/h $\alpha_{oxi08}, \beta_{oxi08}, Hf_{o2,oxi08}, R_{oxi08}$

Authigenic mineralization
$$Dwou_{aut} = \alpha_{aut08} \cdot \exp(\beta_{aut08} \cdot TmpW) \cdot WOU + R_{aut08a} \cdot Dwou_{oxi}$$
$$+ R_{aut08b} \cdot (32/12) \cdot \left(\sum_{i=1}^{3} Dwde_{ami} + \sum_{j=1}^{2} Dwdm_{ami} \right)$$
mg/l/h $\alpha_{aut08}, \beta_{aut08}, R_{aut08a}, R_{aut08b}$

$[Dissolved\ oxygen\ (DO) : 09]$

Biochemical net production / consumption

$$\dot{C}_{wdo} = \frac{32}{12} \cdot Dpp_{pho} - \frac{32}{12} \cdot Dpp_{res} - \frac{32}{12} \cdot Dzp_{exc}$$

$$- \frac{32}{12} \cdot \left(\sum_{i=1}^{3} Dwde_{omi} + \sum_{j=1}^{2} Dwdm_{omj} \right)$$

$$- 2 \cdot \frac{32}{14} \cdot Dwnx_{nit} - Dwou_{oxi}$$

$$- \frac{32}{12} \cdot \frac{1}{\Delta z} \cdot (Dsfb_{exc} + Ddfb_{exc}) \qquad \text{(bottom layer only)}$$

mgO$_2$/l/h

Table A6. Formulation of essential biochemical processes (benthic system)

Biochemical Processes	Formulation $[\min(a, b) = a\ (a > b),$ or $b\ (a < b)\ ;\ g(X, a_{half}) = X / (X + a_{half})\ ]$	Unit	Parameters [see Table 2, Table 3]
[Suspension feeders : 51]			
Biochemical net production / consumption	$\dot{C}_{sfb} = Dsfb_{fee} - Dsfb_{fec} - Dsfb_{exc} - Dsfb_{mor} + Dsfb_{lar}$	$\mu gC/\,cm^2/h$	
Feeding	$Dsfb_{fee} = \min(Lim_{filter}, Lim_{growth})$	$\mu gC/cm^2/h$	
Filter rate limitation	$Lim_{filter} = v_{fee51} \cdot u_{fee51} \cdot (PP + ZP + WDE_1 + WDE_2 + WDE_3)$ $\cdot R_{cor0} \cdot SFB$	$\mu gC/cm^2/h$	
Filter rate	$v_{fee51} = 1.2 \times 10^{-5} \cdot TmpB^{1.25} \cdot A_{Wwet51} {}^{-0.75}/A_{wd51}/A_{cd51}$ $(TmpB > 10)$ $1.2 \times 10^{-5} \cdot 10^{1.25} \cdot A_{Wwet51} {}^{-0.75}/ A_{wd51}/ A_{cd51}$ $(TmpB < 10)$	$ml/h/\mu gC$	$A_{Wwet51}, A_{wd51}, A_{cd51}$
Oxygen saturation limitation	$u_{fee51} = \min(1, do_{sat}/R_{O2mor51})$ do_{sat}: Oxygen saturation of bottom water (calculated)	-	$R_{O2mor51}$
Decreasing by double filtering	$R_{cor0} = (1 - \exp(-COR0)) / COR0$	-	
Double filtering ratio	$COR0 = v_{fee51} \cdot u_{fee51} \cdot SFB \cdot dt / A_{Depq}$	-	A_{Depq}
Growth rate limitation	$Lim_{growth} = (\alpha_{grt51} + \alpha_{bas51}) / (R_{ege51} \cdot (1 - R_{exc51})) \cdot SFB \cdot F_{temp}$ F_{temp}: function of temperature	$\mu gC/cm^2/h$	$\alpha_{grt51}, \alpha_{bas51},$ R_{ege51}, R_{exc51}
Feces	$Dsfb_{fec} = (1 - R_{ege51}) \cdot Dsfb_{fee}$	$\mu gC/cm^2/h$	R_{ege51}
Excretion	$Dsfb_{exc} = R_{exc51} \cdot R_{ege51} \cdot Dsfb_{fee} + \alpha_{bas51} \cdot SFB$	$\mu gC/cm^2/h$	$R_{exc51}, R_{ege51}, \alpha_{bas51}$
Mortality	$Dsfb_{mor} = u_{mor51} \cdot \exp(\beta_{mor51} \cdot TmpB) \cdot SFB$	$\mu gC/cm^2/h$	β_{mor51}
Rate of mortality	$u_{mor51} = \alpha_{mor51a} + \alpha_{mor51b} \cdot (1 - u_{fee51})$	$\mu gC/cm^2/h$	$\alpha_{mor51a}, \alpha_{mor51b}$
Larva input	$Dsfb_{lar} = R_{lar51} \cdot Dsfb_{fee,\,av}$ $Dsfb_{fee,\,av}$: spatial and temporal average of feeding (calculated)	$\mu gC/cm^2/h$	R_{lar51}
[Deposit feeders : 52]			
Biochemical net production / consumption	$\dot{C}_{dfb} = Ddfb_{fee} - Ddfb_{fec} - Ddfb_{exc} - Ddfb_{mor} + Ddfb_{lar}$	$\mu gC/cm^2/h$	
Feeding	$Ddfb_{fee} = \alpha_{fee52} \cdot \exp(\beta_{fee52} \cdot TmpB) \cdot u_{fee52a} \cdot u_{fee52b} \cdot u_{fee52c} \cdot DFB$	$\mu gC/cm^2/h$	$\alpha_{fee52}, \beta_{fee52}$
Food limitation	$u_{fee52a} = 1 - \exp(A_{ivl52} \cdot \min(0, A_{kai52} - Food_{52}))$ $Food_{52}$: average concentration of food in mud. (calculated)	-	A_{ivl52}, A_{kai52}
Cannibalism efficiency	$u_{fee52b} = g(DFB, Hf_{dfb,fee52})$	-	$Hf_{dfb,fee52}$
Oxygen saturation limitation	$u_{fee52c} = \min(1, do_{sat} / R_{O2mor52})$	-	$R_{O2mor52}$
Feces	$Ddfb_{fec} = (1 - u_{fee52}) \cdot Ddfb_{fee}$	$\mu gC/cm^2/h$	
Assimilation efficiency	$u_{fee52} = 1 - R_{undg52} \cdot (1 + g(Food_{52}, Hf_{fod52,fee52}))$	-	$R_{undg52}, Hf_{fod52,fee52}$
Excretion	$Ddfb_{exc} = R_{exc52} \cdot u_{fee52} \cdot Ddfb_{fee}$	$\mu gC/cm^2/h$	R_{exc52}
Mortality	$Ddfb_{mor} = v_{mor52} \cdot u_{mor52} \cdot DFB$	$\mu gC/cm^2/h$	
Temperature dependency	$v_{mor52} = \min(\exp(\beta_{mor52} \cdot TmpB), \exp(\beta_{mor52} \cdot A_{temp,fee52}))$	-	$A_{temp,fee52}, \beta_{mor52}$
Rate of mortality	$u_{mor52} = \alpha_{mor52a} + \alpha_{mor52b} \cdot (1 - u_{fee52c})$	$1/h$	$\alpha_{mor52a}, \alpha_{mor52b}$
Larva input	$Ddfb_{lar} = R_{lar52} \cdot Ddfb_{fee,\,av}$ $Ddfb_{fee,\,av}$: spatial and temporal average of feeding (calculated)	$\mu gC/cm^2/h$	R_{lar52}
[Detritus : 53,i] *[i = 1 (fast-labile), 2 (slow-labile), 3 (refractory/very slow-labile)]*			
Biochemical net production / consumption	$\dot{C}_{det,i} = R_{Zfec51} \cdot R_{FODS1,i} \cdot Dsfb_{fec} + R_{Zmor51} \cdot R_{SFB,\,i} \cdot Dsfb_{mor}$ $+ R_{Zfec52} \cdot R_{FOD52,i} \cdot Ddfb_{fec} + R_{Zmor52} \cdot R_{DFB,i} \cdot Ddfb_{mor}$ $- R_{Zfee52} \cdot R_{FOD52,i} \cdot Ddfb_{fee} + R_{Zmor60} \cdot R_{BAL,i} \cdot Dbal_{mor}$ $- Ddet_{omi,i} - Ddet_{smi,i} - Ddet_{ami,i} - Ddet_{dec,i}$	$\mu gC/cm^3/h$	$R_{SFB,i}, R_{DFB,i}, R_{BAL,i},$ $[i = 1 - 3]$
Oxic mineralization	$Ddet_{omi,i} = m_{53,i} \cdot g(DOO, Hf_{o2b,omi}) \cdot DET_i / G \cdot (1 - \phi)$	$\mu gC/cm^3/h$	$Hf_{o2b,omi}$
Suboxic mineralization	$Ddet_{smi,i} = m_{53,i} \cdot g(HNY, Hf_{no3b,smi})$ $\cdot (1 - g(DOO, Hf_{o2b,smi})) \cdot DET_i / G \cdot (1 - \phi)$	$\mu gC/cm^3/h$	$Hf_{o2b,smi}, Hf_{no3b,smi}$
Anoxic mineralization	$Ddet_{ami,i} = m_{53,i} \cdot (1 - g(HNY, Hf_{no3b,ami}))$ $\cdot (1 - g(DOO, Hf_{o2b,ami})) \cdot DET_i / G \cdot (1 - \phi)$	$\mu gC/cm^3/h$	$Hf_{o2b,ami}, Hf_{no3b,ami}$
Mineralization rate	$m_{53,i} = \alpha_{mi53,i} \cdot \exp(\beta_{mi53,i} \cdot TmpB)$ $G = g(DOO, Hf_{o2b,omi})$ $+ g(HNY, Hf_{no3b,smi}) \cdot (1 - g(DOO, Hf_{o2b,smi}))$ $+ (1 - g(HNY, Hf_{no3b,ami})) \cdot (1 - g(DOO, Hf_{o2b,ami}))$	$1/h$ -	$\alpha_{mi53,i}, \beta_{mi53,i}\ [i = 1 - 3]$ $Hf_{o2b,omi},$ $Hf_{o2b,smi}, Hf_{no3b,smi}$ $Hf_{o2b,ami}, Hf_{no3b,ami}$
Decomposition	$Ddet_{dec,i} = R_{dec53,i} \cdot (Ddet_{omi,i} + Ddet_{smi,i} + Ddet_{ami,i})$	$\mu gC/cm^3/h$	$R_{dec53,i}\ [i = 1 - 3]$
[Dissolved organic matter (DOM): 54,j] *[j = 1 (labile), 2 (refractory)]*			
Biochemical net production / consumption	$\dot{C}_{dom,j} = R_{ext60,j} \cdot Dbal_{ext} + \sum\limits_{i=1}^{3} R_{DOM,ji} \cdot Ddet_{dec,i}$ $- R_{Zfec52} \cdot R_{FOD52,54j} \cdot Ddfb_{fee}$ $+ R_{Zfec52} \cdot R_{FOD52,54j} \cdot Ddfb_{fec}$ $- Ddom_{omi,j} - Ddom_{smi,j} - Ddom_{ami,j}$	$\mu gC/cm^3/h$	$R_{ext60,j}, R_{DOM,ji}$ $[i = 1 - 3, j = 1 - 2]$
Oxic mineralization	$Ddom_{omi,j} = m_{54,j} \cdot g(DOO, Hf_{o2b,omi}) \cdot DOM_j / G \cdot (\phi + \overline{\rho}_s \cdot K_{ads54j} \cdot (1 - \phi))$	$\mu gC/cm^3/h$	$Hf_{o2b,omi},$ $K_{ads54j}\ [j = 1 - 2]$
Suboxic mineralization	$Ddom_{smi,j} = m_{54,j} \cdot g(HNY, Hf_{no3b,smi})$ $\cdot (1 - g(DOO, Hf_{o2b,smi})) \cdot DOM_j / G \cdot (\phi + \overline{\rho}_s \cdot K_{ads54j} \cdot (1 - \phi))$	$\mu gC/cm^3/h$	$Hf_{o2b,smi}, Hf_{no3b,smi},$ $K_{ads54j}\ [j = 1 - 2]$
Anoxic mineralization	$Ddom_{ami,j} = m_{54,j} \cdot (1 - g(HNY, Hf_{no3b,ami}))$ $\cdot (1 - g(DOO, Hf_{o2b,ami})) \cdot DOM_j / G \cdot (\phi + \overline{\rho}_s \cdot K_{ads54j} \cdot (1 - \phi))$	$\mu gC/cm^3/h$	$Hf_{o2b,ami}, Hf_{no3b,ami},$ $K_{ads54j}\ [j = 1 - 2]$
Mineralization rate	$m_{54,j} = \alpha_{mi54,j} \cdot \exp(\beta_{mi54,j} \cdot TmpB)$	$1/h$	$\alpha_{mi54,j}, \beta_{mi54,j}\ [j = 1 - 2]$

Table A.6. (Continued)

[NH₄-N : 55] — *[NH$_4$-N : 55]*

Biochemical net production / consumption

$$\dot{C}_{hnx} = R_{nc60} \cdot Dbal_{res} + Dhnx_{red}$$
$$+ \sum_{i=1}^{3} R_{nc03,i} \cdot (Ddet_{om,i} + Ddet_{am,i} + Ddet_{ni})$$
$$+ \sum_{j=1}^{2} R_{nc04,j} \cdot (Ddom_{om,j} + Ddom_{am,j} + Ddom_{am,j})$$
$$- R_{nc60} \cdot Dbal_{pho} \cdot \frac{R_{pho60,55} \cdot HNX}{R_{pho60,55} \cdot HNX + HNY} - Dhnx_{nit}$$
$$+ R_{Zexc52} \cdot R_{ncFOD52} \cdot Ddfb_{exc}$$

μgN/cm³/h $R_{nc60}, R_{nc03,i}, R_{nc04,j},$ $R_{pho60,55}$ $[i = 1 - 3, j = 1 - 2]$

Nitrification

$$Dhnx_{nit} = \alpha_{nit55} \cdot exp(\beta_{nit55} \cdot TmpB) \cdot g(DOO, Hf_{o2,nit55}) \cdot HNX$$
$$\cdot (\phi + \overline{\rho}_s \cdot K_{ads55} \cdot (1 - \phi))$$

μgN/cm³/h $\alpha_{nit55}, \quad \beta_{nit55}, \quad Hf_{o2,nit55},$ K_{ads55}

Nitrate reduction

$$Dhnx_{red} = (14/12) \cdot (4/(8 - 3 \cdot R_{den56})) \cdot (1 - R_{den56})$$
$$\cdot \left(\sum_{i=1}^{3} Ddet_{am,i} + \sum_{j=1}^{2} Ddom_{am,j} \right)$$

μgN/cm³/h R_{den56}

[NO₃-N : 56] — *[NO$_3$-N : 56]*

Biochemical net production / consumption

$$\dot{C}_{hny} = Dhnx_{nit} - Dhny_{den} - Dhnx_{red}$$
$$- R_{nc60} \cdot Dbal_{pho} \cdot \frac{HNY}{R_{pho60,55} \cdot HNX + HNY}$$

μgN/cm³/h $R_{nc60}, R_{pho60,55}$

De-nitrification

$$Dhny_{den} = (14/12) \cdot (4/(8 - 3 \cdot R_{den56})) \cdot R_{den56} \cdot \left(\sum_{i=1}^{3} Ddet_{am,i} + \sum_{j=1}^{2} Ddom_{am,j} \right)$$

μgN/cm³/h R_{den56}

[PO₄-P : 57] — *[PO$_4$-P : 57]*

Biochemical net production / consumption

$$\dot{C}_{dip} = R_{pc60} \cdot Dbal_{res}$$
$$+ \sum_{i=1}^{3} R_{pc03,i} \cdot (Ddet_{om,i} + Ddet_{am,i} + Ddet_{ni})$$
$$+ \sum_{j=1}^{2} R_{pc04,j} \cdot (Ddom_{om,j} + Ddom_{am,j} + Ddom_{am,j})$$
$$- R_{pc60} \cdot Dbal_{pho} + R_{Zexc52} \cdot R_{pcFOD52} \cdot Ddfb_{exc}$$

μgP/cm³/h $R_{pc60}, R_{pc03,i}, R_{pc04,j}$ $[i = 1 - 3, j = 1 - 2]$

[ODU : 58]

Biochemical net production / consumption

$$\dot{C}_{odu} = (32/12) \cdot \left(\sum_{i=1}^{3} Ddet_{am,i} + \sum_{j=1}^{2} Ddom_{am,j} \right) - Dodu_{aut} - Dodu_{oxi}$$

μg/cm³/h

Oxidation

$$Dodu_{oxi} = \alpha_{oxi58} \cdot exp(\beta_{oxi58} \cdot TmpB) \cdot g(DOO, Hf_{o2,oxi58}) \cdot ODU \cdot \phi$$
$$+ R_{oxi58} \cdot (32/12) \cdot \left(\sum_{i=1}^{3} Ddet_{am,i} + \sum_{j=1}^{2} Ddom_{am,j} \right)$$

μg/cm³/h $\alpha_{oxi58}, \beta_{oxi58}, Hf_{o2,oxi58}$ R_{oxi58}

Authigenic mineralization

$$Dodu_{aut} = \alpha_{aut58} \cdot exp(\beta_{aut58} \cdot TmpB) \cdot ODU \cdot \phi + R_{aut58a} \cdot Dodu_{oxi}$$
$$+ R_{aut58b} \cdot (32/12) \cdot \left(\sum_{i=1}^{3} Ddet_{am,i} + \sum_{j=1}^{2} Ddom_{am,j} \right)$$

μg/cm³/h $\alpha_{aut58}, \beta_{aut58},$ R_{aut58a}, R_{aut58b}

[Dissolved oxygen (DO) : 59]

Biochemical net production / consumption

$$\dot{C}_{doo} = \frac{32}{12} \cdot Dbal_{pho} - \frac{32}{12} \cdot Dbal_{res}$$
$$- \frac{32}{12} \cdot \left(\sum_{i=1}^{3} Ddet_{om,i} + \sum_{j=1}^{2} Ddom_{om,j} \right)$$
$$- 2 \cdot \frac{32}{14} \cdot Dhnx_{nit} - Dodu_{oxi}$$

μg/cm³/h

[Benthic algae : 60]

Biochemical net production / consumption

$$\dot{C}_{bal} = \int (Dbal_{pho} - Dbal_{ext} - Dbal_{res} - Dbal_{mor}) dz$$
$$- R_{FOD52,60} \cdot Ddfb_{fee}$$

μgC/cm²/h

Photosynthesis

$$Dbal_{pho} = v_{pho60} \cdot u_{pho60a} \cdot u_{pho60b} \cdot BAL \cdot R_{Zpho60}$$

μgC/cm³/h

Maximum growth rate

$$v_{pho60} = \alpha_{pho60} \cdot exp(\beta_{pho60} \cdot TmpB)$$

1/h $\alpha_{pho60}, \beta_{pho60}$

Nutrient limitation

$$u_{pho60a} = min[\ g(HNX + HNY, Hf_{n,pho60}), \ g(DIP, Hf_{p,pho60}) \ ]$$

-    $Hf_{n,pho60}, Hf_{p,pho60}$

Light availability

$$u_{pho60b} = \frac{1}{\Delta z} \int_{z}^{z+\Delta z} \frac{I_B}{I_{opt60}} e^{-k_b z} exp\left\{ 1 - \frac{I_B}{I_{opt60}} e^{-k_b z} \right\} dz$$

   I_{opt60}, k_b

Extra-release

$$Dbal_{ext} = R_{ext60} \cdot Dbal_{pho}$$

μgC/cm³/h R_{ext60}

Respiration

$$Dbal_{res} = R_{res60a} \cdot exp(\beta_{pho60} \cdot TmpB) \cdot g(DOO, Hf_{o2,res60}) \cdot BAL \cdot R_{Zres60}$$
$$- R_{res60b} \cdot Dbal_{pho}$$

μgC/cm³/h $R_{res60a}, \beta_{pho60}, Hf_{o2,res60},$ R_{res60b}

Mortality

$$Dbal_{mor} = v_{mor60} \cdot BAL \cdot R_{Zmor60}$$

μgC/cm³/h

Rate of mortality

$$v_{mor60} = \alpha_{mor60} \cdot exp(\beta_{mor60} \cdot TmpB)$$

   $\alpha_{mor60}, \beta_{mor60}$

ACKNOWLEDGMENT

Part of this research was supported by the Program for Promoting Fundamental Transport Technology Research from the Japan Railway Construction, Transport and Technology Agency (JRTT).

REFERENCES

Aoyama, H. & Suzuki, T. (1997). In Situ Measurement of Particulate Organic Matter Removal Rates by a Tidal Flat Macrobenthic Community. *Bull. Japan. Soc. Fish. Oceanogr. 61(3)*, 265–274. (in Japanese with English abstract).

Baretta, J.W., Ebenhöh, W. & Ruardij, P. (1995). The European Regional Seas Ecosystem Model, a complex marine ecosystem model. *Neth. J. Sea Res. 33*, 233–246.

Baretta, J.W., Ruardij, P. & de Wolf, P. (eds.). (1988). *Tidal flat estuaries, simulation and analysis of the Ems Estuary, Ecological Studies 71*, Springer-Verlag.

Baretta-Bekker, J.G. & Baretta, J.W. (1997). Microbial dynamics in the marine ecosystem model ERSEM II with decoupled carbon assimilation and nutrient uptake. *J. Sea Res. 38*, 195–211.

Berg, P., Risgaard-Petersen, N. & Rysgaard, S. (1998). Interpretation of measured concentration profiles in sediment pore water. *Limnol. Oceanogr. 43*, 1500–1510.

Berner, R.A. (1980). *Early Diagenesis – A Theoretical Approach.* Princeton University Press, New Jersey, 241 pp.

Boudreau, B.P. (1996). A method of lines code for carbon and nutrient diagenesis in aquatic sediments. *Computers Geosci. 22(5)*, 479–496.

Boudreau, B.P. & Jørgensen, B.B. (eds.) (2001). *The Benthic Boundary Layer: Transport Processes and Biogeochemistry.* Oxford University Press, Oxford, 440 pp.

Canfield, D.E., Jørgensen, B.B., Fossing, H., Glud, R., Gundersen, J., Ramsing, N.B., Thamdrup, B., Hansen, J.W., Nielsen, L.P. & Hall, P.O.J. (1993). Pathways of organic carbon oxidation in three continental margin sediments, *Mar. Geol. 113*, 27–40.

Chiba Prefectural Fisheries Research Center (2001–2006). *Hypoxia quick information.* (web site http://www.awa.or.jp/home/cbsuishi/04tkhinsanso/04tkhinsansoflame.htm) (in Japanese).

Dedieu, K., Rabouille, G., Gilbert, F., Soetaert, K., Metzger, E., Simonucci, G., Jézéquel, D., Prévot, F., Anschutz, P., Hulth, S.,Ogier, S. & Mesnage, V. (2007). Coupling of carbon, nigrogen and oxygen cycles in sediments from a Mediterranian lagoon: a seasonal prespective. *Mar. Ecol. Prog. Ser. 346*, 45–59.

Furota, T. (1988). Effects of low-oxygen water on benthic and sessile animal communities in Tokyo Bay. *Bull. Coast. Oceanogr. 25(2)*, 104–113. (in Japanese).

Hata K. & Nakata, K. (1998). Evaluation of eelgrass bed nitrogen cycle using an ecosystem model. *Environ. Model. Software 13*, 491–502.

Imao, K., Suzuki, T. & Takabe, T. (2004). New method to predict changes in the structure and function of a macrobenthic community from changes in environmental oxygen concentrations. *Fish. Engineering, 41(1)*, 13–24. (In Japanese with English abstract).

Ishida, M. & Hara, T. (1996). Changes in water quality and eutrophication in Ise and Mikawa Bays. *Bull. Aichi Fish. Res. Inst. 3*, 29–41. (in Japanese with English abstract).

Jørgensen, B.B. (1978). A comparison of methods for the quantification of bacterial sulfate reduction in coastal marine sediments: II Calculations from mathematical models. *Geomicrobiol. J. 1*, 29–47.

Kanagawa Prefectural Fisheries Research Institute. (2005–2006). *Tokyo Bay Dissolved Oxygen Information.* (web site http://www.agri.pref.kanagawa.jp/suisoken/kankyo/sanso/TokyoBayOxInfo.htm). (In Japanese).

Kikuchi, T. (1993). Ecological characteristics of the tidal flat ecosystem and importance of its conservation. *Japan. J. Ecol. 43*, 223–235. (In Japanese).

Koike, K. (2000). *Reclamation of Tokyo Bay and Artificial Beach in Kanto and Ogasawara Areas – Japanese Geography*. University of Tokyo Press. (In Japanese).

Kuwae, T., Kibe, E. & Nakamura, Y. (2003). Effect of emersion and immersion on the porewater nutrient dynamics of an intertidal sandflat in Tokyo Bay. *Estuar. Coast. Shelf S. 57*, 929–940.

Kuwae, T., Inoue, T., Miyoshi, E., Konuma, S., Hosokawa, S. & Nakamura, Y. (2005). Modeling the coastal marine ecosystem coupled with tidal flats based on the study of oxygen cycling in sediments. In: *Report of Program for Promoting Fundamental Transport Technology Research,* Japan Railway Construction, Transport and Technology Agency, pp. 262–423. (In Japanese).

Luff, R. & Moll, A. (2004). Seasonal dynamics of the North Sea sediments using a three-dimensional coupled sediment-water model system. *Cont. Shelf Res. 24*, 1099–1127.

Matsumoto, E. (1983). The sedimentary environment in Tokyo Bay. *Earth Chemistry 17*, 27–32. (In Japanese).

Ministry of the Environment. (2006). *Reference data of a basic principle on the regulation of total amount control for chemical oxygen demand, contained amount of nitrogen and phosphorus.* (in Japanese)

Miyata, M. (2003). Bibliography of Zappai historical sources (*Zappai shiryou kaidai*). Systemized Japanese historical bibliography, *Seisyohdosyoten,* pp. 501.

Nakamura, Y., Nomura, M. & Kamio, K. (2004). Field observation and analysis of benthic-pelagic coupling in Banzu tidal flat and the adjoincent coastal area of Tokyo Bay. *Report of the Port and Airport Res. Inst. 43(2)*, 35–71. (In Japanese with English abstract).

Nakata, K., Horiguchi, F., Taguchi, K. & Setoguchi, Y. (1983a). Three dimensional simulation of tidal current in Oppa Estuary. *Bull. Natl. Res. Inst. Poll. Resources 12(3)*, 17–36. (In Japanese).

Nakata, K., Horiguchi, F., Taguchi, K. & Setoguchi, Y. (1983b). Three dmensional eco-hydrodynamical model in coastal region., *Bull. Natl. Res. Inst. Poll. Resources 13(2)*, 119–134. (In Japanese).

Odum, E.P. (1971). *Fundamentals of Ecology,* 3rd ed. W.B. Saunders, Philadelphia.

Pritchard, D.W. (1967). What is an estuary: physical viewpoint. In G.H. Lauff (Ed.), *Estuaries*, AAAS Publication no. 83, Washington, DC, pp. 3–5.

Revsbech, N. P., Madsen, B. & Jørgensen, B.B. (1986). Oxygen production and consumption in sediments determined at high spatial resolution by computer simulation of oxygen microelectrode data. *Limnol. Oceanogr. 31(2)*, 293–304.

Rysgaard, S. & Berg, P. (1996). Mineralization in a northeastern Greenland sediment : mathematical modeling, measured sediment pore water profiles and actual activities, *Aquatic Microb. Ecol. 11*, 297–305.

Sayama, M. (2005). Modeling the coastal marine ecosystem coupled with tidal flats based on the study of oxygen cycling in sediments. In: *Report of Program for Promoting Fundamental Transport Technology Research, Japan Railway Construction*, Transport and Technology Agency, pp. 424–456. (In Japanese).

Soetaert, K., Herman, P.M.J. & Middleburg, J.J. (1996a). A model of early diagenetic processes from the shelf to abyssal depth. *Geochimica et Cosmochimica Acta 60(6)*, 1019–1040.

Soetaert, K., Herman, P.M.J. & Middelburg, J.J. (1996b). Dynamic response of deep-sea sediments to seasonal variations: A model. *Limnol. Oceanogr. 41(8)*, 1651–1668.

Soetaert, K., Middelburg, J.J., Herman, P.M.J. & Buis, K. (2000). On the coupling of benthic and pelagic biogeochemical models. *Earth-Science Rev. 51*, 173–201.

Sohma, A. (2005). *Development of a multiple coastal ecosystem model including benthic, pelagic and tidal flat ecosystems for ecological evaluation in hypoxic estuary*. Ph. D. thesis, Tokai University, 368pp.

Sohma, A. (2009). Paradigm shift from a clean ocean to a bountiful ocean: An essential vision revealed by ecological modeling of "tidal flats - central bay area coupling" and "benthic-pelagic ecosystem coupling". In: Gertsen, N. & Sønderby, L. (eds.) *Water Purification*. Nova Science Publishers, New York, pp. 55-136.

Sohma, A., Sato, T. & Nakata, K. (2000). New numerical model study on a tidal flat system – seasonal, daily and tidal variation. *Spill Sci. Tech. Bull. 6*, 173–185.

Sohma, A. & Sayama, M. (2002). Modeling for coupled cycle of oxygen, nitrogen, and carbon in a coastal marine sediment: A new ecological model for dynamics in the micro profiles. *Proc. Coast. Engineering, JSCE 49*, 1231–1235. (In Japanese).

Sohma, A. & Sekiguchi, Y. (2003). Development of a new multiple coastal ecosystem model focused on ecological network and benthic vertical mechanisms in the micro scale: application of a hydrodynamics model and benthic ecosystem model in the central bay area of Tokyo Bay. In: *Proceedings of the Advanced Marine Science and Technology Conference in Autumn*, pp. 87–92. (In Japanese).

Sohma, A., Sekiguchi, Y. & Kakio, T. (2005). Development of a new multiple coastal ecosystem model "ZAPPAI" including benthic, pelagic and tidal flat ecosystems for ecological evaluation in hypoxic estuary: autonomous response to the tidal flat creation, dredging, sand capping, load reduction and red tide. *J. Advanced Mar. Sci. Tech. Soc. 11(2)*, 21–52 (In Japanese with English abstract).

Sohma, A., Sekiguchi, Y. & Nakata, K. (2004). Modeling and evaluating the ecosystem of seagrass beds, shallow waters without seagrass, and an oxygen-depleted offshore area. *J. Mar. Syst. 45*, 105–142.

Sohma, A., Sekiguchi, Y., Kuwae, T., Nakamura, Y. (2008). A Benthic-pelagic coupled ecosystem model to estimate the hypoxic estuary including tidal flats: model description and validation of seasonal/daily dynamics. *Ecol. Model. 215*, 10–39.

Sohma, A., Sekiguchi, Y., Yamada, H., Sato, T. & Nakata, K. (2001). A new coastal marine ecosystem model study coupled with hydrodynamics and tidal flat ecosystem effect. *Mar. Poll. Bull. 43*, 187–208.

Suzuki, T., Aoyama, H., Kai, M. & Imao, K. (1998). Effect of dissolved oxygen deficiency on a shallow benthic community in an embayment. *Oceanography in Japan 7(4)*, 223–236. (In Japanese with English abstract).

Suzumura, M., Kokubun, H. & Itoh, M. (2003). Phosphorus cycling at the sediment-water interface in a eutrophic environment of Tokyo Bay, Japan. *Oceanography in Japan 12(5)*, 501–516. (In Japanese with English abstract).

In: Estuaries: Classification, Ecology and Human Impacts
Editor: Steve Jordan

ISBN: 978-1-61942-083-0
© 2012 Nova Science Publishers, Inc.

Chapter 11

MONITORING THE CONDITION OF THE ESTUARIES OF THE UNITED STATES: THE NATIONAL COASTAL ASSESSMENT EXPERIENCE

J. Kevin Summers, Virginia D. Engle, Lisa M. Smith, Linda M. Harwell, John M. Macauley and James E. Harvey
U.S. Environmental Protection Agency, Gulf Ecology Division, Gulf Breeze, FL, US

ABSTRACT

Coastal waters in the United States include estuaries, bays, sounds, coastal wetlands, coral reefs, intertidal zones, mangrove and kelp forests, seagrass meadows, and coastal ocean and upwelling areas. These coastal areas encompass a wide diversity of ecosystems that result from the tidal exchanges between freshwater rivers and saline ocean water that occur within coastal estuaries. Coastal habitats provide spawning grounds, nursery areas, shelter, and food sources critical for the survival of finfish, shellfish, birds, and other wildlife populations that contribute substantially to the economic health of the Nation. Section 305(b) of the Clean Water Act requires that the U.S. Environmental Protection Agency (EPA) report periodically on the condition of the nation's coastal waters. As part of this process, coastal states provide valuable information about the condition of their coastal resources to EPA; however, because the individual states use a variety of approaches for data collection and evaluation, it has been difficult to compare this information among states or on a national basis. In 1999—to better address questions about national coastal condition—EPA, the National Oceanic and Atmospheric Administration (NOAA), and the U.S. Fish and Wildlife Service (FWS), agreed to participate in a multiagency effort to assess the condition of the nation's coastal resources. The agencies chose to assess condition using nationally consistent monitoring surveys to minimize the problems created by compiling data collected using multiple approaches. The results of these assessments were compiled periodically into a series of *National Coastal Condition Reports* (NCCR). This series of reports contains the most comprehensive ecological assessment of the condition of U.S. coastal bays and estuaries.

This chapter describes the planning, execution and communication of the National Coastal Assessments.

INTRODUCTION

A key action of the Clean Water Action Plan (U.S. EPA 1998a) was the development of a coordinated monitoring plan for coastal waters and a comprehensive report on the condition of the Nation's coastal waters. In support of the Clean Water Action Plan, the U.S. Environmental Protection Agency (EPA) initiated the National Coastal Assessment (NCA) which had its roots in earlier research efforts within EPA's Environmental Monitoring and Assessment Program (EMAP; Summers et al. 1995; McDonald et al. 2001). NCA evolved from the coastal and estuarine element of EMAP which began in the early 1990s with an overall goal to develop methods and approaches to assess the condition of the nation's coastal resources.

The objective of the National Coastal Assessment was to create an integrated comprehensive coastal monitoring program across all U.S. coastal states and territories to assess the condition of estuaries at multiple geographic and spatial scales (Paul et al. 2008). NCA was designed to answer the question, "What is the condition of the nation's estuaries and is that condition getting better or worse?" From 1999 through 2006, coastal states collected and analyzed information in partnership with NCA to assess the condition of coastal waters and develop an approach that could be implemented by EPA, coastal states, and coastal tribal governments. The NCA research and monitoring program culminated in a series of National Coastal Condition Reports (NCCRs; U.S. EPA 2001a, 2004, 2008), describing the condition of the nation's estuaries through the period 1999–2006 and creating a framework for this approach to be incorporated into the coastal monitoring strategies of EPA and coastal states and territories.

WHY COASTAL WATERS?

Coastal waters are productive and diverse ecosystems, including estuaries, coastal wetlands, coral reefs, mangrove forests and upwelling areas. Critical coastal habitats provide spawning grounds, nurseries, shelter and food for finfish and shellfish as well as essential nesting, resting, feeding and breeding habitat for waterfowl and other migratory birds. Coastal areas also are the most developed areas in the U. S. This narrow fringe—only 17% of the total conterminous U.S. land area—is home to more than 50% of the U.S. population (Crossett et al. 2004). With an average of 300 people mi^{-2}, U.S. coastal counties are more densely populated than non-coastal counties (average 98 persons mi^{-2}; Crossett et al. 2004). In addition to being popular places to live, U.S. coasts are a source of many other valuable commodities: more than 50% of the U.S. Gross National Product is generated in coastal counties; a significant proportion of commercial fisheries (46% by weight and 68% by value) and recreational fisheries (80%) are dependent on estuaries; and approximately180 million people visit coastal areas for recreation each year (Cunningham and Walker 1996; Lellis-Dibble et al. 2008; NOAA 2011).

For these reasons and because public perception and scientific evidence pointed to a general deterioration of the coastal environment, the need for more comprehensive monitoring of the status and trends in the condition of coastal waters was outlined in the Clean Water Action Plan (U.S. EPA 1998a). In the 1990s we discovered that we could not answer questions such as, "What is the ecological condition of the estuaries of the United States and are they getting better or worse?" despite a plethora of available data (Summers et al. 1995). Our inability did not stem from a lack of data or from a lack of desire to answer the question. There were literally millions of pieces of data concerning various attributes of estuarine ecosystems. Many entities desired the answer for a number of reasons (e.g., wise investment of fiscal resources for environmental and human health protection, "bragging rights" for the nation's cleanest and healthiest shorelines, prudent management of resources and ecosystem services). Our inability to answer the question stemmed from the lack of a comprehensive national monitoring program designed specifically to address the question.

THE IMPORTANCE OF QUESTION

Environmental monitoring programs have been designed for many purposes: to provide information on compliance with regulations, establish the aquatic resource conditions or status, measure the effectiveness of management and regulatory programs, and supply information to policy, planning and decision-making processes. The first point of discussion or clarification when designing a monitoring program, therefore, must be "What question are you trying to answer?" In essence, all aspects of a monitoring design hinge upon this initial question. The downfall of many monitoring programs is failure to clearly articulate the question(s) that must be addressed. Without this clear articulation of the question(s) and adherence to restrictions placed upon the survey design by the question(s), a monitoring program will generally fail to meet its intended needs.

The primary question for NCA was "What is the ecological condition of the estuarine resources of the United States?" This question was supplemented by NCA's state and federal partners to reflect two additional questions based on spatial scale: "What is the ecological condition of estuarine resources in regions of the United States?" and "What is the ecological condition of estuarine resources in each coastal state?" These questions had specific implications for the types of survey designs (probability versus targeted) and response designs (type of indicators and timing of collection) that could be implemented. A number of tertiary questions were later developed regarding individual indicators or indices based on the selected design. It cannot be stressed enough – the primary and secondary questions that a monitoring program needs to answer must be *clearly* determined prior to construction of the survey design and indicator selection to ensure that the design will provide the data necessary to answer the question(s).

The question(s), therefore, determine the monitoring objectives and guide the development of all elements of an environmental monitoring program to ensure data collection at the appropriate spatial and temporal scales to answer the question(s). The critical elements of an environmental monitoring program (U.S. EPA 2003) include monitoring objectives, survey design, response design (including selection of indicators), quality assurance, data management, data analysis, and reporting.

Monitoring Objectives and Survey Design

The key steps to constructing a survey design for monitoring any environmental resource (Olsen et al. 1999) are to:

1. State the monitoring objectives, precisely and quantitatively;
2. Define the target population explicitly and precisely;
3. Construct a sample frame that accurately represents the target population;
4. Decide which survey design will best provide information to meet objectives.

NCA's monitoring objectives were (1) to assess the condition of estuarine waters and (2) to track changes in condition over time at state, regional and national scales. The target population for NCA was U.S. estuaries (i.e., coastal waters that extend from the saltwater-freshwater interface to the mouth of an estuarine drainage basin). We defined this target population explicitly as:

"... all coastal waters of the conterminous U.S. from the head-of-salt to confluence with ocean including inland waterways and major embayments such as Florida Bay and Cape Cod Bay. For the purposes of this study the head-of-salt is generally defined as < 0.5 psu and represents the landward/upstream boundary. The seaward boundary extends out to where an imaginary straight-line intersecting two land features would fully enclose a body of coastal water. All waters within the enclosed area are defined as estuarine, regardless of depth or salinity." (U.S. EPA 2010, p. 1)

A sample frame was constructed to represent this target population in a GIS framework. The National Estuary Dataset was created as a vector dataset from modified USGS 1:100,000 Digital Line Graph (DLG) datasets (Bourgeois et al. 1997). This line work representing the boundary of estuary influence was used to generate new polygons which were identified using NOAA navigational charts. The National Estuary Dataset contained polygons representing all coastal waters of the conterminous U.S., plus Alaska, Hawaii, Puerto Rico, Guam, American Samoa, and the U.S. Virgin Islands.

NCA implemented a probability survey design to obtain unbiased estimates of the condition of estuaries at multiple spatial scales. Probability survey designs provide a scientifically rigorous way to (1) estimate the condition of an entire natural resource by randomly selecting only a subset of that resource and (2) quantify the uncertainty surrounding that estimate (McDonald et al. 2001; Olsen et al. 1999). A generalized random tessellation stratified (GRTS) unequal probability survey design was used to ensure that the sample sites were spatially distributed and that important sub-classes (strata or multi-density categories) were included (Stevens & Olsen 2004). When strata are explicitly defined (e.g., a state or specific region within a state), separate estimates of condition can be made for each stratum with known confidence. The use of multi-density categories allocated a specified proportion of sample sites to different resources within a state to ensure that all resources of interest were included in the design. A GRTS design increases efficiency, improves precision, includes design-based estimates of variance, and is based on a unified statistical theory for all aquatic resources (Stevens & Olsen 2004).

Table 1. National Coastal Assessment 2005–2006 survey designs for coastal states

State	Within State Strata	Multi-Density Category (MDC)	Frame Area (km^2)	% of Sites in each MDC	Total Number of Sites
Maine	North	N/A	1665.8283	50%	50
	South	N/A	1635.5105	50%	
New Hampshire	N/A	N/A	57.8210	100%	50
Massachusetts	N/A	N/A	2241.4568	100%	50
Rhode Island	N/A	Deep Water	193.5269	50%	50
		Shallow Water & Ponds	197.0302	50%	
Connecticut	N/A	Long Island Sound	3129.4490	80%	50
		Tidal Rivers	111.9986	20%	
New York	N/A	Hudson River	273.8935	8%	50
		Estuaries excluding LIS	1282.1788	92%	
New Jersey	N/A	NJ Harbor	226.6679	36%	50
		NJ Coastal Bays	520.9446	64%	
Delaware	DE Bay	DE River Estuary	972.8287	32%	50
		DE Bay (main)	1047.4231	36%	
		DE Bay small estuaries	68.8775	32%	
	Inland Bays	N/A	89.6472	100%	50
Maryland	N/A	Chesapeake Bay (main)	3574.6194	40%	50
		Ches. Bay (tributaries)	2650.6164	40%	
		Coastal Bays	267.1776	20%	
Virginia	N/A	Chesapeake Bay (main)	3845.1494	20%	50
		Ches. Bay (tributaries)	1534.5540	56%	
		Coastal Bays	417.0846	24%	
North Carolina	N/A	Estuaries < 250 km^2	2436.0889	50%	50
		Estuaries ≥ 250 km^2	6238.6760	50%	
South Carolina	N/A	Tidal Creeks		50%	120
		Open Water		50%	
Georgia	N/A	N/A	671.6192	100%	50
Florida	Atlantic Coast	N/A	1370.6884	100%	50
	Gulf Coast	N/A	10471.2536	100%	50

Table 1. (Continued)

State	Within State Strata	Multi-Density Category (MDC)	Frame Area (km^2)	% of Sites in each MDC	Total Number of Sites
Alabama	N/A	N/A	1577.0159	100%	50
Mississippi	N/A	N/A	1713.2051	100%	50
Louisiana	N/A	N/A	12844.2527	100%	50
Texas	N/A	N/A	5839.4188	100%	100
California	North	San Francisco Bay	1148.7702	60%	50
		Rest of N. CA bays	141.7751	40%	
	South	Los Angeles/Long Beach	58.7146	33%	50
		San Diego Bay	45.7073	33%	
		Rest of S. CA bays	31.6530	33%	
Oregon	N/A	N/A	382.9321	100%	50
Washington	Puget Sound	N/A	2968.7308	100%	30
	Coastal Bays	N/A	708.4499	100%	20
Hawaii (2005)	N/A	Estuaries < 0.1 km^2	5.0656	20%	50
		Estuaries 0.1 – 1.0 km^2	26.2629	20%	
		Estuaries 1.0 – 10 km^2	72.8728	20%	
		Estuaries > 10 km^2	120.8485	40%	
Alaska	Aleutian (2005)	N/A	2608.9007	100%	50
	Bering (2006)	N/A	116764.9819	100%	50

Survey designs were constructed for each coastal state and island territory from 2000 to 2006. For the purpose of illustration, we limit this discussion to the last two years of the program (2005–2006). NCA implemented GRTS survey designs with at least 50 sites per state, to be sampled over this two year period. A sample size of 50 was chosen to attain a conservative estimate of precision of ±12% with 90% confidence (Cochran 1987). Although these survey designs were essentially standardized, some survey designs were tailored to meet individual state monitoring needs (see Table 1). For example, Maine needed to sample the northern portion of its coast in 2005 and the southern portion in 2006 for operational and logistical reasons; therefore, separate survey designs were completed with 25 sites allocated to each stratum (northern and southern Maine). States with a single large estuary and many small estuaries (e.g., Maryland, Virginia, Delaware) incorporated multi-density categories into their survey designs to ensure that a proportion of sites were selected within each of these resources (Table 1). Long Island Sound includes open waters shared by New York and Connecticut and small harbors and bays bordering the Sound in both states; the survey design ensured that sites were spatially distributed across the resources of both states (Paul et al. 2008). States with large coastlines (i.e., Florida, Texas, and California) implemented survey designs with 100 sites allocated over the two-year period (Table 1). These examples of variations in state survey designs demonstrate the flexibility and robustness of the GRTS unequal probability survey design which can accommodate many of the difficulties inherent in applying traditional stratified random survey designs to natural resources (Stevens & Olsen 2004).

RESPONSE DESIGN

While the survey design determines which sites will be sampled, the response design determines when, what and how to collect information at the selected sites. NCA conducted surveys during an index period from July through mid-September for most of the U.S., because ecological effects in estuaries related to hypoxia, increased contaminant availability, and increased human usage are most prevalent in late summer (Summers 2001). To help ensure that the NCA collected the appropriate types of data to meet its objectives, the program convened a diverse panel of environmental scientists to formulate a list of core indicators to characterize water quality, sediment quality, and biological condition at each site. Indicators were selected to characterize ecological condition and included in the program if they (1) could be quantified over time, (2) were sensitive to potential impacts, (3) were easily measured and interpreted, and (4) could be benchmarked against reference values (Kurtz et al. 2001). All samples were collected and analyzed in accordance with NCA field and quality assurance guidelines (U.S. EPA 2001b, 2001c; see Quality Assurance section in this chapter).

Water Quality

The first NCCR (U.S. EPA 2001c) presented trophic conditions in the nation's estuaries as summarized data from NOAA's National Estuarine Eutrophication Assessment, an assessment of the "symptoms" of eutrophication based on existing data and expert opinion (Bricker et al. 1999). The need for quantitative data that could be used to assess water quality more comprehensively led to the inclusion into NCA of measures of dissolved nutrients such as dissolved inorganic nitrogen (DIN) and phosphorus (DIP), along with chlorophyll a, dissolved oxygen (DO), and water clarity.

NCA assessed water quality by measuring DO, water clarity, dissolved inorganic nitrogen (DIN) and phosphorus (DIP) and chlorophyll a. These five water quality metrics were used in the calculation of a composite water quality index for coastal resources of the U.S. (U.S. EPA 2004). However, a major challenge in developing a national scale water quality indicator was, and continues to be, accounting for regional differences and recognizing that a "one size fits all" approach is not appropriate for scaling these metrics in national coastal assessments.

Percent Area

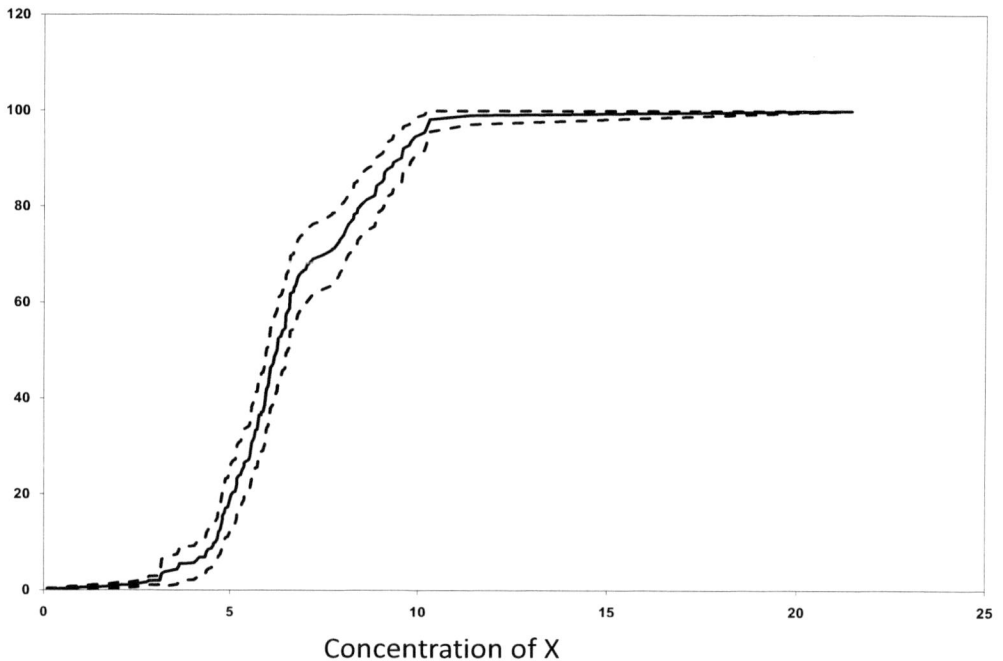

Figure 1. A hypothetical cumulative distribution function showing the concentration of a hypothetical environmental indicator on the X-axis and cumulative percentage of estuarine area on the Y-axis. (The solid line is the cumulative distribution function and the dashed lines are the 95% confidence intervals).

DO is the most commonly measured water quality parameter in aquatic resource monitoring programs. Many states apply a threshold concentration of 4–5 mg l^{-1} to establish water quality standards. Aquatic life use criteria have been developed for dissolved oxygen in fresh and saltwater environments and a threshold at which stress is induced in organisms has been determined as < 2.0 mg l^{-1} (U.S. EPA 2000a). DO concentrations are often lowest in bottom waters where hypoxic conditions directly affect benthic organisms; therefore, bottom DO concentrations are often used for water quality assessments. Based on water quality standards and aquatic life criteria, NCA used threshold values of 2 and 5 mg l^{-1} to evaluate DO concentrations in all estuaries (U.S. EPA 2004, 2008).

Clear waters are valued by society and contribute to the maintenance of healthy and productive ecosystems, particularly in support of submersed aquatic vegetation (SAV). In NCA, water clarity was measured either with light meters that compared the amount and type of light reaching the water surface to the light measured at a depth of 1 meter (transmissivity), or by Secchi disk depth (transparency). Water clarity varies naturally throughout various parts of the nation; therefore a water clarity indicator was created to compare observed clarity to regional reference conditions (Smith et al. 2006). The regional reference conditions were determined by comparing available data for each of the U.S. regions to set thresholds for water clarity based upon three water body types: those with naturally high turbidity; those with average natural turbidity, and those expected to have clear waters that support SAV (U.S. EPA 2004; 2008; Smith et al. 2006).

Dissolved nutrient concentrations were measured in surface water samples collected at NCA sites. DIN and DIP represent the portion of the nutrient pool available for algal growth. Elevated nutrient concentrations in combination with high chlorophyll *a* concentrations, reduced water clarity and low DO can be indicative of eutrophic conditions and poor water quality. A 1:10 ratio of DIN:DIP was used as the starting point for establishing national thresholds for nutrients. Regional thresholds for evaluating nutrients, chlorophyll *a* and water clarity were set based on differences in natural inputs of nutrients as well as aquatic life uses for various water body types (U.S. EPA 2004; 2008). Higher nutrient threshold values were used for the West Coast region, which experiences natural upwelling of nutrients in the summer months (Nelson & Brown 2008). Lower threshold values for nutrients were established for tropical waters (i.e., Florida Bay, Hawaii, island territories) which are expected to be oligotrophic, have naturally clear waters with low algal production, and to support SAV.

High concentrations of chlorophyll *a* indicate the potential for over-production of algae, a symptom of eutrophication. NCA measured surface concentrations of chlorophyll a at each site. National thresholds were modified to evaluate Florida Bay and the tropical waters of Hawaii and the island territories based on criteria similar to those used to establish nutrient thresholds in clear water systems (U.S. EPA 2004).

Each NCA site was rated good, fair, or poor with respect to each water quality metric based on these threshold values (U.S. EPA 2004). A water quality index was developed as a composite of the rating scores assigned to the five water quality metrics (i.e., if no metric was scored poor and only one was fair, the water quality index score was 'good'; if one metric scored poor and two or more were fair, the water quality index score was 'fair'; and if two or more metrics scored poor, the site was scored 'poor' for the water quality index; U.S. EPA 2004, 2008). Individual water quality metrics and the composite water quality index were analyzed using cumulative distribution functions (Figure 1) to estimate the areal percentage of observed conditions within each region. The areal percentages were also evaluated at the national scale to represent the water quality of the nation's estuarine resources (U.S. EPA 2004, 2008).

Sediment Quality

Another issue of major environmental concern in coastal waters is the contamination of sediments with toxic chemicals. A wide variety of metals and organic substances, such as polycyclic aromatic hydrocarbons (PAHs), polychlorinated biphenyls (PCBs), and pesticides, are discharged into coastal waters from urban, agricultural, and industrial sources in a watershed. These contaminants may accumulate and have adverse effects on the benthic community of invertebrates, shellfish, and crustaceans. Indicators of sediment toxicity, sediment contaminants, and sediment total organic carbon (TOC) were combined into a sediment quality index to assess sediment quality. Sediment toxicity was assessed as the percent survival of amphipods exposed to field-collected sediments relative to the survival of amphipods exposed to uncontaminated reference sediment (U.S. EPA 1994). In the absence of quantitative sediment quality criteria, measured concentrations of metals and organics in sediments were compared to published ERL and ERM values[1] (Long et al. 1995) to assess the potential for adverse biological effects. Although TOC exists naturally in coastal sediments as the result of degradation of autochthonous and allochthonous organic materials (e.g., phytoplankton, leaves, twigs, dead organisms), anthropogenic sources (e.g., organic industrial wastes, untreated or only primary-treated sewage) can significantly elevate the concentration of TOC in sediments. As a component of sediment quality, TOC can alter the bioavailability of sediment contaminants and change the benthic community structure.

A sediment quality index was developed as a composite of the rating scores assigned to sediment toxicity, sediment contaminants and TOC. A "poor" sediment quality indicated that any one of these metrics scored poor. If none of the metrics scored poor and the sediment contaminants metric scored fair, the sediment quality index score was 'fair'. If none of the metrics were rated poor and the sediment contaminants metric was rated good, the sediment quality index score was 'good' (U.S. EPA 2004, 2008).

Benthic Condition

Population and community characteristics of aquatic organisms that inhabit sediments (i.e., benthic macroinvertebrates) serve as reliable indicators of coastal environmental quality because they are sensitive to contaminants, hypoxia, salinity fluctuations, and habitat alteration. To distinguish degraded benthic habitats from natural, healthy benthic habitats, NCA applied regional benthic indices of environmental condition (Engle et al. 1994; Weisberg et al. 1997; Engle & Summers 1999; Van Dolah et al. 1999; Hale & Heltshe 2008). These indices reflect changes in benthic community diversity and the abundance of pollution-tolerant and pollution-sensitive species. A 'good' benthic index score means that estuarine sediment samples contain a wide variety of benthic species, as well as a low proportion of pollution-tolerant species and a high proportion of pollution-sensitive species (U.S. EPA 2004, 2008). A 'poor' benthic index score indicates that the benthic communities are less diverse than expected, are populated by more pollution-tolerant species than expected, and contain fewer pollution-sensitive species than expected (U.S. EPA 2004, 2008). Benthic indices vary by region because benthic species assemblages depend on prevailing temperature, salinity, and the silt-clay content of sediments (Engle & Summers 2000, Hale 2010). The benthic index was rated poor at a site when the index values fell below regional thresholds (U.S. EPA 2004, 2008). Regions that have not developed benthic indices (i.e., West Coast, Alaska, Hawaii, American Samoa, Guam, Puerto Rico, and the U.S. Virgin Islands) used benthic community diversity indices or species richness as surrogate measures of benthic condition.

---

[1] ERM (Effects Range Mean) is the concentration of a contaminant, above which adverse biological effects were observed in 50% of published studies. ERL (Effects Range Low) is the concentration of a contaminant above which adverse biological effects were observed in 10% of published studies (Long et al. 1995).

Fish Tissue Contaminants

Chemical contaminants may enter a marine organism through direct uptake from contaminated water, consumption of contaminated sediment, or consumption of previously contaminated organisms and may bioaccumulate in tissues and biomagnify in organisms at higher tropic levels though food chain effects. Tissue contaminant concentrations were measured in demersal (bottom-dwelling) and slower-moving pelagic (water column dwelling) species (e.g., finfish, shrimp, lobster, crab, sea cucumbers; collectively referred to as "fish") that were representative of each geographic region. Contaminant residues were compared with risk-based EPA Advisory Guidance values (U.S. EPA 2000b) for use in establishing fish advisories. A fish tissue contaminants index was used as an indicator of exposure to environmental contaminants as well as an indicator of risk to fish populations and the geographical distribution of contaminants in fish populations (Harvey et al. 2008).

QUALITY ASSURANCE

The worth of a coastal monitoring program is measured by the quality of its data. Quality assurance is a vital process in any data management system. NCA implemented a series of steps to ensure data quality. Prior to implementation of NCA environmental data collection, a detailed Quality Assurance Project Plan (QAPP) was developed to serve as the roadmap for the construction of the program's quality system, as well as a key planning tool for the various elements of the program. The QAPP stated the data quality goals and responsibilities at all levels of NCA operations: experimental design, field collection of environmental samples and data, sample processing and laboratory analyses, data quality reviews and data management (U.S. EPA 2001c).

With the QAPP as the foundation, NCA built a robust quality program to provide guidelines and set standards to ensure that data produced were of known and acceptable quality. A NCA Quality Assurance (QA) Team was appointed to monitor and assess data quality at all levels of the program, from sample collection to final report. The QA Team included QA Coordinators at the national, regional and state levels who ensured that the NCA quality program was in place and followed, facilitated training, provided oversight and conducted audits of field teams, laboratories and contractors, and reviewed and documented data quality.

To ensure that the data collected generated sound estimates of environmental condition, the NCA established Data Quality Objectives (DQOs) and Measurement Quality Objectives (MQOs) to set the overall level of data quality required by management to make informed decisions. The DQO for condition estimates was: "For each indicator of condition, estimate the portion of the resource in degraded condition within ±10%, with 90% confidence based on a complete sampling regime" (U.S. EPA 2001b)[2]. The MQOs established the quality goals for individual measurements, usually expressed in terms of accuracy, precision, and completeness. The MQOs were used as quality control criteria to set the bounds of acceptable measurement error (U.S. EPA 2001c).

---

[2] As discussed in the design section, an error rate of 12% was achieved.

NCA conducted workshops to train state personnel on NCA monitoring concepts and how they could be applied in their state programs and on the use of standardized methods to collect the core environmental samples and data. At least once during the course of the field sampling period, each state field crew was audited to evaluate the crew's compliance with the QAPP guidelines.

NCA analytical chemistry laboratories were required to demonstrate technical capability by either successfully analyzing Standard Reference Materials (SRMs) or submitting results from the National Institute of Standards and Technology (NIST) annual inter-laboratory comparison. To accommodate states who lacked the analytical capabilities mandated by the program, NCA established national contracts with commercial laboratories to perform the required analyses. National contract laboratories were required either to adhere to the requirements of the QAPP or provide a plan with requirements that were equivalent to or more stringent than those of NCA. Management of the contracts, coordination of the shipment of samples, and distribution of resulting data was performed by EPA.

Furthermore, NCA implemented standardized data collection protocols to ensure data quality. Standardized data collection forms were used to capture observational data. Sample analysis progress was followed using standardized sample tracking. Data contributors were required to review the quality of their data prior to submission. All data were examined for standard editing errors, outliers, missing data, and mismatched data and all data corrections were fully documented. Statistical analyses were used to determine if the data met program MQOs.

DATA MANAGEMENT

Coastal monitoring data are inherently challenging to manage. New technologies and partnering strategies are increasing both the volume and complexity of data. This explosion of available data is not only making data harder to use but also more difficult to understand (*The Economist* 2010). Managing these data in an informative and scientifically meaningful way remains an *essential* part of creating effective environmental strategies (Günther 1998).

Making quality data accessible, easy to find, understandable, *and* scientifically useful is essential to the success of coastal monitoring programs (Hale et al. 1998). Data management was incorporated into every monitoring activity in NCA, from design development to data analyses. NCA data management requirements focused less on the structures used to house data and more on the functions of obtaining and preparing data for use in assessments. The NCA data management system was built on an existing information management strategy (Hale et al. 1999) which emphasized (1) distributing data management activities to increase productivity and data quality, (2) using data consistency and formatting standards to help minimize data integration issues, (3) implementing processes to ensure data integrity and quality, and (4) making data of known quality accessible to data users.

As in the case with NCA, it is often necessary and desirable to develop partnerships to acquire the depth and breadth of data necessary to comprehensively describe coastal ecology, but these relationships add another layer of complexity to managing data (Hale et al. 2000, 2003). Key personnel from EPA's Atlantic, Gulf and Western Ecology Divisions supported NCA data management activities. These geographically-focused or regional data hubs worked closely with NCA participants located within their respective geographic areas to resolve region-specific data issues and facilitate the flow of data from the field to NCA's central data repository.

There are many tools available for physically organizing and storing monitoring data; however, there is rarely a "one-size-fits-all" solution, especially within a partnership framework. To ensure continued flow and quality of data, a variety of techniques was used to facilitate the data integration needed to aggregate data in a central data warehouse (Hale et al. 1998). For example, data format, file content, and documentation were standardized to minimize data integration issues, fulfill minimum data requirements and guide development of sustainable data management structures for organizing and storing data (Hale et al. 1998). NCA data centers maintained data using similarly designed, tiered data repository structures to help minimize the impact of implementing new technologies.

The tiered-data approach used by each of NCA's data centers also provided a physical measure of version control during different phases of data processing (e.g., raw data archive, quality assurance (QA), and final assembly). A core set of data standards, along with data quality processing, ensured the integrity and continuity of the data. Because each tier of data was offered with a known measure of quality, potential data users could begin working with data at any stage of the refinement process, with some reasonable expectation of data quality. Final assemblages of qualified NCA data were sent to NCA's central repository where data were prepared for public access.

Metadata are integral to ensuring that the data help a coastal monitoring program meet its needs. NCA metadata were used by data managers to track and organize data while others used the same metadata to determine what data were available or how the data were collected. Three components of metadata (i.e., data directory, catalog and dictionary) formed the nexus that identified, described, and located NCA data.

Data management is at the heart of coastal monitoring. It provides the framework to organize a valuable asset and showcase the quality of a program. The data contribute to the legacy of coastal monitoring. Quality data that are understandable, easy to find, and readily accessible extend the life a program long after the last sample is collected.

DATA ANALYSIS AND REPORTING

The NCA was designed to provide estimates from a subset of estuaries to make inferences about condition of all estuaries in the U.S. These estimates were reported as the percentage of estuarine area in good, fair, or poor condition with respect to the indicators. The process of data analysis to compute the estimates is outlined in detail in Appendix A of the NCCR III (U.S. EPA 2008). Briefly, each site received a good, fair, or poor score for each indicator based on comparison of the data values to established reference values. A cumulative distribution function (CDF) was then used to calculate the percentages of area with good, fair, and poor scores; CDFs were weighted by the proportion of estuarine area in each region, state, within-state stratum, or multi-density category, depending on the spatial scale of interest. A great advantage of using a probability survey design is that statistical theories can be applied to calculating confidence intervals for the estimates (Stevens & Olsen 2003).

The results of data analysis were reported in the series of NCCRs. Geographic regions and the Nation were rated overall as good, fair, or poor based on the percentage of estuarine area that was rated poor or fair for each index (U.S. EPA 2008). In the NCCRs, estimates of condition were presented in charts and maps using a traffic light color scheme (i.e., green = good, yellow = fair, and red = poor).

PARTNERSHIPS, COLLABORATION, AND COMMUNICATION

Partnerships with state resource agencies were necessary to implement NCA successfully. NCA identified state resource agencies that were responsible for reporting on the condition of their coastal resources and interested in developing collaborative partnerships with EPA. Cooperative agreements were established between EPA and state coastal resource agencies. These formal agreements provided EPA funds to build state resource agency capacity to conduct coastal monitoring surveys. The NCA Field Operations Manual and the Quality Assurance Project Plan (U.S. EPA 2001b; 2001c) formed the foundation for the cooperative agreements between state resource agencies and EPA, ensuring that data collected under the NCA would be comparable across all states. Data comparability is essential for estimating regional and national condition and for establishing baseline coastal resource conditions.

Training workshops were conducted (1) to introduce the monitoring and assessment methods to the states and (2) to learn more about each state's needs for monitoring and assessment. Workshops and subsequent dialog with state resource agencies emphasized that to make this type of monitoring program relevant, results had to provide information for making sound and effective science-based decisions and meet the needs of state and local governments, tribes and EPA.

As NCA data became available, EPA scientists worked with state resource agencies to communicate the information to tribes, state and local decision makers, and citizens. Just as EPA produced the NCCRs to provide regional and national assessments of coastal condition, some states and territories also published reports on the condition of their estuarine resources (e.g., Van Dolah et al. 2003; Didonato et al. 2009; Simons & Smith 2009; Trowbridge et al. 2009).

The NCA data currently are available at http://www.epa.gov/emap/nca/html/data/index.html for public use. It is especially valuable for establishing historic baseline coastal conditions. The NCA data were used most recently to establish baseline estuarine condition after Hurricane Katrina in 2005 (Engle et al. 2009; Macauley et al. 2010; Smith et al. 2009) and after the 2010 Deepwater Horizon oil spill. Researchers have also used NCA data to investigate a variety of research questions, e.g., geographical patterns of fish tissue chemical contamination (Harvey et al. 2008); benthic community patterns and biogeography (Garza 2008; Hale 2010); and to develop fish community indices (Jordan et al. 2010). The NCA data also can be used to correlate estimates of condition with stressors at various spatial scales (e.g., percent estuarine area with benthic communities in poor condition, compared to DO concentrations less than 2 mg l^{-1}). These spatial arrays of information help scientists and the public to visually compare the levels of stressors with indicators of condition at state, regional, and national levels.

TECHNICAL TRANSFER AND FUTURE OF NCA MONITORING

Over the past decade, national coastal monitoring programs have consistently adapted to changing national priorities and emerging issues. As demand for coastal and marine resources increases with growing populations and intensified development, ecosystems will be affected by the resulting environmental stresses. The NCA program produced four NCCRs that assessed the condition of U.S. coastal waters. Each consecutive report in the NCCR series presented an expanded geographical extent of monitoring, improved indices of coastal condition, and the current state of coastal monitoring science.

The success of NCA was a direct result of engaging state partners early on and understanding the unique perspectives of decision makers, as well as their challenges, concerns, and technical capability. Sound science is the foundation for effective environmental decisions. The success and defensibility of EPA policy and regulations, and state and local decision making, is strengthened in direct proportion to the amount and quality of scientific support for specific decisions. This support is most effective when the connection between decision support needs and research products is clear and transparent. The value of communication and outreach for creating this connection has been emphasized repeatedly (e.g., NRC 1994, 1996, 2000, U.S. EPA 1998b); NCA has provided a strong example.

The legacy of NCA continues with the National Coastal Condition Assessment (NCCA), under the purview of EPA's Office of Water. The NCCA is part of the National Aquatic Resource Surveys program, an effort to assess the quality of all U.S. aquatic resources, including lakes, rivers and streams, and wetlands[3]. The NCCA will continue to collaborate with the coastal states to provide important information about the condition of the nation's coastal and estuarine resources, and the status of key stressors at national and regional scales. The first NCCA was conducted in 2010, encompassing all of the U.S. coastal and Great Lakes states except Alaska. The results of the 2010 survey will be reported in the fifth NCCR, including, for the first time, measurements of *Enterococcus* spp. as indicators of bacterial contamination and the recreational safety of coastal waters. This expansion of coastal monitoring reflects the evolving priorities of EPA's Office of Water to prioritize human health, link with potential climate change effects, and to perpetually update monitoring programs to answer new questions.

CONCLUSIONS

Estuaries are complex physical, chemical and biological environments that require an understanding of their structure and function to devise a tractable, useful and meaningful monitoring program. Regardless of the type of ecosystem, ecological condition monitoring requires that the sampling design be driven by the question(s) the monitor asks. Too often in the past, monitoring, often deemed "successful", has been a hodgepodge of data collection for the sake of having information. Then, users of this information are surprised when it cannot address specific questions or issues that they have. Monitoring requires knowledge of these questions before a single sample is taken to ensure that the information, collected at some

---

[3] http://www.epa.gov/OWOW/monitoring/nationalsurveys.html.

often not insignificant cost, can be used to address these questions. The NCA is an example of how early planning, a comprehensive and sometimes exhaustive discussion of potential questions, objectives, permissible error, and dedication to detail supplies data that can answer multiple questions at national, regional, state, and to some extent local spatial scales, even in environments as complex as estuaries. No monitoring system can answer all possible questions about a particular ecosystem or resource. Even NCA data, which can be used to address many questions, does not have the spatial or temporal resolution to apply to most local scales, seasonal differences, or tidally-specific scales. The questions posed for NCA simply did not require such fine-scale information, because it was designed to assess worst case conditions at national, regional and state spatial scales, so seasonal and tidal variations had little relevance to the questions that were posed.

Once the monitoring questions have been defined, an understanding of the physical, chemical and biological attributes of the ecosystem is important for developing a response design. Knowledge of what characteristics drive the ecosystem in question and the influence of varying geographical locations and drivers (e.g., off-shore currents, temperature regimes, physical topography) is essential. NCA attempted to enforce consistency in all instances except where the natural functioning of the estuary or estuarine ecosystem differed fundamentally. Examples are contrasts between tidal rivers and coastal embayments, differences in expected nutrient concentrations in eastern, western and tropical estuaries, and threshold values for water clarity adjusted for natural differences in turbidity.

Finally, no monitoring system is likely to be useful without some recognition of environmental policy issues. For NCA, this acknowledgment is demonstrated in the two-tiered approach to determining the proportion of estuarine area that is in unacceptable condition. The most basic element of condition assessment is scientific; i.e. the proportion of area that is above or below some criterion (e.g., dissolved oxygen levels < 2 mg L^{-1}). A second, policy-oriented element, often more controversial, is the labeling of indicator ranges as good, fair, or poor and condition as acceptable or unacceptable. Finally, even deeper into the policy realm, is determining how much of a spatial unit (estuary, state, region, nation) has to be in unacceptable condition before you label the entire spatial unit as in unacceptable condition. For example, if 3% of the estuarine area of the Gulf of Mexico has dissolved oxygen below 2 mg L^{-1} in late summer, is it appropriate to assess the condition of Gulf of Mexico estuaries as unacceptable with respect to dissolved oxygen? Distinctions among good, fair, poor, and acceptable or unacceptable are normative, not scientific, even though rigorously based on consistent interpretations of data, and agreed upon by scientists.

The NCA, over 15 years of development within EPA's Office of Research and Development, demonstrated that a useful and meaningful estuarine monitoring program can address a large suite of questions in a consistent and structured manner. The adoption of this approach by a regulatory program office (EPA's Office of Water) means that consistent information will continue to be generated so that EPA and coastal states can assess the condition of estuaries and how the condition changes over time.

REFERENCES

Bourgeois, P.E., Robb, S.C., Summers, J.K. & Macauley, J.M. (1997). EMAP's Estuary Program: Interagency efforts develop master database. *Geogr. Info. Syst. 7*, 14–18.

Bricker, S.B., Clement, C.G., Pirhalla, D.E., Orlando, S.P. & Farrow, D.R.G. (1999). *National Estuarine Eutrophication Assessment: Effects of Nutrient Enrichment in the Nation's Estuaries.* U.S. Department of Commerce, National Oceanic and Atmospheric Administration, Silver Spring, MD.

Cochran, W.G. (1987). *Sampling Techniques.* John Wiley & Sons, New York.

Crossett, K.M., Culliton, T.J., Wiley, P.C. & Goodspeed, T.R. (2004). *Population Trends along the Coastal United States: 1980–2008.* National Oceanic and Atmospheric Administration, National Ocean Service, Silver Spring, MD.

Cunningham, C. & Walker, K. (1996). Enhancing public access to the coast through the CZMA. *Current: J. Mar. Educ. 14*, 8–12.

DiDonato, G.T., DiDonato, E.M., Smith, L.M., Harwell, L.C. & Summers, J.K. (2009). Assessing coastal waters of American Samoa: territory-wide water quality data provide a critical "big-picture" view for this tropical archipelago. *Environ. Monit. Assess., 150,* 157–165.

Engle, V.D., Hyland, J.L. & Cooksey, C. (2009). Effects of Hurricane Katrina on benthic macroinvertebrate communities along the northern Gulf of Mexico coast. *Environ. Monit. Assess. 150* 193–209.

Engle, V.D. & Summers, J.K. (1999). Refinement, validation, and application of a benthic condition index for northern Gulf of Mexico estuaries. *Estuaries 22,* 624–635.

Engle, V.D. & Summers, J.K. (2000). Biogeography of benthic macroinvertebrates in estuaries along the Gulf of Mexico and western Atlantic coasts. *Hydrobiologia 436,* 17–33.

Engle, V.D., Summers, J.K. & Gaston, G.R. (1994). A benthic index of environmental condition of Gulf of Mexico estuaries. *Estuaries 17,* 372–384.

Garza, C. (2008). Relating spatial scale to patterns of polychaete species diversity in coastal estuaries of the western United States. *Landscape Ecol. 23,* 107–121.

Günther, O. (1998). *Environmental Information Systems.* Springer, Berlin.

Hale, S.S. (2010). Biogeographical patterns of marine benthic macroinvertebrates along the Atlantic coast of the northeastern USA. *Estuar. Coasts 33,* 1039–1053.

Hale, S.S., Bahner, L.H. & Paul, J.F. (2000). Finding common ground in managing data used for regional environmental assessments. *Environ. Monit. Assess. 63,*143–157.

Hale, S.S. & Heltshe, J.F. (2008). Signals from the benthos: Development and evaluation of a benthic index for the nearshore Gulf of Maine. *Ecol. Indic. 8,* 338–350.

Hale, S.S., Hughes, M.M., Paul, J.F., McAskill, R.S., Rego, S.A., Bender, D.R., Dodge, N.J., Richter, T.R. & Copeland, J.L. (1998). Managing scientific data: The EMAP approach. *Environ. Monit. Assess. 51,* 429–440.

Hale, S.S., Miglarese, A.H., Bradley, M.P., Belton, T.J., Cooper, L.D., Frame, M.T., Friel, C.A., Harwell, L.M., King, R.E., Michener, W.K., Nicolson, D.T. & Peterjohn, B.G. (2003). Managing troubled data: Coastal data partnerships smooth data integration. *Environ. Monit. Assess. 81,* 133–148.

Hale, S.S., Rosen, J.S., Scott, D., Paul, J.F. & Hughes, M.M. (1999). *EMAP Information Management Plan: 1998–2001.* EPA/620/R-99/001a. U.S. Environmental Protection Agency, Office of Research and Development, Washington, DC.

Harvey, J., Harwell, L. & Summers, J.K. (2008). Contaminant concentrations in whole-body fish and shellfish from US estuaries. *Environ. Monit. Assess. 137,* 403–412.

Jordan, S.J., Lewis, M.A., Harwell, L.M. & Goodman, L.R. (2010). Summer fish communities in northern Gulf of Mexico estuaries: Indices of ecological condition. *Ecol. Indic. 10,* 504–515.

Kurtz, J.C., Jackson, L.E. & Fisher, W.S. (2001). Strategies for evaluating indicators based on guidelines from the Environmental Protection Agency's Office of Research and Development. *Ecol. Indic. 1,* 49–60.

Lellis-Dibble, K.A., McGlynn, K.E. & Bigford, T.E. (2008). *Estuarine Fish and Shellfish Species in U.S. Commercial and Recreational Fisheries: Economic Value as an Incentive to Protect and Restore Estuarine Habitat.* NMFSF/SPO-90. U.S. Department of Commerce, National Oceanic and Atmospheric Administration, Silver Spring, MD.

Long, E.R., MacDonald, D.D., Smith, S.L., & Calder, F.D. (1995). Incidence of adverse biological effects within ranges of chemical concentrations in marine and estuarine sediments. *Environ. Manage. 19,* 81–97.

Macauley, J.M., Smith, L.M., Harwell, L.C., & Benson, W.H. (2010). Sediment quality in near coastal waters of the Gulf of Mexico: influence of Hurricane Katrina. *Environ. Toxicol. Chem. 29,* 1403–1408.

McDonald, M., Blair, R., Dlugosz, J.J., Hale, S., Hedtke, S., Heggem, D., Jackson, L., Jones, K., Levinson, B., Olsen, A., Paulsen, S., Stoddard, J., Summers, K., & Veith, G. (2001). Environmental Protection Agency's Environmental Monitoring and Assessment Program (EMAP) in the 21st century. *Hydrol. Sci. Tech. 18,* 133–143.

NRC (Natural Resources Council). (1994). *Science and Judgment in Risk Assessment.* National Academy Press, Washington, DC:

NRC (Natural Resources Council). (1996). *Understanding Risk: Informing Decisions in a Democratic Society.* National Academy Press, Washington, DC:.

NRC (Natural Resources Council). (2000). *Clean Coastal Waters: Understanding and Reducing the Effects of Nutrient Pollution.* National Academy Press, Washington, DC:

Nelson, W.G. & Brown, C.A. (2008). Use of probability-based sampling of water-quality indicators in supporting development of quality criteria. *ICES J. Mar. Sci. 65,* 1421–1427.

NOAA. (2011). NOAA's State of the Coast. National Oceanic and Atmospheric Administration, National Ocean Service. Available at http://stateofthecoast.noaa.gov (Accessed January 2011).

Olsen, A.R., Sedransk, J., Edwards, D., Gotway, C.A., Liggett, W., Rathbun, S., Reckhow, K., & Young, L.J. (1999). Statistical issues for monitoring ecological and natural resources in the United States. *Environ. Monit. Assess. 54,* 1–45.

Paul, J.F., Walker, H.A., Galloway, W., Pesch, G., Cobb, D., Strobel, C.J., Summers, J.K., Charpentier, M., & Heltshe, J. (2008). Combining existing monitoring sites with a probability survey design – examples from U.S. EPA's National Coastal Assessment. *The Open Environ. Biol. Monit. J. 1,* 16–25.

Simons, J.D. & Smith, C.R. (2009). Texas National Coastal Assessment (2000–2004): challenges, solutions, lessons learned and future directions. *Environ. Monit. Assess. 150,* 167–179.

Smith, L.M., Engle, V.D., & Summers, J.K. (2006). Assessing water clarity as a component of water quality in Gulf of Mexico estuaries. *Environ. Monit. Assess. 115,* 291–305.

Smith, L.M., Macauley, J.M., Harwell, L.C., & Chancy, C.A. (2009). Water quality in the near coastal waters of the Gulf of Mexico affected by Hurricane Katrina: before and after the storm. *Environ. Manage. 44,* 149–162.

Stevens, D. L. & Olsen, A. R. (2003). Variance estimation for spatially balanced samples of environmental resources. *Environmetrics 14,* 593–610.

Stevens, D.L. & Olsen, A.R. (2004). Spatially balanced sampling of natural resources. *J. Am. Stat. Assoc. 99,* 262–278.

Summers, J.K. (2001). Ecological condition of the estuaries of the Atlantic and Gulf coasts of the United States. *Environ. Toxicol. Chem. 20,* 99–106.

Summers, J.K., Paul, J.F., Robertson, A. (1995). Monitoring the ecological condition of estuaries in the United States. *Toxicol. Environ. Chem. 49,* 93–108.

The Economist. (2010). Data, data everywhere: A special report on managing information. The Economist Newspaper, Limited. London. [http://www.economist.com/] 27 Feb 2010.

Trowbridge, P.R. & Jones, S.H. (2009). Detecting water quality patterns in New Hampshire's estuaries using National Coastal Assessment probability-based survey data. *Environ. Monit. Assess. 150,* 129–142.

U.S. Environmental Protection Agency (U.S. EPA). (1994). *Methods for assessing the toxicity of sediment-associated contaminants with estuarine and marine amphipods.* EPA/600/R-94/025.

U.S. Environmental Protection Agency, Office of Research and Development, Washington, DC.

U.S. EPA. (1998a). *Clean Water Action Plan: Restoring America's Waters.* EPA-840-98-001.

U.S. Environmental Protection Agency, Office of Research and Development, Washington, DC.

U.S. EPA. (1998b). *Guidelines for Ecological Risk Assessment.* EPA/630/R-95/002F. U.S. Environmental Protection Agency, Office of Research and Development, Washington, DC.

U.S. EPA. (2000a). *Ambient Water Quality Criteria for Dissolved Oxygen (Saltwater): Cape Cod to Cape Hatteras.* EPA/822-R-00-012. U.S. Environmental Protection Agency, Office of Water, Washington, DC.

U.S. EPA. (2000b). *Guidance for Assessing Chemical Contaminant Data for Use in Fish Advisories, Volume 2: Risk Assessment and Fish Consumption Limits.* EPA/823-B-00-008. U.S. Environmental Protection Agency, Office of Water, Washington, DC.

U.S. EPA. (2001a). *National Coastal Assessment Field Operations Manual.* EPA/620-R-01-003. U.S. Environmental Protection Agency, Office of Research and Development, Washington, DC.

U.S. EPA. (2001b). *National Coastal Assessment Quality Assurance Project Plan.* EPA/620-R-01-002. U.S. Environmental Protection Agency, Office of Research and Development, Washington, DC.

U.S. EPA. (2001c). *National Coastal Condition Report*. EPA/620-R-01-005. U.S. Environmental Protection Agency, Office of Research and Development and Office of Water, Washington, DC.

U.S. EPA. (2003). *Elements of a State Water Monitoring and Assessment Program*. EPA/841-B-03-003. U.S. Environmental Protection Agency, Office of Wetlands, Oceans and Watersheds, Washington, DC.

U.S. EPA. (2004). *National Coastal Condition Report II*. EPA/620-R-03/002. U.S. Environmental Protection Agency, Office of Research and Development and Office of Water. Washington, DC.

U.S. EPA. (2008). *National Coastal Condition Report III*. EPA/842-R-08/002. U.S. Environmental Protection Agency, Office of Research and Development and Office of Water. Washington, DC.

U.S. EPA (2010). National Coastal Condition Assessment Site Evaluation Guidelines. http://water.epa.gov/type/oceb/upload/ncca-siteeval.pdf Accessed June 21, 2011.

Van Dolah, R.F., Chestnut, D.E., Jones, J.D., Jutte, P.C., Riekerk, G., Levisen, M., & McDermott, W. (2003). The importance of considering spatial attributes in evaluating estuarine habitat condition: the South Carolina experience. *Environ. Monit. Assess. 81,* 85–95.

Van Dolah, R.F., Hyland, J.L., Holland, A.F., Rosen, J.S., & Snoots, T.T. (1999). A benthic index of biological integrity for assessing habitat quality in estuaries of the southeastern USA. *Mar. Environ. Res. 48,* 269–283.

Weisberg, S.B., Ranasinghe, J.A., Dauer, D.D., Schnaffer, L.C., Diaz, R.J., & Frithsen, J.B. (1997). An estuarine benthic index of biotic integrity (B-IBI) for Chesapeake Bay. *Estuaries 20,* 149–158.

In: Estuaries: Classification, Ecology and Human Impacts   ISBN: 978-1-61942-083-0
Editor: Steve Jordan   © 2012 Nova Science Publishers, Inc.

Chapter 12

IMPACTS OF FLOODS AND SEA-LEVEL RISE ON COASTS AND ESTUARIES: A PROBABILISTIC BAYESIAN MODELING STRATEGY

M. Rajabalinejad
Delft University of Technology, Delft, the Netherlands
Z. Demirbilek
U.S. Army Engineer R&D Center, Coastal and Hydraulic Laboratory, Vicksburg, MS, US

ABSTRACT

The ability of coasts to withstand floods and sea-level rise is investigated using a probabilistic Bayesian method. Fundamental principles of recently developed probabilistic computational methods for monotonic models are presented. Computational efficiency of these models and their coupling with a family of Monte Carlo (MC) methods are investigated for coastal estuaries and inter-connecting coasts. These integrated state-of-the-art computational modeling techniques are used in strategic planning and engineering management decision tools to evaluate impacts of storm hazards on estuaries and coasts in development of flood protection plans and assessment of damage to infrastructures.

INTRODUCTION

A probabilistic Bayesian modeling strategy is presented in this chapter for investigating the strength and resiliency of coasts against floods and sea-level rise (SLR). An accurate and reliable assessment of this ability is paramount to coastal, civil and hydraulics engineering communities. The need for good predictive technology became apparent after Hurricane Katrina's devastation of New Orleans (USA) and nearby communities. The failure of coastal defense systems built to protect the city of New Orleans endangered humans and the environment, resulting in significant casualties, loss of life and large economic damage

(Demirbilek 2010a,b, Rajabalinejad 2008). The tragic events in New Orleans caused by Hurricane Katrina were seen worldwide, exemplifying types of damages, losses and human suffering that could occur in exposed coastal regions throughout the world. Global warming and SLR are expected to lead to higher storm surges and increasing flood risks (van Gelder et al. 2007, Demirbilek et al. 2009, Mark et al. 2007). More resilient and reliable flood defenses will be required to meet increasing risks for failures of systems caused by episodic extreme meteorological and atmospheric (metocean) events.

The focus in this chapter is on introduction of theory and applications of some recently developed novel probabilistic methods for assessing the safety of coastal estuaries and inter-connecting coasts. A method is proposed an its implementation is demonstrated through a step-by-step procedure for two real-world practical applications at New Orleans and the Scheldt estuary (Netherlands). The method is used with other computational models to evaluate impacts of sea-level rise and floods on estuaries of different sizes and shapes. Results of this integrated state-of-the-art computational modeling techniques are incorporated into strategic decision tools to assess potential damages caused by tropical and extra-tropical storms to estuaries, coasts and infrastructure (Rajabalinejad 2008, Starr 2009, Haimes 2004, Thompson & Graham 1996).

The mathematical method demonstrated employs the monotonic concept described in Rajabalinejad (2009) and Rajabalinejad et al. (2010a,b). However, it is not limited to monotonicity; it also uses Bayes' theory to incorporate prior knowledge from the solution of a problem. This is called the Bayesian Monte Carlo method (BMC), which integrates informative priors to optimize a system's load-response relationship in the simulations. The basic theoretical principles of these methods were developed in a recent PhD study (Rajabalinejad 2009), and further information about BMC and dynamic bounds (DB) methods are presented in the dissertation. The emphasis in this chapter is on application of these methods for evaluating the safety and functional performance of estuarine and coastal defense structures. The safety assessment of monotonic models is investigated at three levels. The first level assessment is based on the DB method, briefly described here with references given for interested readers. At the second level, the DB method is integrated with the BMC method by using the Gaussian error function. This method improves accuracy of estimates and is highly efficient. The third level of the computational method involves an integration of BMC with the beta distribution. This method fully considers the monotonic property, but it is comparatively more complex than other methods mentioned. Therefore, in this chapter we describe state-of-the-art tools for monotonic models with different levels of complexity for determining failure probability of flood defence systems.

The proposed numerical procedure allows modelers to consider different types of prior information. The numerical model is computationally efficient, with a speed up factor ranging from 50 to 1000 for the standard MC simulations, depending on the number of controlling parameters and linear (or nonlinear) characteristics of problems being solved. This efficient new computational procedure is demonstrated for assessment of safety measures for two estuaries in different parts of the world: Scheldt estuary in The Netherlands and New Orleans in the USA.

DYNAMIC BOUNDS

The DB method was introduced in Rajabalinejad (2009) and Rajabalinejad et al. (2010c), as a method for reliability analysis of engineering systems with monotonic limit state equations (LSEs), which have a limited number of influential variables. The application of this method to a complex flood defense structure is presented in Jaynes (2003). The DB method takes into account the monotonicity and typical prior information of LSEs in order to reduce the very large number of calculations required in the MC process. The DB method divides the range of an LSE into three regions: stable, unstable and unqualified. This procedure speeds up the MC process (as depicted in Figure 1) to improve the calculation efficiency by gradually shrinking the size of the "unqualified" region. The details of the algorithm for this computational technique are described in Rajabalinejad (2009) and Rajabalinejad et al. (2010c). For completeness, because the DB method is very effective for monotonic LSEs, a brief discussion of monotonicity and the underlying principles of the Bayesian Monte Carlo method will be described next..

PRINCIPLES OF THE BAYESIAN MONTE CARLO (BMC) METHOD

Consider a continuous function U that we wish to estimate at a number of discrete points. We define the u_i set of discrete points by a vector $\vec{u}$ assigned to discrete points (pixels). The elements of observed data points d_i are defined by vector $d = [d_1, \cdots, d_n]$, and their locations are stored in a n-dimensional vector. Let $P(u_j \mid D, I)$ be the univariate probability (P) density function (pdf) for an arbitrary pixel u_j. The data D and informational context I can be found from the simulations and model, respectively. The global uncertainty σ (e.g., global standard deviation) was first used in Rajabalinejad (2009)—see also Rajabalinejad et al. (2008c), Rajabalinejad & Mahdi (2010), and Rajabalinejad et al. (2011)—to define a nuisance parameter. With marginalization, global uncertainty can be written as

$$P(u_j \mid D, I) = \int P(u_j, \sigma \mid D, I)\, d\sigma. \tag{1}$$

By application of Bayes theorem, we find

$$P(u_j \mid D, I) = \int P(u_j \mid \sigma, D, I)\, P(\sigma \mid D, I)\, d\sigma. \tag{2}$$

With the global uncertainty (σ) defined, we can estimate the value of the limit state equation (LSE) at any arbitrary point x_i from an interpolation function (model) f using information about its neighboring points. Let the estimate be $\hat{u}_i$. In this model, the value of

u_i is estimated by its neighboring points. There are $m+1$ neighboring points for each arbitrary location among these points. The model f_m is defined using Equation (2), where the index m is the order of model (function). We can estimate value of LSE at the point x close to the middle of its neighboring points by

$$f_m(x_i) = \sum_{k=0,k\neq r}^{m} u_{i-r+k} L_{i-r+k}(x_i), \tag{3}$$

where $r = \text{int}\,(\dfrac{m+1}{2})$, u_i is the LSE response assigned to point x_i, and L_k is the i-th fundamental polynomial defined as

$$L_k(x_i) = \prod_{k\neq l} \left(\frac{x_i - x_l}{x_k - x_l} \right). \tag{4}$$

where $L_k(x) = 1$ at $x = x_k$ and $L_k(x) = 0$ at all other data points x_l for $k \neq l$. The linear response is given by

$$\hat{u}_i = f_1(u_{i+1}, u_{i-1}) = u_{i-1} + \frac{u_{i+1} - u_{i-1}}{x_{i+1} - x_{i-1}}(x_i - x_{i-1}). \tag{5}$$

For the m-th order model f_m, the error e_i becomes

$$e_i = u_i - \hat{u}_i = u_i - f_m(x_i) = u_i - \sum_{k=0,k\neq r}^{m} u_{i-r+k} L_{i-r+k}(x_i), \tag{6}$$

where r and m are defined by Equation (3).

In this section, we will show the extension of Equation (2) to higher dimensions. In this case, we define a vector of global uncertainties as $\vec{\sigma} = [\sigma_1, \sigma_2, \ldots, \sigma_n]$, where each uncertainty is associated with its respective dimension $\vec{x} = [x_1, x_2, \ldots, x_n]$ of the LSE $G(\vec{x})$. The exact value of $G(\vec{x})$ at any point $\vec{x}_j$ is u_j. Using Equation (2), we have

$$P(u_j \mid D, I) = \int \cdots \int P(u_j, \vec{\sigma} \mid D, I) d\vec{\sigma}. \tag{7}$$

For instance, a two-dimensional problem involves two-dimensional global uncertainty given by $\vec{\sigma} = [\sigma_1, \sigma_2]$. We treat these uncertainties as two "nuisance parameters" (Rajabalinejad 2009, Sivia & Skilling 2006) that can be integrated to yield

$$P(u_j \mid D, I) = \iint P(u_j, \vec{\sigma} \mid D, I) \, d\vec{\sigma}, \tag{8}$$

Equation (8) may be re-written using the Bayes theorem in the following form:

$$P(u_j \mid D, I) = \iint P(u_j \mid \vec{\sigma}, D, I) P(\vec{\sigma} \mid D, I) \, d\vec{\sigma}. \tag{9}$$

The uncertainty estimation for two and higher dimensions is straight forward and is described elsewhere (Bojanov et al. 1993, DeBoor & Ron 1990). Two extensions of the BMC to monotonic models either with a Gaussian or Beta distribution for estimation of modeling error are described in the next two sections. These are followed by description of the computational algorithm and two practical applications demonstrating the application of the integrated modeling system.

BMC for Monotonic Models with a Gaussian Distribution

We need to define the first term $P(u_j \mid \sigma)$ in the right hand side of Equation 2, which is equivalent to Equation (7) if the global uncertainty (σ) has only one element. The value of error with a zero mean could be positive or negative and its unknown variance is σ_j^2. Assume that the standard deviation of error is proportional to the shortest distance from its neighboring data points. This allows us to use the Gaussian density function for the model error in the standard error form (Jaynes 2003). The Gaussian error is defined as

$$P(e_j \mid \sigma_j) = \frac{1}{\sqrt{2\pi}\sigma_j} \exp\left\{ -\frac{1}{2\sigma_j^2} e_j^2 \right\}, \tag{10}$$

where e_j is the error and σ_j^2 is the unknown variance. By changing the variable from e_j to u_j, we find the following multivariate pdf for the pixels u_j as:

$$P(u_j \mid \sigma_j) = \frac{1}{\sqrt{2\pi}\sigma_j} \exp\left\{ -\frac{1}{2\sigma_j^2} \left(u_j - f_m(x_j) \right)^2 \right\}. \tag{11}$$

Following the approach used in Rajabalinejad (2009) and Rajabalinejad et al. (2008a) to associate the global and local uncertainties, we define

$$\sigma_j = \alpha_j \sigma, \tag{12}$$

where

$$\alpha_j = \sqrt[q]{\min\left\{\left|x_{j-r+k} - x_j\right| : k = 1 \cdots m\right\}}, \tag{13}$$

m is the order of model and $r = abs\left(\dfrac{m+1}{2}\right)$, as defined in Equation (3) and q is the model response order as discussed in Rajabalinejad et al. (2010 a, c).

To determine the second term $P(\sigma \mid D, I)$ in the right hand side of Equation (2), we apply the Gaussian distribution in Equation (10) to the error of the i-th interval e_i having an unknown variance of σ_i^2 and a mean value $\hat{u}_i$ obtained from Equation (3). There are n non-overlapping data points, and the error of the model at the location of any observed data point may be defined as:

$$e_i = u_i - d_i, \tag{14}$$

where $i = 1, \ldots, n$. Assuming the error in Equation (14) has a zero mean (e.g., the error may have positive or negative values) with an unknown variance σ_i^2, we apply the principle of Maximum Entropy to define its probability density function as follows:

$$P(\sigma_i \mid e_i) = \frac{1}{\sqrt{2\pi}\sigma_i} \exp\left\{-\frac{1}{2\sigma_i^2} e_i^2\right\}. \tag{15}$$

Substituting Equation (14) into (15) and changing the variable from the error e_i to data d_i, the function can be re-written as

$$P(\sigma_i \mid u_i, d_i) = \frac{1}{\sqrt{2\pi}\sigma_i} \exp\left\{-\frac{1}{2\sigma_i^2}\left(d_i - f_m(x_i)\right)^2\right\}, \tag{16}$$

where $i = 1, \ldots, n$. By assuming logical independence between the errors, and making appropriate substitutions, we obtain the following function

$$P(\sigma_1, \ldots, \sigma_n \mid u_1, \ldots, u_n, d_1, \ldots, d_n)$$

$$= \frac{1}{(2\pi)^{n/2}(\sigma_1 \ldots \sigma_n)} \exp\left\{-\sum_{i=1}^{n} \frac{1}{2\sigma_i^2}\left(d_i - f_m(x_i)\right)^2\right\}. \tag{17}$$

In Equation (17), σ_i is the standard deviation of error. Assuming that the error is proportional to the shortest distance to neighboring points, we apply Equations (35) and (36) to relate local standard deviations σ_i to a global standard deviation σ. Equation (17) can then be written as

$$P(\sigma \mid u_1,\ldots,u_n,d_1,\ldots,d_n)$$
$$= \frac{1}{(2\pi)^{n/2}(B\sigma^n)}\exp\left\{-\frac{1}{2\sigma^2}\sum_{i=1}^{n}\frac{\left(d_i - f_m(x_i)\right)^2}{\alpha_i^2}\right\}, \tag{18}$$

where B is

$$B = \alpha_0 \cdot \alpha_1 \cdots \alpha_{v+1}. \tag{19}$$

Equation (18) may be expressed as

$$P(\sigma \mid D,I) \propto \frac{1}{\sigma^n}\exp\left\{-\frac{1}{2\sigma^2}\sum_{i=1}^{n}\frac{\left(d_i - f_m(x_i)\right)^2}{\alpha_i^2}\right\}. \tag{20}$$

Multiplication of Equations (11) and (12) according to Equation (2) leads to the following equation:

$$P(u_j \mid D,I) \propto$$
$$\int \frac{1}{\sigma_j \sigma^n}\exp\left\{-\frac{1}{2\sigma_j^2}\left(u_j - f_m(x_j)\right)^2 - \frac{1}{2\sigma^2}\sum_{i=1}^{n}\frac{\left(d_i - f_m(x_i)\right)^2}{\alpha_i^2}\right\}d\phi. \tag{21}$$

Equation (21) can also be expressed as

$$P(u_j \mid D,I) \propto$$
$$\int \frac{1}{\sigma^{n+1}}\exp\left\{-\frac{1}{2\sigma^2}\left(\frac{\left(u_j - f_m(x_j)\right)^2}{\alpha_j^2} - \sum_{i=1}^{n}\frac{\left(d_i - f_m(x_i)\right)^2}{\alpha_i^2}\right)\right\}d\sigma. \tag{22}$$

Integration of Equation (22) over σ gives the posterior $P(u_j \mid D,I)$ as

$$P(u_j \mid D, I) \propto \left[\frac{\dfrac{\left(u_j - f_m\left(x_j\right)\right)^2}{\alpha_j^2}}{\displaystyle\sum_{i=1}^{n} \left\{ \dfrac{\left(d_i - f_m\left(x_i\right)\right)^2}{\alpha_i^2} \right\}} + 1 \right]^{-\frac{1}{2}n} . \tag{23}$$

BMC FOR MONOTONIC MODELS USING A BETA DISTRIBUTION

We need to define the first term $P(u_j \mid \sigma)$ in the right hand side of Equation (2). The value of error with a zero mean could be positive or negative and its unknown variance is σ_j^2. Assume the standard deviation of error is proportional to the shortest distance from its neighboring data points. We use the generalized beta (GB) density function for the error (Equation 6) to assure monotonic constraint of the model. The GB distribution ensures that $\hat{u}_j = f(x_j)$ is bounded between u_{j-1} and u_{j+1} and its density function for error is a suitable choice. The GB density is defined as

$$P(x \mid c, d) = \frac{(x-c)^{p-1}(d-x)^{q-1}}{B(p,q)(d-c)^{p+q-1}}, \tag{24}$$

for $c \leq x \leq d$; $B(p,q)$ is the beta function where p and q are the Beta parameters. The mean and standard deviation of this distribution are defined in Equations (25) and (26), respectively as

$$E(x) = c + \frac{p}{p+q}(d-c), \tag{25}$$

$$Var(x) = \frac{pq}{(p+q)^2(p+q+1)}(d-c)^2. \tag{26}$$

Using the GB distribution at the interval of $[u_{j-1} \ u_{j+1}]$ and assuming $u_{j-1} < u_{j+1}$, we have

$$P(x \mid u_{j-1}, u_{j+1}) = \frac{(x-u_{j-1})^{p-1}(u_{j+1}-x)^{q-1}}{B(p,q)(u_{j+1}-u_{j-1})^{p+q-1}}, \tag{27}$$

where $u_{j-1} \leq x \leq u_{j+1}$. The estimate of the pixel value is $u_j = x$ and the error function is defined as

$$e_j = x - u_{j-1},$$ (28)

which is equivalent to

$$e_j = u_j - u_{j-1}.$$ (29)

Substituition of Equation (28) into Equation (27) gives

$$P(u_j \mid u_{j-1}, u_{j+1}, p_j, q_j) = \frac{(u_j - u_{j-1})^{p_j-1}(u_{j+1} - u_j)^{q_j-1}}{B(p_j, q_j)(u_{j+1} - u_{j-1})^{p_j+q_j-1}}.$$ (30)

The first and second moments of Equation (30) are:

$$E(u_j) = \hat{u}_j = u_{j-1} + \frac{p_j}{p_j + q_j}(u_{j+1} - u_{j-1}),$$ (31)

$$Var(u_j) = \sigma_j^2 = \frac{p_j q_j}{(p_j + q_j)^2 (p_j + q_j + 1)}\left(u_{j+1} - u_{j-1}\right)^2.$$ (32)

The model estimate $\hat{u}_j$ is the expected value of the pdf in Equation (30), while its standard deviation is unknown. With these definitions, we define the parameters of GB density in Equation (27) as

$$p_j = \frac{(u_{j-1} - \hat{u}_j)(-\hat{u}_j u_{j+1} + u_{j+1} u_{j-1} + \sigma_j^2 + \hat{u}_j^2 - \hat{u}_j u_{j-1})}{(u_{j+1} - u_{j-1})\sigma_j^2},$$ (33)

$$q_j = \frac{-\hat{u}_j u_{j+1}^2 + u_{j-1} u_{j+1}^2 + u_{j+1}\sigma_j^2 + 2\hat{u}_j^2 u_{j+1} - 2\hat{u}_j u_{j-1} u_{j+1} - \sigma_j^2 \hat{u}_j + \hat{u}_j^2 u_{j-1} - \hat{u}_j^3}{(u_{j+1} - u_{j-1})\sigma_j^2},$$ (34)

where σ_j is the standard deviation of error. Following the approach used in Rajabalinejad (2009) and Rajabalinejad et al. (2008a), we find

$$\sigma_i = \alpha_i \sigma,$$ (35)

where

$$\alpha_i = \sqrt[q]{\min\left\{|x_{i-r+k} - x_i| : k = 1 \cdots m\right\}}, \tag{36}$$

where m is the model order, $r = abs\left(\dfrac{m+1}{2}\right)$ as indicated in Equation (3), and q is the expected response order of the model as discussed in Rajabalinejad et al. (2010 a,c).

To calculate the second term $P(\sigma \mid D, I)$ in the right hand side of Equation (2), we apply the beta distribution in Equation (27) to the error of the i-th interval e_i having an unknown variance of σ_i^2 and a mean value $\hat{u}_i$. We obtain

$$P(e_i \mid \sigma_i, \hat{u}_i) \propto \frac{(e_i - (u_{i-1} - u_i))^{p_i - 1}((u_{i+1} \quad u_i) \quad e_i)^{q_i - 1}}{B(p_i, q_i)(u_{i+1} - u_{i-1})^{p_i + q_i - 1}}. \tag{37}$$

Substitute Equation (29) into Equation (37), and change the variable from the error e_i to the data d_i to obtain the likelihood function as

$$P(\sigma_i \mid d_i, \hat{u}_i) \propto \frac{(d_i - u_{i-1})^{p_i - 1}(u_{i+1} - d_i)^{q_i - 1}}{B(p_i, q_i)(u_{i+1} - u_{i-1})^{p_i + q_i - 1}}. \tag{38}$$

Assuming logical independence between the errors of n data points, the likelihood function can be written as

$$P(\sigma_1, \ldots, \sigma_n \mid u_1, \ldots, u_n, d_1, \ldots, d_n) \propto \prod_{i=1}^{n}\left(\frac{(d_i - u_{i-1})^{p_i - 1}(u_{i+1} - d_i)^{q_i - 1}}{B(p_i, q_i)(u_{i+1} - u_{i-1})^{p_i + q_i - 1}}\right). \tag{39}$$

where σ_i is the standard deviation of error. Assuming that this uncertainty is also associated with the global uncertainty in Equations (35) and (36), we find

$$\sigma_i = \alpha_i \sigma, \tag{40}$$

Equation (40) can be written as

$$P(\sigma \mid u_1, \ldots, u_n, d_1, \ldots, d_n) \propto \prod_{i=1, i \in c}^{n}\left(\frac{(d_i - u_{i-1})^{p_i - 1}(u_{i+1} - d_i)^{q_i - 1}}{B(p_i, q_i)(u_{i+1} - u_{i-1})^{p_i + q_i - 1}}\right), \tag{41}$$

where

$$p_i = \frac{(u_{i-1} - \hat{u}_i)(-\hat{u}_i u_{i+1} + u_{i+1} u_{i-1} + \alpha_i^2 \sigma^2 + \hat{u}_i^2 - \hat{u}_i u_{i-1})}{(u_{i+1} - u_{i-1})\alpha_i^2 \sigma^2}, \tag{42}$$

$$q_i = \frac{-\hat{u}_i u_{i+1}^2 + u_{i-1} u_{i+1}^2 + u_{i+1} \alpha_i^2 \sigma^2 + 2\hat{u}_i^2 u_{i+1} - 2\hat{u}_i u_{i-1} u_{i+1} - \alpha_i^2 \sigma^2 \hat{u}_i + \hat{u}_i^2 u_{i-1} - \hat{u}_i^3}{(u_{i+1} - u_{i-1}) \alpha_i^2 \sigma^2}. \tag{43}$$

The final pdf is obtained by multiplying Equations (30) and (41), which yields

$$P(u_j \mid D, I) = \int \prod_{i=1, i \in c}^{n} \left(\frac{(d_i - u_{i-1})^{p_i - 1} (u_{i+1} - d_i)^{q_i - 1}}{B(p_i, q_i)(u_{i+1} - u_{i-1})^{p_i + q_i - 1}} \right) \\ \times \frac{(u_j - u_{j-1})^{p_j - 1} (u_{j+1} - u_j)^{q_j - 1}}{B(p_j, q_j)(u_{j+1} - u_{j-1})^{p_j + q_j - 1}} \, d\sigma \tag{44}$$

where p_j, q_j, p_i and q_i are obtained from Equations (33), (34), (42) and (43), respectively.

COMPUTATIONAL ALGORITHM

The implementation of an algorithm for the MC method is illustrated here for a two-dimensional example, followed by a description of mathematical formulations and computations. In Figure 1, the two-dimensional LSE $G(x_1, x_2) = 0$ is depicted with the contours of the joint probability density function $f(x_1, x_2)$ for two variables (X_1, X_2). The limit state equation $G(x_1, x_2)$ is assumed to be monotonically increasing in both variables. The data points may be exactly on the LSE or beneath or above the LSE surface. This means that $G(x_1, x_2) < 0$ for those points below the LSE surface, defined here as the *unstable region* of points corresponding to a failure of the system. For points which lie above the LSE, the function is positive ($G(x_1, x_2) > 0$), and this is called the *stable region*.

The first random point, $\vec{x}^{(1)} = (x_1^{(1)}, x_2^{(1)})$, is generated from the joint pdf of f. This is depicted by a black square labeled by 1 in Figure 3, for which $G(x_1^{(1)}, x_2^{(1)}) < 0$, corresponding to a failure in the unstable region. From the monotonicity of G it is inferred that $G(x_1, x_2) < 0$ for all points (x_1, x_2) in the quadrant to the left of and below $(x_1^{(1)}, x_2^{(1)})$. Next, a second point $(x_1^{(2)}, x_2^{(2)})$ is generated from f, for which $G(x_1^{(2)}, x_2^{(2)}) > 0$, indicating this point is in the stable region with other points in the right-upper quadrant in the stable region.

This process is repeated for all other points. The result for a small number of iterations is shown in Figure 3. The shaded regions constitute approximations to the stable and unstable regions used to obtain bounds on the probability of failure (p_f). However, the location of the LSE-curve is not known in practical applications, and becomes gradually more visible since it is sandwiched between approximations to stable and unstable regions. The dynamic upper and lower bounds of p_f become tighter and well-defined as more points are generated. It is not necessary to evaluate the LSE for each point generated because the status of a point, stable or unstable, can be determined from other neighboring points.

For a general mathematical description, now we revert to n-dimensional space. Define the stable and unstable sets S and U as

$$S = \{\vec{x} : G(\vec{x}) \geq 0\} \quad \text{and} \quad U = \{\vec{x} : G(\vec{x}) < 0\}. \tag{45}$$

If two points $\vec{x} = (x_1, \ldots, x_n)$ and $\vec{y} = (y_1, \ldots, y_n)$ satisfy the relationship $x_i \leq y_i$ for $i = 1, \ldots, n$, we say that $\vec{x}$ is *less stable than* $\vec{y}$, or $\vec{y}$ is *more stable than* $\vec{x}$. If $\vec{x} \in S$, $\vec{y} \in S$ follows. A similar observation applies to U.

Consider the k-th iteration of the MC process, in which a number of stable points (e.g., $\vec{s}^{(1)}$, ..., $\vec{s}^{(p)}$) and a number of unstable points ($\vec{u}^{(1)}$, ..., $\vec{u}^{(q)}$) were generated. The approximation to the stable region S is the union of the p-orthants (generalizing the quadrants in Figure 1) such that

$$H_i = \{\vec{x} : \overline{x}\text{i's more stable than } \vec{s}^{(i)}\}, \quad i = 1, \ldots, p. \tag{46}$$

If $\vec{s}^{(i)}$ is more stable than $\vec{s}^{(j)}$ for any i and j, its orthant H_i would have to be completely contained in H_j. In this case, there would be no loss of information if $\vec{s}^{(i)}$ were dropped from the list of data points. Likewise, the approximation for the unstable region U is the union of

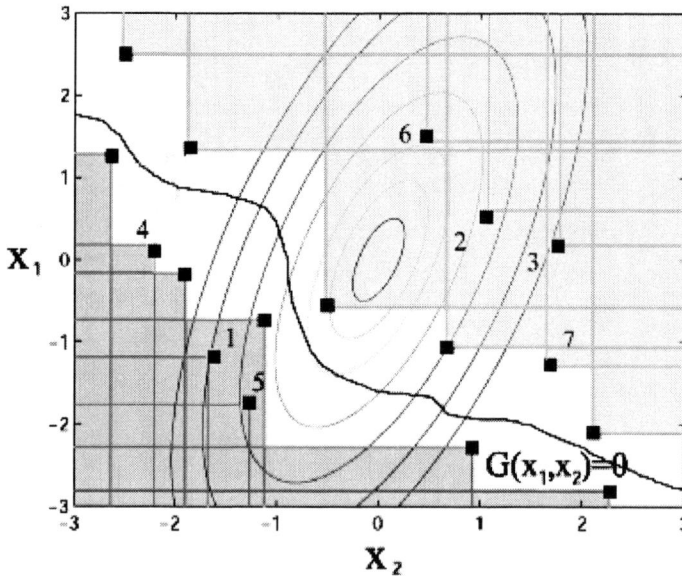

Figure 1. Illustration of the dynamic bounds algorithm. The response of a two dimensional LSE is divided into stable, unstable and unknown regions.

$$L_i = \{\vec{x} : \bar{x}\text{i's less stable than } \bar{u}^{(i)}\}, \quad i = 1, \ldots, q. \tag{47}$$

From now on, we shall assume that S_k and U_k are the corresponding approximations to S and U, and are defined as

$$S_k = \cup_{i=1}^{p} H_i \quad \text{and} \quad U_k = \cup_{i=1}^{q} L_i. \tag{48}$$

When the next random point $\vec{X}^{(k+1)}$ is generated from f, there are three possibilities concerning its location in relation to stable and unstable regions. The first possibility is that $\vec{X}^{(k+1)} \in S_k$, such that the point is located in the stable region. The second possibility is that $\vec{X}^{(k+1)} \in U_k$, where the point is located in the unstable region, increasing the number of failures by one. The third possibility is that $\vec{X}^{(k+1)} \notin S_k \cup U_k$, and in this case the point is located in the unqualified region between S_k and U_k, which means $G(\vec{X}^{(k+1)})$ must be evaluated with the Bayesian integration method using Equation (9). This equation can be evaluated using the Gaussian error function of Equation (23) or beta error function of Equation (44). The joint probability density function is assigned to this point if the following condition is satisfied:

$$\mathbf{1}[P(G(\mathbf{X}_i) < 0 \mid \mathbf{X}_i, I) \geq \alpha] \quad \text{or} \quad \mathbf{1}[P(G(\mathbf{X}_i) \geq 0 \mid \mathbf{X}_i, I) \geq \alpha], \tag{49}$$

The evaluation of $G(\vec{X}^{(k+1)})$ is not necessary. In Equation (43), $\mathbf{1}[C]$ is 1 if condition C is true and 0 otherwise, and I is the prior and α is the acceptance criterion. The parameter α is selected for the desired accuracy level, and we use $\alpha = 0.90$ as the accuracy level corresponding to the MC stop rule described in Rajabalinejad (2010c):

$$N \geq 400(\frac{1}{\hat{p}_f} - 1), \tag{50}$$

where $\hat{p}_f$ is the estimated failure probability. It is important to note that if $\alpha = 1.00$, the DB method is applied.

If $\mathbf{1}[P(G(\mathbf{X}_i) \geq 0 \mid \mathbf{X}_i, I) \geq \alpha]$ is true, $\vec{X}^{(k+1)}$ is added to the collection of known stable points. If $\mathbf{1}[P(G(\mathbf{X}_i) < 0 \mid \mathbf{X}_i, I) \geq \alpha]$ is true, $\vec{X}^{(k+1)}$ is added to the collection of unstable points and a similar update is performed. If condition (49) is not satisfied, $G(\vec{X}^{(k+1)})$ is evaluated.

Note that values of p and q vary during the simulation and depend on the iteration number k.

The key algorithmic steps can be summarized as follows:

1. Determine S_0 and U_0. These could be empty sets or determined from threshold points. Set $k = 0$ and $n_f = 0$.

2. Increase k by 1 to generate $\vec{X}^{(k)}$ from f. If $\vec{X}^{(k)} \in U_{k-1}$, increase n_f by 1, and update U_{k-1} to obtain U_k. If $\vec{X}^{(k)} \nsubseteq S_{k-1} \cup U_{k-1}$, Then

   2.1. For the DB method, go to 3.

   2.2. For the BMC with Gaussian error, evaluate $G(\vec{X}^{(k)})$ using the two dimensional form of Equation (23).

   2.3 For the BMC with beta error, evaluate $G(\vec{X}^{(k)})$ using the two dimensional form of Equation (44).

   If $\mathbf{1}[P(G(\mathbf{X}_i) < 0 \mid \mathbf{X}_i, I) \geq \alpha]$ is true, increment n_f by 1 and update U_{k-1} to obtain U_k. If $\mathbf{1}[P(G(\mathbf{X}_i) \geq 0 \mid \mathbf{X}_i, I) \geq \alpha]$, update S_{k-1} to obtain S_k. If Equation (9) is not fulfilled, evaluate $G(\vec{X}^{(k)})$ and if it is negative, add 1 to n_f and update U_{k-1} to obtain U_k; otherwise, update S_{k-1} to obtain S_k.

3. Repeat until $k = N$.

4. Calculate $\hat{p}_f = n_f / N$ as an MC estimate for p_f.

Figure 2. A plan view of the Scheldt Estuary in The Netherlands. Tthe white area is called "Zeeuwsch Vlaannderen" and is counted as Dike Ring 32 in the protection system of the Scheldt Estuary (Rajabalinejad 2010a).

The estimate $\hat{p}_f$ is as good as the ordinary MC estimate based on N independent samples $1[G(\vec{X}^{(1)}) < 0],...,1[G(\vec{X}^{(N)}) < 0]$, but requires evaluation of G for only a small fraction of all data points available. Additionally, the simulation yields upper and lower bounds of p_f as follow:.

If $U_N \subset U$ and $S_N \subset S$, we have

$$\hat{p}_u := P\left(\vec{X} \in U_N\right) \leq p_f \quad \text{and} \quad \hat{p}_s := P\left(\vec{X} \in S_N\right) \leq 1 - p_f, \tag{51}$$

where $\vec{X}$ is an independent draw from f. In this case, the bounds on p_f are

$$\hat{p}_u \leq p_f \leq 1 - \hat{p}_s. \tag{52}$$

EXAMPLE APPLICATIONS TO FLOOD DEFENSES

Scheldt Estuary in the Netherlands

Located in the Netherlands, the Scheldt estuary is a typical North Sea area protected against coastal flooding by various flood defense structures (Figure 2). Flooding is the dominant hazard in the Netherlands, and most of the areas are protected by forelands, levees, dikes, dunes, barriers, and other flood defenses. The Western Scheldt forms the entrance to the harbor of Antwerp (Belgium). Water levels are influenced by storm surges on the North Sea, as well as river discharges from the Scheldt. Among the elements of the protection system, the dike rings play a key role because of their total length, function and their contribution to the system. Dike rings are divided into two categories, river dikes and sea dikes, which protect large cities like Amsterdam and Rotterdam.

There are four surrounding dike rings along the Western Scheldt. These dike rings are numbered 29–32 (Figure 3). The reliability assessment of this site is defined by utilizing LSEs. A large number of LSEs for various failure scenarios and failure modes were considered (Sivia & Skilling 2006) and their integration is addressed in Rajabalinejad et al. (2010a). These outcomes form the building blocks of this enormous project. In these studies, attempts have been made to define and apply LSEs for various failure modes of a usual flood defense structure, and most of the presented LSEs can be utilized for safety assessment. Among failure modes, sliding is one of the more important, influential, and difficult to model. This failure mode can be investigated by analytical methods or finite element (FE) models. Because reliability assessment of FE models is a complicated and time consuming process (U.S. ACE 2006), analytical models can be used to compute failure probabilities. The outcome is used in a reliability tool that takes into account effects of various parameters to assess safety of the entire dike ring. This process will improve the safety assessment system of dike rings.

The DB method was used for safety assessment of this dike (Rajabalinejad et al. 2010a). Table 1 presents a comparison between the results of the classical MC method and the DB method. The results are based on 95% accuracy for the relative standard error (see Rajabalinejad et al. 2010a). This table shows that the number of simulations is dramatically reduced using the DB method.

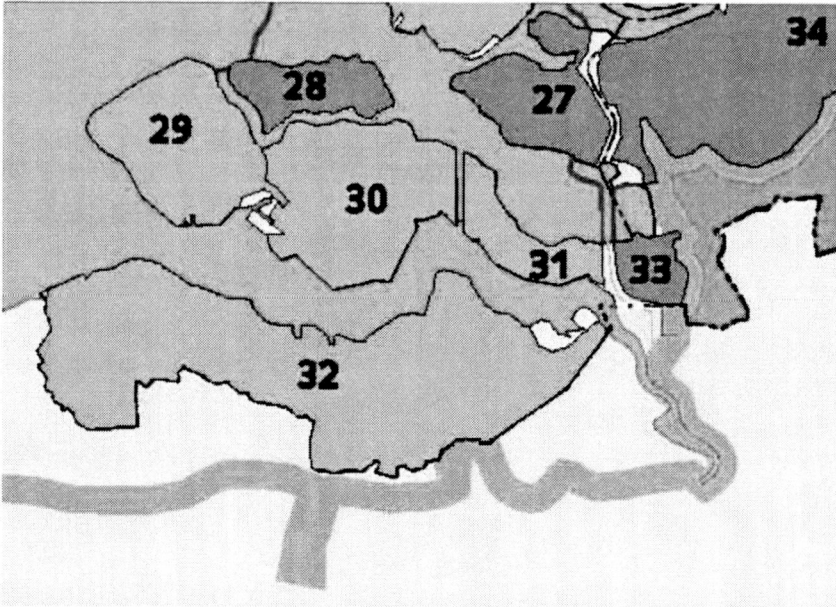

Figure 3. Dike ring areas in the Scheldt Estuary. Dike Ring 32 protects "Zeeuwsch Vlaannderen" highlighted in the previous figure (see Figure 2).

Table 1. Results of the DB method in different dimensions based on the number of variables contributing to failure of dike in the Scheldt Estuary

Dimensions of DB	Number of DB simulations	Equivalent MC simulations
1	3	4500
2	29	4500
3	115	4500
4	280	4500
5	397	4500

17th Street Flood Wall in New Orleans

The computational methods described in this chapter have also been applied to the 17th Street flood wall in New Orleans, USA. This recent structural failure problem has been investigated by many researchers (see Demirbilek 2010a,b, Rajabalinejad et al. 2010c, U.S. ACE 2006, Seed et al. 2006). The 17th Street flood wall is an I-wall located on the east side of the 17th Street Canal. It was breached during Hurricane Katrina when the surge level exceeded 8.0 feet, a typical failure elevation for several flood walls in New Orleans. All data and information about the geometry and material properties of the 17th Street flood wall were obtained from published materials on internet websites (U.S. ACE 2006, Seed et al. 2006).

As shown in Rajabalinejad et al. (2010c, 2008b) the first three influential variables for the 17th Street flood wall sufficed to provide the desired level of accuracy using an FE model as in Rajabalinejad (2010c). The same model is used here to describe implementation of the Bayesian interpolation method for three influential variables to investigate fail-safe characteristics of the 17th Street flood wall and to develop probability of failure estimates. The candidates for the first three influential variables are shown in Table 3 in Rajabalinejad et al. (2010c). It is important to point out that different ranking criteria would produce different outputs[1]. The product moment correlation (r) criterion for the first three influential variables is used here to obtain the probability estimates shown in Table 2. In this analyis, parameters for soil number 3, 8, and 2 (Marsh, Clay and Clay)are indeed the controlling variables for failure of the flood wall[2] according to Rajabalinejad et al. (2010c).

Table 2. The first three influential variables for stability of the 17th Street flood wall based on product moment correlation (r) for Water Level +8 ft (Rajabalinejad et al. 2010c)

Material	Soil No.	Variable No.	r
Marsh Under Levee	3	v_2	0.359
Gray Clay	8	v_3	0.242
Gray Clay	2	v_1	0.194

RESULTS OF THE DYNAMIC BOUNDS METHOD

The DB method was applied to the 17th Street flood wall (Rajabalinejad et al. 2010c). Tables 3 and 4 show the results of applying DB to the 17th Street flood wall using various ranking criteria. A suitable ranking criterion is essential for DB as described in Rajabalinejad (2009) and Rajabalinejad et al. (2010b,c).

Table 3. The calculated probabilities of failure with the dynamic bounds (DB) method for three influential variables using the product moment correlation. The probability of failure (p_f) is the number of failures divided by number of equivalent MC simulations

Water level (ft.)	DB	Failures	Stables	Equivalent MC	p_f
+4 (1.2 m)	190	424	1076	1500	0.28
+6 (1.8 m)	231	670	830	1500	0.45
+8 (2.4 m)	221	781	719	1500	0.52
+10 (3.0 m)	52	162	32	194	0.84
+12 (3.6 m)	23	102	6	108	0.94

---

[1] An efficient criterion is essential to rank variables for the IDB method.
[2] The rank correlation shows that variables 3, 8, and 4 are the most influential variables and this sequence may change when the structure's response becomes nonlinear at a high water level, W.L.=+8 ft (2.4 m).

Table 4. The calculated probability of failure with the dynamic bounds (DB) method for three influential variables according to the rank correlation. The probability of failure is obtained by number of failures divided by number of equivalent MC simulations

Water level (ft.)	DB	Failures	Stables	Equivalent MC	p_f
+4 (1.2 m)	171	377	1123	1500	0.25
+6 (1.8 m)	207	526	939	1465	0.36
+8 (2.4 m)	202	322	438	1500	0.21
+10 (3.0 m)	66	157	37	194	0.81
+12 (3.6 m)	29	102	8	110	0.93

RESULTS OF BMC WITH GAUSSIAN ERROR FUNCTION

The proposed BMC method is demonstrated here for the 17th Street flood wall. Our goal is to develop probability of failure estimates. The controlling variables of the 17th Street Flood Wall $v_1, ..., v_3$ are shown in Table 2. For each randomly generated data point (pixel) in the limit state equation, we develop the joint pdfs of variables from $P(u_j \mid v_1, v_2, v_3)$ as a three-dimensional problem. The integration of joint pdf over the stable or unstable regions determines the location of target pixels (or data points). As the simulation progresses, the accuracy of estimates improves with increasing size of the ensemble population (data points) that reduces the predictive errors. Figure 4 shows a typical comparison between the estimated joint pdf of a two-dimensional problem (v_1, v_2) for 5 and 20 data points. Figure 4 shows that the accuracy of predicted estimates improves with the progression of MC simulations. The proposed approach takes advantage of these numerical characteristics to save enormous computational time in the MC simulations.

The number of simulations required for the BMCM method for the 17th Street I-Wall application for water level +8 ft is shown in Table 5. Results are provided for the BMCM, classical MC, and DB methods. Results show that only a fraction of MC simulation would be required if the BMCM coupled method were used. The BMCM method proposed in this paper requires monotonic constraint and also a Gaussian density function for error estimation. An uncertainty model association is used to relate the local uncertainty σ_i to the global uncertainty σ in the form of $\sigma_i = \alpha_i \sigma$, where α_i corresponds to the cube root of the distance of x_i with its closest neighbor.

RESULTS OF BMC WITH BETA ERROR FUNCTION

Predicted variable estimates are dependent on the variables shown in Table 2, and different estimates would be obtained with a different set of variables. For each randomly generated data point (pixel) in the LSE, we developed the joint pdfs of variables from $P(u_j \mid v_1, v_2, v_3, I)$ as a four-dimensional problem. The integration of a joint pdf over the stable or unstable regions determines the location of the target pixels (or data points). As the simulation progresses, the accuracy of estimates will improve with increasing size of the ensemble population (data points) that reduces the predictive errors. Figure 4 shows a comparison between the estimated joint pdf of a two-dimensional problem (v_1, v_2) for 5 versus 20 data points. Results in Figure 4 shows that the accuracy of predicted estimates improves with the progression of MC simulations. Taking advantage of this characteristic saves enormous computational time in the MC simulations.

The number of simulations required for the Bayesian interpolation method for the 17th Street I-Wall application for water level +8 ft is shown in Table . Results are provided for the BMCM, classical MC, and DB methods, showing that only a fraction of MC simulation is required when the BMCM coupled method is used. The DB method is briefly described here since it has been fully described in previous publications (see References). It is used in the comparisons shown here with a monotonic model. The BMCM method proposed in this paper is also subjected to this requirement, and the GB density function is used for this purpose. An uncertainty model association is used to relate the local uncertainty σ_i to the global uncertainty σ in the form of $\sigma_i = \alpha_i \sigma$, where α_i corresponds to the cubic root of the distance of x_i with its closest neighbor. The cube root relation for the third order model response is used (Rajabalinejad et al. 2010b).

Table 5. The calculated probabilities of failure for the 17th Street I-Wall structure obtained with the Bayesian Monte Carlo (BMCM), Monte Carlo (MC) and dynamic bounds (DB) methods using product moment correlation (Rajabalinejad et al. 2010c) for the three most influential variables. $\mathcal{G}(\hat{p}_f)$ represents the relative standard error of $\hat{p}_f$.

WL: water level

WL (ft.)	BMCM	$\hat{p}_f$	$\mathcal{G}(\hat{p}_f)$	DB	MC	$\hat{p}_f$	$\mathcal{G}(\hat{p}_f)$
+8 (2.4 m)	36	51.2	0.042	221	1500	52	0.024

CONCLUSIONS

The theoretical formulation and numerical implementation details of a new Bayesian interpolation method for a monotonic models are provided in this paper. The proposed modeling approach is an integrated model that employs a Bayesian interpolation, a Gaussian density function and the Dynamic Bounds (DB) method to efficiently calculate an unbiased estimate of the failure probability of complex structural systems. The proposed modeling approach also includes a new concept that relates relationship between global and local uncertainties (Rajabalinejad 2009). With these features, this novel approach makes it possible to obtain an unbiased estimate of failuare probabilities of complex flood protection systems using the Bayesian Monte Carlo (BMC) methods. This integrated approach preserves fundamental properties of the classical MC method, while it greatly improves the computational efficiency by using prior information in a monotonic LSE.

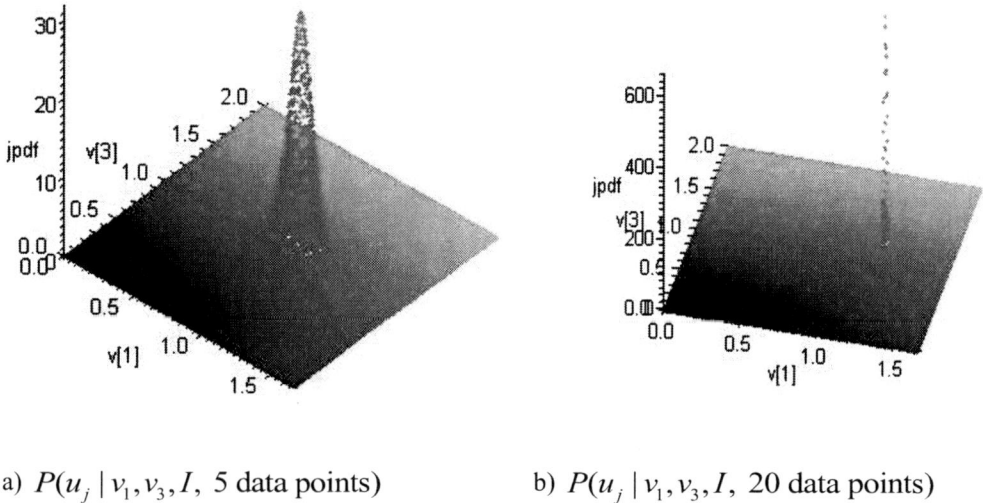

a) $P(u_j \mid v_1, v_3, I,$ 5 data points)

b) $P(u_j \mid v_1, v_3, I,$ 20 data points)

Figure 4. Two-dimensional pdfs assigned to the response estimate u_j for 5 and 20 simulations, respectively in (a) and (b).

The proposed hybrid modeling approach employs Bayesian interpolation and the generalized beta density function to estimate modeling error. Principles of the suggested method are described here, including introduction of global and local uncertainties (Rajabalinejad 2009). The integrated method was applied to investigate load-response characteristics of the 17th Street flood wall. The method also was applied to river dikes and sea dikes designed to protect areas adjacent to the Scheldt Estuary from flooding (extreme water level variations) caused by (1) sea level rise and storm surges on the North Sea, and (2) river discharges from the Scheldt River. An increasing level of analyses (one to five-dimensional) was used to consider potential influence of the controlling parameters. Table 1 demonstrates a remarkable computational efficiency of the DB method as compared to the standard Monte Carlo method. These results show that the DB method is able to speed up the classical Monte Carlo method simulations by factors that range from 100 to 1500. Results of previous numerical realizations (simulations), termed here the prior information, are used in the current simulation with the Bayes theory and the BMCM. This novel integrated technique preserves fundamental properties of the classical MC method, and greatly improves the computational efficiency by using prior information and the monotonic LSE property. The prior information is obtained from previously completed MC simulations. The priors data are constructed based on logical dependence between neighboring data points, which are termed here as "pixels", and using their locations in regions classified as stable or unstable areas of the limit state equation. Separation of data points in this manner avoids unnecessary simulations in the MC method and reduces computational burdens.

ACKNOWLEDGMENT

The information about the 17th Street flood wall project used in this research was obtained from publications available from various Internet sources, including study reports downloaded from different websites. The second author wishes to thank the Coastal Inlets Research Program (CIRP) for the support of his role in this international research activity.

REFERENCES

Bojanov, B., Hakopian, H. & Saaki, A. (1993). *Spline Functions and Multivariate Interpolations,* Kluwer Academic Publ.

De Boor, C. & Ron, A. (1990). On multivariate polynomial interpolation. *Constr. Approx. 6,* 287–302.

Demirbilek, Z. (2010a). A Forensic Analysis of Hurricane Katrina's Impact: Methods and Findings. Special Issue of *Ocean Engineering*, Volume 37, Issue 1 (ed. Demirbilek), Elsevier.

Demirbilek, Z. (2010b). Hurricane Katrina and Ocean Engineering Lessons Learned, Ocean Engineering, Volume 37 No.1, p.1-3.

Demirbilek, Z., Lin, L. & Mark, D.J. (2009). Chesapeake Bay Storm Flooding Frequency-of-Occurrence Studies. ERDC Letter Report for FEMA, U.S. Army Engineer Research and Development Center, Vicksburg, MS.

Haimes, Y.Y. (2004). *Risk Modeling, Assessment, and Management.* John Wiley & Sons, Hoboken, NJ.

Jaynes, E.T. (2003). *Probability Theory, the Logic of Science,* Cambridge University Press, Cambridge, U.K.

Mark, D.J., Demirbilek, Z. Lin, L. & Carson, F.C. (2007). Cursory-level storm surge frequency-of-occurrence analysis of the Lower Potomac River. ERDC Letter Report for FEMA, U.S. Army Engineer Research and Development Center, Vicksburg, MS.

Rajabalinejad, M. (2008). A Systematic Approach to Risk Mitigation. In *The Research Agenda of Risk and Design,* Delft University of Technology, Delft, The Netherlands.

Rajabalinejad, M. (2009). *Reliability Methods for Finite Element Models.* IOS Press, Amsterdam, the Netherlands, 132 pp.

Rajabalinejad, M., Demirbilek, Z. & Mahdi, T. (2010a). Determination of failure probabilities of flood defence systems with improved dynamic bounds method. *Nat. Hazards 55,* 95–109.

Rajabalinejad, M. & Mahdi, T. (2010). The inclusive and simplified forms of Bayesian interpolation for general and monotonic models using Gaussian and generalized beta distributions with application to Monte Carlo simulations. *Nat. Hazards 55,* 29–49.

Rajabalinejad, M., Meester, L.E., Van Gelder, P.H.A.J.M. & Vrijling, J.K. (2011). Dynamic bounds coupled with Monte Carlo simulations. *Reliab. Eng. Syst. Safe. 96,* 278-285.

Rajabalinejad, M., van Gelder, P. & Mahdi, T. (2010b). Stochastic methods for safety assessment of the flood defence system in the Scheldt Estuary of the Netherlands. *Natural Hazards 55,* 123–144.

Rajabalinejad, M., van Gelder, P.H.A.J.M., Demirbilek, Z., Mahdi, T. & Vrijling, H.K. (2010c). Application of dynamic bounds in the safety assessment of flood defences, a case study: 17th Street flood wall, New Orleans. *Georisk 4,* 157-173.

Rajabalinejad, M., van Gelder, P. & van Erp, N. (2008c). Application of Bayesian Interpolation in Monte Carlo Simulation. In: Martorell, S., Soares, C.G. & Barnett, J. (Eds.), *Safety, Reliability and Risk Analysis,* Taylor and Francis Group, London, UK, pp. 705–713.

Rajabalinejad, M., van Gelder, P. & Vrijling, J.K. (2008a). Improved dynamic limit bounds in Monte Carlo simulations. *49th AIAA/ASME/ASCE/AHS/ASC Structures, Structural Dynamics, and Materials Conference.* American Institute of Aeronautics and Astronautics.

Rajabalinejad, M., van Gelder, P.H.A.J.M. & Vrijling, J.K. (2008b). Probabilistic finite elements with dynamic limit bounds: a case study: 17th Street flood wall, New Orleans. In: *6th International Conference on Case Histories in Geotechnical Engineering and Symposium in Honor of Professor James K. Mitchell,* Rolla, Missouri USA, Missouri University of Science and Technology.

Starr, C. (2008). Risk management, assessment, and acceptability. *Risk Anal. 5,* 97–102.

Seed, R.B. & 34 co-authors (2006). *Investigation of the Performance of the New Orleans Flood Protection Systems in Hurricane Katrina.* Report of the Independent Levee Investigation Team, University of California at Berkeley.

Sivia, D. & Skilling, J. (2006). *Data Analysis: a Bayesian Tutorial,* Oxford University Press, USA.

Thompson, K.M. & Graham, J.D. (1996). Going beyond the single number: using probabilistic risk assessment to improve risk management. *Human Ecol. Risk Assess. 2,* 1008-1034.

U.S. ACE (2006). *Orleans and Southeast Louisiana Hurricane Protection System, Volume V The Performance Levees and Floodwalls,* U.S. Army Corps of Engineers, Report of the Interagency Performance Evaluation Task Force.

van Gelder, P., Wang, W. & Vrijling, J.K. (2007). Statistical estimation methods for extreme hydrological events. In: Vasiliev, O.F.; van Gelder, P.H.A.J.M., Plate, E.J. & Bolgov, M.V. (Eds.), *Extreme Hydrological Events: New Concepts for Security.* NATO Science Series: IV: Earth and Environmental Sciences, vol. 78, pp. 199-252.

In: Estuaries: Classification, Ecology and Human Impacts   ISBN: 978-1-61942-083-0
Editor: Steve Jordan   © 2012 Nova Science Publishers, Inc.

Chapter 13

RESTORING AND MANAGING GULF OF MEXICO FISHERIES: A PATH TOWARD CREATIVE DECISION-MAKING

John F. Carriger and William H. Benson

U.S. Environmental Protection Agency, National Health and Environmental Effects
Research Laboratory, Gulf Ecology Division, 1 Sabine Island Drive, Gulf Breeze, FL, US

ABSTRACT

This chapter introduces decision analysis concepts with examples for managing fisheries. Decision analytic methods provide useful tools for structuring environmental management problems and separating technical judgments from preference judgments to better weigh the prospects from decisions. First, an introduction to decision analysis methods will be given. To illustrate the concepts, a decision context of restoring and sustaining fisheries for the Natural Resource Damage Assessment process after the 2010 Gulf of Mexico oil spill is investigated and objectives derived. The fundamental objectives, measures for the achievement of fundamental objectives, and the means to achieve the fundamental objectives are selected. Additional topics important to decisions such as weighing trade-offs with decision-makers or stakeholders, considering uncertainty, and the value of information for monitoring data and other information that can assist in making a decision also are introduced and discussed. The decision analysis field offers tools that can integrate ecological and socioeconomic research to better understand problems and create improved opportunities. The processes described in this chapter might be useful ones for assessing restoration and recovery tasks as well as for providing more rigorous understanding of the opportunities available to better manage fisheries in the Gulf of Mexico and beyond.

INTRODUCTION

Formal decision analysis tools and lessons from the decision science literature contain well-founded advice for structuring environmental decision problems and making more informed decisions. Managers facing complex decisions for estuaries often have to consider multi-dimensional prospects and varied information, including opinions from stakeholders, historical monitoring data, and technical model output. Bringing this information together to weigh the trade-offs from decisions and to focus future data gathering activities can preclude entrenched positions, bias in decisions, and undesirable outcomes.

Beyond environmental concerns, to fully appraise estuarine strategies prior to implementation, it should be considered how policies can affect what is important to stakeholders and their prospects for economic and social well-being. Decision analysis frameworks allow managers to bring together those who have a stake in estuarine problems (e.g., fishers, residents, tourism vendors) and consider how they specify the features that are valuable to them and how they weigh different attributes used to distinguish among decisions. Stakeholder input should be obtained in the beginning of the project and their values should be used to update the project as it moves on. Measures using familiar units or robust constructed scales can be considered by stakeholders, experts, or decision-makers in an estuarine decision analysis task. In addition, examining the trade-offs with stakeholders and decision-makers can decompose a complex situation and provide equivalent analyses, such as how much wetland habitat restored from a management decision is worth a certain monetary investment, or how much various stakeholders would be willing to spend on a project that could better the employment prospects in a watershed. Some techniques for implementing these considerations will be discussed in this chapter.

Implementing structured decision analysis for problems involving estuaries can educate decision-makers and stakeholders on the trade-offs and create better decision opportunities (Gregory et al. 2001). The information extracted from these interactions should be advantageous for estuarine management tasks and obtained in a timely fashion that minimizes the use of resources. Constructive workshop procedures have been in development with protocols for providing understandable information of high quality to stakeholders, tackling complexity, splitting high-dimensional problems into tractable dimensions, and assembling stakeholder concerns to form a holistic value model that will be useful to appraise decisions (Gregory & Slovic 1997; Payne et al. 1999). Sometimes satisfaction of stakeholders or decision-makers is used to describe whether a structured process is successful or not. By itself, this is a sub-optimal measure. The judgment and decision-making literature contains a wealth of advice and scientifically tested theories about how biases in choices and selection might occur (Gregory et al. 2001). Successfully implementing a method that avoids these biases would be one way to determine the acceptability of an elicitation process. Essentially, a value model for stakeholders is constructed in decision analysis; whether this model is useful and proper is something that should be examined in detail. Additionally, the decision process can be tested scientifically to examine various facets of its success. These facets might include whether or not the decision-making process was improved by implementing a method, whether sources of contention between stakeholders were identified and alleviated, whether learning brought about improved objectives, whether communication between technical experts and stakeholders with less technical backgrounds was successful, or whether the decisions were more successful in achieving desired outcomes. Clemen (2008) provides excellent advice for evaluating a decision process based on its ability to select better choices when implementing the decisions (weakly effective) or achieve objectives after implementing the decisions (strongly effective). The purposes of different phases of the decision process should be considered and used to select measures of effectiveness.

Table 1. Generic steps in a decision analysis process

Decision context	The reason for a decision opportunity
Objectives	Expressions of what is important in the decision opportunity
Alternatives	The choices needed to fulfill the objectives
Prospects	The potential outcomes from the decisions and their uncertainties
Trade-offs	The willingness of stakeholders to accept more or less of an objective for another
Recommendations	The optimal strategy for achieving the objectives

There are several steps, outlined in Table 1, that are common to most decision analysis frameworks (Gregory et al. 2006a). The first step consists of selecting a context for the decisions. This step frames the subsequent analysis and helps to choose whether the problem requires a broad focus or a narrow one. The next step is to choose the objectives for the decision. This step requires considering what is valuable to the stakeholders and the attributes that will be used to define and analyze how the decision context affects the objectives. The next step is to choose alternatives for achieving objectives. Normally, there will be several existing alternatives, including a no-action and/or status quo alternative that must be assessed. However, the objectives chosen in the previous step can be used to find new alternatives (Keeney 1992). The next step is to analyze how the alternatives will contribute to achieving (or not achieving) the objectives. The change in attributes for the objectives given different alternatives is especially important for this purpose. Technical analysis can be done using Bayesian networks or other appropriate and defensible quantitative models that propagate uncertainties. The next step is exploring the trade-offs that stakeholders are willing to make among the objectives (e.g., how much of one objective they would be willing to sacrifice to achieve more of another). The final step is selecting strategies for achieving the objectives that are consistent with values and preferences of stakeholders. Each of these steps is iterative; reframing the context or objectives might be necessary as one learns more about the problem and the values of those involved. These steps are illustrated in greater detail throughout this chapter as we explore how managers might assess the outcomes from fisheries decisions in the process of recovery and restoration in the GOM.

TECHNICAL AND VALUE-BASED JUDGMENTS IN MANAGEMENT

In order to assess a decision problem, a model must be available to represent how a decision-maker or expert group envisions that a management response will affect the objectives or the processes that could influence the objectives (Conroy et al. 2008). This requires technical judgments but often, gaining more technical information is viewed as the answer to environmental management issues when, in fact, it is not appropriate (Gregory et al. 2006a). Value systems of affected parties might need to be investigated (using scientific principles) to better understand which objectives should be focused on and how trade-offs should be made. Decision analysis gives us tools to distinguish value-based judgments from technically-based judgments and to minimize confusion about which of these is being discussed and applied. An example of the difference between the two is presented in a hypothetical example in Figure 1 (adapted from Keeney 2002), where an indifference curve (black) is displayed for two management objectives, i.e. restoration cost and acres of lost habitat. These curves consider the trade-offs a decision-maker or stakeholder may find acceptable between decreasing habitat losses and the management costs. The indifference curve is based on notions of trade-off acceptance. In actuality, several indifference curves would be constructed to examine a range of trade-offs throughout the consequence space. The shape of the indifference curve, whether it is linear or concave upward or downward, can tell us properties about the *marginal rate of substitution* between attributes (Clemen & Reilly 2001). For example, a linear curve would allow us to calculate the rate (e.g., $x per wetland acre) a stakeholder would be willing to exchange one attribute for another based on the slope. For Figure 1, more complicated mathematics would be necessary to formally quantify marginal rates of substitution for different consequences (Clemen & Reilly 2001) but general patterns can be easily discerned for different levels of each of the attributes. A red curve is also shown in Figure 1 to indicate a range of potential outcomes from decisions under consideration. This curve would be based on a technical examination of policies, a meaning different from that of the indifference curve in the figure. Considerations of how the potential management outcomes might be selected can be made by combining the technical and value-based judgments. The best decision in this simple example is one where the technical curve is tangent to the values-based indifference curve (Keeney 1992, Keeney 2002).

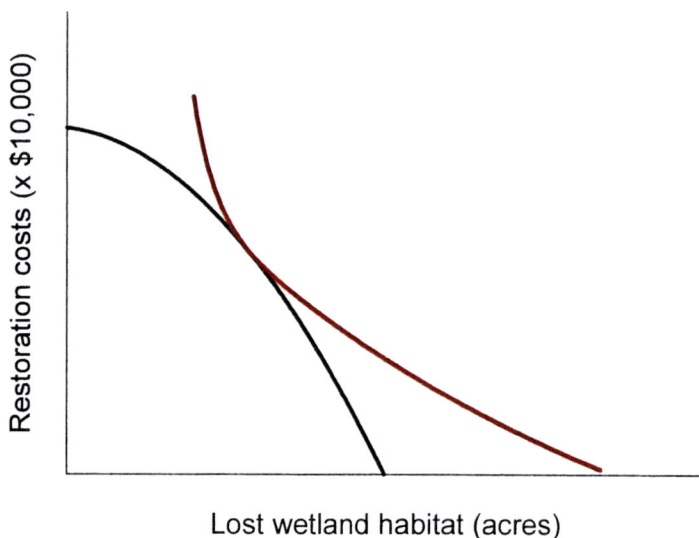

Figure 1. Indifference curve (black) expressing the amount of restoration costs equivalent to acres of lost habitat. The black curve indicates a range that a decision-maker would consider equally acceptable. The restoration curve in red expresses the potential outcomes for various alternatives; adapted from Keeney (2002).

Understanding values means understanding what people care about (Keeney 1992). Eliciting stakeholder opinion can help us understand the values of those who stand to gain or lose from management decisions, assist our understanding in the potential adverse outcomes or rewards from decisions, make the decision process transparent and inclusive, resolve difficulties when contentions arise, and clarify the reasons for contentions among stakeholders or between stakeholders, experts, and decision-makers. From this learning, better decision-making should result.

Decision analysis tools are available to assist with understanding the values, the sources of contention, and communicating in an optimal manner to facilitate better understanding on the part of the elicitors and the stakeholders. Ignoring the will of stakeholders can be a source of contention and actually inhibit the goals of management rather than facilitate them. If stakeholders do not understand the considerations of management and if management does not consider the wishes of stakeholders, the two groups could end up with very different stances, leading to confrontations (Arvai et al. 2008).

THEORETICAL PERSPECTIVES

As described in French et al. (2009), *normative decision analysis* gives a rigid recommendation on how to achieve final states of value or more of what stakeholders might care about given a rational framework. *Descriptive decision analysis* examines how decisions are made including decision behaviors that deviate from a rational framework. *Prescriptive decision analysis* brings the two together to advise a decision-maker on a potential strategy and to facilitate the decision process in an optimum manner. In this chapter, we draw from all three fields of decision analysis but focus primarily on prescriptive approaches for achieving better decisions. The primary aspect of decision-making described is the value-focused approach, as developed by Keeney (1992) and in other work by Keeney and his colleagues. In this approach, understanding what is valued by stakeholders in the decision is prioritized and alternatives are selected in order to achieve what is valuable. This method can be contrasted with alternative-focused approaches where the options are the primary focus and values are secondary. Alternative-focused thinking encourages limited analysis, because identifying values first and then focusing on ways to achieve things of value will likely bring about better decisions than focusing on existing alternatives. After experimentally implementing value-focused and alternative-focused workshops with different stakeholder groups, Arvai et al. (2001) observed that the value-focused group had greater confidence in their choices, felt that their choices were in better alignment with their concerns, felt more informed about the subject matter, and had a broader range of topics in their discussions than the alternative-focused group. However, the value-focused group did not have a significant difference in choice satisfaction from the alternative-focused group (Arvai et al. 2001).

DECISION ENVIRONMENT – PROBLEM FORMULATION

The first step in formulating the decision problems and opportunities is choosing a decision context (Keeney 1992). For example, a decision context might be related to an environmental agency's overall programmatic goals, or it might be focused on rehabilitating a single site damaged by a specific stressor, or it might pertain to the restoration of an entire watershed. This step might seem simple but it can be tricky and will have dramatic implications for any subsequent analyses. Selecting the wrong decision context has been called a Type III error (Clemen & Reilly 2001). Selecting a context that is too narrow will rule out important objectives (and alternatives) that should otherwise be considered. Selecting a context that is too broad will waste time and resources on assessing a more difficult problem than necessary. Assessing a context that is focused on the wrong problem will discredit any subsequent analysis.

Roles

Initially, the roles of the people involved with the problems or opportunities related to a decision context should be investigated. Commonly, there will be a group of stakeholders, decision-makers, experts, and analysts (French et al. 2009). Sometimes, these roles will overlap (French et al. 2009). For example, stakeholders might also be experts in certain processes related to the decision task (stakeholders will always be experts with respect to their own values). Decision-makers should hold a stake in the outcomes. Additional roles could possibly include policy makers, policy implementers, negotiators, and facilitators. Investigating the roles of the decision environment will also help elucidate the decision context. Understanding the decision-makers' capabilities will help clarify the potential objectives and alternatives. Knowledge of the stakeholders will be useful for eliciting values and choosing objectives. As the decision analysis develops, potential options to achieve objectives will be better clarified and might help select additional experts and analysts to examine the potential uncertainties and risks.

Definitions of stakeholders might be situation-specific and temporary (Winn 2001). Within a context, one label of a stakeholder, e.g., a company shareholder, might be different from that in another context, e.g., a company board member. In addition, groups of stakeholders are not monolithic entities; a decision will affect stakeholders in a group differently, and they will not necessarily share the same knowledge (Winn 2001). How the issues affect the stakeholders should be better understood as the elicitations develop, but preliminary data gathering can help decision analysts partition the stakeholder population into a useful sampling frame for building value models in later tasks.

Characterizing the Environment

The decision environment for managing an estuary can be characterized by the regulatory requirements and capabilities of those involved with the context under consideration. Ultimately, we wish to delineate what types of options are most feasible given the policies and strategies that can be implemented. For example, the Gulf States Marine Fisheries Commission has regulatory authority (through the states) for managing interstate fisheries in inshore waters of the Gulf of Mexico (GOM), and can establish annual catch limits or restrictions on harvesting. They do not have the authority to prevent residents from using a chemical on their lawns that might run off into an estuary and harm fish, although they might act in an advisory capacity to urge action by a state environmental agency or the U. S. Environmental Protection Agency (U.S. EPA). Thus, knowing what is feasible for the decision-makers in a decision context provides a foundation for understanding what can be achieved and what means are possible for achieving it.

Likewise, the regulatory environment will be based on previous laws and legislation in place as well as mandates that give decision-makers their authorities. Understanding the decision-making capabilities of those involved with an estuarine management problem allows a better understanding of the fundamental objectives, the means objectives, and the alternatives that might be available to achieve them. Several ways of intervening when issues arise might be available. For example, if a water quality violation is identified, what is the state regulatory agency allowed to do in response? Can polluters be held liable? Or is follow-up primarily informational and through incentives to change behavior? Incentive structures might be emphasized in a regulatory body or a command and control system might be available. An inter-agency context may provide greater leeway for managing a problem or creating new opportunities.

To facilitate better communication with stakeholders and scientific analysis for decision-making, there are several decision analytic structures. Von Winterfeldt & Edwards (2007) review eight common structures that can be applied as generic templates. Some are more specific than others, and often it is best to prepare a mixture of decision analysis structures as demonstrated by von Winterfeldt & Fasolo (2009). The structures discussed in von Winterfeldt & Edwards (2007) are *objectives hierarchies*, *means-ends networks*, *belief networks*, *consequence tables*, *influence diagrams*, *decision trees*, *fault trees*, and *event trees* (Table 2). Influence diagrams, which illustrate cause-effect relationships between management decisions, environmental variables, and objectives, can be helpful in most estuarine management programs. They are intuitive to lay stakeholders and can be precursors to data gathering and uncertainty reduction for management learning opportunities.

Decision trees are related to influence diagrams but distinctly represent the decisions at various time periods and the chance events that follow from decisions. Each possible decision is represented as a branch in the tree and the diagrams can get quite complex. However, their use in describing and analyzing the temporal events in the decision process can be invaluable and there are several schemes for simplifying communication and representation. A decision tree developed by Sheehy & Vik (2002) for the Exxon Valdez oil spill Natural Resource Damage Assessment (NRDA) process was employed to examine the effectiveness of various monitoring regimes for salmon stocks. Figure 2 is a partial decision tree for a problem of fish tissue contamination. Branches are not fully elucidated, probabilities and utilities are not included, and only one path is displayed for demonstration purposes.

Table 2. Eight decision analytic structures, adapted from von Winterfeldt & Edwards (2007)

Problem type	Structure	Description	Primary uses for steps in Table 1
Evaluation	Means-ends network	Identifies and differentiates the end objectives (what is valuable in a problem) and related means objectives (important for achieving what is valuable)	Objectives, alternatives
	Objectives hierarchy	Defines the fundamental objectives and their measures	Objectives, alternatives, trade-offs
	Consequence table	A table relating consequences of alternatives on each objective	Prospects, trade-offs
Decision under uncertainty	Decision tree	Analyzes the chronological relationship between decisions and uncertain processes	Prospects, recommendations
	Influence diagram	Examines the conditional relationships between decisions, states of nature, and objectives in a problem	Prospects, recommendations
Probabilistic inference	Event tree	Inductive approach for examining the outcomes following an initiation event	Prospects, recommendations
	Fault tree	Deductive approach for examining the possible causes of an initiating event	Prospects, recommendations
	Belief network	Inference structure showing the conditional relationships between states of nature and objectives	Prospects, recommendations

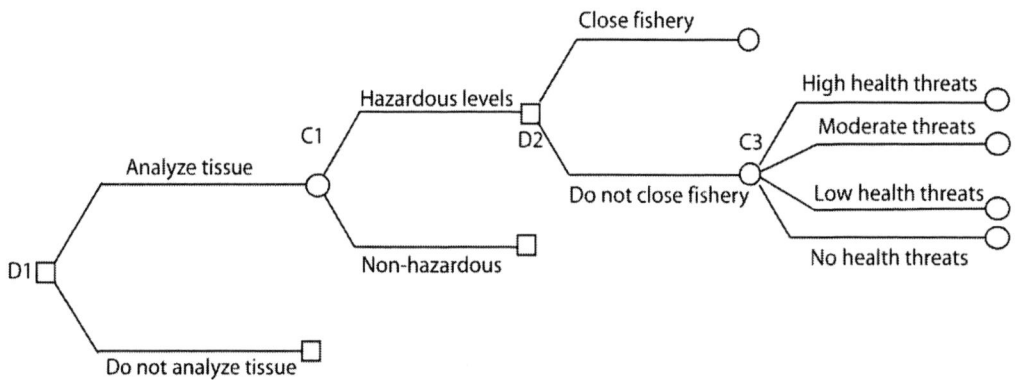

Figure 2. Example of a decision tree structure for mitigating human health effects by testing for and potentially closing fisheries impacted by an oil spill.

Fault trees map the events or faults that lead up to an initiating event. Their usage might be important in describing causes that could lead to future disasters. In fact, fault trees were constructed for deepwater drilling prior to the BP Deepwater Horizon (DWH; Gulf of Mexico, April 2010) spill by engineers for technical risk assessment work (Goldsmith & Ericson 2000). Fault trees are commonly used in engineered system failure and manufacturing quality control scenarios. Hayes (2002) applied fault trees to examine the causal events leading to an invasive species establishment at a port. The Hayes paper also discussed some of the advantages and limitations in applying fault trees to ecological problems. Event trees are similar to fault trees, but describe the cause-effect chain of events for the performance of a system once an initiating event occurs – thus sketching the problem from initiating event to fault (von Winterfeldt & Edwards 2007). Like fault trees, event trees are more commonly used in engineering system reliability issues. Often, they are coupled to fault trees.

Objectives hierarchies and means-ends networks are ways of organizing a problem conceptually and are helpful in most decision analysis tasks (von Winterfeldt & Edwards 2007). A potential application of objectives hierarchies and means-ends networks for a hypothetical example involving fisheries management for restoration and recovery tasks in the GOM is described below in more detail. A decision tree differs from a means-ends network in that the interrelationships and sequence between decisions and chance events are explicitly related in a decision tree. A means-ends network maps out the relationship of means objectives to fundamental objectives based on an importance order for what is valued in the context of the problem. Explicit decisions are not always represented in a means-ends network, and important chance events or states of nature are not represented in the means-ends network, but are considered in the development of the objectives.

MEASURING ENVIRONMENTAL VALUES

Knowledge of what people care about provides better opportunities for satisfactory outcomes and preventing adverse impacts (Gregory 1999). Hajkowicz et al. (2000) and Gregory (1999) give excellent summaries and typologies for valuation methods used in environmental management decision-making. Values include direct use or consumptive values, e.g., catching fish; nonconsumptive use, e.g., hiking, swimming; indirect use, e.g., reading about the ecology of the GOM; and non-use, e.g., existence value of a fish species or ability to catch fish in the future (Gregory 1999). Incorporating the latter values and other non-market values is especially difficult given many of the applied valuation tools that are used frequently (Gregory 1999).

When appraising environmental values, one should choose one or more than one approach that will work best for the situation at hand (Gregory 1999). The decision sciences approach to valuation is discussed here because of its ability to facilitate communication, integrate multi-disciplinary measures, provide procedures for trade-off analyses, and incorporate probabilistic methods for assessing uncertainties in decision outcomes (Gregory 1999). Assessing the trade-offs from a decision analysis among a range of different objectives will give a better perspective on the impacts to stakeholders and the quality and acceptability of decisions (McDaniels & Trousdale 2005). Therefore, it gives more of an indication of what is at stake in a decision problem.

In deciding whether to use a certain method or not, it is useful to think in terms of what Keeney & von Winterfeldt (2007) describe as practical value models. Practical value models weigh the rigor of the value model with required preference depictions; costs, labor and deadlines; communication facets; and usefulness for inferences. Value models are "good enough" in an analysis if increasing details will not improve selection of alternatives or the assessment of how well the alternatives achieve positive outcomes. Value models are "good enough" in terms of a process if they fail only in acceptance by decision-makers or are difficult to understand.

CASE STUDIES- MANAGING FISHERIES IN THE GULF OF MEXICO

To illustrate several decision analysis methods in more detail, case study applications pertaining to fisheries management in the GOM are described below. The first case study is an examination of how decision analysis can organize what is valuable in a problem through the specification of objectives. This study focuses on a decision context of restoring fish communities for tasks after the DWH oil spill in the GOM. Fundamental objectives are differentiated from means objectives and measures are chosen for the fundamental objectives. Additional objectives are introduced and illustrated through fisheries management in the GOM. Then, methods for quantitatively examining the alternatives using the objectives and some applications of multi-attribute utility theory (MAUT) or multi-attribute value theory (MAVT) are described. Finally some concepts pertaining to analyzing uncertainties on decision outcomes are given, along with an example of value of information in fisheries management using influence diagrams.

GULF OF MEXICO BACKGROUND

The GOM, a semi-enclosed sea, covers more than 617,600 square miles (1.6 million km^2), connecting with several major river systems and many estuaries. Gulf coastal estuaries, bays, and sub-estuaries cover 10,643 square miles (27,565 km^2), framed by an extensive system of barrier islands. Gulf Coast estuaries and wetlands are critical feeding, spawning, and nursery habitats for a rich assemblage of fish and wildlife, including essential habitat for shorebirds, colonial nesting birds, and migratory waterfowl. The GOM is a shared resource at risk. Increasing population pressures translate into increased use of and demands on Gulf resources. From 1970 to 2000, the population of coastal zone counties in the Gulf Coast region nearly doubled in size (Colgan 2003). Concomitant with this increase in population growth, there are increased pressures on the natural resources of the Gulf Coast.

On April 20, 2010, the DWH oil rig exploded in the GOM 41 miles (66 km) southeast of the Louisiana coast. The explosion killed 11 people, and by the time the well was capped, it is estimated that the well had released more than 4.9 million barrels ($\sim 8 \times 10^8$ L) of oil into the Gulf, resulting in significant impacts on Gulf communities, ecosystems and economic activity. The long-term environmental, economic and social impacts of the spill may not be known fully for several years.

Structured decision analysis can assist in ongoing and future restoration tasks and enhance the recovery of the GOM. A hypothetical case study is examined here to illustrate how decision analysis can assist in organizing the information needed and make defensible restoration decisions. We explore the restoration of fisheries habitat from impacts due to the oil spill and response efforts as an example decision context. This example will demonstrate how decision analytical structures and objective determination can improve the judgment of regulators and decision-makers in estuarine management problems.

DECISION CONTEXT: NATURAL RESOURCE DAMAGE ASSESSMENT IN THE GULF OF MEXICO

As authorized in the Oil Pollution Act of 1990, public trustees from state, federal and tribal governments construct a NRDA after an oil spill to examine the compensation value of resources damaged by a spill as well as the restoration tasks needed to return resources to their levels of quality prior to a spill (NOAA 1996c). Once a spill occurs, the U.S. EPA and other authorized agents work to remove the oil so as to minimize ongoing hazards to humans and wildlife (NOAA 2011c). After this initial cleanup, the NRDA works to ensure that restoration of pre-spill services is complete and to incorporate public values regarding loss of the services. Comprehending these aspects of damaged resources may take many years after large spills (NOAA 2011c). In a NRDA assessment, trustees are charged with developing the restoration plans, including invited input from stakeholders and responsible parties. The responsible parties are accountable for payment; the stakeholders can give valuable insights as to the damages that accrued to them as well as the facets of the resource restoration that are important.

The NRDA process is, as of this writing, being implemented in the GOM region to evaluate the impact on pre-spill resources. As illustrated in Figure 3, the NRDA process has three generic steps: pre-assessment, restoration planning, and restoration implementation (NOAA 1996a). The pre-assessment phase examines whether impacts have occurred. Conservative retrodictive risk models might be applied in this phase of the assessment to augment the judgments from data. The injury assessment places value on the loss of resources if damage is ruled to have occurred in the first phase. A restoration plan is developed in this phase based on the injuries and the loss to the public. The first phase of the NRDA has already been completed in the GOM, damages are currently being assessed, and initial restoration plans are being developed. Public compensation restoration options for interim losses prior to recovery are included in this restoration plan. In the restoration phase, the restoration is implemented and monitored. The responsible parties pay for and/or implement the restoration actions. Ultimately, the final products of the NRDA process are restoration plans for compensation of the public for loss of use of ecological services and "restoring, replacing, rehabilitating or acquiring the equivalent" of affected resources (NOAA 1996a).

Fisheries contribute a large portion of the economic benefits to the GOM region. Approximately 28% of fish harvested in the U.S. is supplied by GOM fishers (Thibodeaux et al. 2011). Part of the NRDA evaluation for the GOM will look at impacts on fisheries. In

Escambia County, Florida, officials have been submitting proposals for regaining and enhancing access to fisheries and improving fish habitat as well as soliciting public input for project ideas (Escambia County 2011). In this chapter, we investigate a decision context for restoring fisheries in the GOM after the DWH spill.

Stakeholder meetings are being held in each GOM state as the value of lost resources and the restoration plans and priorities are contemplated. Public trustees, through government agencies, are in charge of determining the costs to be paid by the responsible parties for compensating the public as well as holding public meetings to consider the public's perspectives on the impacts and the value of those impacts to them. There is no mandated time constraint for the NRDA process, so it can go on for years, but restoration may be implemented at any time and compensatory restoration is appraised from the time of injury until baseline conditions are re-established. Currently (2011) a bill is before the U.S. House of Representatives and the U.S. Senate to accelerate the NRDA process for restoring fisheries in the GOM.

Efforts to elicit stakeholder opinions throughout the Gulf Coast were made following the spill. The inherent difficulty of assessing the economic dislocation from the spill was discussed in the Recovery report (Mabus 2010). Similar to the ecological systems of the GOM, the human communities are diverse, with many distinctive characteristics (Mabus 2010). Moreover, the economic impacts from the spill will differ among these communities (Mabus 2010). Management plans that are inadequate for addressing impacts of such large scope can be inhibitory and counterproductive. The distribution of actual impacts, the definitions of what constitutes risk, and how risks should be measured, make interpreting how stakeholders perceive and value decisions even more difficult for structured elicitation processes. Unlike many natural disasters, the economic damage to infrastructure from the spill was minimal (Mabus 2010). However, the loss to service and seafood industries was very real and difficult to assess. Chemical risk perceptions provide challenges, and large oil spills like the DWH illustrate that notions of risk are a social construct. See Slovic (1999) for a discussion of risk perceptions.

Knowing the objectives for a decision opportunity and differentiating between *fundamental* or *ends objectives* (what is ultimately cared about in a decision context) and *means objectives* (objectives important for achieving the ends objectives) can enhance the NRDA process and focus the concerns of stakeholders as well as the nature of projects that are being evaluated. For example, a stakeholder group might feel that improving water quality in an open ocean system where fish spawn is important, whereas another might think that cleaning an estuarine feeding ground should receive higher priority. A decision analysis would evaluate what the end objectives are for each situation (e.g., economic value of increased fish to fisheries, employment and income from restoration) and the means for achieving those (e.g., restoring polluted habitat, offering restoration employment to displaced fisheries workers) to further elucidate the issues and focus on what is valuable and how to achieve better outcomes.

The NRDA process includes stakeholder input to better understand the costs that they bore from an oil spill and to allow trustees to develop restoration plans that properly return and compensate for lost value. As in the NRDA process, constructing meaningful dialogue with the stakeholders should be an objective of any formal interaction between policy-makers and stakeholders. Similar to sampling of biological or physical variables in the natural environments, rigorous protocols and implementation procedures should be applied that take into consideration judgmental processes of stakeholders and the meaningfulness of information from stakeholder interactions (Gregory et al. 2001).

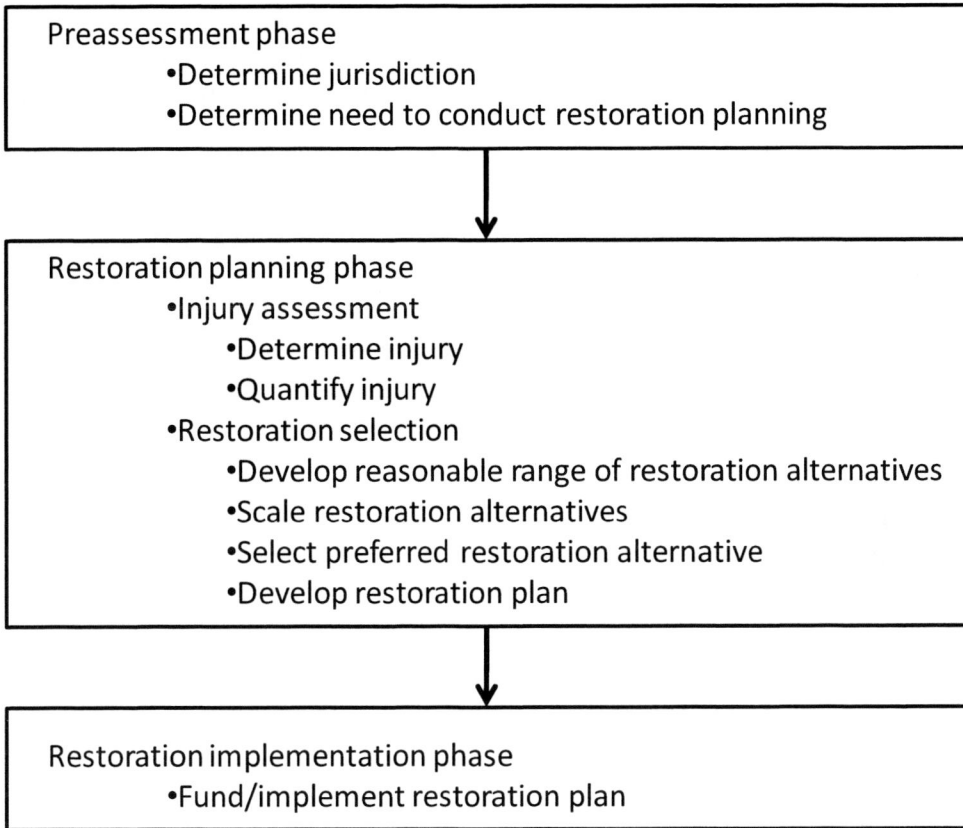

Figure 3. Natural resource damage assessment process for the Oil Pollution Act of the United States from NOAA (1996b).

IDENTIFYING OBJECTIVES

In order to ensure, in a decision context, that we are assessing the final end values and not a means for achieving those values, we differentiate between two types of objectives – fundamental objectives and means objectives (Keeney 1992, McDaniels 2000). Fundamental objectives are the end values that we care about achieving in a decision problem. Means objectives are important only for their ability to achieve the fundamental objectives. Two additional types of objectives can also be useful. One is strategic objectives –the ultimate end objectives that would be important for any decision that might be made in a broader context than the one under consideration. Thus, fundamental objectives would be means for achieving strategic objectives. Also, process objectives might also be considered. These are objectives related to improving the decision process itself.

Prior to choosing the fundamental and means objectives, a decision context must be selected. In the following example, we will be selecting alternatives to restore services from fisheries affected by the DWH oil spill. As described in the Oil Pollution Act for the NRDA, "Services (or natural resource services) means the functions performed by a natural resource for the benefit of another natural resource and/or the public" (OPA regulations § 990.30). This definition will help us to focus the fundamental objectives on the services from fish themselves along with how they would benefit other natural resources or the public.

Success in the NRDA process is defined as the ability of a natural resource to "maintain its normal function and services" without human intervention (NOAA 1996c). A goal is set to return a resource to baseline conditions, but comparison of a resource's structure and function to categorically and spatially comparable resources (which could include baseline data) is also recommended (NOAA 1996c). Effectiveness in restoration is measured by rate of recovery (NOAA 1996c). The Magnuson-Stevens Fishery Conservation and Management Act (MSFMA; as amended through January 12, 2007) discusses a series of optimizations between recreational and commercial use and rebuilding of stocks. The MSFMA prescribes consideration of recreational, food supply, employment, commercial, and community impacts in setting optimum or maximum sustainable yield levels. Benefits of fisheries in the MSFMA are considered at a national level. Thus, Congress acknowledges the economic, health, recreational, and sustenance value of fisheries and contributions to these aspects should be captured in the NRDA objectives.

The damages in a NRDA are based on the costs to restore resources, compensatory restoration for public losses until recovery to baseline conditions, and costs for assessing the risk. Dollars recovered from responsible parties are used by trustees for restoring the injured resources (Restoration Plan 1997). Services lost and replaced should be "of the same type and quality and of comparable value to the injury" (NOAA 1996b). Restoration is scaled to the quantity and quality of resources or services lost, normally, but economic valuation for resources or services might be used when restored services are different. Dollar valuation of lost and replacement services are used if it is not feasible to obtain services of similar value based on type and quality (NOAA 1996b). In the Oil Pollution Act, cost-benefit analysis might be used to appraise restoration alternatives but it is not required, particularly in cases where there are qualitative concerns (§ 1502.23). For all cases, all factors relevant and important to a decision must be clearly stated in a NRDA.

Fundamental Objectives

Fundamental objectives in decision problems are the end objectives that we are concerned with and that we wish to achieve when making decisions (Keeney 1992). They reflect what we care about and what we wish to evaluate in trade-offs when selecting from a package of management options. Each fundamental objective may have sub-objectives that explain in more detail what the fundamental objective signifies. There must be at least two sub-objectives and they should be collectively exhaustive with respect to the overlying objectives. The group of fundamental objectives chosen should have no overlap and none should be a means to another (to prevent double counting). An objective has the structure: [direction of preference] + [item of concern] + [decision context] (Dunning et al. 2000, Keeney 2007, Mollaghasemi & Pet-Edwards 1997). For example, an objective might state "Maximize economic benefits from restoring wetland habitat." The direction of preference is to maximize or increase, the item of concern is economic benefits and the decision context pertains to restoring wetland habitat. Sub-objectives would specify what we mean by economic benefits (e.g., more fish for fisheries, greater tourism revenues, recreation contributions to the economy).

Some examples of fundamental objectives for a decision context of mitigating, restoring, and recovering fisheries from the DWH oil spill are given in Table 3. The overall objective and context is to maximize ecosystem services from restoring fish stocks. The objectives are primarily focused on the beneficiaries of these fish stocks, and reducing unwanted effects from restoration activities such as causing risks to restoration workers or overspending on inefficient tasks. This hierarchy might be pertinent to a group of stakeholders that would include restoration decision-makers, trustees, recreational fishers, commercial fisheries representatives, non-governmental organizations, public health groups, and sustenance fishers. However, it is given for example purposes and an actual hierarchy should be appropriately elicited and considered with an actual group of decision-makers, stakeholders, and domain experts.

The first objective (I.) pertains to human health impacts from restoration tasks. The first sub-objective (I.A.) describes one aspect of what stakeholders might consider for human health impacts from NRDA restoration tasks (i.e., minimizing human health impacts from consuming tainted seafood). The second branch (I.B.) concerns minimizing health threats from the restoration tasks themselves. Each of these sub-objectives has associated attributes pertaining to acute and chronic injuries that are used to fully define the objectives and how they will be measured.

The second fundamental objective (II.) pertains to non-use value of fish stocks. The first two sub-objectives (II.A. and II.B.) further elucidate the objective as providing fish and shellfish for consumption by marine mammals and birds. Marine mammals would include whale species such as the endangered sperm whale (*Physeter macrocephalus*) and dolphins. Through 2010, at least five species of dolphins and two species of whales had been identified in the vicinity of the oil spill from aerial surveys (SEFSC 2010). Survival advantage for different habitat types was discussed in Jordan et al. (2009) and could be adapted as a measure of the effectiveness of restoration. As Beck et al. (2011) describe, fish using a habitat should be considered, not just the ones that reside in it. The third branch (II.C.) concerns existence values held by the public for healthy fish and shellfish populations. An appropriate measure for stakeholders might be fish diversity (evenness and richness), trophic function to indicate the importance of the impacted species in food webs, and abundance (U.S. EPA 2000). While the sub-objective in II.C. pertains to non-endangered species, the last sub-objective (II.D.) describes the non-use value for threatened or endangered fish and shellfish such as the Gulf sturgeon (*Acipenser oxyrhynchus desotoi*). The population growth rate resulting from restoration tasks is used as an example measure for assessing the population dynamics of endangered fish and shellfish. To assess a broad measure like this, the effects of fishing on endangered species would be considered, including competition from stocks of harvested fish and bycatch mortality. Focused measures might be more appropriate and would have to be weighed against the extra information for a broader assessment.

The next objective (III.) penalizes the restoration tasks if they require funding above what is required for baseline restoration. In the NRDA process, compensation from responsible parties is allowed up to the scale of the incident (NOAA 1996b). A restoration plan with costs exceeding the compensation may still be applied if funding from other sources can be obtained. The focus of an NRDA is restoring a resource to its pre-spill condition, but this might be viewed as a goal in that restoration would be achieved once the pre-spill condition is reached. Giving the objectives constraint-free properties prevents them from being used as goals for achieving specific levels of an objective (Keeney 1992). For purposes of this chapter, we assume that the objectives are constraint-free and the feasible tasks selected should be implemented based on their optimal effectiveness in achieving the objectives. Placing constraints on objectives can be problematic in that they might limit the selection of options that score high, but are excluded because they do not meet a pre-defined constraint (Keeney 1992). If constraints are desired for a decision context, they can be indicated next to an objective and considered more fully in quantitative evaluation of alternatives. Likewise, constraints on alternatives can cause similar problems, (e.g., only consider habitat restoration options if the area was affected by the oil spill), but might be required or helpful in certain contexts.

Table 3. Fundamental objectives and attributes specifying the overall objective of maximizing ecological services from fish stocks through mitigation, restoration and recovery from impacts by the DWH oil spill and related response activities

Fundamental objective	Attribute(s)
I. Ensure human health and safety in fisheries restoration tasks	
A. Minimize human health threats from consuming fish and shellfish	
1. For subsistence fishers	Acute/chronic morbidity or mortality
2. For commercial fish consumption	Acute/chronic morbidity or mortality
B. Minimize injuries to restoration and cleanup crews	Acute/chronic morbidity or mortality
II. Maximize non-use value of fish stocks	
A. Maximize availability of fish and shellfish for consumption by marine mammals	Survival advantage (%) per community component
B. Maximize availability of fish and shellfish for consumption by birds that depend on fish species (e.g., brown pelican)	Survival advantage (%) per community component
C. Maximize the public's existence value for non-endangered fish and shellfish	Fish diversity, trophic function, abundance
D. Maximize recovery of threatened/endangered fish and shellfish species (e.g., Gulf sturgeon)	Population growth rates
III. Minimize damage penalties beyond baseline restoration	$ above restoration baseline
IV. Minimize ineffective costs in the NRDA process	
A. Minimize fisheries damage assessment costs	$
B. Minimize fisheries restoration costs	$
C. Minimize fisheries replacement costs	$
D. Minimize cost accrual from delay in recovery	$
E. Minimize costs from fish habitat cleanup activities	$
V. Maximize funds from NRDA for fishery restoration	$

Fundamental objective	Attribute(s)
tasks	
VI. Maximize in-season availability of fish for commercial fishing	Annual recruitment
VII. Maximize long-term stock health from restoration	Juvenile or early life stage survival rate, population structure
VIII. Maximize cultural benefits/minimize cultural costs related to fisheries	
A. Minimize loss of traditional usage of fisheries	# of fishing boats out of service in traditional fishing communities
B. Minimize adverse impacts on traditional fishing cultural patterns	Constructed scale
C. Promote equity	Distribution of benefits/costs
IX. Maximize recreational opportunities in restored sites	# and quality of recreational visitor days related to fisheries protection
X. Minimize losses to recreational opportunities from fishery closures	Days of recreational opportunities lost, quality of lost recreation opportunities from a constructed scale
XI. Incorporate public input to restoration plans	# of stakeholder groups involved throughout project
XII. Maximize future management restoration capabilities	Constructed scale for significance of information in improving future restoration tasks

In the case of the Gulf oil spill, it can be assumed that rehabilitation of resources beyond the restoration of baseline levels is necessary. Some fisheries are likely to be operating at unsustainable levels to maintain the demand in the U.S. and abroad (NOAA 2009b). Currently, 84% of the seafood consumed in the U.S. is imported from other countries (NOAA 2009a). Thus, the objectives might be helpful in a broader decision context, and could inform management about the value of restoration beyond baseline conditions.

Other objectives for reducing unwanted effects from management activities pertain to inefficient and cost-ineffective options (IV.). Cost effectiveness also was used to develop criteria for screening sites to restore in the NRDA restoration plan for Commencement Bay (Restoration Plan 1997). The sub-objectives in IV.A. to IV.D. have a natural scale in dollars which can be based on 2011 dollar values for changes in worth in the future. Collectively they should describe the inefficiencies that are measured in fundamental objective IV. without overlapping each other. Fundamental objective V. pertains to available funds for restoration if obtaining funding support for projects needs additional consideration in appraising alternatives.

Fundamental objectives VI. and VII. highlight the importance of fish availability to fisheries from the restoration tasks. One is used to describe the in-season fish availability at the time of the restoration tasks. Recruitment potential is utilized in the NRDA preassessment process to evaluate the extent of damages (NOAA 1996a) and this might be an appropriate measure for short-term (in season) objectives. The other fundamental objective is for long-term stock health beyond immediate fishery requirements. Most reef restoration tasks following hurricanes in the GOM are measured based on short-term harvests (Beck et al. 2011). Building for long-term sustainable harvests and resiliency should be a priority following the DWH spill in the recovery and restoration efforts (Beck et al. 2011, McDaniels 1995). Securing the long-term health of fish populations for resilience to future disturbances was underscored by Holly Bins of the Pew Environment Group's testimony before the Gulf of Mexico Fishery Management Council (Bins 2010).

For the fisheries objectives, the types of fish required should be assessed based on the impacts from oil and restoration tasks. More than 1,500 species of fish inhabit the GOM (Tunnell 2011). Populations affected by the oil or response activities would be of concern in the current context. For the NRDA restoration process, season- and species-specific variables should be considered when formulating the decision analysis. For example, NOAA (2011a) describes fish species that were highly susceptible to oil spill exposure because of the location of their habitat.

The next fundamental objective looks at the cultural impacts of restoration decisions. Cultural losses would be assessed through the loss of traditional fishing activities (measure: number of boats in traditional fishing communities that are annually out of service) and for affected fishers who remain in service, adverse impacts on traditional fishing patterns in the GOM (a quality measure). The final cultural sub-objective pertains to equitable benefits from management decisions. Restoration should ensure that disadvantaged groups (e.g., low income stakeholders, ethnic groups) are fairly compensated. As in Keeney et al. (1996) and Keeney and von Winterfeldt (1994), this sub-objective can be further defined to consider spatial and inter-generational inequities to groups of people from management decisions.

Fundamental objectives IX. and X. pertain to recreational gains from restoration and recreational losses from fishery closures. Most recreationally important fishes are offshore though some use estuarine habitat for spawning or critical life stage requirements, e.g., spotted seatrout (*Cynoscion nebulosus*) and red drum (*Sciaenops ocellatus*) (U.S. EPA 1999). The measures for recreational fishermen in this context are not focused on revenues from restoring fish stocks but on the opportunities available to catch fish and the quality of the recreational experience from catching fish.

Fundamental objective XI. considers how effectively restoration plans incorporate stakeholder input in the process. Including stakeholders throughout the NRDA restoration planning and implementation phases would be a measure of how well the plans consider stakeholder beliefs and employ their assistance in achieving better and more procedurally equitable outcomes. This objective might best be assessed as a process objective and not a fundamental one (see below).

The final fundamental objective pertains to learning and improving the management restoration process through learning. The ability to achieve the other fundamental objectives might have to be weighed against the cost, time and effort required for adaptive management or scientific decision-making (McDaniels & Gregory 2004). Thus, anticipated benefits from learning should be considered when selecting restoration strategies. For objective VI., Atlantic bluefin tuna (*Thunnus thynnus*) might have been affected because their spawning time coincided with the spill (Bjorndal et al. 2011). However, knowledge about the bluefin's life history in the GOM is inadequate for judgments and should be assessed in the learning objective (XII.).

Learning should also consider baseline data necessary to appraise decisions. A recent call for more baseline data for assessing protected species in the GOM was made by Bjorndal et al. (2011) in *Science*. Understanding the ecological changes from the massive fishery closures would be an important learning task for the restoration activities, beyond the direct impacts of oil on fish populations. For example, higher escapement from fishery closures can induce reductions in prey species available for larval fish with potentially undesirable consequences (Seeb et al. 1997 and references). Besides considering direct mortality from oil and indirect mortality from food web impacts, impacts of reduced fishing effort as well as population growth rate changes from changes in population size (density dependence) should be examined (Schuhmann & Schwabe 2004).

Learning objectives should consider adaptive management opportunities in management decisions. Adaptive management is a means for testing management decisions and their achievement of objectives. It follows principles of good scientific analysis as applied to decision-making. Like uncertainty assessments, all estuary management should consider adaptive methods to reduce uncertainty and to learn from the decision-making task. Fundamental objectives can be created related to adaptive management to weigh the extra costs it might require vs. implementing a strategy without an experiment of any kind to compare the effectiveness of the strategy (McDaniels & Gregory 2004). Additional considerations for adaptive management implementation are discussed in Gregory et al. (2006b).

Means Objectives

Means objectives are objectives that we care about only because of their influence on the fundamental objectives (Keeney 1992). This is not to say that means objectives are of lesser significance. In fact, one means objective can have considerable impacts on several fundamental objectives. Means objectives also are important to ensure that we have specified the right set of fundamental objectives and we are assessing ends over means. If there is a means to achieving another fundamental objective in the objectives hierarchy, then we are double counting, because this means objective is assessed at least twice (on its own and with the objectives it influences). To ensure that we are assessing ends over means – ask "why is that important?" (Keeney 1992). If the objective is important because it contributes to something more valuable in the decision context, it should be considered a means objective. In this process, we also need to make certain that we are not moving into strategic objectives or objectives that would be important for any decision in the estuary, not just the ones we are concerned with in a specific decision context, i.e. restoring fish communities from an oil spill.

Figure 4 displays an example means-ends network for restoring, rehabilitating, and replacing habitat. As we move away from the fundamental objectives, we ask "How do I achieve that?" to construct additional means (Keeney 1992). From Figure 4, we might ask, "How do I obtain greater viability of endangered fish populations and in-season fish stocks?" We may decide that restoring and replacing habitat would be two means for this objective. Then we may ask, "How would I replace oil-damaged habitat?" We decide that creating new habitats and acquiring natural resource equivalents are two ways of achieving this. We then ask, "How do I create new habitats?" and decide that reconstructing oyster reef substrates would be one way of doing this. We may also decide that transplanting seagrass to denuded areas or creating artificial refuges for downstream fishery locations are other means.

Weathered components of oil washed ashore around shoreline habitat, which is essential for fish refugia, breeding, and feeding. Chapter 5 discusses the important aspects of habitat restoration that should be considered including the actual contribution of the habitat to survival. The NRDA is focused on restoring habitats in the GOM that were affected by the spill. The habitats that were affected are important to species and components that contribute to use and non-use values held by the public. Habitats include deep water habitat (e.g., deepwater coral), offshore habitats (e.g., shellfish reefs, water column habitat, bottom sediment), near-shore habitats (e.g., coral reefs, muddy sediments, seagrasses), and coastal habitats (e.g., mangroves, mudflats, wetlands, sandy beaches) (NOAA 2011b). Sediment in marine areas is expected to be the largest repository of oil; ingestion of sediment or contaminated invertebrate infauna is being considered as the most likely route for exposure to oil degradation products.

For each region that was found to be affected by oil, habitat types should be selected and targeted for their essential services to stakeholders. Habitat types and regions should not be viewed as closed systems but as dynamic entities with potential exchange capabilities. Thus, the restoration should not proceed in a piecewise, fragmented fashion but rather one that integrates the spatial and temporal dimensions in connectivity and use by fishes throughout their life cycles.

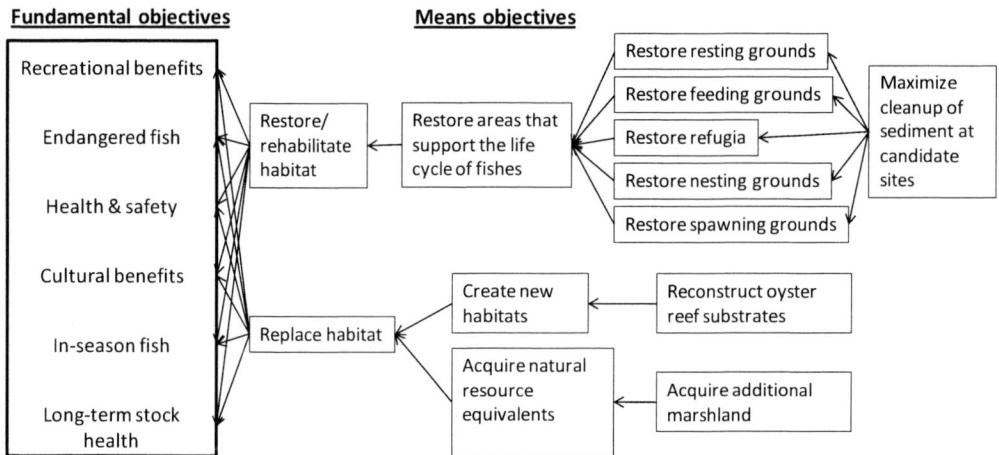

Figure 4. Partial network of means objectives for restoring and rehabilitating habitat and replacing habitat, and their potential impacts on categories of fundamental objectives.

Additional means should be evaluated for their ability in achieving fundamental objectives. According to Figure 4, the decisions for restoring habitat would be focused on cleaning sediment. If this was the primary focus in a decision process, the means-ends network could be used to illustrate that and discuss additional means for restoration. For example, in situ burning might also be important for restoration in coastal wetlands; it was found not to damage plant systems in Louisiana following a Hurricane Katrina-related oil spill (Baustian et al. 2010). Following some initiatives taken by NOAA through funding from the American Recovery and Reinvestment Act, reef reconstruction might be considered to protect coastline, SAV and marsh habitat as well as for providing fish habitat within the reefs themselves. When selecting from a package of options, a variety of habitat types should be provided for fish species (Restoration Plan 1997) and decisions should focus on their restoration or replacement. Along with the values structured in the fundamental objectives, good technical information and modeling should be used to appraise the superiority of some means over others.

All relevant ongoing or planned tasks should be incorporated into the means-ends relationships to account for possibilities of achieving each of the objectives and the uncertainties in achieving them. Creative brainstorming with experts and stakeholders should also be used to derive new strategies or means of achieving the objectives (Keeney 1992). Separating the means and fundamental objectives can allow managers to contemplate how best to achieve objectives, whether the measures being evaluated truly reflect a fundamental objective in a problem, and how objectives important to the problem might have a set of new alternatives that can better achieve them.

Strategic and Process Objectives

Strategic objectives would be the end objectives for any decision process, particularly for an agency or an organization (Keeney 1992). In some decision contexts, strategic objectives would be too broad and difficult to consider. For example, in an estuarine management problem focused on recovery of a biological community from wastewater pollutants impairing an estuarine system, a fundamental objective might be to maximize the abundance of a susceptible fish species in a receiving water body. The contribution of the abundance of the fish species to the gross regional product of adjacent coastal communities might be too broad in this decision context, but it is ultimately important to consider for inter-agency contexts. Differentiating fundamental objectives for a project or program from strategic objectives for an agency or inter-agency context in a decision problem can help prevent the assessment from being too broad and focused on endpoints that might not be susceptible to management decisions. Getting this balance right can be difficult but is essential to ensure that the analytical work is focused on the right problem. Ultimately, restoring a watershed or implementing sustainable initiatives would require greater thinking than a focused context or even a single government agency would be capable of. Identifying strategic objectives would be useful for a broad mandate, such as the recovery of the GOM from recent natural and anthropogenic disasters, which considers economic growth, human well-being and ecological integrity in a watershed or region.

The purpose of process objectives is to improve the decision-making process, not to appraise the consequences of the decisions in a certain context. Figure 5 illustrates the relationship between strategic, fundamental, means, and process objectives in a broader context of managing fisheries in the GOM. In Figure 5, a decision context of restricting or enhancing access to fishery resources is faced by an organization like the Gulf of Mexico Fisheries Management Council. The means objectives pertain to increasing accessibility to fishing sites and establishing spatial, temporal or categorical restrictions on certain fisheries. The fundamental objectives deal with long-term stock health (McDaniels 1995), wildlife benefits from fish consumption, minimizing bycatch mortality for species like the Kemp's ridley sea turtle (*Lepidochelys kempii*), and present stock health to economic benefits from fisheries in employment and income. The strategic objectives that would benefit from achievement of the fundamental objectives would be focused on ecological integrity of the entire GOM, economic benefits to coastal regions and the nation, social well-being, and human health. Process objectives would pertain to how we can make better decisions or improving the decision process. In this case, minimizing conflicts between different users of fisheries, the inclusion of stakeholders in management decisions, finding better data, and ensuring the ability to enforce decisions would all contribute to a better decision-making process.

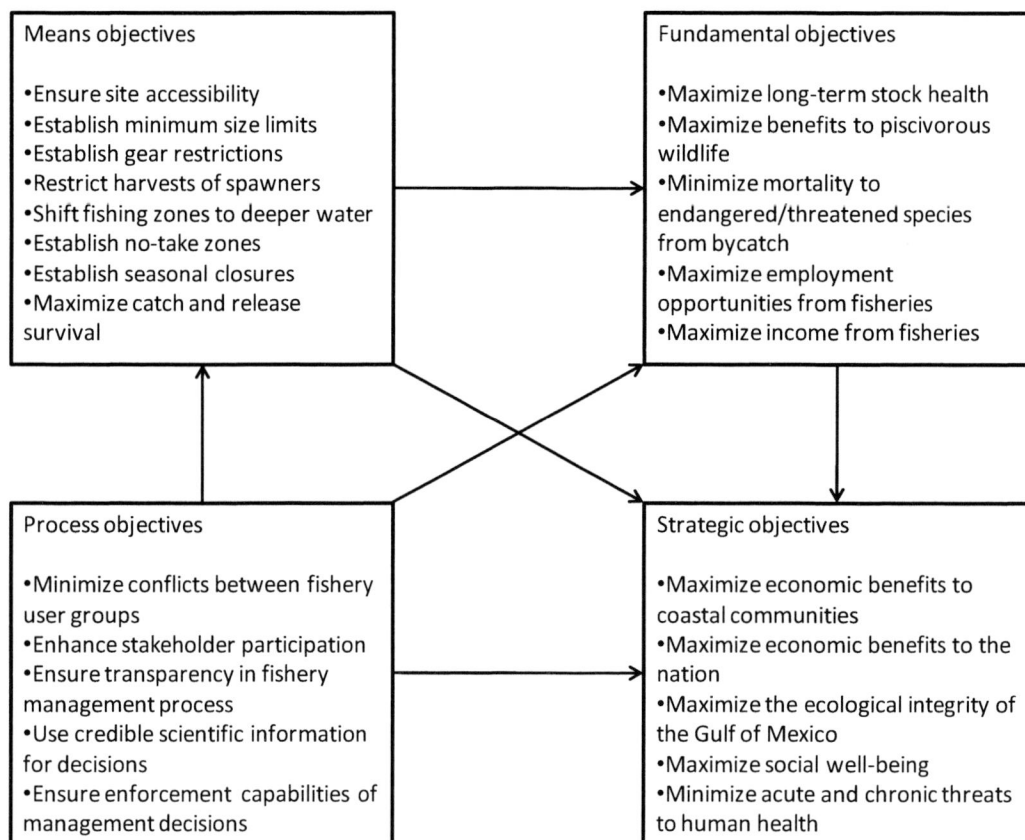

Figure 5. Relationship between means, fundamental, process, and strategic objectives for a decision context focused on restricting or enhancing access to fish resources (adapted from Keeney 2007).

IDENTIFYING ATTRIBUTES TO MEASURE THE OBJECTIVES

Determining attributes is one of the most important aspects of setting up a decision problem. Attributes should be unambiguous, comprehensive, direct, operational, and understandable (Table 3; Keeney & Gregory 2005). Attributes should be selected that distinguish among the alternatives under consideration or generated during the decision process (Failing & Gregory 2003). Thus, they should not attempt to completely encompass every facet of value in estuarine resources. Often, there will be a larger list of attributes than necessary and further analysis might find that some are uncontrollable or unchangeable within the management context. In those cases, it might be useful to place restrictions on attributes that make them more meaningful and measurable. For example, a group of stakeholders may be interested in the feasibility of management options for protecting human lives. The attributes might be predicted numbers of mortal and non-mortal injuries, but lifestyle choices outside the management purview could increase or decrease mortalities, thus the attributes would be insensitive to management options. In this case, lives lost to flooding and illnesses from exposure to hazardous algal blooms would be better attributes for evaluating management options.

There are natural, constructed and proxy attributes (Keeney 1992). Keeney & Gregory (2005) recommend choosing a natural measure that is comprehensive, direct and operational if available. If not, the fundamental objective should be broken into additional sub-objectives to obtain a natural measure. Alternatively, constructing an attribute is recommended. If the constructed attribute is not proper, or it is infeasible to make one, than a proxy attribute should be selected. Sometimes natural attributes might not be available for such objectives as aesthetic impacts to an estuarine resource. In this case, constructed attributes might be used.

Constructed attributes typically are scaled to indicate relative changes to an objective. When using constructed attributes, the method should be justified, the scale measurable by data collection, and the values relevant to evaluating trade-offs (Keeney & Gregory 2005). Thus, it is good to ensure that decision outcomes are represented by the scales (Keeney & von Winterfeldt 2007). Though normally discrete (nominal or ordinal scales), constructed attributes can be made continuous by constructing uniform changes between integer levels (Keeney 1981). In any case, the values of constructed attributes should clearly differentiate changes meaningful for stakeholders (Keeney 1981). Table 4 gives an example of a constructed scale for the attribute of lost recreational opportunities, from the objectives hierarchy in Table 3. Each level of the constructed attribute should be understandable to stakeholders so that they are certain of the meaning of quality fishing days and what a slightly lower quality recreational fishing site would be like.

Proxy attributes are sometimes used when other attributes are not available for an objective (Keeney 1992). A chosen proxy attribute should have a natural scale available. Complexity in assessing direct attributes for the objective often leads to the use of proxies (Keeney & Gregory 2005). For example, the objective "minimize injuries to restoration crews" has a natural attribute of injuries and lives lost from acute and chronic threats but could also be partially implied by the proxy "sea state conditions in open water habitats during restoration events." A measure for fish kills might be "number of hypoxic events" instead of the actual number of fish kills. The former may be chosen because it is easier to measure hypoxic events than fish kills. Essentially, selecting a proxy attribute is akin to using a measure for a means objective instead of one for a fundamental objective (Keeney 1992).

Several issues arise in the use of proxy attributes (Keeney 1992). First, double counting the objectives is more likely, because a means objective is likely to affect more than one fundamental objective. Second, weighting a proxy attribute is likely to create bias. Stakeholders might have a difficult time inferring how changes in a proxy attribute indicate changes in the object of value for a fundamental objective. As Fischer et al. (1987) observe, stakeholders might focus on concerns specified in the actual objective rather than consider the functional relationship between the proxy and the objective. Overweighting a proxy is more likely as the range of outcomes for the proxy would be greater than for a probabilistic outcome for a more fundamental concern (Fischer et al. 1987). For deterministic relationships, this might not be the case, but a deterministic relationship between a proxy attribute and a fundamental objective is less likely. Subjects might also weight the proxy attribute based on a believable worst case scenario. In any case, these simplifications are likely to introduce bias in the assessment of decision prospects. The more fundamental attribute should be chosen for valuation purposes despite the increased difficulty in modeling and assessment to relate a fundamental attribute to a more easily observed proxy (Smith & von Winterfeldt 2004).

Table 4. Constructed scale for the quality of lost recreational opportunities attribute (adapted from Keeney and von Winterfeldt 1994)

Level	Representative recreational quality impact
0	No recreational impact
1	Minor recreational disruption at a local site for potential site visitors with other local recreational opportunities of slightly lower quality
2	Moderate recreational disruption that affects over 1,000 quality fishing days for potential site visitors with no other local recreational opportunities of equal quality
3	Major recreational disruption to 10,000 quality fishing days for potential site visitors with other local recreational opportunities of slightly lower quality
4	Major recreational disruption to 100,000 quality fishing days for potential site visitors with no other local recreational opportunities of equal quality

Complicated attributes and objectives can sometimes come from requests by stakeholders to represent more aspects of the objective that they feel knowledgeable about (von Winterfeldt & Fasolo 2009). The result can impede the analysis as more considerations must be accounted for in the modeling exercises (von Winterfeldt & Fasolo 2009). To balance the requests of stakeholders with the tractability of the analysis, the elicitation protocol should include useful criteria for attributes that the stakeholders might agree on, for example, highly correlated attributes of the same objective might be condensed.

BEHAVIORAL DECISION RESEARCH

Behavioral decision research is largely descriptive, giving us information on how we make decisions. Important information can be found in the behavioral decision research on how biases that were previously unknown arise in the elicitation and decisionmaking process. As Keeney (1992) notes, biases cannot be completely eliminated. The values or preferences that are identified and elicited cannot be completely separated from the elicitation process (Slovic 1995). However, biases cannot be ignored and minimizing the biases is an objective of any decision analysis.

Early utility maximization theories considered decision-makers to be rational, leading to utility axioms such as transitivity and complete preferences (Slovic 1995). From these considerations, some valuation methods assume that for valuing final assets stakeholders will consider the magnitude of a loss in the same manner as they value a gain (Gregory 1999). However, the behavioral decision sciences have illustrated that this is not the case (Tversky & Kahneman 1981). Perceptions of values in losses or gains can influence how a policy decision is considered; anticipated outcomes might be over- or underweighted as a result of such perceptions, while policymakers may not recognize this assumption (Gregory 1999).

Value in a NRDA can be established by the amount of services, goods, or currency an individual is willing to accept to lose the use of a good or service or the amount they are willing to pay for a good or service (NOAA 1996b). However, if residents of the GOM states are asked for the maximum willingness to pay to prevent a future spill similar to the DWH disaster, their values would likely be lower than if asked for their willingness to accept another spill (Gregory 1999). Another example would occur if restoration alternatives were presented as giving stakeholders gain vs. redressing losses (Gregory 1999). The latter will tend to be weighted higher than the former even if the expected benefits or costs will provide the same final state of wealth. To avoid these issues, a constructive approach to valuation with stakeholders would consider how gains and losses are depicted and their potential influence on a problem (Slovic 1995). For example, the implications of loss of fish habitat could be described, but in combination with the positive implications of remaining habitat to minimize the impacts of focusing solely on a loss. Bias can be reduced in some cases by using gains and losses together when presenting issues. Additional procedural biases that could influence the plausibility of normative assumptions and the elicited preferences are summarized in French et al. (2009).

Arvai et al. (2008 and citations) discuss some additional heuristics that can impede complex decision-making tasks such as *availability*- judgments are focused on ability to recall similar events, *representativeness*- the likelihood of events is judged based on similarities to known events (which can lead to conjunction fallacies), and *anchoring*- cues can lead decision-makers or stakeholders to anchor on a reference point and incrementally adjust their judgments around it. The above-mentioned potential biases and others are indicative of the care that must be taken when eliciting judgments and values from decisionmakers and stakeholders (Arvai et al. 2008).

CONSTRUCTING MULTI-ATTRIBUTE UTILITY AND VALUE FUNCTIONS

Once an objectives hierarchy is constructed and alternatives to achieve the objectives are identified, a multi-attribute utility or measurable value function can be used to quantify the impacts of the alternatives on the objectives. The basic steps of applying a multi-attribute utility or value function are (1) to assess the value or utility for individual objectives and attributes, (2) determine an appropriate multi-attribute utility or value equation through independence checks, and (3) assess scaling functions. Whether a utility or value function is appropriate depends on whether there are uncertainties in the problem (Keeney 1992). If the outcomes for attributes are expressed as distributions, then a utility function should be constructed. If the decision-makers are certain that outcomes will be achieved by implementing their decisions, a value function might be more appropriate. Keeney & von Winterfeldt (2007) discuss how value and utility functions can be related to one another through monotonic functions and when they might be considered equivalent.

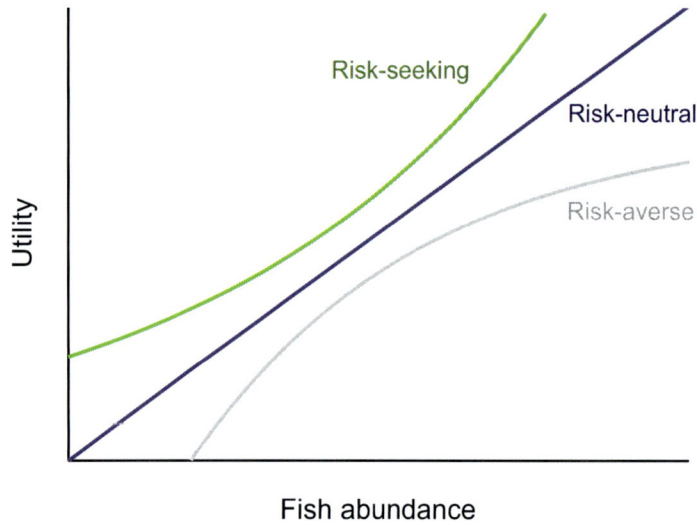

Figure 6. Three individual utility functions showing different risk attitudes to gains in fish abundance. The green curve illustrates risk-seeking behavior, in which utility increases at a faster rate at higher levels of the objective. The blue curve illustrates risk-neutral behavior. The grey curve indicates risk-averse behavior, in which utility increases at a slower rate at higher levels of the objective.

Typically, an additive form is used to bring together the different attributes (Gregory 2000). Additive utility implies very strict independence between the attributes and is demonstrated below:

$$U(x_1, \ldots, x_m) = k_1 U_1(x_1) + \cdots + k_m U_m(x_m)$$
$$= \sum_{i=1}^{m} k_i U_i(x_i)$$

where $U(x_1, \ldots, x_m)$ represents the utility of an outcome on m attributes for an alternative, $k_1, \ldots, k_m$ are weights for the m attributes (their assessment is discussed in the next section), and $U_m(x_m)$ are individual utility functions for each attribute m with outcome degree x_m (Clemen & Reilly 2001). (For an additive value function, substitute a V for U). For each alternative, the overall impact on the objectives can be compared using an additive function if the assumptions are properly met. For information on calculating overall value or utility for a complex hierarchy with sub-objectives, see French et al. (2009).

Individual value functions can be constructed using a value increment procedure for the strength of preference between subinterval attribute levels as demonstrated by Merrick et al. (2005) for an objectives hierarchy constructed for a watershed in the Chesapeake Bay. Individual utility functions can be judged or elicited so as to model the risk behaviors of the stakeholders. A simple example (Keeney & von Winterfeldt 2007): stakeholders might be risk neutral to losses of fish abundance over a region where a species of fish is abundant but risk averse when a fish species is rare, threatened, or endangered. Figure 6 displays the general shapes of risk neutral, risk prone, and risk averse utility functions. For a risk neutral case, the measurable value function is equivalent to the utility function. In order to capture the preferences of stakeholders under conditions of uncertainty, lotteries (e.g., certainty equivalent, probability equivalent, risky-risky choice) might be given in some contexts to elicit individual utility points. These lotteries can be difficult for lay stakeholders, and even experienced decision-makers, to comprehend and express preferences with because of misunderstandings of the implications of probabilities, or difficulties choosing among outcomes in simple lotteries. The difficulty of creating single attribute utility functions with lotteries can be circumvented by inferring single attribute utility functions (e.g., risk neutral, risk averse, risk prone) from stakeholder attitudes (Keeney & von Winterfeldt 2007). Sensitivity analysis should be used on the resulting value model's weights or risk tolerance parameters to choose aspects that should be examined in more detail. Bounding the parameters to plausible ranges would also assist in the sensitivity analysis stage (Keeney & von Winterfeldt 2007).

WEIGHING TRADE-OFFS WITH STAKEHOLDERS

Trade-offs should be examined in light of the expected range of consequences for each of the fundamental objectives and their attributes. Two processes for assessing trade-offs are discussed here. The first is indifference or cost-equivalence analysis. The second is swing weighting (von Winterfeldt & Edwards 1986). Indifference curves are very useful for establishing the amount one would be willing to give up in exchange for receiving more or less of another objective. Indifference curves can be constructed easily when all individual value or utility functions are linear. Figure 1 depicts an example of an indifference curve.

A related way of considering the trade-off behavior of stakeholders is through an equivalent cost procedure (Keeney & von Winterfeldt 2007). This procedure can be used if the attributes are scaled numerically, the individual value functions are close to linear, the multi-attribute value function is additive, and cost is an important factor in the problem. Thus, costs of losing or gaining individual attributes can be calculated, such as the cost of losing an acre of seagrass. This method could be useful in a NRDA assessment. Trustees utilize a resource-to-resource or service-to-service procedure to appraise the extent of resource replacement (NOAA 1996b). Comparing the value of services lost to those returned is done when service-to-service or resource-to-resource comparisons cannot be made. Equivalent value is the standard in this case. The Oil Pollution Act (§990.53) allows for selecting a restoration action using cost equivalence of lost services based on a dollar value when resource-to-resource or valuation scaling are unfeasible.

For multiple objectives and attributes, the attribute outcome levels for each alternative can be multiplied by their per-unit equivalent cost for this calculation. Cost-benefit considerations are similar but the model for cost-equivalence may be based on trade-offs elicited from stakeholders (Keeney & von Winterfeldt 2007). As Keeney & von Winterfeldt (2007) describe, equivalent cost can be calculated using the equation:

$$V(x_1, x_2, \ldots, x_i \ldots, x_n, c) = c + \sum w_i x_i$$

where V is the multi-attribute value function which will be expressed as an equivalently valued consequence in dollar costs across the attributes for an alternative, c is cost, x_i is the outcome level for attribute i, and w_i is the equivalent cost for each unit of x_i.

Equivalent costs can be used to compare management alternatives based on how well each of the attribute categories performs on the option. Thus, an equivalent cost of $90,000 dollars for the loss in recreational opportunities from closing a fishery and one of $45,000 for the loss in recreational opportunities from not closing a fishery would indicate that not closing the fishery would perform better on recreational opportunities than closing it. In their nuclear waste repository site selection analysis, Merkhofer & Keeney (1987) demonstrate how attributes might be made equivalent to costs through weighting functions and meeting the assumptions of multi-attribute additive models. Equivalent cost analysis using societal values for cost expenditures for different impacts on health, property, social, and environmental objectives with combinations of probabilistic and risk modeling for assessing outputs is given in von Winterfeldt et al. (2004).

Swing weighting is useful when not all of the individual value functions are linear (von Winterfeldt & Edwards 1986, Keeney & von Winterfeldt 2007). In swing weighting, stakeholders are given a hypothetical alternative where each of the objectives is at their worst consequence. Then, they choose which objective they would like to swing first or move from the worst level to the best level. They repeat this procedure until every objective has been swung from the worst to the best possible outcome. After this exercise, an ordered list of objectives will be available. The stakeholders then take this list and write how much more they prefer one objective that was swung before another using a numerical scale or distributed numbers.

An example swing weighting procedure is given in Table 5. From the table, a group of stakeholders or decisionmakers would individually be given a range of consequences on each attribute's units to contemplate. They would then be given a hypothetical situation where each attribute is at its worst possible level. They would be asked, "Which attribute would you swing from the worst to the best level first?" The stakeholders would continue this process until each attribute is swung from the worst to the best level to establish a rank order. Next, they would be given a list of each of their attributes in the order they were swung and asked to indicate the relative importance of each attribute on a scale, starting at 100 for the first attribute swung. They may be permitted to weight some attributes as zero, effectively eliminating them from consideration in the objectives.

The scaling constants estimated from swing weighting also can be used to establish equivalent costs. Swing weighting could be useful if one does not wish to emphasize exchanges between non-monetary objectives and dollar values as observed in McDaniels and Trousdale (2005) for the Metis people's analysis of trade-offs. As illustrated in McDaniels and Trousdale (2005), financial equivalents can be derived from the scaling constants and an additive value function to construct a value model that "prices out" the outcomes. For example, the abundance of an endangered species could be transferred to a dollar value based on relative priorities determined in swing weighting results with a monetary objective. Because swing weighting the objectives allows one to consider how much a change in one objective might be worth relative to a change in another, a non-monetary objective's relative worth in dollars can be easily estimated with an additive function, when the individual value functions are linear, and if at least one of the objectives has a measure in dollars. Keeney (1980) gives examples for calculating cost equivalents for more complex multi-attribute problems.

Table 5. Example fundamental objective attribute ranges from management decisions for restoring sites that might be considered in evaluating trade-offs

Attribute	Worst level	Best level	Units
Worker injuries	150	0	# of acute morbid cases
Survival advantage for prey fish species for mammals and birds	0	500	%
Management costs	80	10	million U.S. dollars (2011)
Traditional fishing impacts	0	8	constructed impact significance scale
Recreational catch per person/trip	0	1180	kg of fish caught per day trip
Public acceptance	0	4	public acceptance scale

Swing weighting is a commonly applicable procedure that allows analysts and decision-makers to contemplate stakeholder trade-off behavior using real ranges of attributes. Simple relative importance rating of objectives could be nonsensical if the implicit ranges are not reflective of real outcomes (Keeney 1992). For example, directly rating importance would not allow stakeholders to consider such factors as the relative worth in costs of an acre of seagrass. It could turn out in the analysis that the weighting by stakeholders indicates that spending over $1 billion is worth less than an acre of seagrass if stakeholders rank without consideration of ranges. Thus, it renders blanket statements without consideration of attribute levels operationally useless. Assessing trade-off behaviors in a manner that is useful is one of the potential benefits of a decision analysis and an important insight that can be gleaned.

Consistency checks should be performed after any weighting or value model building procedure. Questions might include, "Your analysis indicates that 100 tons of spawning stock biomass are worth about $10,000 in management costs, does this seem reasonable?" (Keeney & von Winterfeldt 2007). Increases and decreases in utility should also be checked for consistency. For example, what percentage gain (loss) in survival advantage for a fish population would produce an increase (decrease) of one unit of utility? The value models would give us the ability to determine this and examine the overall reasonability of the outcomes of the process with stakeholders. For additional information on weighing objectives, Keeney (2002) describes considerations that should be understood prior to implementing trade-off analyses with stakeholders.

EVALUATING ALTERNATIVES

With the additive model, one can evaluate alternatives using a portfolio approach (where a subset of options is chosen based on their ability to achieve the objectives and subject to any necessary constraints) or a ranking approach (where a single option is chosen with the highest overall value or lowest cost equivalent). Each has similar evaluation procedures but the portfolio approach will also evaluate a no action alternative as a baseline to evaluate the incremental value of implementing each option (Clemen & Smith 2009). The ranking approach will focus on the best possible option to pursue from a set of options. A decision-consequence table is exemplified in Table 6. The objectives and attributes in Table 6 were taken from the ones in Table 5. Overall values are displayed in the last row of Table 6 for each alternative. For each alternative, the attribute scores from the individual value functions would be multiplied by the attribute's scaling constant and summed with the other weighted attribute scores to obtain an overall value estimate. Thus, in Table 6, worker injuries might have a higher weight over the attribute ranges than survival advantage which would have a higher weight than public acceptance. From Table 6, Alternative 3 is clearly dominant while Alternative 2 is preferred to Alternative 1.

Table 6. Decision-consequence table for assessing performance of NRDA objectives under different restoration alternative scenarios and overall value (V(x)) for each alternative

Attribute	Units	Alternative 1	Alternative 2	Alternative 3
Worker injuries (x1)	# of acute morbid cases	150	60	0
Survival advantage for prey fish species for mammals and birds (x2)	%	20	0	500
Management costs (x3)	Million U.S. dollars (2011)	10	80	70
Traditional fishing impacts (x4)	Constructed impact significance scale	8	0	4
Recreational catch per trips (x5)	Kg of fish caught per person/day trip	0	276	1180
Public acceptance (x6)	Public acceptance scale	0	1	4
V(x)		0.37	0.59	0.86

Alternatives in estuarine management problems are uncertain, so the appropriate value model should generally follow the structure of a utility function. However, Keeney & von Winterfeldt (2007) discuss important considerations for when to use value functions in lieu of utility functions and how to properly relate the two to each other through relative risk aversion functions. Individual attributes will have single utility functions that are brought together using an MAU function. For simplicity in use and application, an additive utility function for multiple attributes should be used for properly structured problems that meet additive assumptions. The tests for the applicability of various multi-attribute functions are preferential, weak-difference, utility, and additive independence tests (Keeney 1992). Formal tests for additivity can be difficult. Although the additive function requires more stringent assumptions, if the objectives are structured so that none are missing and the fundamental objectives are truly ends objectives and not means of achieving them, then the additive model usually will be appropriate (Keeney & von Winterfeldt 2007).

In order to calculate expected utility with the technical uncertainty analysis conducted on the prospects, the probability distribution for each alternative, p_j, must be calculated and combined with the utility function, U(x) (Keeney 1980). For continuous attributes, an integration of probabilities multiplied by utilities for each possible consequence must be done, whereas summation is used for the discrete case. A general equation for the continuous case is shown below:

$$E_j(u) = \int_x p_j(x)U(x)dx$$

where $E_j(u)$ is the expected utility of alternative j and U(x) is the utility function (Keeney 1980). Additional information on uncertainty analysis is discussed in the following section.

If expected utility was used instead of measurable value, the impacts on the attributes from the alternatives would be considered as best estimates, expected, or average values as listed in Table 6. Distribution parameters (e.g., standard deviations for normal distributions) used in calculations should also be reported for continuous variables along with interval probabilities for the ranges in constructed scales. Sensitivity analysis should be used on the predicted rankings based on changes in inputs from the utility models as well as changing parameters from the distributions over the attributes.

UNCERTAINTY ANALYSIS

There are many different definitions of uncertainty and risk. Following Hubbard (2009), uncertainty means there is more than one prospect. Risk is a detrimental outcome that might or might not occur. Uncertainty is measured as a group of prospects with corresponding probabilities, and risk is measured as a group of prospects with probabilities and magnitudes of loss of money, satisfaction, health, etc. (Hubbard 2009). Risk is an inherent consideration in NRDA because injuries are estimated from uncertain knowledge. There are many ways to account for uncertainty, but the most common and useful is probabilistic analysis. Deterministic models follow rules to establish relationships between variable configurations. Probabilistic models allow the specification of uncertainty about relationships of variables. Considering the uncertainty in a problem brings more reliable decision recommendations with less of a chance of false positive and negatives. It can also bring more understanding of the decision environment and sources of contention about decision outcomes.

Causality is of primary importance in NRDA decisions, as analysts evaluate damage and potential restoration successes from an oil spill. A causal decision problem can be addressed with either a top-down or a bottom-up approach (Pauker & Wong 2005). In the top-down approach, the model is conceptualized in an influence diagram, conditional independence network, or some type of cognitive map, while the quantitative data necessary to specify relationships between variables in the map is employed at a later stage. The bottom-up approach begins with considering what knowledge or evidence is available and what this evidence tells us about the decision problem. Neither of these approaches is mutually exclusive; they might be blended depending on the needs of analysts and decision-makers.

Given the uncertainties and likely disagreements among interested parties about the effectiveness of different options, proper analytical models should be developed to address uncertainties and the potential range of impacts from decisions (Keeney 2009). In a fisheries NRDA problem, processes that would be modeled include various uncertainties in restoration outcomes in open marine water or an estuarine marsh. Also, spatial and temporal uncertainties about the sensitivity of life stages (e.g., young-of-year, spawning adults) to a spill-related stressor, the sensitivity of various fish species to oil, and their likely responses to restoration could be accounted for if it was necessary for appraising decisions.

The decision context and the objectives should include spatial and temporal dimensions so that these elements of uncertainty can be properly weighted in the modeling process. Discussions of how spatial scales apply to quantifying benefits in estuaries can be found in Chapter 5. A combination of empirical and mechanistic modeling might be used, limited by constraints on predictions from data for the empirical aspect of models, the scientific understanding needed for the mechanistic aspect of models, and the strength of data used to build the models (Strayer et al. 2003). Model outputs are used to make inferences based on inputs that are assumed to be accurate (Pourret et al. 2008). Bayesian belief networks (BBNs) or influence diagrams (IDs) can be useful for combining mechanistic and empirical models and for measuring the uncertainty in quantitative predictions from each. Relationships in BBNs are described by conditional probabilities to explicitly represent the uncertainty in data and mechanisms, and structures can be formally tested with data and expert knowledge (Wooldridge & Done 2004).

Another consideration is the amount of damage that has occurred to fishery resources as a result of the oil spill. Data documenting the spill have been posted on web sites by U.S. Fish and Wildlife Service and NOAA[1]. Fish kills have been observed and nearly 1000 miles (1600 km) of oiled shorelines have been documented, but the number of unobserved incidents of fish kills and the uncertain impacts of oiled shorelines need to be evaluated to appraise impacts of restoration decisions on objectives. Given the life cycles of some long-lived fish, the impacts of the spill will take years to fully understand. Communicating these uncertainties should be considered during model selection phases. How uncertainty and risk are defined by different groups of experts or stakeholders would be made explicit to enhance insights. Competing models for different stakeholder groups can be developed to examine beliefs in unknown decision outcomes as Conroy et al. (2008) did for conservationists and fisherfolk in assessing gill net entanglement risks to Hector's dolphin (*Cephalorynchus hectori*).

A NRDA plan is currently in effect to investigate fish kills from the DWH oil spill (Powers and Blanchet 2010). The plan requires documented incidents of fish kills potentially related to the oil spill as well as baseline data for fish kills that occurred over the past ten years. GIS layers are being constructed to document fish kills prior to, during, and after the spill. Baseline data are included primarily for relating causes to effects probabilistically. In the past, evidence that fish kills have been caused by specific stressors has been weakened by probabilistic reasoning that did not consider the likelihood of a fish kill in the absence of the stressor, or the probability of the stressor's presence (Stow 1999, Newman et al. 2007). These assessments estimated the probability of a stressor given a fish kill instead of the more

---

[1] http://www.noaa.gov/sciencemissions/bpoilspill.html.

rigorous determination of the probability of a fish kill given a stressor, which can be estimated from the above factors using Bayes' theorem.

Expert Opinion

Expert opinion might also be used in certain cases. Often in estuarine problems, the relationships between options, intermediate variables, and objectives are complex and data for specific aspects of these relationships might be sparse. For example, upland land use, forested buffers, agricultural practices, etc. usually are expected to affect the quality and quantity of water flowing into estuaries. Variations in water quality characteristics such as nutrient and suspended solids concentrations also can be influenced by hydrologic variables outside of the decision-maker's control. All of these factors might be relevant to evaluating impacts of decisions on fish habitats and populations even if data are lacking. With expert opinion, we can better understand the uncertainty in a decision problem, weigh evidence for what might happen from decisions in the future, and what might have happened in the past, and establish prior beliefs for updating as new data are gathered. Individual experts might be commissioned and a subjective probability distribution elicited for an important process, or a panel of experts might weigh the evidence for causes of a potential impairment in an estuary using Bayes' theorem and rigorous accounting. Hesitancy about using expert judgment might stem from unfamiliarity with the methods, but uncertainty about important processes and outcomes that arise from limited data might be better appraised using robust protocols for eliciting judgments from experts (Gregory et al. 2006a).

Hora (2007) provides an overview of protocols for eliciting opinions from individual experts. Clemen & Winkler (2007) give an overview of procedures for aggregating expert probability distributions from subjectively elicited ones. The method(s) chosen should be based on the type of problem being investigated as well as the qualities of the elicitations.

Value of Information

Although the pre-assessment phase of the NRDA has been completed for the GOM, information gathering continues as the restoration plans are being developed and stakeholder values considered. Probabilistic assessments in combination with measures of value can be applied to augment the data gathering process, resolve differences in stakeholder or expert opinions, and focus research efforts. Conroy et al. (2008) provide just such a structure, using influence diagrams and stakeholder workshops in a problem pertaining to gill net entanglement regulations proposed for reducing dolphin mortalities. Sheehy & Vik (2002) conducted a *value of information* analysis that demonstrated to managers, based on their willingness to pay for the information, the importance of monitoring data for fish stocks. The use of an influence diagram constructed in Netica™ (Norsys 2010; Figure 7) is shown here to illustrate how valuation and technical data can be used to better consider the expected value of perfect and imperfect information.

In an influence diagram, there are three types of nodes (boxes) related to variables in a problem. *Chance nodes* pertain to random variables, *decision nodes* pertain to actions under contemplation, and *utility nodes* contain a utility function that relates the outcomes of decisions to their expected utilities (Shachter 1986). Each chance node is exhaustive (i.e. covers the possible range of the variable) and comprised of two or more states that are mutually exclusive (non-overlapping) as to the outcomes they describe. Each decision node contains discrete states related to potential options a manager might consider. The nodes are connected by arcs or arrows that indicate some type of a conditioning influence when connected to chance nodes, an informational link when connected to decision nodes, and an influence on utility in the problem when connected to utility nodes. The expected utility for each management option is displayed in the decision node and is calculated as a weighted summation of the probability of each of the outcomes times their utility if they were to occur. For each closure decision and each stock level of fish when the decision is made, there is an associated utility score.

Influence diagrams are helpful in structuring and analyzing decision problems, because they display intuitively how decisions affect processes of importance and the potential impacts of decisions on value. Given this potential value, decision scenarios can be selected to maximize the expected utility for a problem and provide a normative approach to recommendations.

In Figure 7, the decision is whether to institute a closure of high intensity, low intensity, or not to close the fisheries at all. The stock level of the fish is either above optimum yield, supporting optimum yield, or below it. It is assumed that a high fishery closure will bring about optimum yield within a required time period, and a low closure will better maintain optimum yield if it is currently supporting it. The utility node contains a utility function based on the impacts from the fishery closure and the status of the fish stock. If the wrong decision is made, for example, a high closure when the stock level is above optimum yield, then the expected utility is lower. Also, any closure will cause lower utility through short-term impacts on fishing yields, and this would be reflected in the utility scale. The marginal distribution of stock level indicates that it is highly likely that the stock level is below optimum yield. In the decision node, the higher expected utility for the high closure decision indicates that this is the best choice given the evidence in the ID. In this example, we assume that the multi-attribute utility function represented in the utility node was discovered to be equivalent to a measurable value function and differences in preference strength between options can be assessed directly. This will allow us to impart meaning in the difference between expected utilities for a value of information example described below.

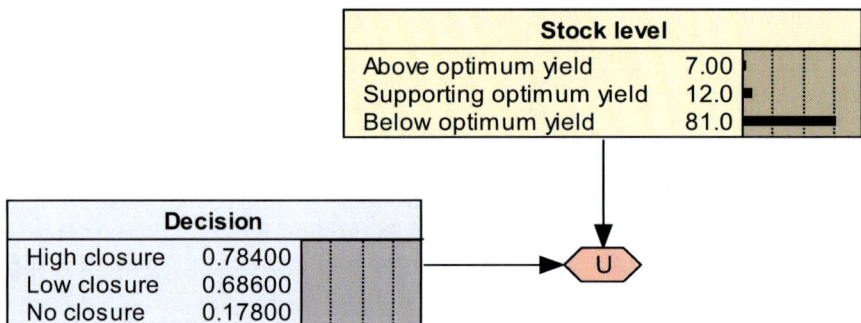

Figure 7. Influence diagram for a fishery closure decision based on the current level of fish stock in a hypothetical system. Adapted from Clemen and Reilly (2001).

In Figure 8a, the influence diagram from Figure 7 is shown, but an arc is drawn between stock level and the decision node, indicating that the stock level is known prior to a decision. Figures 8b-d illustrate how the expected values change based on different outcomes for this information. We can see that knowledge of the best decision is highly dependent on the information about the stock level. From Figures 7 and 8, the expected value of perfect information (knowledge of stock level prior to a decision) can be calculated. First the expected value for Figures 8b-d is found by multiplying the highest value for the best decision in each diagram by its probability of occurrence. These values are summed and the expected value of the decision if this information is not known (Figure 7) is subtracted from it (0.907-0.784). This gives an additional value of 0.123. This number is the additional value of knowing the stock level prior to making a decision. From the trade-off analysis, we can convert values to relative costs and know the equivalent dollar worth in management costs that this additional information is worth. It would be hard to judge this from the measurable value scale alone. As illustrated in McDaniels (1995), seemingly low or insensitive utility or measurable value ranges can sometimes translate into equivalent high dollar benefits or management costs.

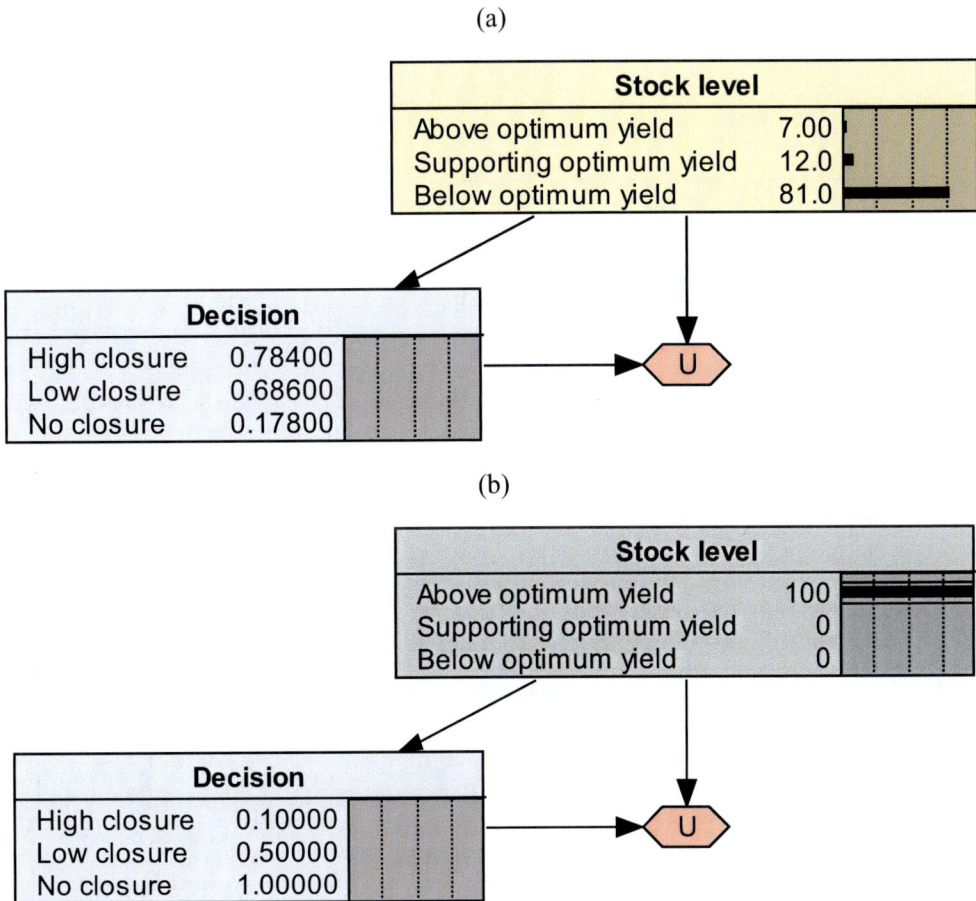

(a)

(b)

Figure 8. (Continued)

(c)

Stock level	
Above optimum yield	0
Supporting optimum yield	100
Below optimum yield	0

Decision	
High closure	0.40000
Low closure	0.70000
No closure	0.90000

U

(d)

Stock level	
Above optimum yield	0
Supporting optimum yield	0
Below optimum yield	100

Decision	
High closure	0.90000
Low closure	0.70000
No closure	0

U

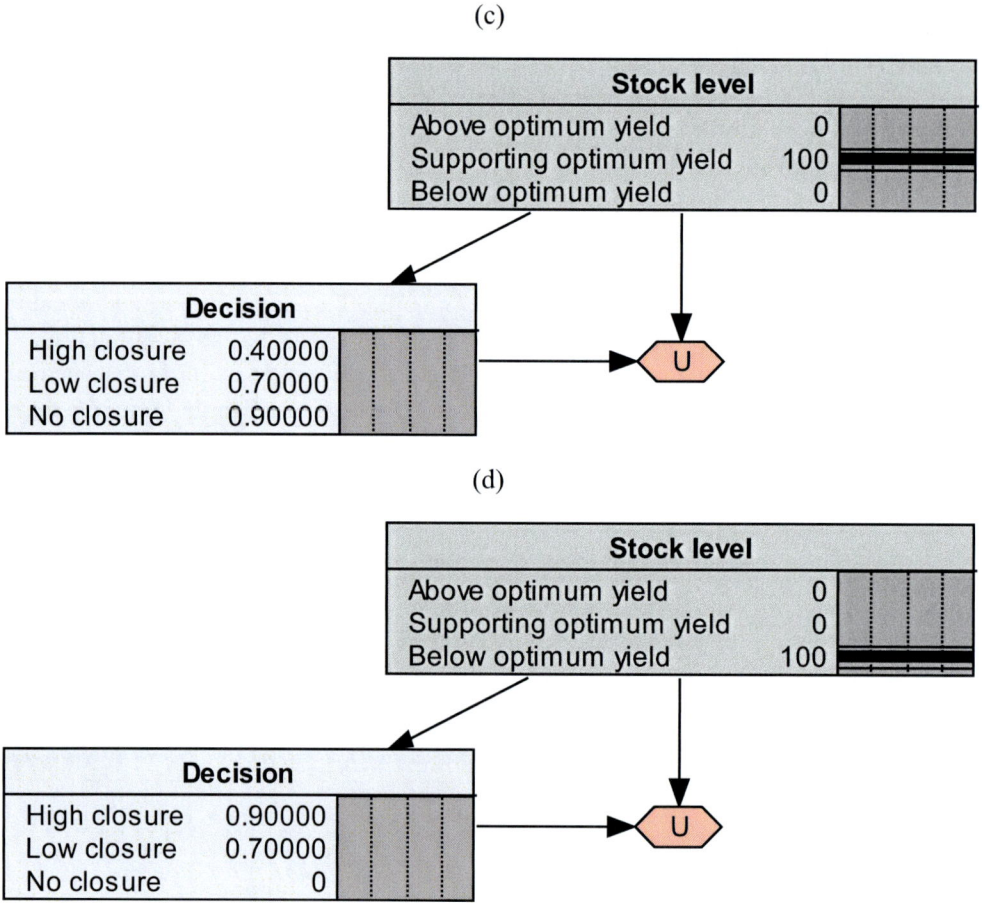

Figure 8. Influence diagram for a fishery closure decision based on the current level of fish stock in a hypothetical system when stock level is known with certainty prior to the decision (a), with knowledge that stock level is above optimum yield (b), with knowledge that stock level is supporting optimum yield (c), and with knowledge that stock level is below optimum yield (d).

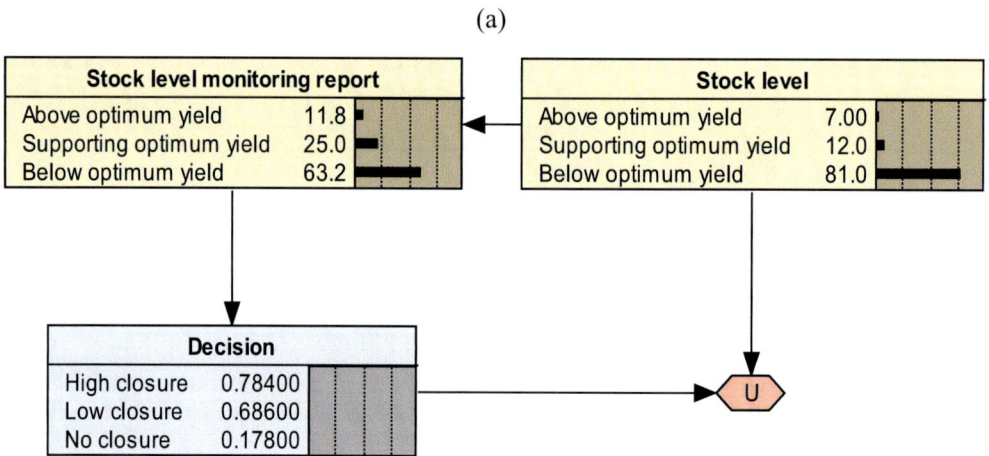

(a)

Stock level monitoring report	
Above optimum yield	11.8
Supporting optimum yield	25.0
Below optimum yield	63.2

Stock level	
Above optimum yield	7.00
Supporting optimum yield	12.0
Below optimum yield	81.0

Decision	
High closure	0.78400
Low closure	0.68600
No closure	0.17800

U

(b)

(c)

(d)

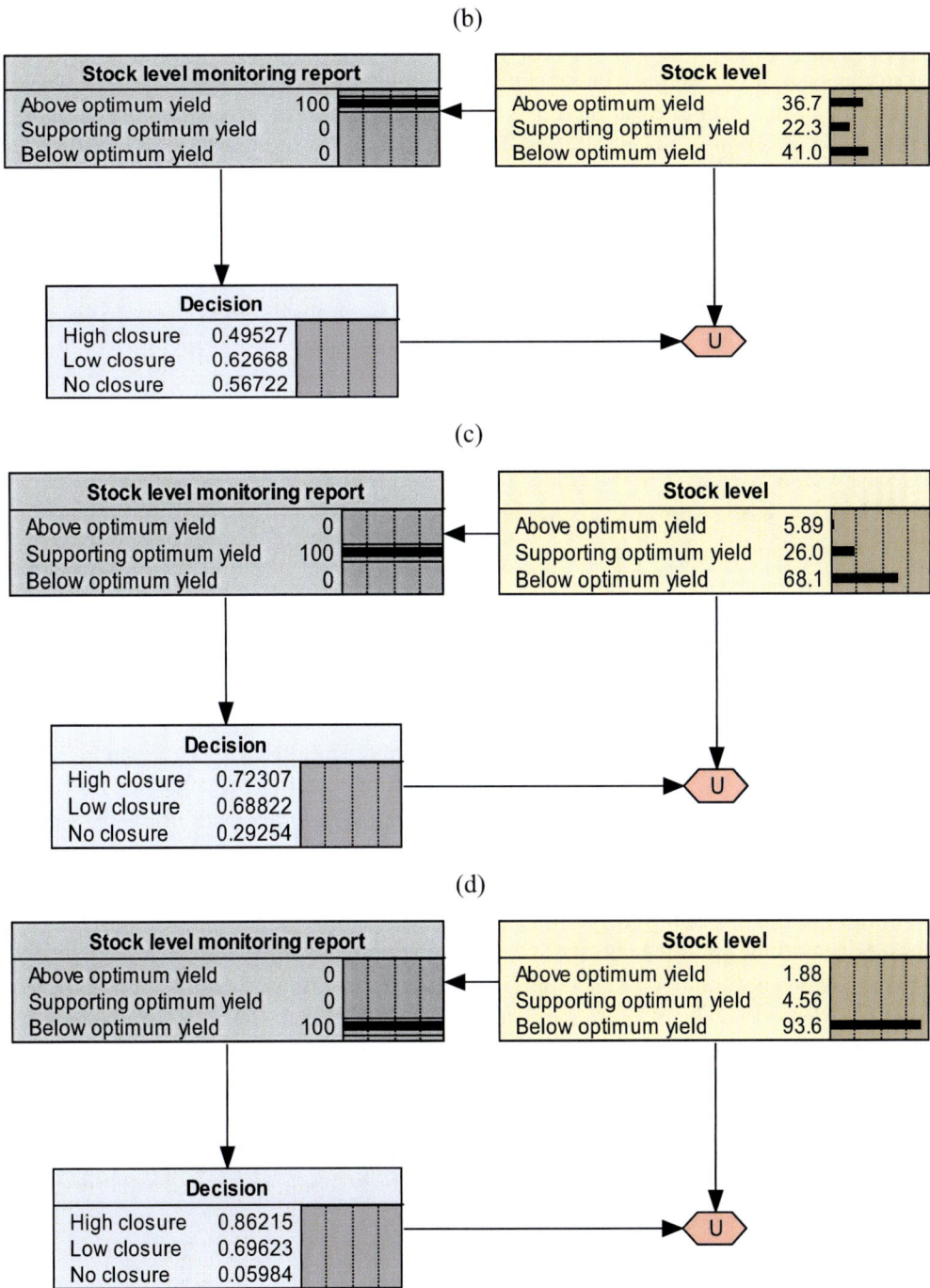

Figure 9. Influence diagram for a fishery closure decision based on the current level of fish stock in a hypothetical system when uncertain information on stock level is available prior to the decision (a), with uncertain knowledge that stock level is above optimum yield (b), with uncertain knowledge that stock level is supporting optimum yield (c), and uncertain knowledge that stock level is below optimum yield (d).

Figure 9 displays the same problem but an additional chance node is added to indicate that the stock level can only be known based on monitoring data that has some uncertainty. As can be seen in Figure 9a, there is uncertainty in the stock level monitoring report given that the stock level is potentially in a certain state. This additional chance node would be useful for examining the impact of the sensitivity and specificity of a monitoring or experimental procedure on a decision. The expected values for each of the scenarios are calculated for Figure 9 as was done in Figure 8. The expected value of imperfect information is found to 0.016 value units. In a more complex influence diagram, this value could be compared to the value of information for other certain or uncertain evidence to prioritize data gathering (Conroy et al. 2008). Figures 8 and 9 also illustrate that having perfect or imperfect information could potentially change the recommended decision. If the recommended decision could not be changed from additional information, then it would not be useful for improving the decision problem (Shachter 1990). For more complicated structures, decision trees are often used to solve value of information problems in decision analysis as illustrated by Clemen & Reilly (2001) and Korb & Nicholson (2011).

CONCLUDING REMARKS

The GOM recovery plans are described in several in-depth investigations and reports examining the impacts of local, state, and federal agencies in the recovery effort. The importance of restoring human health and economic benefits by management efforts in the GOM was described in the Mabus report (Mabus 2010). Sustainability issues in the restoration of the GOM, including the establishment of restoration science hubs, were further emphasized in the *Beyond Recovery* report (Gordon et al. 2011). An executive order signed by President Barack Obama in 2010 directed all federal agencies to coordinate their efforts for the restoration and the recovery of the Gulf's social well-being and lifestyles. In responding to the spill, federal, state, local, and community agencies are partners in fulfilling the social and economic goals set forth in previous discussions. Restoration of the GOM will not be focused solely on mitigating the impacts from the DWH oil spill. Rather, more comprehensive restoration and recovery are being undertaken to remedy harms caused by a long series of calamities, including hurricanes, coastal land loss, and failures of infrastructure. Interacting with stakeholders and experts in fulfilling these tasks is considered key to the success of restoration initiatives in the GOM. To implement these interactions optimally, knowledge supplied by the science of decision analysis should be applied extensively.

The decision sciences comprise a broad field with ongoing development in qualitative and quantitative procedures, applicable to a variety of situations. We have presented an overview of some of the methods that can be used and some important considerations for technical and value-based evaluations in managing estuaries. Describing all potentially valuable approaches available in the behavioral, preference management and construction, risk analysis, and other decision science literature in one chapter would have been impossible. Better discrimination among decision analysis frameworks and better evaluation of approaches will become possible as we gain understanding of what works and does not work in different contexts, problems, and opportunities.

There are many competing ideas about how best to approach management problems. Public decisions often can be controversial, with adversarial debates about the usefulness of projects and the anticipated impacts on groups of stakeholders (Gregory and Keeney 1994). The decision analysis literature gives a wealth of information about how to properly structure these problems by identifying the involved parties, considering their preferences, and analyzing the uncertainty related to outcomes for things of value. The usefulness of a decision analysis arises partly from logically structuring the preferences or values of stakeholders. From these values, better alternatives may be found. Without the structure and rigor supplied by decision science, decision-makers and other key players in the problem may focus on the wrong problem entirely, and thus have little basis for understanding sources of contention in finding solutions to problems of managing and restoring estuaries and other natural resources and ecosystems.

ACKNOWLEDGMENTS

Thank you to Drs. Jason Merrick, Thomas Stockton, and Brian Dyson for reviewing drafts of this chapter and providing constructive criticism. These individuals do not necessarily agree or disagree with any of this chapter. This research was supported by the U.S. EPA and while reviewed according to EPA guidelines, it does not necessarily reflect EPA policy. Mention of trade names or commercial products does not constitute endorsement by the U.S. EPA. This is contribution 1436 from the Gulf Ecology Division.

REFERENCES

Arvai, J.L., Gregory, R., & McDaniels, T.L. (2001). Testing a structured decision approach: Value-focused thinking for deliberative risk communication. *Risk Anal. 21,* 1065–1076.

Arvai, J., Gregory, R., & Zaksek, M. (2008). Improving management decisions. In: W.E. Martin, C. Raish, & B. Kent (eds.), *Wildfire Risk: Human Perceptions and Management Implications.* Resources for the Future, Washington, DC.

Baustian, J., Mendelssohn, I., Lin, Q., & Rapp, J. (2010). In situ burning restores the ecological function and structure of an oil-impacted coastal marsh. *Environ. Manage. 46,* 781–789.

Beck, M.W., Brumbaugh, R.D., Airoldi, L., Carranza, A., Coen, L.D., Crawford, C., Defeo, O., Edgar, G.J., Hancock, B., Kay, M.C., Lenihan, H.S., Luckenbach, M.W., Toropova, C.L., Zhang, G.,& Guo, X. (2011). Oyster reefs at risk and recommendations for conservation, restoration, and management. *Bioscience 61,* 107–116.

Bins, H. (2010). Testimony before the Gulf of Mexico Fishery Management Council. Holly Binns, Pew Environment Group, June 16, 2010. http://www.pewenvironment.org/uploadedFiles/PEG/Publications/Other_Resource/testimony_06_16_10.pdf

Bjorndal, K.A., Bowen, B.W., Chaloupka, M., Crowder, L.B., Heppell, S.S., Jones, C.M., Lutcavage, M.E., Policansky, D., Solow, A.R., & Witherington, B.E. (2011). Better science needed for restoration in the Gulf of Mexico. *Science 331,* 537–538.

Clemen, R.T. (2008). Improving and measuring the effectiveness of decision analysis: Linking decision analysis and behavioral decision research. In: T. Kugler, J.C. Smith, T. Connolly, & Y.-J. Son (Eds.), *Decision Modeling and Behavior in Complex and Uncertain Environments.* Springer, New York, NY.

Clemen, R.T., & Reilly, T. (2001). *Making Hard Decisions with DecisionTools®.* Duxbury, Pacific Grove, CA.

Clemen, R.T., & Smith, J.E. (2009). On the choice of baselines in multiattribute portfolio analysis: A cautionary note. *Decision Analysis 6(4),* 256–262.

Clemen, R.T., & Winkler, R.L. (2007). Aggregating probability distributions. In: W. Edwards, R.F. Miles, Jr., & D. von Winterfeldt (Eds.), *Advances in Decision Analysis: From Foundations to Applications.* Cambridge University Press, New York, NY.

Colgan, C.S. (2003). Living near… and making a living from… the nation's coasts and oceans. Prepared for the United States Commission on Ocean Policy. http://www.oceancommission.gov/

Conroy, M.J., Barker, R.J., Dillingham, P.W., Fletcher, D., Gormley, A.M., & Westbrooke, I.M. (2008). Application of decision theory to conservation management: Recovery of Hector's dolphin. *Wildlife Res. 35,* 93–102.

Dunning, D.J., Ross, Q.E., & Merkhofer, M.W. (2000). Multiattribute utility analysis for addressing Section 316(b) of the Clean Water Act. *Environ. Sci. Policy 3,* S7–S14.

Escambia County (2011). Natural resource damage assessment process. Online resource-http://www.co.escambia.fl.us/Bureaus/CommunityServices/NRDAProjects.html, accessed April 6, 2011.

Failing, L., & Gregory, R. (2003). Ten common mistakes in designing biodiversity indicators for forest policy. *J. Environ. Manage. 68,* 121–132.

Fischer, G.W., Damodaran, N., Laskey, K.B., & Lincoln, D. (1987). Preferences for proxy attributes. *Manage. Sci. 33,* 198–214.

French, S., Maule, J., & Papamichail, N. (2009). *Decision Behaviour Analysis and Support.* Cambridge University Press, New York, NY.

Goldsmith, R., & Ericson, R. (2000). Lifetime cost of subsea production systems. Prepared for Subsea JIP, System Description & FMEA Rev. 2, Det Norske Veritas, Minerals Management Service.

Gordon, K., Buchanan, J., Singerman, P., Madrid, J., & Busch, S. (2011). Beyond recovery: moving the Gulf Coast toward a sustainable future. Center for American Progress/Oxfam America, Washington, DC.

Gregory, R.S. (1999). Identifying environmental values. In: V.H. Dale & M.R. English (Eds.), *Tools to Aid Environmental Decision Making.* Springer, New York, NY.

Gregory, R.S. (2000). Valuing environmental policy options: a case study comparison of multiattribute and contingent valuation survey methods. *Land Econ. 76,* 151–173.

Gregory, R.S., & Keeney, R.L. (1994). Creating policy alternatives using stakeholder values. *Manage. Sci. 40,* 1035–1048.

Gregory, R., & Slovic, P. (1997). A constructive approach to environmental valuation. *Ecol. Econ. 21,* 175–181.

Gregory, R., Arvai, J., & McDaniels, T. (2001). Value-focused thinking for environmental risk consultations. *Res. Soc. Prob. Public Policy 9,* 249–273.

Gregory, R.S., Failing, L., Ohlson, D., & McDaniels, T.L. (2006a). Some pitfalls of an overemphasis on science in environmental risk management decisions. *J. Risk Res. 9*, 717–735.

Gregory, R.S., Ohlson, D., & Arvai, J. (2006b). Deconstructing adaptive management: criteria for applications to environmental management. *Ecological Applications 16*, 2411–2425.

Hajkowicz, S., Young, M., Wheeler, S., Hatton, D.H., & Young, D., (2000). Supporting decisions: understanding natural resource management assessment techniques. CSIRO Land and Water, Policy and Economic Research Unit, Adelaide, SA, Australia.

Hayes, K.R. (2002). Identifying hazards in complex ecological systems. Part 1: fault-tree analysis for biological invasions. *Biological Invasions 4*, 235–249.

Hubbard, D.W. (2009). *The Failure of Risk Management.* John Wiley & Sons, Inc., Hoboken, NJ.

Hora, S.C. (2007). Eliciting probabilities from experts. In: W. Edwards, R.F. Miles, Jr., D. von Winterfeldt (Eds.), *Advances in Decision Analysis: From Foundations to Applications.* Cambridge University Press, New York, NY.

Jordan, S. J., Smith, L.M., & Nestlerode, J.A. (2009). Cumulative effects of coastal habitat alterations on fishery resources: toward prediction at regional scales. *Ecol. Soc. [On-line serial] 14 (1)*, 16 http://www.ecologyandsociety.org/vol14/iss1/art16/.

Keeney, R.L. (1980). *Siting Energy Facilities.* Academic Press, New York, NY.

Keeney, R.L. (1981). Measurement scales for quantifying attributes. *Behav. Sci. 26*, 29–36.

Keeney, R.L. (1992). *Value-Focused Thinking: A Path to Creative Decisionmaking.* Harvard University Press, Cambridge, MA.

Keeney, R.L. (2002). Common mistakes in value trade-offs. *Oper. Res. 50*, 935–945.

Keeney, R.L. (2007). Developing objectives and attributes. In: W. Edwards, R.F. Miles, Jr., D. von Winterfeldt (Eds.), *Advances in Decision Analysis: From Foundations to Applications.* Cambridge University Press, New York, NY.

Keeney, R.L. (2009). The foundations of collaborative group decisions. *Int. J. Collab. Eng. 1*, 4–18.

Keeney, R.L., & Gregory, R.S. (2005). Selecting attributes to measure the achievement of objectives. *Oper. Res. 53*, 1–11.

Keeney, R.L., & von Winterfeldt, D. (1994). Managing nuclear waste from power plants. *Risk Anal. 14*, 107–130.

Keeney, R.L., & von Winterfeldt, D. (2007). Practical value models. In: W. Edwards, R.F. Miles, Jr., D. von Winterfeldt (Eds.), *Advances in Decision Analysis: From Foundations to Applications.* Cambridge University Press, New York, NY.

Keeney, R.L., McDaniels, T.L., & Ridge-Cooney, V.L. (1996). Using values in planning wastewater facilities for metropolitan Seattle. *Water Resour. Bull. 32*, 293–303.

Korb, K., & Nicholson, A. (2011). *Bayesian Artificial Intelligence*, 2nd, ed. CRC Press, Boca Raton, FL.

Mabus, R. (2010). America's Gulf Coast: a long term recovery plan after the Deepwater Horizon oil spill. www.restorethegulf.gov.

McDaniels, T.L. (1995). Using judgment in resource management: A multiple objective analysis of a fisheries management decision. *Oper. Res. 43*, 415–426.

McDaniels, T.L. (2000). Creating and using objectives for ecological risk assessment and management. *Environ. Sci. Policy 3*, 299–304.

McDaniels, T.L., & Gregory R. (2004) Learning as an objective within a structured risk management decision process. *Environ. Sci. Technol. 38(7)*, 1921-1926.

McDaniels, T.L., & Trousdale, W. (2005). Resource compensation and negotiation support in an aboriginal context: using community-based multi-attribute analysis to evaluate non-market losses. *Ecol. Econ. 55*, 173–186.

Merkhofer, M.W., & Keeney, R.L. (1987). A multiattribute utility analysis of alternative sites for the disposal of nuclear waste. *Risk Anal. 7*, 173–194.

Merrick, J.R.W., Parnell, G.S., Barnett, J., & Garcia, M. (2005). A multiple-objective decision analysis of stakeholder values to identify watershed improvement needs. *Decision Anal. 2*, 44–57.

Mollaghasemi, M., & Pet- Edwards, J. (1997). *IEEE Computer Society Technical Briefing: Making Multiple-Objective Decisions*. IEEE Computer Society Press, Los Alamitos, CA.

Newman, M.C., Zhao, Y., & Carriger, J.F. (2007). Coastal and estuarine ecological risk assessment: The need for a more formal approach to stressor identification. *Hydrobiol. 577*, 31–40.

NOAA (1996a). *Preassessment phase: guidance document for Natural Resource Damage Assessment under the Oil Pollution Act of 1990*. National Ocean and Atmospheric Administration, Damage Assessment and Restoration Program, Silver Spring, MD.

NOAA (1996b). *Restoration planning: guidance document for Natural Resource Damage Assessment under the Oil Pollution Act of 1990*. National Ocean and Atmospheric Administration, Damage Assessment and Restoration Program, Silver Spring, MD.

NOAA (1996c). *Primary restoration: guidance document for Natural Resource Damage Assessment under the Oil Pollution Act of 1990*. National Ocean and Atmospheric Administration, Damage Assessment and Restoration Program, Silver Spring, MD.

NOAA (2009a). *Fisheries of the United States: 2009*. National Oceanic and Atmospheric Administration, National Marine Fisheries Service, Office of Science and Technology, Silver Spring, MD.

NOAA (2009b). *Fishery Management Plan for Regulating Offshore Aquaculture in the Gulf of Mexico*. National Oceanic & Atmospheric Administration, National Marine Fisheries Service, Southeast Regional Office, St. Petersburg, FL.

NOAA (2011a). Affected Gulf resources. Online resource, http://www.gulfspillrestoration.noaa.gov/oil-spill/affected-gulf-resources/, accessed April 21, 2011.

NOAA (2011b). Co-trustees. Online resource, http://www.gulfspillrestoration.noaa.gov/about-us/co-trustees/, accessed April 6, 2011.

NOAA (2011c). Damage assessment. Online resource, http://www.gulfspillrestoration.noaa.gov/assessment/, accessed April 6, 2011.

Norsys. 2010. Netica™ 4.16. Norsys Software Corp., Vancouver, BC.

Pauker, S.G., & Wong, J.B. (2005). The influence of influence diagrams in medicine. *Decision Anal. 2*, 238–244.

Payne, J.W., Bettman, J.R., & Schkade, D.A. (1999). Measuring constructed preferences: Towards a building code. *J. Risk Uncertain. 19*, 243–270.

Pourret, O. (2008). Introduction to Bayesian networks. In: O. Pourret, P. Naim, & B. Marcot (Eds.), *Bayesian Networks: A Practical Guide to Application*. John Wiley & Sons, Ltd., Hoboken, NJ.

Powers, S.P., & Blanchet, H. (2010). Investigative plan for fish and invertebrate kills in the Northern Gulf of Mexico. For the Mississippi Canyon 252 Trustees, http://losco-dwh.com/viewWorkPlans.aspx

Restoration Plan (1997). Commencement Bay Natural Resource Restoration: Restoration Plan. http://www.darrp.noaa.gov/northwest/cbay/admin.html.

Schuhmann, P.W., & Schwabe, K.A. (2004). Modeling the dynamics of fishery stock recovery: Implications for the assessment of natural resource damages. *Nat. Resour. Model. 17*, 191–212.

Seeb, L.W., Templin, W.D., Tarbox, K.E., Davis, R.Z., & Seeb, J.E. (1997). *Exxon Valdez* oil spill restoration project final report: Kenai River Sockeye Salmon Restoration. Restoration Project 96255-2, Final Report, Alaska Department of Fish and Game, Division of Commercial Fisheries Management and Development, Anchorage, AK.

SEFSC (2010) Work Plan for the Collection of Data to Determine Impacts of the Deepwater Horizon Mississippi Canyon 252 Incident on Endangered and Protected Marine Mammals in the Northern Gulf. Southeast Fisheries Science Council, Signed Offshore Whale Plan Rev. 5 June 11, 2010.

Shachter, R.D. (1986). Evaluating influence diagrams. *Oper. Res. 34(6)*, 871–882.

Shachter, R.D. (1990). An ordered examination of influence diagrams. *Networks, 20*, 535–563.

Sheehy, D., & Vik, S. (2002). Applying decision analysis methods to NRDA restoration planning. In: P.J. Cannizzaro (Ed.), *Proceedings 29th Annual Ecosystem Restoration and Creation Conference*. Tampa, FL, pp 113–127.

Slovic, P. (1995). The construction of preference. *Am. Psychol. 50*, 364–371.

Slovic, P. (1999). Trust, emotion, sex, politics, and science: surveying the risk-assessment battlefield. *Risk Anal. 19*, 689–701.

Smith, J.E., & von Winterfeldt, D. (2004). Decision analysis in Management Science. *J. Manage. Sci. 50*, 561–574.

Stow, C.A. (1999). Assessing the relationship between *Pfiesteria* and estuarine fishkills. *Ecosystems 2*, 237–241.

Strayer, D.L., Beighley, R.E., Thompson, L.C., Brooks, S., Nilsson, C., Pinay, G., & Naiman, R.J. (2003). Effects of land cover on stream ecosystems: roles of empirical models and scaling issues. *Ecosystems 6*, 407–423.

Thibodeaux, L.J., Valsaraj, K.T., John, V.T., Papadopoulos, K.D., Pratt, L.R., & Pesika, N.S. (2011). Marine oil fate: knowledge gaps, basic research, and development needs: a perspective based on the Deepwater Horizon spill. *Environ. Eng. Sci. 28*, 87–93.

Tunnell Jr., J.W. (2011). An expert opinion of when the Gulf of Mexico will return to pre-spill harvest status following the BP Deepwater Horizon MC 252 oil spill. http://masgc.org/gmrp/documents/oiltunnell.pdf

Tversky, A., & Kahneman, D. (1981) The framing of decisions and the psychology of choice. *Science 211 (4481)*, 453–458.

U.S. EPA (1999). The ecological condition of estuaries in the Gulf of Mexico. EPA 620-R-98-004, United States Environmental Protection Agency, Office of Research and Development, Washington DC.

U.S. EPA (2000). Evaluation guidelines for ecological indicators. EPA/620/R-99/005, United States Environmental Protection Agency, Office of Research and Development, Washington DC.

Von Winterfeldt, D., & Edwards, W. (1986). *Decision Analysis and Behavioral Research.* Cambridge University Press, Cambridge UK.

Von Winterfeldt, D., & Edwards, W. (2007). Defining a decision analytic structure. In: W. Edwards, R.F. Miles, Jr., D. von Winterfeldt (Eds.), *Advances in Decision Analysis: From Foundations to Applications.* Cambridge University Press, New York, NY.

Von Winterfeldt, D., & Fasolo, B. (2009). Structuring decision problems: a case study and reflection for practitioners. *Eur. J. Oper. Res. 199*, 857–866.

Von Winterfeldt, D., Eppel, T., Adams, J., Neutra, R., DelPizzo, V. (2004). Managing potential health risks from electric powerlines: a decision analysis caught in controversy. *Risk Anal. 24*, 1487–1502.

Winn, M.I. (2001). Building stakeholder theory with a decision modeling methodology. *Bus. Soc. 40*, 133–166.

Wooldridge, S., & Done, T. (2004). Learning to predict large-scale coral bleaching from past events: a Bayesian approach using remotely sensed data, in-situ data, and environmental proxies. *Coral Reefs 23*, 96–108.

In: Estuaries: Classification, Ecology and Human Impacts   ISBN: 978-1-61942-083-0
Editor: Steve Jordan   © 2012 Nova Science Publishers, Inc.

Chapter 14

Synthesis: Estuaries under Stress

Stephen J. Jordan

U.S. Environmental Protection Agency, Gulf Ecology Division, Gulf Breeze, FL, US

Abstract

Estuaries are subject to stresses that originate from intensive human activities in the proximal estuarine environment and in watersheds that drain to the coastal zone. These stresses are summarized here, with examples from this book and the scientific literature. To sustain the many benefits that estuaries provide to society, and to restore those that have been lost, will require better decisions than have been made in the past. Management and policy will need to be more proactive and less reactive, deal with upstream causes to relieve downstream effects, and account for a full range of ecosystem services in the economic balance of costs and benefits.

Keywords: Watershed(s), Gulf of Mexico, Chesapeake Bay, pollution, economics, decision science

Estuaries under Stress

At one time, I lived with my wife and newborn son in a cabin in the mountains of eastern Kentucky. Many stories could be told, but the relevant one is this: There was no trash disposal service in this remote area and no public dumping facility in the entire county. So I asked my landlord, who had lived his whole life in the area, what we should do with our trash. He replied, "Put it down there on the creek bank – come a tide[1] it will take it all away." I learned that this was the standard method of rubbish disposal in this part of the world – everything from disposable diapers to large appliances and automobile carcasses was left along creek banks to be taken all away.

How is this little anecdote relevant to estuaries? Our creek flowed into Quicksand Creek, a tributary of the Kentucky River, in turn a tributary of the Ohio River, which flows into the

---

[1] "Come a tide" in the local dialect meant "when the creek floods."

Mississippi River, which flows into estuaries and coastal waters of the Gulf of Mexico. Although it is unlikely that our bits of trash would have made it as far as the Gulf (>1000 km by waterway), the story illustrates that estuaries are downstream from almost everywhere, and susceptible to whatever happens upstream. The watershed of the Mississippi River, the principal freshwater source for the Gulf of Mexico, has been so extensively altered by human activities that it is difficult to imagine its original condition. Deforestation, agriculture, mining, heavy industry, urban development, and a vast array of water control structures (dams, locks, levees, canals etc.) have changed the volume, timing, spatial distribution, and quality of freshwater and sediment flows (Kesel 1986, Day et al. 2000), while the river contributes excessive quantities of nutrients to coastal waters (Mitsch et al. 2001, Scavia et al. 2003).

Despite all of these insults, plus the disastrous Deepwater Horizon oil well blowout of 2010 (Suran 2011), the chronic development of a huge hypoxic zone each summer in the north-central Gulf of Mexico (Rabalais et al. 2002), and precipitous losses of coastal wetlands (Kesel 1986), the waters of the Gulf coast continue to support thriving fisheries, tourism, recreation, industry, and shipping. The ecosystem is resilient, but to what extent can it continue to absorb such affronts and remain vibrant? So far, we are not able to predict when one more environmental disaster, the steady accumulation of habitat losses, or the combination will push the system beyond a tipping point, followed by a possibly unrecoverable decline to a less desirable stable state. This phenomenon is thought to have been a major reason for the rapid decline and painfully slow recovery of Chesapeake Bay, where catastrophic freshwater flooding and sediment loading from Hurricane Agnes (1972) capped more than a century of disregard for uncontrolled exploitation and the environmental costs of pollution (Boesch et al. 2001).

Estuaries around the world face similar stresses. Asian, European and North American estuaries are represented in this volume; estuaries in Africa (e.g., Turpie et al. 2002, Ewa-Oboho & Oladimej 2007), Australia (Deeley & Paling 1999), and South America (e.g., Perillo et al. 2005, Lacerda 2006) are subject to the same kinds of anthropogenic pressures, with variations in emphasis.

Themes of pollution, habitat loss, and restoration pervade this book, interwoven with the sub-plot of variability that is so important in estuaries (Geyer et al. 2000). Kurtz et al. (Chapter 2) demonstrate how variation among estuaries in responses to a stressor (nitrogen loading) complicates the problem of classification for the purposes of management and regulation. Caçador and Duarte (Chapter 3), Jordan and Peterson (Chapter 5), and Seaver (Chapter 7) illustrate various aspects of variability in chemical or biological properties over decades. Mukherjee and Ray (Chapter 6), Bartolomé et al. (Chapter 8) and Gredilla et al. (Chapter 9) describe and analyze shorter-term temporal and spatial variability. Given the many threats from stressors and the inherent variability of estuaries, it is plain that solid data are needed to comprehend the status of estuarine ecosystems, how they are changing, and how they might change in the future. In concert, the studies presented here could hardly make a stronger case for the systematic, comprehensive, long-term monitoring described by Summers et al. (Chapter 11).

DECIDING THE FUTURE OF ESTUARIES

Our Kentucky landlord probably had no idea nor even curiosity about where a flood would take the trash, except "away." The same is likely the case for most of the inland residents of estuarine watersheds. Their decisions about land uses and waste disposal have been made largely in ignorance or disregard for the downstream consequences. Currently, there is great controversy over proposed criteria for the state of Florida that would limit discharges of nutrients in inland waters in order to protect estuaries and coastal waters from eutrophication.[2] A similar problem, at a vastly larger scale, applies to attempts to reduce hypoxia in the northern Gulf of Mexico. Hypoxia is strongly related to nutrient loads from the Mississippi River, which drains all or parts of 31 U.S. states. None of these states has adopted regulatory nutrient limitations specifically to protect Gulf of Mexico waters – the emphasis has been on a patchwork of voluntary programs and incentives to reduce nutrient discharges (Mississippi River/Gulf of Mexico Watershed Nutrient Task Force 2010).

Even when we confine our concerns to the immediate coastal zone, the problems of conserving and restoring the values of estuaries, while managing their multiple uses, are complex. Sohma (Chapter 10) presents an intriguing case study where tidal flats in Tokyo Bay were reclaimed (i.e. elevated with fill) for development. His simulations indicate that loss of the tidal flats reduced the capacity of the estuary to process nutrients productively, leading to poor water quality and hypoxia. Reducing nutrient loads to the bay could improve water quality, but without restoring the productivity that would produce a "bountiful ocean." On the other hand, restoring the tidal flats could alleviate the water quality problems and restore a system productive of desirable aquatic life. These findings required an elaborate, state-of-the-art mathematical model, which unfortunately was developed long after the decisions were made to reclaim the tidal flats.

Many decisions with similar import for estuaries are being made now and will be made in the future, and like the effects of reclaiming the Tokyo Bay tidal flats, the unwanted consequences of these decisions may be difficult or impossible to reverse. A pressing concern is for how coastal property owners and communities will respond to rising sea level (Rajabalinejad et al., Chapter 12). Extensive armoring of shorelines with bulkheads, seawalls, dikes and revetments will protect properties, but also destroy or degrade shallow water habitats that are critical for aquatic vegetation and many estuarine animals. In turn, the loss of habitats will be detrimental to fisheries (Jordan & Peterson, Chapter 5; Jordan et al. 2009), tourism, recreation, the aesthetic values of natural shorelines, and possibly water quality and other beneficial properties of coastal ecosystems.

It is relatively easy to assign economic values to such actions as protecting property from floods and storm surges, dredging to maintain shipping channels, or development of businesses, housing and infrastructure in sensitive coastal areas. Conversely, methods for confidently estimating the composite values of ecosystems and ecosystem services are only now being developed, have large uncertainties, and can be controversial (Simpson 2010). It is even more difficult to estimate how these values might change as the result of decisions about where and what to develop, how to respond or adapt to sea level rise, how to recover from disasters such as hurricanes and large oil spills, or the best uses of dredged sediment. In the

---

[2] http://water.epa.gov/lawsregs/rulesregs/florida_index.cfm.

balance, explicit monetary benefits or cost savings can outweigh the (apparently) less tangible benefits of functioning coastal ecosystems and productive estuarine environments.

In recent years, ecologists have begun working with economists and social scientists to improve (1) the quality of information available for making decisions that jointly affect the environment and society; and (2) the processes used by society to analyze and facilitate complex, collective decisions. In the first realm, models of ecological production functions and economic valuation methods are being combined to estimate the marginal values of environmental change in novel ways (e.g., Jordan et al. 2010). In at least one case, these models are being employed to evaluate how alternative futures for the entire watershed of a large estuary will affect the production, delivery and value of ecosystem services (Russell et al. 2011, http://www.epa.gov/ged/tbes/). In the realm of decision analysis, Carriger & Benson (Chapter 13) illustrate how decision science could be used to address the highly complex alternatives and trade-offs involved in minimizing the impacts of an event such as the 2010 Gulf of Mexico oil well blowout and discharge of >800 million L of crude oil into the Gulf. As another example, a comprehensive, scientific approach to decision-making in ecosystem management for coastal areas of northern Greece was reported by Pavlikakis & Tsihrintzis (2003).

Within the next several years, we will have the tools to make more proactive and informed decisions about the future of the world's estuaries and coastal ecosystems. The decision process will be able to account for a full spectrum of ecosystem services and benefits to society, as well as society's needs and preferences. In the place of narrow economic considerations—incomes, tax revenues, property values and the like—we will be able to carry out true economic analyses, accounting holistically for changes in public welfare and the sustainability of coastal communities. Having the tools is not enough, though. If they are to be used widely and wisely to conserve and restore estuaries, to balance uses and trade-offs, and to foster sustainability, everyone, from those who make decisions locally in the coastal zone, to the people far upstream, will need to be better informed about the issues described here and about the value of estuaries as functional ecosystems.

DISCLAIMER

The views expressed in this chapter are those of the author and do not necessarily reflect the views or policies of the U.S. Environmental Protection Agency. I appreciate thoughtful reviews by Michael Lewis and John Carriger. This is contribution 1434 from the Gulf Ecology Division, and a product of EPA's Ecosystem Services Research Program.

REFERENCES

Boesch, D.F., Brinsfield, R.B. & Magnien, R.E. (2001). Chesapeake Bay eutrophication: Scientific understanding, ecosystem restoration, and challenges for agriculture. *J. Environ. Qual. 30*, 303–320.

Day, J.W., Jr., Britsch, L.D., Hawes, S.R., Shaffer, G.P., Reed, D.J. & Cahoon, D. (2000). Pattern and process of land loss in the Mississippi Delta: A spatial and temporal analysis of wetland habitat change. *Estuaries 23*, 425–438.

Deelley, D.M. & E.I. Paling (1999). Assessing the ecological health of estuaries in Australia. Land and Water Resources Research and Development Corporation Occasional Paper 17/99, Canberra. http://au.riversinfo.org/archive/?doc_id=9.

Ewa-Obo, I. & Oladimej, O. (2007). Studies on the short-term effects of the Mobil Idoho oil spill on the littoral biota of southeastern Nigeria. *W. Afr. J. Applied Ecol. 11*, http://www.wajae.org/volume11.html.

Geyer, W.R., Morris, J.T., Prahl, F.G. & Jay, D.A. (2000). Interaction between physical processes and ecosystem structure: A comparative approach. In: Hobbie, J.E. (ed.). *Estuarine Science: A synthetic approach to research and practice.* Island Press, Washington, DC, pp. 177–210.

Jordan, S. J., Hayes, S. E., Yoskowitz, D., Smith, L.M., Summers, J.K., Russell, M. & Benson, W.H. 2010. Accounting for natural resources and environmental sustainability: linking ecosystem services to human well-being. *Environ. Sci. Technol. 44,* 1530–1536.

Jordan, S.J., Smith, L.M. & Nestlerode, J.A. (2009). Cumulative effects of coastal habitat alterations on fishery resources: toward prediction at regional scales. *Ecol. Soc. 14,* Article 16. http://www.ecologyandsociety.org/vol14/iss1/art16/

Kesel, R. (1986). The decline in the suspended load of the lower Mississippi River and its influence on adjacent wetlands. *Environ. Geol. 11*, 271–281.

Lacerda, L.D. (2006). Inputs of nitrogen and phosphorus to estuaries of northeastern Brazil from intensive shrimp farming. *Braz. J. Aquat. Sci. Technol. 10*, 13–27.

Mississippi River/Gulf of Mexico Watershed Nutrient Task Force (2010). *Gulf Hypoxia Annual Report 2010.* http://water.epa.gov/type/watersheds/named/msbasin/upload/ Hypoxia-Task-Force-FY10-Annual-Report_508.pdf.

Mitsch, W.J., Day, J.W., Jr., Gilliam, J.W., Groffman, P.M., Hey, D.L., Randall, G.W., & Naiming, W. (2001). Reducing nitrogen loading to the Gulf of Mexico from the Mississippi River basin: Strategies to counter a persistent ecological problem. *Bioscience 51*, 373–388.

Pavlikakis, G.E. & Tsihrintzis, V.A. (2003). A quantitative method for accounting human opinion, preferences and perceptions in ecosystem management. *J. Environ. Manage. 68*, 193–205.

Perillo, G.M.E, Pérez, D.E., Piccolo, M.C., Palma, E.D. & Cuadrado D.G. (2005). Geomorphologic and physical characteristics of a human impacted estuary: Quequén Grande River Estuary, Argentina. *Estuar. Coast. Shelf S. 62*, 301–312.

Rabalais, N.N., Turner, R.E. & Wiseman, W.J., Jr. (2002). Gulf of Mexico hypoxia, a.k.a "The Dead Zone." *Annu. Rev. Ecol. Syst. 33*, 235–263.

Russell, M., Rogers, J., Jordan, S., Dantin, D., Harvey, J., Nestlerode, J. & Alvarez, F. (2011). Prioritization of ecosystem services research: Tampa Bay demonstration project. *J. Coast. Conserv.* DOI 10.1007/s11852-0158-z.

Scavia, D., Rabalai, N.N., Turner, R.E., Justić, D. & Wiseman, W.W., Jr., (2003). Predicting the response of Gulf of Mexico hypoxia to variations in Mississippi River nitrogen load. *Limnol. Oceanog. 4,* 951–956.

Simpson, R.D. (2010). The "ecosystem service framework." In: N. Hanley (ed.), *Valuation of Regulating Services of Ecosystems*, University of Stirling, UK.

Suran, M. (2011). Natural healing: How does nature repair itself after an oil spill? *EMBO Reports 12*, 27–30.

Turpie, J.K., Adams, J.B., Joubert, A., Harrison, T.D., Colloty, B.M., Maree, R.C., Whitfield, A.K., Wooldridge, T.H., Lamberth, S.J., Taljaard, S. & Van Niekerk, L. (2002). Assessment of the conservation priority status of South African estuaries for use in management and water allocation. *Water SA 28*, 191–206.

ABOUT CONTRIBUTORS

Gorka Arana (Chapter 9) is an assistant professor in the Analytical Chemistry Department at the University of the Basque Country. He received a Ph.D. in Chemistry at the University of the Basque Country in 1996. From 1996-1998 he was a postdoctoral fellow at the Institute of Reference Materials and Measurements (JRC-IRMM) of the European Commission in Geel (Belgium) working on a setup for the nuclear activation analysis of short lived isotopes. Since 1999 he has been a lecturer in analytical chemistry at the University of the Basque Country, working on environmental analysis and on the diagnosis of built cultural heritage. He is also involved in chemical metrology as he is the Spanish national team leader of the TrainMiC (Training in Metrology in Chemistry) Programme coordinated by the JRC-IRMM.

Luis Bartolomé (Chapter 8) got his Ph.D. in Analytical Chemistry in 2007 after a study about the *Prestige* oil tanker spill impact along the Bay of Biscay. Most of his publications are related to occurrence and distribution of organic micro-contaminants between compartments (biota, sediment and water) in estuaries and coastal areas. Since 2008, Dr Bartolomé has been a Specialist Technician in the Central Service of Analysis at the Basque Country and works as a chromatographer developing new methodologies for environmental analysis.

William H. Benson (Chapter 13) serves as Director of the National Health and Environmental Effects Research Laboratory's Gulf Ecology Division within the U.S. Environmental Protection Agency's (EPA) Office of Research and Development. Dr. Benson's research activities have been directed towards assessing the influence of environmental stressors on health and ecological integrity. He has conducted research in the areas of metal and pesticide bioavailability, reproductive and developmental effects in aquatic organisms, endocrine disrupting chemicals, and use of indicators in assessing health and ecological integrity. In addition, he has published over 100 scientific publications focusing on environmental toxicology and chemistry. Dr. Benson obtained a B.S. in Biology from Florida Institute of Technology and his M.S. and Ph.D. degrees in Toxicology were obtained from the University of Kentucky. While in graduate school he was the first recipient of the Society of Environmental Toxicology and Chemistry (SETAC) Pre-Doctoral Fellowship sponsored by The Procter & Gamble Company. Dr. Benson is a Past President of SETAC and has served on the International Council of SETAC. He was elected as a Fellow of the American Association for the Advancement of Science, and is active in several other professional societies, including the Society of Toxicology.

Isabel Caçador (Chapter 3) is a Professor at the Faculty of Sciences of the University of Lisbon. Presently, she is also an Investigator of the Marine Botany Laboratory at the Center of Oceanography of Lisbon. The main areas of study are salt marsh ecology, primary production, carbon cycling, coastal marsh stability and sedimentation processes, with special interests in pollution, heavy metal speciation, biogeochemical cycling and soil-plant interactions. Recently,

Dr. Caçador has been actively involved in research dealing with environmental concerns associated with wetlands. These include heavy metals retention, eutrophication, carbon cycling in salt marshes, and responses of salt marshes to sea-level rise.

John F. Carriger (Chapter 13) is a post-doctoral ecologist at the Gulf Ecology Division within U.S. EPA's Office of Research and Development. Dr. Carriger has a Ph.D. in marine science from the College of William & Mary's Virginia Institute of Marine Science, with M.S. and B.S. degrees from Florida International University. Dr. Carriger joined the U.S. EPA in 2010; his work is currently focused on the development of decision support frameworks for coral reef management in the U.S. Virgin Islands and Puerto Rico. Prior to joining the U.S. EPA, Dr. Carriger had post-doctoral experience investigating ecological risks from contaminants for the Everglades restoration effort. His primary research interests are qualitative and quantitative approaches for environmental decision-making and probabilistic approaches for representing and analyzing risks and uncertainties in environmental management problems.

Alberto de Diego (Chapter 9) is an assistant professor in the Analytical Chemistry Department at the University of the Basque Country. He received a Ph.D. in Chemistry at the University of the Basque Country in 1996. In 1996 and 1997 he was a postdoctoral fellow at the Laboratoire de Chimie Analytique Bio-Inorganique et Environnement, UMR CNRS 5034 (Université de Pau et des Pays de l'Adour), working on metal speciation analysis of environmental samples. Since 1998 he has been lecturing in analytical chemistry at the University of the Basque Country and working on environmental analytical chemistry and chemometrics.

Z. Demirbilek (Chapter 12). No biographical information provided.

Bernardo Duarte (Chapter 3) is presently an investigator at the Marine Botany Laboratory within the Center of Oceanography of Lisbon. The main areas of investigation in the recent years have been in heavy metal contamination in salt marshes and estuaries, microbial enzymes and decomposition processes, biogeochemical and metal cycling, salt marsh plants ecophysiology, and heavy metal stress. Nowadays his investigations mostly focus on halophyte photosynthesis mechanisms, carbon cycles, CO_2 emissions, dynamics in salt marshes, and climate change.

Virginia D. Engle (Chapter 11) is a research ecologist at the Gulf Ecology Division of U.S. EPA's Office of Research and Development in Gulf Breeze, FL. Ms. Engle received a B.A. in Biology and Mathematics at Goucher College in 1986. She has worked at the Gulf Ecology Division since 1990, primarily in the Environmental Monitoring and Assessment Program and the National Coastal Assessment. Her principal research interests are in coastal monitoring and assessment, estuarine ecology, benthic ecology, statistical designs and analysis, and, more recently, coastal wetlands and ecosystem services. Ms. Engle is a long-time member of the Coastal and Estuarine Research Federation and a recent member of the Society of Wetland Scientists. She has received numerous awards and citations from the U.S. Environmental Protection Agency, including seven Scientific and Technological Achievement Awards and three bronze medals.

Nestor Etxebarria (Chapter 8) received a Ph.D. from The University of the Basque Country, Bilbao, in 1993 with research on the hydrolysis of several transition metals. From 1994 to 1996 he worked at the Institute of Reference Materials and Measurements (IRMM, Geel, Belgium) under the supervision of Dr. Piotr Robouch with the application of instrumental neutron activation analysis for the production control of reference materials. In 1996 Dr. Etxebarria was promoted to associate professor at the Department of Analytical Chemistry, University of the Basque Country. His research is essentially devoted to environmental analytical chemistry, in the development of new analytical methodology and in the analysis of the chemical behavior of pollutants in the environment. Dr. Extebarria works in an interdisciplinary environment with other chemists, cell biologist, toxicologist, geologist, etc. In this framework, he participates in two postgraduate programs: Environmental Contamination and Toxicology and Marine Environmental Resources.

Silvia Fdez-Ortiz de Vallejuelo (Chapter 9) is a researcher in the Department of Analytical Chemistry of the University of the Basque Country. She received a Ph.D. in Chemistry at the University of the Basque Country in 2008. Since 2005 she is involved in issues related to analysis of environmental samples, mainly sediments, water samples and biota. She also works on the diagnosis of built Cultural Heritage by application of non-destructive techniques such as Raman and IR spectroscopy.

Phani Bhusan Ghosh (Chapter 6) is a senior scientist at the Institute of Environmental Studies and Wetland Management. Dr. Ghosh received his Ph.D. from Calcutta University. He has been engaging in research in the field of the chemical environment of mangrove and estuarine systems.

Ainara Gredilla (Chapter 9) is a Ph.D. student in the Department of Analytical Chemistry of the University of the Basque Country. Her thesis, "Development of analytical and chemometric tools for the quantification and prediction of trace element contamination in estuaries," is scheduled to be presented in September 2011. She obtained her B.S. in Analytical Chemistry in 2007, and her M.Sc. in Environmental Pollution and Toxicology in 2008. She has been involved in several research projects to study trace element contamination in selected estuaries of the Basque Coast, and spent four months in the Faculty of Life Sciences of the University of Copenhagen working on the chemometric analysis of environmental data.

James D. Hagy III (Chapter 2) is a research ecologist at the Gulf Ecology Division of U.S. EPA's Office of Research and Development (ORD) in Gulf Breeze, Florida. Dr. Hagy received a Ph.D. in Marine, Estuarine and Environmental Science at the University of Maryland in 2001. From 2001 through the present, he has served the U.S. EPA, conducting research that addresses impacts and management of anthropogenic nutrient enrichment in estuaries, coastal waters, and their watersheds. His research has addressed water quality issues in Chesapeake Bay, the northern Gulf of Mexico, and Florida estuaries, often applying empirical methods at a large scale. Most recently his work has addressed methods for developing numeric water quality criteria for nutrients. Dr. Hagy is a member of the Coastal and Estuarine Research Federation.

James E. Harvey (Chapter 11) is a research biologist at the Gulf Ecology Division of U.S. EPA's Office of Research and Development in Gulf Breeze, FL. Mr. Harvey received an M.S. in Biology at the University of New Orleans in 1985 with a focus on molecular toxicology and fate of Cd and Hg in mammals. He received his B.S. in Zoology from Southeastern Louisiana University in 1982. His current research interest is the ecology of large coastal ecosystems and the delivery of ecosystem services through the landscape at different scales.

Linda C. Harwell (Chapter 11) is an information technology specialist at the Gulf Ecology Division of U.S. EPA's Office of Research and Development in Gulf Breeze, FL. From 1985-1993, Ms. Harwell worked in the private sector as a computer programmer and data systems analyst. Since 1993, she has provided the Gulf Ecology Division with data management and data analysis expertise supporting large-scale coastal monitoring efforts such as the Environmental Monitoring and Assessment Program and the National Coastal Assessment. Ms. Harwell has received a number of awards and citations from U.S. EPA, including two Scientific and Technological Achievement Awards and two bronze medals.

Stephen J. Jordan (Chapters 1, 5, and 14; Editor) is a senior research ecologist and Special Assistant to the Director at the Gulf Ecology Division of U.S. EPA's Office of Research and Development in Gulf Breeze, Florida. Dr. Jordan received a Ph.D. in Marine, Estuarine and Environmental Science at the University of Maryland in 1987. From 1985-2002, he was employed by the Maryland Department of Natural Resources, serving from 1992-2002 as Director of the Cooperative Oxford Laboratory. His principal research interests are in the ecology of large coastal ecosystems and applications of ecological information in management. Dr. Jordan is a long-time member of the American Fisheries Society and the Coastal and Estuarine Research Federation. He has received awards and citations from the Chesapeake Bay Program, the Maryland Department of Natural Resources, the Governor of Maryland, the U.S. Senate, the National Fish Habitat Board, Sigma Xi, and U.S. EPA.

Janis C. Kurtz (Chapter 2) is a research biologist at the Gulf Ecology Division of U.S. EPA's Office of Research and Development (ORD) in Gulf Breeze, Florida. Dr. Kurtz received a Ph.D. in Environmental Biology and Public Policy at George Mason University in 1994. From 1986 to the present she has served with the U.S. EPA, as Science Officer for Environmental Effects, Transport and Fate with the Science Advisory Board until 1989, followed by service as a research biologist for ORD. From 1997-1998 she was a Visiting Research Scientist at the University of Missouri, Columbia in the Department of Biochemistry. Her research interests include the impacts of nutrients and pollutants on estuarine and marine communities and processes, especially phytoplankton and microbial communities, and comparisons between estuarine and coastal ecosystems as susceptibility varies based on the physical and hydrological properties of watersheds and receiving waters. She is an editor of the journal *Ecological Indicators*, a member of the Coastal and Estuarine Research Federation, and has received numerous awards for outstanding service and exemplary publications during her tenure with U.S. EPA.

John M. Macauley (Chapter 11) is a research biologist at the Gulf Ecology Division of U.S. EPA's Office of Research and Development in Gulf Breeze, FL. Mr. Macauley received a B.S. in Biological Sciences (1979) and an M.S. in Biology (1989) from The University of West Florida. He joined the Gulf Ecology Division 1979 and assisted in conducting toxicity tests on aquatic organisms. His work has included ecological effects studies of pesticide runoff from agricultural development adjacent to estuaries and developing monitoring designs and methodology for the Environmental Monitoring and Assessment (EMAP) and National Coastal Assessment programs. His principal research interests are in estuarine plant ecology with emphasis on submerged aquatic vegetation. He has received several Scientific and Technological Achievement Awards from the U.S. EPA for journal publications and was awarded a Gold Medal for Exceptional Service from U.S. EPA for work on the EMAP Program.

Juan Manuel Madariaga (Chapter 9) is a full professor in the Analytical Chemistry department at the University of the Basque Country. He received a Ph.D. in Chemistry at the University of the Basque Country in 1983. Since 1984 he has been lecturing on analytical chemistry at the University of the Basque Country and working on environmental analytical chemistry and chemometrics. He is the leader of a research group that received the award of excellence in 2002, working on characterization and diagnosis of environmental impacts on both natural and cultural heritage, including development of laboratory and hand-held instrumentation for problem-solving cases.

Joyita Mukherjee (Chapter 6) is a senior research scholar in the Ecological Modelling Laboratory, Department of Zoology, Visva-Bharati University, Santiniketan, India. She has been working toward a Ph.D. thesis on organic carbon cycling through phytoplankton in estuarine system under the supervision of Professor Santanu Ray.

Mark S. Peterson (Chapters 4 and 5) is a fish ecologist and Professor in the Department of Coastal Sciences at the University of Southern Mississippi's Gulf Coast Research Laboratory (GCRL). Dr. Peterson received his Ph.D. in Biological Sciences in 1987 at the University of Southern Mississippi followed by a Post-Doctoral Fellowship (1987-1989) at the Harbor Branch Oceanographic Institution. From 1989-1994 he was an Assistant (1989-1994) and then Associate Professor (1994) in Biological Sciences at Mississippi State University; in 1994 he relocated to the USM GCRL (1994-1998) as an Associate Research Scientist. In 1998, the Department of Coastal

Sciences was formed and he became an Associate Professor (1998-2001) and then Professor. His principal research interests center around resource ecology of coastal nekton and how natural and anthropogenic impacts modify these relationships. Dr. Peterson is a long-time member of the American Fisheries Society, American Society of Ichthyologists and Herpetologists, and the Coastal and Estuarine Research Federation. He has been President of the Mississippi Chapter of AFS, Southern Division of the ASIH, and Gulf Estuarine Research Society. He has served on the National Fish Habitat Initiative Scientific Committee, two National Research Council committees, and the National Sea Grant College Program Novel Bay and Estuary Ecosystems working group.

Mohammad Rajabalinejad (Chapter 12). No biographical information provided.

Chet F. Rakocinski (Chapter 4) is Professor at the Gulf Coast Research Laboratory (GCRL) in Ocean Springs, MS, within the Department of Coastal Sciences of the University of Southern Mississippi (USM). Dr. Rakocinski received a Ph.D. in Biology with a Zoology emphasis from USM in 1986. His postdoctoral experience included stints as a temporary Assistant Professor at the University of Mississippi, as a Research Associate within the Coastal Fisheries Institute at Louisiana State University, and as an Associate Research Scientist at the USM GCRL. His research broadly centers on the aquatic ecology of fishes and macrobenthic invertebrates. Research interests include responses of macrobenthic communities to environmental change and anthropogenic alteration, recruitment ecology of early life-stages of estuarine dependent species, and trophic relationships of fishes. He embraces an eclectic approach using a combination of descriptive field studies, field and laboratory experiments, and simulation modeling. Dr. Rakocinski formerly served as Associate Editor for the journal *Estuaries*, as a panel member for the U.S. EPA STAR and STAR Fellowship programs, and as a member of technical review panels for New York Sea Grant and the West Coast and Polar Undersea Research Center. He has also served as a reviewer of numerous proposals for various funding agencies and of 95 manuscripts for 51 journals.

Juan Carlos Raposo (Chapter 8) received his doctorate in 2001 in Analytical Chemistry at the University of the Basque Country. His principal research interests are in the ecology of natural systems, and interpretation and applications of ecological information in management. Since 2005, Dr. Raposo has been a Specialist Technician in the Central Service of Analysis (CSA) at the Basque Country as a part of the vice-chancellor's office. The CSA seeks to help promote, develop and disseminate multidisciplinary technological and scientific research in order to foster the advance of knowledge and cultural, social and economic development. Dr. Raposo is also involved in training the research staff of the University of The Basque Country and in dissemination of scientific information.

Santanu Ray (Chapter 6) is a professor in the Ecological Modelling Laboratory, Department of Zoology, Visva-Bharati University, Santiniketan, India. Dr. Ray received his Ph.D. from Calcutta University in 1988. He did his postdoctoral research at the Czech Academy of Science, University of South Bohemia, Czech Republic; University of Copenhagen, Denmark; and Odum's School of Ecology, University of Georgia, USA. Since 2001 he has been serving Visva-Bharati University, engaging in research in the field of theoretical ecology, especially integrated ecosystem theories, and systems ecology of mangrove and estuarine ecosystems. He is associate editor of the journal *Ecological Modelling – an International Journal of Ecological Modelling and Systems Ecology*, published by Elsevier. He is also a member of the editorial board of the journals *Computational Ecology and Software* and *Network Biology*.

Madhumita Roy (Chapter 6) is an assistant professor, South Calcutta Girls College, Kolkata. She has been working toward a Ph.D. thesis on the role of mangrove litter in the detritus food chain of estuarine ecosystems, under the supervision of Professor Santanu Ray.

Anghuman Sarkar (Chapter 6) is an assistant professor of the Department of Statistics, Visva-Bharati University. Dr. Sarkar received his Ph.D. from Visva-Bharati University. He has been engaging in research in the field of sampling design and statistical modeling of ecological systems.

George Seaver (Chapter 7) is a research scientist in physical estuary processes, in particularly in nutrient transport through precipitation, groundwater, estuaries and into bays at climatic scales. He is also a principal in SeaLite Engineering of Cataumet, Massachusetts. Beginning in 1987 and continuing until the present, nutrient and other environmental measurements from several estuaries, their watersheds and downstream bays on Cape Cod, Massachusetts have been collected to analyze their trends and cycles. From this data an understanding of when the seasonal and climatic scales are independent of each other and when they are dependent has been explored. Seaver has also developed optical sensors for use in physical oceanography, as well as innovative techniques to inhibit the biofouling of these sensors. He has 10 U.S. and foreign patents issued on these techniques, as well as 20 papers in professional journals (physical oceanography, instrumentation, estuaries). Seaver received his Ph.D. from Harvard University in Applied Physics in 1973, was a post doctoral fellow at Harvard and then a Research Associate at MIT until 1976. He subsequently formed SeaLite Engineering to develop optical sensors and devices for use in oceanography.

Lisa M. Smith (Chapter 11) is a research biologist at the Gulf Ecology Division of U.S. EPA's Office of Research and Development in Gulf Breeze, FL. Ms. Smith received a B.S. in Biological Sciences from Clemson University in 1993 and an M.S. in Ecology and Coastal Zone Studies from the University of West Florida in 1995. She joined the Gulf Ecology Division's Coastal Ecology Branch in 1997 and began working with the National Coastal Assessment in 1999. Her principal research interests are in estuarine ecology with emphasis in estuarine chemistry and estuarine habitat evaluations. Current research is focused on the relationships between ecosystem services and aspects of human well-being. She has received several Scientific and Technological Achievement Awards from the U.S. EPA for journal publications and was awarded a Scientific Achievement Award by the National Fish Habitat Board.

Akio Sohma (Chapter 10) is a Principal Researcher at Mizuho Information & Research Institute, Japan. He received his M.S. in Physics from Aoyama Gakuin University and a Ph.D. in Marine Science from Tokai University. From 1994–2003, he worked on marine environmental modeling and assessment at the Fuji Research Institute. In 2004, he was a Guest Researcher at the Bergen Center for Computational Science, University of Bergen. Since 2005, he has served in Mizuho Information & Research Institute. His principal research interests are in modeling ecological chains from lower to higher trophic levels, in evaluating ecological balance among ecosystem components, and applications of ecological models in environmental management. Dr. Sohma has experienced many ocean modeling and assessment projects as a leader. He received the Hotta Memorial incentive award from the Advanced Marine Scienceand Technology Society in 2007.

J. Kevin Summers (Chapter 11) is the Associate Director of Science for U.S. EPA's Gulf Ecology Division. He received his doctorate in 1979 from the University of South Carolina for his work in developing ecosystem models of carbon and energy flow in coastal and estuarine ecosystems. His focus for the past 30 years has been systems ecology, modeling, monitoring and assessment, and ecosystem services. While at Martin Marietta Environmental Systems, Dr. Summers with Drs. Douglas Heimbuch and Kenneth Rose developed an approach for times series categorical regressions (CATREG) which permits the parsing of the contribution of stock dependency, natural environmental events and anthropogenic factors from the long-term trajectories of individual fish stocks and their fisheries. In 2000, Dr. Summers initiated the

National Coastal Assessment, a U.S.-wide monitoring program to assess the ecological condition of the Nation's estuaries culminating in a series of reports, the National Coastal Condition Reports, for which he was the senior author, publishing published collaboratively with NOAA and USGS. At present, Dr. Summers spearheads a portion of the Ecosystem Services Research Program that is examining the relationship among ecosystem services and human well-being. This research is targeted toward the development of an Index of Well-Being for the United States. Dr. Summers has received numerous awards and citations from U.S. EPA and several awards from outside the agency for his research in estuarine monitoring and condition assessment.

Deborah N. Vivian (Chapter 4) is a biologist at the Gulf Ecology Division of U.S. EPA's Office of Research and Development in Gulf Breeze, FL. Ms. Vivian received a master's degree in Coastal Sciences at the University of Southern Mississippi's Gulf Coast Research Laboratory (GCRL) in 2005. She worked as a technician for the Rutgers University Marine Field Station from 1997-2001 on numerous projects pertaining to estuarine ecology and benthic habitats. Her principal research interests continue to focus on the ecology of larval and juvenile estuarine resident fishes. She has received awards from the Mississippi-Alabama Sea Grant Program and U.S. EPA.

INDEX

D

E

G

H

I

S

T

U

DATE DUE